DISCARDED

NAS 1.21: 4012/v.6

NASA SP-2000-4012

NASA HISTORICAL DATA BOOK
Volume VI

NASA Space Applications, Aeronautics and Space Research and Technology, Tracking and Data Acquisition/Support Operations, Commercial Programs, and Resources 1979—1988

Judy A. Rumerman

The NASA History Series

National Aeronautics and Space Administration
NASA History Office
Office of Policy and Plans
Washington, D.C.   2000

Library of Congress Cataloguing-in-Publication Data
(Revised for Vol. 6)

NASA historical data book.

(The NASA historical series) (NASA SP ; 2000-4012)
Vol. 1 is a republication of: NASA historical data book, 1958–1968. / Jane Van Nimmen and Leonard C. Bruno.
Vol. 6 in series: The NASA history series.
Includes bibliographical references and indexes.
Contents: v. 1 NASA resources, 1958–1968 / Jane Van Nimmen and Leonard C. Bruno — v. 2. Programs and projects, 1958–1968 / Linda Neuman Ezell — v. 3. Programs and projects, 1969—1978 / Linda Neuman Ezell — v. 4. NASA resources, 1969–1978 / Ihor Gawdiak with Helen Fedor — v. 5. NASA launch systems, space transportation, human spaceflight, and space science, 1979–1988 / Judy A. Rumerman — v. 6. NASA space applications, aeronautics and space research and technology, tracking and data acquisition/support operations, commercial programs, and resources, 1979–1988 / Judy A. Rumerman.

1. United States. National Aeronautics and Space Administration—History. I. Van Nimmen, Jane. II. Bruno, Leonard C. III. Ezell, Linda Neuman. IV. Gawdiak, Ihor. V. Rumerman, Judy A. VI. Rumerman, Judy A.
VII. Series. VIII. Series: NASA SP ; 2000-4012.

For sale by the U.S. Government Printing Office
Superintendent of Documents, Mail Stop: SSOP, Washington, DC 20402-9328
ISBN 0-16-050266-7

CONTENTS

List of Figures and Tables v

Preface and Acknowledgments xxi

Chapter One: Introduction 1

Chapter Two: Space Applications 9

Chapter Three: Aeronautics and Space Research and Technology 173

Chapter Four: Tracking and Data Acquisition/Space Operations 297

Chapter Five: Commercial Programs 353

Chapter Six: Facilities and Installations 381

Chapter Seven: NASA Personnel 461

Chapter Eight: Finances and Procurement 495

Index 601

About the Compiler 613

The NASA History Series 615

LIST OF FIGURES AND TABLES

Chapter Two: Space Applications

Figure 2–1	Office of Space and Terrestrial Applications	19
Figure 2–2	Office of Space Science and Applications	20
Figure 2–3	OSTA-1 Payload Location	22
Figure 2–4	Shuttle Imaging Radar-A	23
Figure 2–5	OSTA-1 Payload Ground Coverage	24
Figure 2–6	Mission Peculiar Equipment Support Structure (MPESS)	24
Figure 2–7	OSTA-2 Integrated Payload	26
Figure 2–8	OSTA-3 Payload Configuration With FILE, MAPS, and SIR-B	27
Figure 2–9	SAGE Orbit Configuration	29
Figure 2–10	Components of the Earth Energy Budget	30
Figure 2–11	Relative Sizes of Nimbus-7, Landsat D, and UARS	33
Figure 2–12	View of the UARS Spacecraft	33
Figure 2–13	NOAA 6 and NOAA 7 Spacecraft Configuration	36
Figure 2–14	NOAA 9 Spacecraft Configuration	38
Figure 2–15	NOAA 10 Orbit	39
Figure 2–16	GOES Satellite Configuration	40
Figure 2–17	VISSR Atmospheric Sounder	41
Figure 2–18	Thematic Mapper Configuration	43
Figure 2–19	Landsat 4 Flight Configuration	44
Figure 2–20	Magsat Orbital Configuration	46
Figure 2–21	Basic Concept of Satellite Search and Rescue	49
Figure 2–22	Galaxy Components	52
Figure 2–23	RCA Satcom 3, 3R, and 4	53
Figure 2–24	SBS Satellite Features	54
Figure 2–25	Westar 4, 5, and 6 Configuration	56
Figure 2–26	Growth of Intelsat Traffic (1975–1990)	58
Figure 2–27	Intelsat 5 Spacecraft	59
Figure 2–28	Fltsatcom Coverage Areas	60
Figure 2–29	Fltsatcom Spacecraft	61
Figure 2–30	NATO III Spacecraft	63
Figure 2–31	Anik C Configuration	65
Figure 2–32	Anik D Geographical Coverage at 104 Degrees West Longitude	66
Figure 2–33	Arabsat Coverage Area	67
Table 2–1	Applications Satellites (1979–1988)	72
Table 2–2	Science and Applications Missions Conducted on the Space Shuttle	75
Table 2–3	Total Space Applications Funding History	76
Table 2–4	Programmed Budget by Major Budget Category	77

Table 2–5	Resource Observations/Solid Earth Observations Funding History	78
Table 2–6	Landsat D/Landsat 4 Funding History	79
Table 2–7	Magnetic Field Satellite Funding History	79
Table 2–8	Shuttle/Spacelab Payload Development Funding History	79
Table 2–9	Extended Mission Operations Funding History	80
Table 2–10	Geodynamics Funding History	80
Table 2–11	Geodynamics Research and Data Analysis Funding History	81
Table 2–12	AgRISTARS Funding History	81
Table 2–13	Environmental Observations Funding History	82
Table 2–14	Upper Atmospheric Research Program Funding History	84
Table 2–15	Upper Atmospheric Research and Data Analysis Funding History	84
Table 2–16	Interdisciplinary Research and Analysis Funding History	84
Table 2–17	Shuttle/Spacelab Resource Observations Payload Development Funding History	85
Table 2–18	Operational Satellite Improvement Program Funding History	85
Table 2–19	Earth Radiation Budget Experiment Funding History	85
Table 2–20	Halogen Occultation Experiment Funding History	86
Table 2–21	Halogen Occultation Extended Mission Operations Funding History	86
Table 2–22	National Oceanic Satellite System Funding History	86
Table 2–23	Nimbus 7 Funding History	86
Table 2–24	Upper Atmosphere Research Satellite Experiments and Mission Definition Funding History	87
Table 2–25	Atmospheric Dynamics and Radiation Research and Analysis Funding History	87
Table 2–26	Oceanic Processes Research and Development Funding History	87
Table 2–27	Space Physics/Research and Analysis Funding History	88
Table 2–28	Tethered Satellite System Funding History	88
Table 2–29	Scatterometer Funding History	88
Table 2–30	Ocean Topography Experiment Funding History	88
Table 2–31	Airborne Science and Applications Funding History	88
Table 2–32	Applications Systems Funding History	89
Table 2–33	Airborne Instrumentation Research Program Funding History	89
Table 2–34	Shuttle/Spacelab Mission Design and Integration Funding History	89
Table 2–35	NASA Integrated Payload Planning Funding History	90
Table 2–36	Materials Processing in Space Funding History	90
Table 2–37	Materials Processing Research and Data Analysis Project Funding History	91
Table 2–38	Shuttle/Spacelab Materials Processing Payload Development Funding History	91
Table 2–39	Materials Processing Experiment Operations (Microgravity Shuttle/Space Station Payloads) Funding History	91

Table 2–40	Technology Transfer Funding History	92
Table 2–41	Applications Systems Verification and Transfer Funding History	92
Table 2–42	Regional Remote Sensing Applications Funding History	92
Table 2–43	User Requirements and Supporting Activities Funding History	93
Table 2–44	Civil Systems Funding History	93
Table 2–45	Space Communications Funding History	94
Table 2–46	Space Communications Search and Rescue Mission Funding History	95
Table 2–47	Space Communications Technical Consultation and Support Studies Funding History	95
Table 2–48	Space Communications Research and Data Analysis Funding History	95
Table 2–49	Communications Data Analysis Funding History	96
Table 2–50	Applications Data Service Definition Funding History	96
Table 2–51	Advanced Communications Technology Satellite Funding History	96
Table 2–52	Information Systems Program Funding History	97
Table 2–53	Data Systems Funding History	97
Table 2–54	Information Systems Funding History	97
Table 2–55	OSTA-1 Payload	98
Table 2–56	OSTA-2 Experiments	99
Table 2–57	OSTA-2 Instrument Module Characteristics	99
Table 2–58	SAGE (AEM-2) Characteristics	100
Table 2–59	ERBS Instrument Characteristics	101
Table 2–60	Earth Radiation Budget Satellite Characteristics	101
Table 2–61	UARS Instruments and Investigators	102
Table 2–62	UARS Development Chronology	104
Table 2–63	NOAA Satellite Instruments (1978–1988)	107
Table 2–64	NOAA 6 Characteristics	108
Table 2–65	NOAA B Characteristics	109
Table 2–66	Advanced Very High Resolution Radiometer Characteristics	110
Table 2–67	NOAA 7 Characteristics	110
Table 2–68	NOAA 8 Characteristics	111
Table 2–69	NOAA 9 Characteristics	112
Table 2–70	NOAA 10 Characteristics	114
Table 2–71	NOAA 11 Characteristics	115
Table 2–72	VISSR Atmospheric Sounder Infrared Spectral Bands	115
Table 2–73	GOES 4 Characteristics	116
Table 2–74	GOES 5 Characteristics	118
Table 2–75	GOES 6 Characteristics	119
Table 2–76	GOES G Characteristics	120
Table 2–77	GOES 7 Characteristics	121
Table 2–78	Landsat 4 Instrument Characteristics	122
Table 2–79	Landsat 4 Characteristics	123
Table 2–80	Landsat 5 Characteristics	124
Table 2–81	Magsat Characteristics	125

Table 2–82	Magsat Investigations	126
Table 2–83	ASC-1 Characteristics	128
Table 2–84	Comstar D-4 Characteristics	128
Table 2–85	Telstar 3-A Characteristics	129
Table 2–86	Telstar 3-C Characteristics	129
Table 2–87	Telstar 3-D Characteristics	130
Table 2–88	Galaxy 1 Instrument Characteristics	130
Table 2–89	Galaxy 2 Characteristics	131
Table 2–90	Galaxy 3 Characteristics	131
Table 2–91	Satcom 3 Characteristics	132
Table 2–92	Satcom 3-R Investigations	133
Table 2–93	Satcom 4 Characteristics	134
Table 2–94	Satcom 5 Characteristics	135
Table 2–95	Satcom 6 Characteristics	136
Table 2–96	Satcom 7 Characteristics	137
Table 2–97	Satcom K-2 Characteristics	138
Table 2–98	Satcom K-1 Characteristics	138
Table 2–99	SBS-1 Characteristics	139
Table 2–100	SBS-2 Characteristics	140
Table 2–101	SBS-3 Characteristics	140
Table 2–102	SBS-4 Characteristics	141
Table 2–103	Westar Satellite Comparison	142
Table 2–104	Westar 3 Characteristics	143
Table 2–105	Westar 4 Characteristics	143
Table 2–106	Westar 5 Characteristics	144
Table 2–107	Westar 6 Characteristics	144
Table 2–108	Intelsat Participants	145
Table 2–109	International Contributors to Intelsat	147
Table 2–110	Intelsat V F-2 Characteristics	148
Table 2–111	Intelsat V F-1 Characteristics	148
Table 2–112	Intelsat V F-3 Characteristics	149
Table 2–113	Intelsat V F-4 Characteristics	149
Table 2–114	Intelsat V F-5 Characteristics	150
Table 2–115	Intelsat V F-6 Characteristics	151
Table 2–116	Intelsat V F-9 Characteristics	151
Table 2–117	Intelsat V-A F-10 Characteristics	152
Table 2–118	Intelsat V-A F-11 Characteristics	152
Table 2–119	Intelsat V-A F-12 Characteristics	153
Table 2–120	Fltsatcom 2 Characteristics	154
Table 2–121	Fltsatcom 3 Characteristics	154
Table 2–122	Fltsatcom 4 Characteristics	155
Table 2–123	Fltsatcom 5 Characteristics	155
Table 2–124	Fltsatcom 7 Characteristics	156
Table 2–125	Fltsatcom 6 Characteristics	156
Table 2–126	Leasat 2 Characteristics	157
Table 2–127	Leasat 1 Characteristics	157
Table 2–128	Leasat 3 Characteristics	158
Table 2–129	Leasat 4 Characteristics	158
Table 2–130	NATO IIID Characteristics	159

Table 2–131	Anik D-1 Characteristics	159
Table 2–132	Anik C-3 Characteristics	160
Table 2–133	Anik C-2 Characteristics	160
Table 2–134	Anik D-2 Characteristics	161
Table 2–135	Anik C-1 Characteristics	161
Table 2–136	Arabsat-1B Characteristics	162
Table 2–137	Aussat 1 Characteristics	162
Table 2–138	Aussat 2 Characteristics	163
Table 2–139	Insat 1A Characteristics	164
Table 2–140	Insat 1B Characteristics	165
Table 2–141	Morelos 1 Characteristics	165
Table 2–142	Morelos 2 Characteristics	166
Table 2–143	Palapa B-1 Characteristics	166
Table 2–144	Palapa B-2 Characteristics	167
Table 2–145	Palapa B-2P Characteristics	167
Table 2–146	UoSat 1 Characteristics	168
Table 2–147	UoSat 2 Characteristics	169
Table 2–148	NOVA 1 Characteristics	169
Table 2–149	NOVA 3 Characteristics	170
Table 2–150	NOVA 2 Characteristics	170
Table 2–151	SOOS-I (Oscar 24/Oscar 30) Characteristics	171
Table 2–152	SOOS-2 (Oscar 27/Oscar 29) Characteristics	171
Table 2–153	SOOS-3 (Oscar 23/Oscar 30) Characteristics	172
Table 2–154	SOOS-4 (Oscar 25 and Oscar 31) Characteristics	172

Chapter Three: Aeronautics and Space Research and Technology

Figure 3–1	Office of Aeronautics and Space Technology (as of October 5, 1983)	185
Figure 3–2	Office of Aeronautics and Space Technology (as of August 14, 1984)	186
Figure 3–3	Percentage of R&D and SFC&DC Budget Allocated for Office of Aeronautics and Space Technology Activities	188
Figure 3–4	The Propfan	193
Figure 3–5	Basic Propeller Installation Configurations	194
Figure 3–6	Modified Gulfstream II Aircraft Used for Propfan Test Assessment	196
Figure 3–7	Laminar Flow Control Through Suction	200
Figure 3–8	Factors Affecting Laminar Control	200
Figure 3–9	Tilt-Rotor Aircraft	209
Figure 3–10	Increased Turning Capability of HiMAT Compared With Other Aircraft	215
Figure 3–11	HiMAT Dimensions	217
Figure 3–12	Forward-Swept Wing	218
Figure 3–13	Relative Thickness of Conventional, Supercritical and Thin Supercritical Wing Designs	219
Figure 3–14	National Aerospace Plane Program Schedule and Milestones	223

Figure 3–15	Four Generic X-30 Designs With Fully Integrated Engines and Airframes	226
Figure 3–16	Proposed National Aerospace Plane	227
Figure 3–17	NASA Transport Systems Research Vehicle (TSRV)	230
Figure 3–18	Artist's Depiction of the Effect of Wind Shear on an Aircraft	234
Figure 3–19	On-Board Wind Shear Warning Systems	236
Figure 3–20	Aerodynamic Coefficient Identification Package (ACIP) Experiment	243
Figure 3–21	Catalytic Surface Effects (CSE) Experiment (Lower Surface View)	244
Figure 3–22	Dynamics, Acoustics, and Thermal Environment (DATE) Experiment	244
Figure 3–23	Induced Environment Contamination Monitor (IECM)	245
Figure 3–24	Tile Gap Heating (TGH) Effects	245
Figure 3–25	OAST-1 Payload Elements	246
Table 3–1	Total Office of Aeronautics and Space Technology Program Funding	259
Table 3–2	Major Budget Category Programmed Budget History	260
Table 3–3	Total Aeronautical Research and Technology Program Funding	261
Table 3–4	Aeronautics Research and Technology Base Funding History	262
Table 3–5	Aerodynamics (Fluid and Thermal Physics) Research and Technology Funding History	263
Table 3–6	Propulsion (and Power) Systems Research and Technology Funding History	263
Table 3–7	Materials and Structures (Aeronautics) Research and Technology Funding History	264
Table 3–8	General Aviation Research and Technology Funding History	264
Table 3–9	Applied Aerodynamics Research and Technology Funding History	264
Table 3–10	Low Speed (Subsonic) Aircraft Research and Technology Funding History	265
Table 3–11	High Speed (High-Performance) Aircraft Research and Technology Funding History	265
Table 3–12	Rotorcraft Research and Technology Funding History	265
Table 3–13	Avionics and Flight Control (Aircraft Controls and Guidance) Research and Technology Funding History	266
Table 3–14	Human Factors Research and Technology Funding History	266
Table 3–15	Multidisciplinary Research and Technology Funding History	266
Table 3–16	Transport Aircraft Research and Technology Funding History	267
Table 3–17	Computer Science and Applications (Information Science) Research and Technology Funding History	267

Table 3–18	Flight Systems Research and Technology Funding History	267
Table 3–19	System Studies Funding History	268
Table 3–20	Systems Technology Program Funding History	268
Table 3–21	Aircraft Energy Efficiency Funding History	270
Table 3–22	Materials and Structures Systems Technology Funding History	270
Table 3–23	Low Speed (Subsonic) Aircraft Systems Technology Funding History	271
Table 3–24	High Speed (High-Performance) Aircraft Systems Technology Funding History	271
Table 3–25	Propulsion Systems Technology Funding History	271
Table 3–26	Avionics and Flight Control Systems Technology Funding History	272
Table 3–27	Transport Aircraft Systems Technology Funding History	272
Table 3–28	Advanced Propulsion Systems Technology Funding History	272
Table 3–29	Aeronautical System Studies Technology Funding History	273
Table 3–30	Numerical Aerodynamic Simulation Funding History	273
Table 3–31	Advanced Rotorcraft Technology Funding History	273
Table 3–32	Experimental Programs Funding History	274
Table 3–33	Space Research and Technology Funding History	274
Table 3–34	Space Research and Technology Base Funding History	275
Table 3–35	Materials and Structures (Space) Research and Technology Funding History	276
Table 3–36	Space Power and Electric Propulsion (Space Energy Conversion) Research and Technology Funding History	276
Table 3–37	Platform Systems (Systems Analysis) Research and Technology Funding History	277
Table 3–38	Information Systems (Space Data and Communications) Research and Technology Funding History	277
Table 3–39	Computer Sciences and Electronics (Information Sciences) Research and Technology Funding History	277
Table 3–40	Electronics and Automation Research and Technology Funding History	278
Table 3–41	Transportation Systems (Space Flight) Research and Technology Funding History	278
Table 3–42	(Chemical) Propulsion Research and Technology Funding History	278
Table 3–43	Spacecraft Systems Research and Technology Funding History	279
Table 3–44	Fluid Physics (Aerothermodynamics) Research and Technology Funding History	279
Table 3–45	Control and Human Factors (Controls and Guidance) Research and Technology Funding History	279
Table 3–46	Human Factors Research and Technology Funding History	280
Table 3–47	System Studies (Space) Funding History	280

Table 3–48	Systems Technology Program (Civil Space Technology Initiative) Funding History	281
Table 3–49	Space Systems Studies Funding History	282
Table 3–50	Information (Systems) Technology Funding History	282
Table 3–51	Space Flight Systems Technology Funding History	282
Table 3–52	Spacecraft Systems Technology Funding History	283
Table 3–53	Automation and Robotics Funding History	283
Table 3–54	(Chemical) Propulsion Systems Technology Funding History	283
Table 3–55	Vehicle Funding History	283
Table 3–56	Large Structures and Control Funding History	284
Table 3–57	High-Capacity Power Funding History	284
Table 3–58	Experimental Programs Funding History	284
Table 3–59	Standards and Practices Funding History	284
Table 3–60	Energy Technology Applications Funding History	285
Table 3–61	Transatmospheric Research and Technology Funding History	285
Table 3–62	Helicopter and Compound RSRA Configurations	286
Table 3–63	HiMAT Characteristics	286
Table 3–64	X-29 Characteristics	287
Table 3–65	Boeing 737 Transport Systems Research Vehicle Specifications	287
Table 3–66	Experiments of the Orbiter Experiments Program	288
Table 3–67	Long Duration Exposure Facility (LDEF) Mission Chronology	290
Table 3–68	Long Duration Exposure Facility (LDEF) Characteristics	291

Chapter Four: Tracking and Data Acquisition/Space Operations

Figure 4–1	OSTDS Organizational Configuration (Pre–Space Network)	302
Figure 4–2	OSTDS Configuration During the Space Network Period	303
Figure 4–3	The Deep Space Network Ground Communications Facility Used Land Mines, Microwave Links, Satellite Cables, and Communications Satellites to Link the Network's Elements	308
Figure 4–4	Twenty-Six-Meter Antenna at Goldstone	309
Figure 4–5	Very Long Baseline Interferometry Radio Navigation	310
Figure 4–6	TDRSS Coverage Area	315
Figure 4–7	User Data Flow	316
Figure 4–8	TDRSS Elements	316
Figure 4–9	White Sands Ground Terminal	317
Table 4–1	Total Office of Space Tracking and Data Systems Funding	322
Table 4–2	Major Budget Category Programmed Funding History	324
Table 4–3	Operations Funding	325
Table 4–4	Spaceflight Tracking and Data Network (STDN) Funding History	325
Table 4–5	Deep Space Network Funding History	326

Table 4–6	Aeronautics and Sounding Rocket Support Operations Funding History	326
Table 4–7	Communications Operations Funding History	326
Table 4–8	Data Processing Operations Funding History, 1979–1982	327
Table 4–9	Systems Implementation Funding History	327
Table 4–10	Spaceflight Tracking and Data Network (STDN) Implementation Funding History	328
Table 4–11	Deep Space Network Implementation Funding History	328
Table 4–12	Aeronautics and Sounding Rocket Support Systems Implementation Funding History	328
Table 4–13	Communications Implementation Funding History	329
Table 4–14	Data Processing Systems Implementation Funding History, 1979–1982	329
Table 4–15	Advanced Systems Funding History	330
Table 4–16	Initial TDRSS Funding History	330
Table 4–17	Space Network Funding History	331
Table 4–18	Tracking and Data Relay Satellite System Funding History	332
Table 4–19	Space Network Operations Funding History	333
Table 4–20	Systems Engineering and Support Funding History	333
Table 4–21	TDRS Replacement Spacecraft Funding History	334
Table 4–22	Second TDRS Ground Terminal Funding History	334
Table 4–23	Advanced TDRSS Funding History	334
Table 4–24	Ground Network Funding History	335
Table 4–25	Spaceflight Tracking and Data Network (STDN) Systems Implementation Funding History	335
Table 4–26	Spaceflight Tracking and Data Network (STDN) Operations Funding History	336
Table 4–27	Deep Space Network Systems Implementation Funding History	336
Table 4–28	Deep Space Network Operations Funding History	337
Table 4-29	Aeronautics, Balloon, and Sounding Rocket Support Systems Implementation Funding History	337
Table 4–30	Aeronautics, Balloon, and Sounding Rocket Support Operations Funding History	337
Table 4–31	Communications and Data Systems Funding History	338
Table 4–32	Communications Systems Implementation Funding History	338
Table 4–33	Communications Operations Funding History	339
Table 4–34	Mission Facilities Funding History	339
Table 4–35	Mission Operations Funding History	340
Table 4–36	Data Processing Systems Implementation Funding History, 1982–1988	340
Table 4–37	Data Processing Operations Funding History, 1982–1988	341
Table 4–38	Tracking and Data Acquisition Stations (1979–1988)	342
Table 4–39	Single-Access Link Summary	350
Table 4–40	Multiple-Access Link Summary	350
Table 4–41	TDRS Characteristics	351

Chapter Five: Commercial Programs

Table 5–1	Total Commercial Programs Funding	363
Table 5–2	Major Budget Category Programmed Funding History	364
Table 5–3	Technology Dissemination Funding	365
Table 5–4	Program Control and Evaluation Funding	365
Table 5–5	Technology Applications Funding	366
Table 5–6	Technology Utilization Funding	366
Table 5–7	Product Development Funding History	367
Table 5–8	Acquisition, Dissemination, and Network Operations Funding History	367
Table 5–9	Program Development, Evaluation, and Coordination Funding History	367
Table 5–10	Industrial Outreach Funding History	367
Table 5–11	Commercial Use of Space Funding	368
Table 5–12	Commercial Applications R&D Funding	368
Table 5–13	Commercial Development Support Funding	368
Table 5–14	Centers for the Commercial Development of Space	369
Table 5–15	Cooperative Agreements Between NASA and the Private Sector	370
Table 5–16	SBIR Funding by Fiscal Year	377
Table 5–17	Proposals and Contract Awards From Each Annual Solicitation	377
Table 5–18	NASA SBIR Awards by Phase and Topic Area (1983–1988)	378
Table 5–19	Cumulative NASA SBIR Funding by Topic Area (1983–1988)	379

Chapter Six: Facilities and Installations

Figure 6–1	NASA Facilities (1980)	384
Figure 6–2	NASA Installations (1979–1988)	385
Table 6–1	NASA Centers' Major Programs and Mission Areas	404
Table 6–2	Property: In-House and Contractor-Held (FY 1979–1988)	405
Table 6–3	Value of Real Property Components as a Percentage of Total Real Property: In House and Contractor-Held (FY 1979–1988)	406
Table 6–4	NASA Facilities Total Investment Value (FY 1979): In-House and Contractor-Held	407
Table 6–4A	NASA Facilities Total Investment Value (FY 1980–1982): In-House and Contractor-Held	408
Table 6–4B	NASA Facilities Total Investment Value (FY 1983–1985): In-House and Contractor-Held	410
Table 6–4C	NASA Facilities Total Investment Value (FY 1986–1988): In-House and Contractor-Held	412
Table 6–5	Land Owned by Installation and Fiscal Year in Acres: In-House and Contractor-Held	414

Table 6–6	Number of Buildings Owned by Installation and Fiscal Year: In-House and Contractor-Held	415
Table 6–7	Number of Square Feet of Buildings Owned by Installation and Fiscal Year: In-House and Contractor-Held	416
Table 6–8	Total Real Property Value by Installation and Fiscal Year: In-House and Contractor-Held	417
Table 6–9	Land Value by Installation and Fiscal Year: In-House and Contractor-Held	418
Table 6–10	Building Value by Installation and Fiscal Year: In-House and Contractor-Held	419
Table 6–11	Other Structures and Facilities Value by Installation and Fiscal Year: In-House and Contractor-Held	420
Table 6–12	Capitalized Equipment Value by Installation and Fiscal Year: In-House and Contractor-Held	421
Table 6–13	Land Value as a Percentage of Total Real Property Value by Installation and Fiscal Year: In-House and Contractor-Held	422
Table 6–14	Building Value as a Percentage of Total Real Property Value by Installation and Fiscal Year: In-House and Contractor-Held	423
Table 6–15	Other Structures and Facilities Value as a Percentage of Total Real Property Value by Installation and Fiscal Year: In-House and Contractor-Held	424
Table 6–16	Real Property Value of Installations Ranked as a Percentage of Total Real Property Value: In-House and Contractor-Held	425
Table 6–17	Capitalized Equipment Value of Installations Ranked as a Percentage of Total Capitalized Equipment Value: In-House and Contractor-Held	426
Table 6–18	NASA Tracking and Data Acquisition Stations	427
Table 6–19	Distribution of Research and Development and Space Flight Control and Data Communications Budget Plan by Installation and Program Office: FY 1988	428
Table 6–20	NASA Headquarters Major Organizations	429
Table 6–21	Headquarters Capitalized Equipment Value	430
Table 6–22	Headquarters Personnel	430
Table 6–23	Headquarters Funding by Fiscal Year	431
Table 6–24	Headquarters Total Procurement Activity by Fiscal Year	431
Table 6–25	Ames in-House and Contractor-Held Property	432
Table 6–26	Ames Value of Real Property Components as a Percentage of Total	432
Table 6–27	Ames Personnel	433
Table 6–28	Ames Funding by Fiscal Year	434
Table 6–29	Ames Total Procurement Activity by Fiscal Year	434
Table 6–30	Dryden in-House and Contractor-Held Property	435
Table 6–31	Dryden Value of Real Property Components as a Percentage of Total	435
Table 6–32	Dryden Personnel	436

Table 6–33	Dryden Funding by Fiscal Year	436
Table 6–34	Dryden Total Procurement Activity by Fiscal Year	436
Table 6–35	Goddard in-House and Contractor-Held Property	437
Table 6–36	Goddard Value of Real Property Components as a Percentage of Total	438
Table 6–37	Goddard Personnel	438
Table 6–38	Goddard Funding by Fiscal Year	439
Table 6–39	Goddard Total Procurement Activity by Fiscal Year	439
Table 6–40	JPL in-House and Contractor-Held Property	440
Table 6–41	JPL Value of Real Property Components as a Percentage of Total	441
Table 6–42	JPL Funding by Fiscal Year	441
Table 6–43	JPL Total Procurement Activity by Fiscal Year	441
Table 6–44	Johnson in-House and Contractor-Held Property	442
Table 6–45	Johnson Value of Real Property Components as a Percentage of Total	442
Table 6–46	Johnson Personnel	443
Table 6–47	Johnson Funding by Fiscal Year	443
Table 6–48	Johnson Total Procurement Activity by Fiscal Year	444
Table 6–49	Kennedy in-House and Contractor-Held Property	444
Table 6–50	Kennedy Value of Real Property Components as a Percentage of Total	445
Table 6–51	Kennedy Personnel	445
Table 6–52	Kennedy Funding by Fiscal Year	446
Table 6–53	Kennedy Total Procurement Activity by Fiscal Year	446
Table 6–54	Langley in-House and Contractor-Held Property	447
Table 6–55	Langley Value of Real Property Components as a Percentage of Total	447
Table 6–56	Langley Personnel	448
Table 6–57	Langley Funding by Fiscal Year	449
Table 6–58	Langley Total Procurement Activity by Fiscal Year	449
Table 6–59	Lewis in-House and Contractor-Held Property	450
Table 6–60	Lewis Value of Real Property Components as a Percentage of Total	450
Table 6–61	Lewis Personnel	451
Table 6–62	Lewis Funding by Fiscal Year	452
Table 6–63	Lewis Total Procurement Activity by Fiscal Year	452
Table 6–64	Marshall in-House and Contractor-Held Property	453
Table 6–65	Marshall Value of Real Property Components as a Percentage of Total	454
Table 6–66	Marshall Personnel	454
Table 6–67	Marshall Funding by Fiscal Year	455
Table 6–68	Marshall Total Procurement Activity by Fiscal Year	455
Table 6–69	National Space Technology Laboratories/Stennis in-House and Contractor-Held Property	456
Table 6–70	National Space Technology Laboratories/Stennis Value of Real Property Components as a Percentage of Total	456

Table 6–71	National Space Technology Laboratories/ Stennis Personnel	457
Table 6–72	National Space Technology Laboratories/ Stennis Funding by Fiscal Year	458
Table 6–73	National Space Technology Laboratories/ Stennis Total Procurement Activity by Fiscal Year	458
Table 6–74	Wallops in-House and Contractor-Held Property	459
Table 6–75	Wallops Value of Real Property Components as a Percentage of Total	459
Table 6–76	Wallops Personnel	460
Table 6–77	Wallops Funding by Fiscal Year	460
Table 6–78	Wallops Total Procurement Activity by Fiscal Year	460

Chapter Seven: NASA Personnel

Figure 7–1	Civil Service, Contractor, and Total Workforce Trend	463
Figure 7–2	Growth in Professional Occupational Groups From 1979 to 1988	464
Figure 7–3	Comparison of Percentage of Educational Levels of Permanent Employees (1979 and 1988)	464
Figure 7–4	Average Salary Trends of Permanent Employees by Pay Plan	465
Figure 7–5	Growth in Minority Employment as a Percentage of NASA Total Employees (1979–1988)	465
Figure 7–6	Growth in Percentage of Female Permanent Employees (1979–1988)	466
Table 7–1	Total NASA Workforce	468
Table 7–2	Accessions and Separations of Permanent Employees	468
Table 7–3	Permanent Employees by NASA Occupational Code Group: Number on Board	469
Table 7–4	Average Annual Salaries of Permanent Employees by Pay Plan	470
Table 7–5	Educational Profile of Permanent Employees: Number on Board	471
Table 7–6	Educational Profile of Permanent Employees: Percentage	471
Table 7–7	Paid Employees by NASA Installation: Number on Board (Permanent and Other)	472
Table 7–8	Paid Employees by NASA Installation: Percentage of NASA Total	473
Table 7–9	Paid Employees by NASA Installation: Changes in Number on Board	474
Table 7–10	Permanent Employees by NASA Installation: Number on Board	475
Table 7–11	Temporary Employees by NASA Installation: Number on Board	476
Table 7–12	NASA Excepted and Supergrade Employees by NASA Installation: Number on Board	477

Table 7–13	NASA Excepted and Supergrade Employees by NASA Installation: Percentage of NASA Total	478
Table 7–14	Scientific and Technical Permanent Employees (Occupational Code Group 200, 700, and 900) by NASA Installation: Number on Board	479
Table 7–15	Technical Support Permanent Employees (Occupational Code Group 300) by NASA Installation: Number on Board	480
Table 7–16	Trades and Labor Permanent Employees (Occupational Code Group 100) by NASA Installation: Number on Board	481
Table 7–17	Clerical and Professional Administrative Permanent Employees (Occupational Code Group 600 and 500) by NASA Installation: Number on Board	482
Table 7–18	Minority Permanent Employees: Number on Board	483
Table 7–19	Minority Permanent Employees by NASA Occupation Code Group: Number on Board	484
Table 7–20	Minority Permanent Employees by Grade Range: Number on Board	485
Table 7–21	Average GS Grade Level of Minority and Nonminority Permanent Employees by Occupational Code Group (1979–1988)	486
Table 7–22	Minority Permanent Employees by NASA Installation: Number on Board (1979–1988)	487
Table 7–23	Minorities as a Percentage of Permanent Employees by NASA Installation (1979–1988)	488
Table 7–24	Female Permanent Employees by NASA Occupation Code Group: Number on Board	489
Table 7–25	Female Permanent Employees by Grade Range: Number on Board	490
Table 7–26	Average GS Grade Level of Male and Female Permanent Employees by Occupational Code Group (1979–1988)	491
Table 7–27	Female Permanent Employees by NASA Installation: Number on Board (1979–1988)	492
Table 7–28	Females as a Percentage of Permanent Employees by NASA Installation (1979–1988)	493
Table 7–29	Age Profile of Permanent Employees by Grade Range: Number on Board	494

Chapter Eight: Finances and Procurement

Figure 8–1	NASA Appropriations	498
Figure 8–2	Percentage of NASA Obligations Devoted to Procurement Actions	499
Figure 8–3	Total Percentage of Procurement Actions by Kind of Contractor (1979–1988)	500
Figure 8–4	Total Number of Procurement Actions by Year	500

Figure 8–5	Percentage of Procurement Award Value by Kind of Contractor	501
Figure 8–6	Total Value of Awards by Fiscal Year	501
Figure 8–7	Distribution of Prime Contract Awards by Region	502
Figure 8–8	Total Value of Business and Educational and Nonprofit Awards by Fiscal Year	503
Table 8–1	NASA's Budget Authority as a Percentage of the Total Federal Budget	507
Table 8–2	NASA Appropriations by Appropriation Title and Fiscal Year and Percentage Change	507
Table 8–3	NASA's Budget History (1979–1988)	508
Table 8–4	Authorizations and Appropriations Compared With Budget Requests	510
Table 8–5	Budget Requests, Authorizations, Appropriations, and Obligations	512
Table 8–6	Research and Program Management Funding by Installation	513
Table 8–7	Research and Development Funding by Installation	514
Table 8–8	Space Flight Control and Data Communications Funding by Installation	515
Table 8–9	Construction of Facilities Funding by Facility	516
Table 8–10	Research and Development Appropriation by Program	517
Table 8–11	Space Flight Control and Data Communications Appropriation by Program	519
Table 8–12	Research and Development Funding by Program	520
Table 8–13	Space Flight Control and Data Communications Funding by Program	522
Table 8–14	NASA Budget Authority in Millions of Real-Year Dollars and in Equivalent FY 1996 Dollars	522
Table 8–15	Total Number of Procurement Actions by Kind of Contractor: FY 1979–1988	522
Table 8–16	Number of Procurement Actions by Kind of Contractor and Fiscal Year	523
Table 8–17	Number of Procurement Actions Awarded to Small and Large Business Firms by Fiscal Year	524
Table 8–18	Total Procurement Award Value by Kind of Contractor and Method of Procurement: FY 1979–1988	525
Table 8–19	Value of Awards by Kind of Contractor and Fiscal Year	526
Table 8–20	Value of Awards to Small and Large Business Firms by Fiscal Year	527
Table 8–21	Value of Awards to Business Firms by Kind of Procurement and Fiscal Year	528
Table 8–22	Value and Percentage of Direct Awards to Business Firms by Contract Pricing Provision: 1979–1988	529
Table 8–23	Number and Percentage of Procurement Action in Direct Awards to Business Firms by Contract Pricing: 1979–1988	530

Table 8–24	Distribution of Prime Contract Awards by State: 1979–1988	531
Table 8–25	Distribution of Prime Contract Awards by Region: FY 1979–1988	536
Table 8–26	Value and Percentage of Total of Awards by Installation	538
Table 8–27	Ranking of NASA's Top Ten Contractors by Fiscal Year	540
Table 8–28	Top Fifty Contractors: FY 1979	541
Table 8–29	Top Fifty Contractors: FY 1980	544
Table 8–30	Top Fifty Contractors: FY 1981	547
Table 8–31	Top Fifty Contractors: FY 1982	550
Table 8–32	Top Fifty Contractors: FY 1983	553
Table 8–33	Top Fifty Contractors: FY 1984	556
Table 8–34	Top Fifty Contractors: FY 1985	559
Table 8–35	Top Fifty Contractors: FY 1986	562
Table 8–36	Top Fifty Contractors: FY 1987	565
Table 8–37	Top Fifty Contractors: FY 1988	568
Table 8–38	Top Fifty Educational and Nonprofit Institutions: FY 1979	571
Table 8–39	Top Fifty Educational and Nonprofit Institutions: FY 1980	574
Table 8–40	Top Fifty Educational and Nonprofit Institutions: FY 1981	577
Table 8–41	Top Fifty Educational and Nonprofit Institutions: FY 1982	580
Table 8–42	Top Fifty Educational and Nonprofit Institutions: FY 1983	583
Table 8–43	Top Fifty Educational and Nonprofit Institutions: FY 1984	586
Table 8–44	Top Fifty Educational and Nonprofit Institutions: FY 1985	589
Table 8–45	Top Fifty Educational and Nonprofit Institutions: FY 1986	592
Table 8–46	Top Fifty Educational and Nonprofit Institutions: FY 1987	595
Table 8–47	Top Fifty Educational and Nonprofit Institutions: FY 1988	598

PREFACE AND ACKNOWLEDGMENTS

In 1973, NASA published the first volume of the *NASA Historical Data Book,* a hefty tome containing mostly tabular data on the resources of the space agency between 1958 and 1968. There, broken into detailed tables, were the facts and figures associated with the budget, facilities, procurement, installations, and personnel of NASA during that formative decade. In 1988, NASA reissued that first volume of the data book and added two additional volumes on the agency's programs and projects, one each for 1958–1968 and 1969–1978. NASA published a fourth volume in 1994 that addressed NASA resources for the period between 1969 and 1978. Earlier in 1999, NASA published a fifth volume of the data book, containing mostly tabular information on the agency's launch systems, space transportation, human spaceflight, and space science programs between 1979 and 1988.

This sixth volume of the *NASA Historical Data Book* is a continuation of those earlier efforts. This fundamental reference tool presents information, much of it statistical, documenting the development of several critical areas of NASA responsibility for the period between 1979 and 1988. This volume includes detailed information on the space applications effort, the development and operation of aeronautics and space research and technology programs, tracking and data acquisition/space operations, commercial programs, facilities and installations, personnel, and finances and procurement during this era.

Special thanks are owed to the student research assistants who gathered and input much of the tabular material—a particularly tedious undertaking. Additional gratitude is owed to Ashok Saxena of Informatics, Inc., who provided the resources to undertake this project.

There are numerous people at NASA associated with historical study, technical information, and the mechanics of publishing who helped in myriad ways in the preparation of this historical data book. Stephen J. Garber helped in the management of the project and handled final proofing and publication. M. Louise Alstork edited and prepared the index of the work. Nadine J. Andreassen of the NASA History Office performed editorial and proofreading work on the project; and the staffs of the NASA Headquarters Library, the Scientific and Technical Information Program, and the NASA Document Services Center provided assistance in locating and preparing for publication the documentary materials in this work. The NASA Headquarters Printing and Design Office developed the layout and handled printing. Specifically, we wish to acknowledge the work of Jane E. Penn, Jonathan L. Friedman, Joel Vendette, and Kelly L. Rindfusz for their editorial and design work. In addition, Michael Crnkovic, Stanley Artis, and Jeffery Thompson saw the book through the publication process. Thanks are due them all.

CHAPTER ONE
INTRODUCTION

CHAPTER ONE
INTRODUCTION

During the period between 1979 and 1988, NASA experienced many trials and triumphs. The Space Shuttle flew its first orbital mission in April 1981, suffered the *Challenger* accident in January 1986, and made a heroic return to flight in September 1988. The two Voyager space probes encountered both Jupiter and Saturn and sped outward to Uranus and Neptune. President Ronald Reagan, in his State of the Union address in 1984, announced that the United States would build a space station and set a goal of completing it within a decade.

Throughout this decade, moreover, the President increased funding for NASA. In response to new initiatives and the operational commitment to the Space Shuttle, the NASA budget rose from $4.96 billion in fiscal year 1979 to $9.06 in 1988. The increases allowed the agency not only to sustain the Space Shuttle program and to begin the construction of a space station, but also to bring to fruition a series of important science, aeronautics, and space applications programs.

Space Applications

In the area of space applications, there were several significant activities. For example, NASA began in the 1970s to build and launch Earth resource mapping satellites, the first of which was the Landsat series. Landsat 1, launched on July 23, 1972, as the Earth Resources Technology Satellite (ERTS) and later renamed, changed the way in which Americans looked at the planet. It provided data on vegetation, insect infestations, crop growth, and associated land-use information. Two more Landsat vehicles were launched in January 1975 and March 1978, performed their missions, and exited service in the 1980s. Landsat 4, launched on July 16, 1982, and Landsat 5, launched on March 1, 1984, were "second-generation" spacecraft, with greater capabilities to produce more detailed land-use data. The system enhanced the ability to develop a worldwide crop forecasting system. Moreover, Landsat imagery has been used to devise a strategy for deploying equipment to contain oil spills, to aid navigation, to monitor pollution, to assist in water management, to site new power plants and pipelines, and to aid in agricultural development.[1]

Aeronautics and Space Research and Technology

From 1979 to 1988, NASA aeronautics and space research and development programs moved forward on a variety of fronts. The National Aeronautics and Space Act of 1958 gave NASA a broad mandate to "plan, direct, and conduct aeronautical and space activities," to involve the nation's scientific community in these activities, and to disseminate widely information about them. The most significant aeronautics endeavors of the era revolved around the effort to improve the efficiency of aircraft. For instance, in 1987, the NASA-industry advanced turboprop team at Lewis Research Center received the Robert J. Collier Trophy for the development of a new fuel-efficient turboprop propulsion system. The National Aeronautic Association has given this award every year since 1911 "for the greatest achievement in aeronautics and astronautics in America."[2]

Until 1970, NASA included basic aeronautics research as one of its major activities. The results of basic research added to the pool of knowledge and did not apply to any ongoing project. This effort was divided into four sections: fluid dynamics, electrophysics, materials, and applied mathematics. The aeronautics function also addressed the problems that vehicles might encounter during launch, ascent through the atmosphere, and atmospheric reentry. For instance, NASA conducted research in the areas of lifting-body research and planetary entry research. NASA also worked at improving the operational electronics systems, while reducing their size, weight, cost, and power requirements. Several NASA centers directed a variety of projects with this goal in mind. Especially important was work in aeronautical operating systems, aerodynamics research, aeronautical propulsion, and special efforts in short takeoff and landing (STOL) aircraft and experimental transport aircraft. The aeronautics effort also conducted projects in the areas of general aviation, environmental factors, vertical/STOL aircraft, supersonic/hypersonic aircraft, and military support.

---

[1]Roger D. Launius, *NASA: A History of the U.S. Civil Space Program* (Malabar, FL: Krieger Publishing Co., 1994), p. 104; Pamela E. Mack, *Viewing the Earth: The Social Construction of Landsat* (Cambridge, MA: MIT Press, 1990).

[2]Mark D. Bowles and Virginia P. Dawson, "The Advanced Turboprop Project: Radical Innovation in a Conservative Environment," in Pamela E. Mack, ed., *From Engineering Science to Big Science: The NACA and NASA Collier Trophy Research Project Winners* (Washington, DC: NASA Special Publication (SP)-4219, 1998), pp. 321–43.

Tracking and Data Acquisition/Space Operations

Another central mission of the agency throughout this period was tracking and data acquisition. The Deep Space Network, charged with communication with space missions beyond Earth orbit, was rapidly changing. Its Mark III Data System implementation task had been completed. The new capabilities required for the Voyager Jupiter encounter were in operation throughout the Deep Space Network, and operations teams were trained in their use. The Viking mission had been "extended" through May 1978 and then further "continued" through February 1979. Both Pioneer Venus missions had been successfully completed in 1978, and both Voyager spacecraft were approaching Jupiter and expected to carry out a full program of science experiments and imaging sequences during their brief encounters with Jupiter in March and July 1979.

For their success, each of the Voyager encounters depended not only on operable spacecraft, but also on ever-greater significant enhancement of the uplink and downlink Deep Space Network capabilities. At the same time, a heavy expenditure of Deep Space Network operational resources in personnel, training, and facilities was required simply to maintain a viable science data return from existing missions. Toward the end of the Voyager era, a truly international cooperative mission made its appearance. The Venus-Balloon mission in mid-1985 involved the Soviet, French, and North American space agencies. Although of very short duration, it presented a complex engineering and operational challenge for the Deep Space Network. Its successful completion established a basis for future relationships between these agencies in the area of tracking and data acquisition support for deep space missions.[3]

Commercial Programs

Very early in the 1980s, the United States developed an official policy to apply the resources of the nation to preserve the role of the country as a leader in space science and technology and their applications. Brought on by the emergence of the Space Transportation System (STS) as a space vehicle, many people began to believe that the dawn of an era of widespread commercial activities in space was at hand. Ensuring that

---

[3]William R. Corliss, "A History of the Deep Space Network," NASA CR-151915, 1976, NASA Historical Reference Collection, NASA History Office, NASA Headquarters, Washington, DC; N.A. Renzetti, *et al.,* "A History of the Deep Space Network from Inception to January 1, 1969," Technical Report 32-1533, Volume 1, Jet Propulsion Laboratory, September 1971. A more detailed account of many of these topics are contained in the unfinished and unpublished notes on the early (prior to 1962) history of the Deep Space Network compiled by Craig Waff at the Jet Propulsion Laboratory in 1993. The Waff notes are held in the Jet Propulsion Laboratory's Archives, Pasadena, CA.

national leadership would require the support and expansion of commercial space activities.[4]

The President's National Space Policy of July 4, 1982, directed NASA to expand U.S. private-sector investment and involvement in civil space and space-related activities. In light of this directive and because substantial portions of the U.S. technological base and motivation reside in the U.S. private sector, NASA will invigorate its efforts to take necessary and proper actions to promote a climate conducive to expanded private-sector investment and involvement in space by U.S. domestic concerns.[5] To more effectively encourage and facilitate private-sector involvement and investment in civil space and space-related activities, beginning in the early 1980s, NASA directed a portion of its space research and development activities toward supporting the research, development, and demonstration of space technologies with commercial application. To further support this objective, NASA would directly involve the private sector in initiatives that are consistent with NASA program objectives and that support commercial space activity.[6] Those initiatives included:

- Engaging in joint arrangements with U.S. domestic concerns to operate on a commercial basis facilities or services that relieve NASA of an operational responsibility
- Engaging in joint arrangements with U.S. domestic concerns to develop facilities or hardware to be used in conjunction with the STS or other aspects of the U.S. space program
- Entering into transactions with U.S. concerns designed to encourage the commercial exploitation of space

In addition to making available the results of NASA research, principal NASA incentives included:

- Providing flight time on the STS on appropriate terms and conditions as determined by the NASA Administrator
- Providing technical advice, consultation, data, equipment, and facilities to participating organizations
- Entering into joint research and demonstration programs in which each party funds its own participation

---

[4] W.D. Kay, "Space Policy Redefined: The Reagan Administration and the Commercialization of Space," *Business and Economic History* 27 (Fall 1998): 237–47.

[5] "Remarks at Edwards Air Force Base, California, on Completion of the Fourth Mission of the Space Shuttle *Columbia*," July 4, 1982, in *Public Papers of the Presidents, Ronald Reagan, 1982* (Washington, DC: U.S. Government Printing Office, 1983), p. 892.

[6] "NASA Policy to Enhance Commercial Investment in Space," September 13, 1983, NASA Historical Reference Collection.

In making the necessary determination to proceed under this policy, the NASA Administrator will consider the need for NASA-funded support or other NASA action to commercial endeavors and the relative benefits to be obtained from such endeavors. The primary emphasis of these joint arrangements will be to provide support to ventures that result in or facilitate industrial activity in space when such activity would otherwise be unlikely to occur because of high technological or financial risk. Other ventures involving new commercial activities in space will also be supported. In either case, private capital must be at risk. The major areas that NASA pursued emphasized:

- Effect of the private-sector activity on NASA programs
- Enhanced exploitation of NASA capabilities, such as the STS
- Contribution to the maintenance of U.S. technological superiority
- Amount of proprietary data or background information to be furnished by the concern
- Rights to be granted the concern in consideration of its contribution
- Impact of NASA sponsorship on a given industry
- Provision for a form of exclusivity in special cases when needed to promote innovation
- Recoupment of the contribution under appropriate circumstances
- Support of socioeconomic objectives of the government
- The willingness and ability of the proposer to market any resulting products and services

Facilities and Resources

During the ten years between 1979 and 1988, NASA's facilities and resources underwent significant alterations. Personnel, budgets, finances, procurement, and many other resource issues rose in response to the increased emphasis placed on spaceflight and NASA during the decade. This volume concludes with a discussion of these issues.

CHAPTER TWO
SPACE APPLICATIONS

CHAPTER TWO
SPACE APPLICATIONS

Introduction

From NASA's inception, the application of space research and technology to specific needs of the United States and the world has been a primary agency focus. The years from 1979 to 1988 were no exception, and the advent of the Space Shuttle added new ways of gathering data for these purposes. NASA had the option of using instruments that remained aboard the Shuttle to conduct its experiments in a microgravity environment, as well as to deploy instrument-laden satellites into space. In addition, investigators could deploy and retrieve satellites using the remote manipulator system, the Shuttle could carry sensors that monitored the environment at varying distances from the Shuttle, and payload specialists could monitor and work with experimental equipment and materials in real time.

The Shuttle also allowed experiments to be performed directly on human beings. The astronauts themselves were unique laboratory animals, and their responses to the microgravity environment in which they worked and lived were thoroughly monitored and documented.

In addition to the applications missions conducted aboard the Shuttle, NASA launched ninety-one applications satellites during the decade, most of which went into successful orbit and achieved their mission objectives. NASA's degree of involvement with these missions varied. In some, NASA was the primary participant. Some were cooperative missions with other agencies. In still others, NASA provided only launch support. These missions are identified in this chapter.

Particularly after 1984, NASA's role in many applications missions complied with federal policy to encourage the commercial use of space and to privatize particular sectors of the space industry, while keeping others under government control.[1] Congress supported President Ronald

---

[1]See Title VII, "Land Remote-Sensing Commercialization Act of 1984," Public Law 98–365, 98th Cong., 2d sess., July 17, 1984; "National Space Strategy," White House Fact Sheet, August 15, 1984.

Reagan's proposal to move land remote sensing (Landsat) to the private sector but insisted that meteorological satellite activities remain a government enterprise. Legislation spelled out intentions of Congress in these areas.

This chapter discusses the applications missions that were launched from 1979 through 1988 in which NASA had a role. It also addresses other major missions that NASA developed during the decade but were not launched until later.

The Last Decade Reviewed (1969–1978)

From 1969 to 1978, NASA added monitoring the state of the environment to its existing applications programs in advanced communications and meteorology research. Geodetic research was a fourth responsibility. The Office of Applications divided these areas of responsibility into four program areas (called by different names during the decade): weather, climate, and environmental quality; communications; Earth resources survey; and Earth and ocean dynamics.

Meteorology

NASA conducted advanced research and development activities in the field of meteorology and served as launch vehicle manager for the fleet of operational satellites of the National Oceanic and Atmospheric Administration (NOAA). In addition, NASA actively participated in the Global Atmospheric Research Program, an international meteorological research effort.

NASA's major meteorology projects consisted of TIROS (Television Infrared Observation Satellite), the Synchronous Meteorological Satellites (SMS), and Nimbus. TIROS began with the ESSA 9 polar-orbiting satellite in 1969. The decade ended with the 1978 launch of TIROS N, a new TIROS prototype. This satellite preceded the group of NOAA satellites that NASA would launch in the following decade. The advantage of SMS over TIROS was its ability to provide daytime and nighttime coverage from geostationary orbit. NASA funded and managed the SMS project but turned it over to NOAA for its operations. Following SMS 1 and 2, this operational satellite was called Geostationary Operational Environmental Satellite (GOES). Three GOES satellites were launched through 1978.

Communications

NASA's research and development activities during this decade were limited to the joint NASA-Canadian Communications Technology Satellites (CTS) and experiments flown on Applications Technology Satellites (ATS). CTS demonstrated that powerful satellite systems could bring low-cost television to remote areas almost anywhere on the globe.

The remaining fifty-eight communications satellites NASA launched were operational satellites that provided commercial communications, military network support, or aids to navigation. NASA provided the launch vehicles, the necessary ground support, and initial tracking and data acquisition on a reimbursable basis. During this period, NASA expanded its communications satellite launching service to include foreign countries, the amateur ham radio community, and the U.S. military. The International Telecommunications Satellite Organization (Intelsat), established in August 1964, was the largest user of NASA communications launch services.

Applications Technology Satellites

The ATS program investigated and flight-tested technology common to a number of satellite applications. NASA launched six ATS spacecraft during the 1970s. These spacecraft carried a variety of communications, meteorology, and scientific experiments. ATS 1 and ATS 3, launched in 1966 and 1967, respectively, provided service into the 1980s.

Earth Observations

The Earth Observations program emphasized the development of techniques to survey Earth resources and changes to those resources and to monitor environmental and ecological conditions. It consisted of three projects: (1) Skylab; (2) the Earth Resources Survey program, consisting of specially equipped aircraft that tested cameras and remote-sensing equipment; and (3) the Earth Resources Technology Satellite (ERTS) program, later renamed Landsat. ERTS and Landsat spacecraft were the first satellites devoted exclusively to monitoring Earth's resources.

The Skylab project was a series of four orbital workshops that were occupied by astronaut crews. A primary objective was to study the long-term effects of weightlessness on humans. In addition, crew members conducted experiments in many discipline areas, providing investigators with hundreds of thousands of images, photographs, and data sets.

An ERTS/Landsat-type program was first conceived in the 1960s. The program grew with input from the Department of Agriculture, the U.S. Geological Survey, NASA, the Department of the Interior, the Department of Commerce, and academia. NASA's efforts focused on sensor development, and the agency launched ERTS 1 in 1972, followed by three Landsat satellites—all of which surpassed their predicted operational lifetimes. Investigators applied satellite data obtained from sensors aboard these satellites to agriculture, forestry, and range resources; cartography and land use; geology; water resources; oceanography and marine resources; and environmental monitoring.

Other Earth Observation Activities

NASA launched five other Earth-observation-type missions during the 1970s: Seasat 1, a satellite designed to predict ocean phenomena; the Laser Geodynamics Satellite (LAGEOS), which demonstrated the capability of laser satellite tracking techniques to accurately determine the movement of Earth's crust and rotational motions; GEOS 3, which studied Earth's shape and dynamic behavior; TOPO 1 for the U.S. Army Topographic Command; and the Heat Capacity Mapping Mission, which was the first in a series of applications explorer missions. All were successful except Seasat 1, which failed 106 days after launch.

Space Applications (1979–1988)

As in the previous decade, most of the applications missions that NASA launched from 1979 to 1988 were commercial missions or missions that were managed by other government agencies. Table 2–1 lists all of the applications satellites that NASA launched during this decade. Only the Stratospheric Aerosol and Gas Experiment (SAGE or AEM-2) and the Magnetic Field Satellite (Magsat or AEM-C), both of which were part of the Applications Explorer Mission (AEM), and the Earth Radiation Budget Satellite (ERBS) were NASA satellites. NASA's other applications missions took place aboard the Space Shuttle. Table 2–2 lists these missions. Additional applications experiments conducted on the Shuttle are discussed under the appropriate STS mission in Chapter 3, "Space Transportation/Human Spaceflight," in Volume V of the *NASA Historical Data Book*.

Environmental Observations

NASA launched two satellites as part of its Applications Explorer Mission. SAGE, launched from Wallops Flight Facility, Virginia, in February 1979, profiled aerosol and ozone content in the stratosphere. The satellite observed the violent eruptions of the volcano La Soufriere in the Caribbean in April 1979, the Sierra Negra volcanic eruption on the Galapagos Islands, and the eruption of Mount St. Helens. Magsat, launched later in 1979, was part of NASA's Resource Observations program.

NASA's other environmental observations missions consisted of two series of meteorological satellites that were developed, launched, and operated in conjunction with NOAA. The new polar-orbiting series of satellites succeeded the TIROS system. This two-satellite weather satellite system obtained and transmitted morning and afternoon weather data. The GOES series continued the group of geosynchronous satellites that began with SMS in the 1970s. Also intended to operate with two satellites, one located near the east coast of the United States and the other near the west coast, GOES provided almost continuous coverage of large areas.

In addition, Nimbus 5, launched in 1972, continued to operate until April 1983. Nimbus 6, launched in 1975, ceased operations in September 1983. Nimbus 7, launched in October 1978, provided useful data until the end of 1984. Its Total Ozone Monitoring System (TOMS) provided the first global maps of total ozone with high spatial and temporal resolution. This was the first time investigators could study short-period dynamic effects on ozone distribution. A series of these measurements provided information related to long-term, globally averaged ozone changes in the atmosphere of both natural and human origin.

NASA also continued to participate in the Global Weather Experiment as part of the Global Atmospheric Research Program. The goal of the program was to devise a way to improve satellite weather forecasting capabilities.

In 1984, NASA launched the ERBS, the first part of a three-satellite system comprising the Earth Radiation Budget Experiment (ERBE). (Other ERBE instruments flew on NOAA 9 and NOAA 10.) Part of NASA's climate observing program, ERBS data allowed scientists to increase their understanding of the physical processes that governed the interaction of clouds and radiation.

The effects of ozone on the upper atmosphere received increasing attention during the 1980s. The Nimbus series of satellites continued to provide data on ozone levels from its backscatter ultraviolet instrument. The Upper Atmospheric Research Satellites (UARS) program, which NASA initiated with an Announcement of Opportunity in 1978, also moved ahead. The program would make integrated, comprehensive, long-term measurements of key parameters and would improve investigators' abilities to predict stratospheric perturbations.

NASA reported to Congress and the U.S. Environmental Protection Agency in January 1982 (as required by the Clean Air Act Amendments of 1977) its assessment of what was known about key processes in the stratosphere, especially about the effect of human-produced chemicals on the ozone layer. This assessment was developed from the findings of a workshop sponsored by NASA and the World Meteorological Organization, in which approximately 115 scientists from thirteen countries participated. The scientists concluded that a continued release of chlorofluorocarbons 11 and 12 (Freon-11 and -12) at 1977 rates would decrease total global ozone by 5 to 9 percent by about the year 2100, but the effects of other changes in atmospheric composition could modify that result.

During 1984, Congress approved the UARS mission, and work began on the observatory and ground data-handling segments of the program. UARS, initially scheduled for launch in late 1989 and later moved to 1991, would be the first satellite capable of simultaneous measurements of the energy input, chemical composition, and dynamics of the stratosphere and mesosphere. The discovery of an Antarctic ozone hole in 1985 and Arctic ozone depletion in 1988 further emphasized the urgency of the mission.

Resource Observations

NASA launched the Magsat satellite in October 1979. Magsat was part of the Applications Explorer Mission and the first spacecraft specifically designed to conduct a global survey of Earth's vector magnetic field. Placed into a significantly lower orbit than previous magnetic field-measuring satellites, it provided more detailed and precise information about the nature of magnetic anomalies within Earth's crust than earlier missions and improved large-scale models of crustal geology.

Data obtained through remote sensing from space attracted a growing number of government and private-sector users during this decade. New ground stations were brought on-line and began receiving data transmitted from the Landsat satellites. Remote-sensing techniques were also used for geologic mapping as part of the NASA-Geosat Test Case Project, a joint research project with private industry. The results indicated that an analysis of remote-sensing measurements could yield geological information not commonly obtained by conventional field mapping.

President Jimmy Carter announced in 1979 that NOAA would manage all space-based operational civilian remote-sensing activities. NASA would continue its involvement in these activities, centered primarily in the Landsat program, through the launch and checkout of the spacecraft. The Land Remote-Sensing Commercialization Act of 1984, passed during the Reagan administration, moved remote-sensing activities from the public to the private sector. In accordance with this legislation, the Earth Observation Satellite Company (EOSAT) was chosen to begin operating the Landsat system. EOSAT initiated the development of a satellite-receiving center and an operations and control center that captured and processed data and flight control for the next-generation Landsat 6 and future spacecraft.

NASA launched Landsat 4 and Landsat 5 in 1982 and 1984, respectively. The Thematic Mapper instrument aboard these satellites, developed by NASA, provided data in several additional spectral bands and had better than twice the resolution of the Multispectral Scanner, which was the instrument used on earlier Landsat spacecraft. The satellites were turned over to NOAA following their checkout and to EOSAT after it assumed operation of the system.

Congress approved the AgRISTARS project in 1979. This multi-agency project—NASA, the Department of Agriculture, the Department of the Interior, NOAA, and the Agency for International Development—was to develop and test the usefulness of remote sensing for providing timely information to the Department of Agriculture. NASA was responsible for the selected research and development, exploratory and pilot testing, and support in areas in which it had specialized capabilities. It served as the lead agency for the Supporting Research project and the Foreign Commodity Production Forecasting project, both of which involved using remote-sensing techniques related to crop production and development. In

1982, Congress reduced the scope of AgRISTARS to focus it primarily on the Department of Agriculture's priority needs. NASA phased out its participation in 1984, but the space agency also conducted investigations in geodynamics and materials processing during this period.

Communications

From 1979 to 1988, NASA's role in the communications satellite field was primarily as a provider of launch services. The agency launched sixty-five operational communications satellites. Operational satellites included: ten Intelsat, four Westar, eight RCA Satcom, four Satellite Business Systems (SBS), one Comstar, three Telstar, five Anik/Telesat (Canada), one Arabsat (Saudi Arabia), two Morelos (Mexico), and two Aussat (Australia). The government of India reimbursed NASA for the launch of two Insat satellites, and the Republic of Indonesia paid for the launch of three Palapa satellites. NASA launched one NATO defense-related communications satellite. For the U.S. Department of Defense (DOD), NASA launched six Fleet Satellite Communications (Fltsatcom) satellites (U.S. Navy and Air Force) and four Leasat/Syncom satellites. In addition, NASA launched seven navigation satellites for the U.S. Navy: four SOOS and three Nova satellites. It also launched four other DOD communications satellites with classified missions.

These commercial missions enabled NASA to use some of its launch capabilities for the first time. SBS-1 was the first to use the Payload Assist Module (PAM) in place of a conventional third stage. The launch of SBS-3 marked the first launch from the Shuttle's cargo bay.

NASA's communications activities centered around its Search and Rescue Satellite-Aided Tracking system (SARSAT), its development of the Advanced Communications Technology Satellite (ACTS), its continued work on its mobile satellite program, and its development of an information systems program to handle the huge quantities of data returned from space missions. In addition, NASA's ATS program carried over into the 1980s. ATS 1, launched in 1966, and ATS 3, launched in 1967, continued to provide important communications services, especially in areas unreachable by more traditional means. ATS 1 operated until it was shut down in October 1985; ATS 3 was still operating into 1996.

SARSAT was an ongoing international project that used satellite technology to detect and locate aircraft and vessels in distress. The United States, the Soviet Union, Canada, and France developed the system. Norway, the United Kingdom, Sweden, Finland, Bulgaria, Denmark, and Brazil were other participants. The Soviet Union contributed a series of COSPAS satellites, beginning with the launch of COSPAS 1 in 1982. This was the first spacecraft that carried instruments specifically to determine the position of ships and aircraft in distress. It was interoperable with the SARSAT equipment on U.S. satellites and ground stations. During the 1980s, the United States operated instruments on NOAA's polar-orbiting spacecraft. The first was NOAA 8, which launched in March 1983.

The system became fully operational in 1984 and succeeded in saving more than 1,000 lives during the 1980s.

Work on NASA's ACTS began in 1984. ACTS was to allow large numbers of U.S. companies, universities, and government agencies to experiment with spot beams, hopping beams, and switchboard-in-the-sky concepts that were to enter the marketplace by the mid-1990s. The mission was originally planned to launch in 1988 but was delayed until September 1993. The program was canceled and resurrected several times; it was restructured in 1988 in response to congressional direction to contain costs.

The joint mobile satellite program among NASA, U.S. industry, and other government agencies was to provide two-way, satellite-assisted communication with a variety of vehicles in the early 1990s. As of the close of 1988, international frequencies had been allocated, and licensing approval by the Federal Communications Commission was expected shortly.

NASA's information systems program, which had become part of the newly formed Communications and Information Systems Division in 1987, operated large-scale computational resources used for data analysis. It also worked with specialized programs to establish data centers for managing and distributing data and developed computer networks and exploited advanced technologies to access and process massive amounts of data acquired from space missions. NASA established the National Science Space Data Center at the Goddard Space Flight Center to archive data from science missions and coordinate management of NASA data at distributed data centers.

Management of the Applications Program at NASA

From 1971, NASA managed applications missions independently from science missions, first through the Office of Applications and then, from 1977, through the Office of Space and Terrestrial Applications (OSTA). In November 1981, OSTA and the Office of Space Science merged into the Office of Space Science and Applications (OSSA).

OSTA's objective was to "conduct research and development activities that demonstrate and transfer space-related technology, systems and other capabilities which can be effectively used for down-to-earth practical benefits."[2] It was divided into divisions for materials processing in space, communications and information systems, environmental observation, research observation, and technology transfer (Figure 2–1). Anthony J. Calio, who had assumed the position of associate administrator in October 1977, continued leading OSTA until the new OSSA was formed. John Carruthers led the Materials Processing in Space Division until mid-1981, when Louis R. Testardi became acting division director. John

---

[2]"Office of Space and Terrestrial Applications," *Research and Development Fiscal Year 1981 Estimates, Budget Summary* (Washington, DC: NASA, 1981).

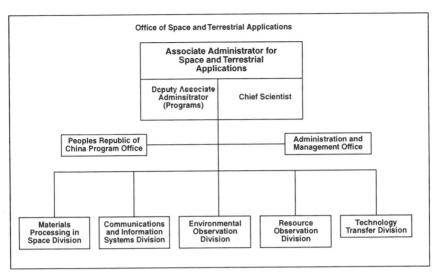

Figure 2–1. Office of Space and Terrestrial Applications

McElroy served as director of the Communications Division until late 1980, when Robert Lovell became division chief. Pitt Thome led the Resource Observation Division, Floyd Roberson served as director of the Technology Transfer Division, and Lawrence Greenwood led the Environmental Observation Division.

Andrew Stofan, who had been head of the Office of Space Science, became associate administrator of the new OSSA until he was replaced by Burton Edelson in February 1982. Edelson remained at the post until he resigned in February 1987. Lennard A. Fisk was appointed to the position in April of that year.

Initially, two OSSA divisions and two offices handled applications—the Environmental Observation and Communications Divisions and the Information Systems and Materials Processing Offices (Figure 2–2). The Information Systems Office was responsible for NASA's long-term data archives, institutional computer operations in support of ongoing research programs, and advanced planning and architecture definition for future scientific data systems. Anthony Villasenor served as acting manager of this office until Caldwell McCoy, Jr., assumed the position of manager in 1983. McCoy held the post until the office merged with the Communications Division in 1987.

Robert Lovell led the Communications Division until he left in early 1987. The division director position remained vacant until Ray Arnold became acting division director later that year. He was appointed permanent director of the division, which had merged with the Information Systems Office in September 1987 to become the new Communications and Information Systems Division. This new division handled all the communications and data transmission needs of OSSA.

Shelby G. Tilford led the Environmental Observation Division until it was disestablished in January 1984. He then assumed leadership of the

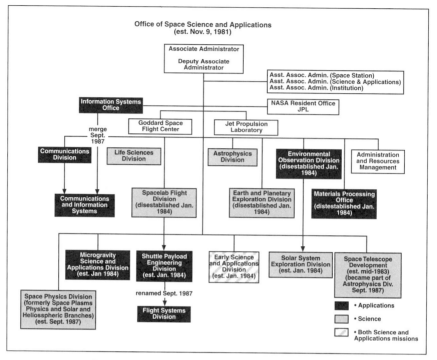

Figure 2–2. *Office of Space Science and Applications*

newly established Earth Science and Applications Division. He remained at that post throughout the decade.

Louis R. Testardi managed the Materials Processing Office through 1982, when he left the position. The post remained vacant until Richard Halpern became manager in the first half of 1983. He led the office until it was disestablished in January 1984 and then led the new Microgravity Sciences and Applications Division, where he remained until mid-1986. The position of director of the Microgravity Science and Applications Division then remained vacant until Kathryn Schmoll became acting director in early 1987. Robert Naumann assumed the post of division director in early 1988 and remained until later that year, when Frank Lemkey replaced him as acting division director.

The Shuttle Payload Engineering Division evolved from the Spacelab Flight Division, which had managed the science-related elements of the Spacelab missions. The new division had responsibility for developing and integrating all science- and applications-related Space Shuttle payloads. Michael Sander led the new Shuttle Payload Engineering Division until late 1985, when Robert Benson became acting director of the division. Benson became permanent division director in 1987 and continued leading the renamed Flight Systems Division.

Money for Space Applications

Budget data (request or submission, authorization, and appropriation) for the major budget categories are from the annual Budget Chronological Histories. Request or submission data for the more detailed budget items come from the annual budget estimates produced by NASA's budget office. No corresponding authorization or appropriations data were available. All programmed (actual) figures come from NASA's budget estimates. It should be noted that the amounts in this section reflect the value of the funds at the time that they were submitted; inflation has not been added. The funding histories of NASA applications from 1979 through 1988 appear in Tables 2–3 through 2–54.

Applications Programs

Space Shuttle Payloads

As with NASA's science missions, the Space Shuttle was a natural environment for many applications investigations. NASA conducted three on-board applications missions under the management of OSTA: OSTA-1 in 1981, OSTA-2 in 1983, and OSTA-3 in 1984. It also participated in the Spacelab missions described in Chapter 4, "Space Science," in Volume V of the *NASA Historical Data Book* and in OAST-1, which was managed by the Office of Aeronautics and Space Technology and is addressed in Chapter 3, "Aeronautics and Space Research and Technology," in this volume.

OSTA-1

OSTA-1 flew on STS-2, the second Space Shuttle test flight. It was the Space Shuttle's first science and applications payload. The objectives of OSTA-1 were to:

- Demonstrate the Shuttle for scientific and applications research in the attached mode
- Operate the OSTA-1 payload to facilitate the acquisition of Earth's resources, environmental, technology, and life science data
- Provide data products to principal investigators within the constraints of the STS-2 mission

The experiments selected for the OSTA-1 payload emphasized terrestrial sciences and fit within the constraints of the STS-2 tests. Experiments relating to remote sensing of Earth resources, environmental quality, ocean conditions, meteorological phenomena, and life sciences made up the payload. Five of the seven experiments were mounted on a Spacelab pallet in the Shuttle payload bay (Figure 2–3); two were carried in the Shuttle cabin. The Spacelab Program Office at the Marshall

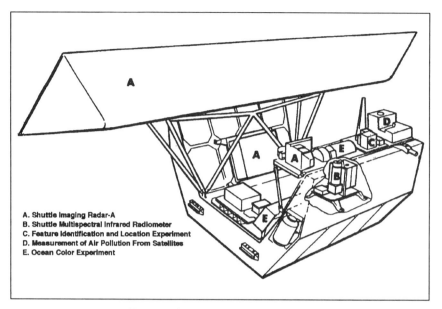

Figure 2–3. OSTA-1 Payload Location

Space Flight Center was responsible for the design, development, and integration of the overall orbital flight test pallet system. Table 2–55 lists the principal investigators and a description of the experiments, including the first Shuttle Imaging Radar (SIR-A), which is depicted in Figure 2–4.

During the flight, *Columbia* assumed an Earth-viewing attitude called Z-axis local vertical, in which the instruments carried in the payload bay were aimed at Earth's surface. Figure 2–5 shows the payload ground coverage and ground resolution of each instrument.

Although most investigation objectives were accomplished, certain conditions affected the quantity and quality of some of the data. During the first twenty-eight hours of the mission, experiment data collection was affected by the loss of one fuel cell and the crew's focus on the orbiter power situation. Instrument operations were restricted to minimize orbiter power usage, and some targets were missed. In addition, the final orbiter maneuvering system burn was delayed for one orbit because of power considerations, which caused the time over specific Earth locations to change and the need to develop new instrument on/off times.

The delay in launch of two hours, forty minutes changed solar illumination conditions along the ground track and the Sun elevation angle, which affected the Ocean Color Experiment, the Shuttle Multispectral Infrared Radiometer, and the Feature Identification and Location Experiment. Cloud cover also affected the Ocean Color Experiment and Shuttle Multispectral Infrared Radiometer targets.

In addition, the shortened mission and intense crew activity limited opportunities for the crew to operate the Nighttime/Daylight Optical Survey of Thunderstorm Lightning (NOSL) experiment. The limited amount of data collected did not allow this experiment to achieve its

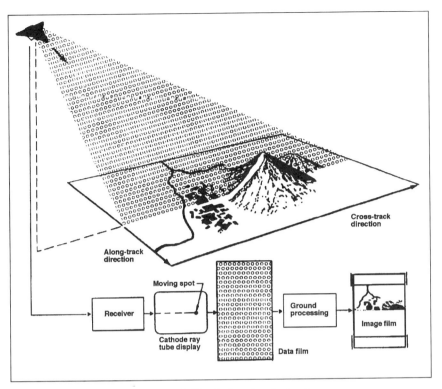

Figure 2–4. Shuttle Imaging Radar-A
(The beam of the SIR-A side-looking radar hit the ground at an angle, giving the resultant image perspective and showing vertical objects in shadowed relief. The intensity of the echoes from the target surface controlled the brightness of a spot tracing a line across a cathode ray tube. An overlapping succession of these lines was recorded on a strip of photographic film moving past the cathode ray tube at a rate proportional to the speed of the Shuttle. Thus, the terrain echo was recorded on the data film with the cross-track dimension across the width of the film. Complex ground processing transformed the data film into an image of the terrain.)

objective of surveying lightning and thunderstorms from space, but the data collected did demonstrate the feasibility of collecting thunderstorm data with the equipment used on this mission. The experiment was reflown on STS-6. The shortened mission also did not allow sufficient time for the Heflex Bioengineering Test to achieve its objective of determining plant growth as a function of initial soil moisture. A mission duration of at least four days was required to permit sufficient growth of the seedlings. This experiment was successfully reflown on STS-3.

OSTA-2

OSTA-2 flew on STS-7. It was the first NASA materials processing payload to use the orbiter cargo bay for experimentation and the initial flight of the Mission Peculiar Equipment Support Structure

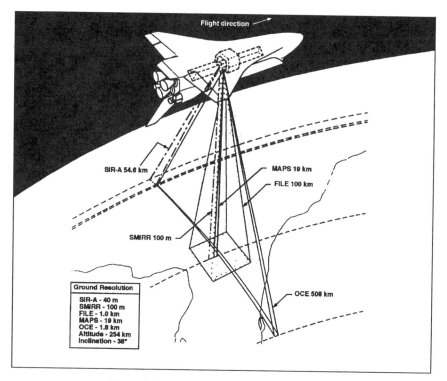

Figure 2–5. OSTA-1 Payload Ground Coverage

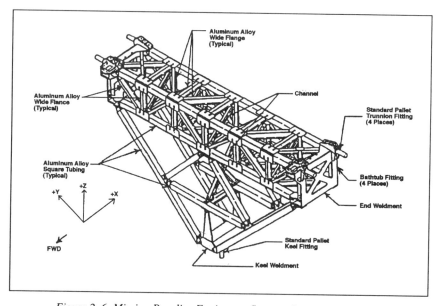

Figure 2–6. Mission Peculiar Equipment Support Structure (MPESS)

(MPESS) carrier (Figure 2–6) and the Materials Experiment Assembly (MEA) payload.

OSTA-2 was a cooperative payload with the Federal Republic of Germany and included three German Project MAUS payloads sponsored by the German Ministry for Research and Technology.[3] The Marshall Space Flight Center developed the NASA facility, and the German facility was developed under the management of the German Aerospace Research Establishment. The primary objectives of OSTA-2 were engineering verifications of the following:

- The MEA facility for the conduct of materials processing experiments
- Materials processing experiment furnaces and apparatus
- The Mission Peculiar Equipment Support Structure system as a carrier of attached payloads

One secondary objectives was to obtain MEA materials science experiment specimens processed in a low-gravity space environment and flight experiment data for scientific investigation. Another secondary objective was to exchange results from MEA and MAUS data analysis between NASA and the German Ministry for Research and Technology.

The elements of the OSTA-2 payload were located on an MPESS in the orbiter carrier bay. In addition to mechanical support, the MPESS provided a near-hemispherical space view for the MEA payload thermal radiator. Payload on/off command switches were activated by the Shuttle crew. Figure 2–7 shows the location of the payload on the MPESS.

The NASA payload, the MEA, was a self-contained facility that consisted of a support structure for attachment to the MPESS and thermal, electrical, data, and structural subsystems necessary to support experiment apparatus located inside experiment apparatus containers. The MEA contained three experiment apparatus that were developed for the Space Processing Applications Rocket project and modified to support OSTA-2 MEA experiments. Two of the three experiment furnaces in the MEA were successfully verified, and scientific samples were processed for analysis. The MEA experiments were selected from responses to an Announcement of Opportunity issued in 1977. The MEA flew again with the German D-1 Spacelab mission on STS 61-A in 1985. The payload demonstrated and verified a cost-effective NASA-developed carrier system. In addition, it demonstrated the reuse of materials processing experiment hardware on the Shuttle that had been developed for suborbital, rocket-launched experiments.

The MAUS experiments were part of the German materials science program, which was established, in part, by the opportunity to fly in Get Away Special (GAS) canisters on a low-cost, space-available basis. The three containers had autonomous support systems, and each container had

---

[3]The acronym MAUS stands for the German name: Materialwissenschaftliche Autonome Experimente unter Schwerelosigkeit.

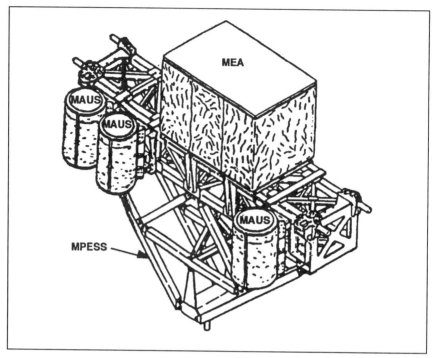

Figure 2–7. OSTA-2 Integrated Payload

its own service module containing experiment hardware, electrical power, experiment control, data acquisition, and storage, as well as housekeeping sensors. Two of the Get Away Special canisters contained identical experiments. The first operated for almost the full programmed duration of approximately eighty hours and shut down automatically. The second shut down prematurely following the first experiment processing cycle. The MEA and MAUS experiments are identified in Table 2–56.

OSTA-3

OSTA-3 was the second in a series of Earth observation payloads that flew on the Shuttle. It flew on STS 41-G. The mission objectives were to:

- Evaluate the utility of advanced remote-sensing systems for various types of Earth observations
- Use remote observations of Earth's surface and its atmosphere to improve current understanding of surficial processes and environmental conditions on Earth

The OSTA-3 payload consisted of four experiments: SIR-B, the Large Format Camera, Measurement of Air Pollution From Satellites (MAPS), and Feature Identification and Landmark Experiment (FILE). All except the Large Format Camera had flown on OSTA-1 on STS-2. SIR-B, MAPS, and FILE were mounted on a pallet carrier (Figure 2–8).

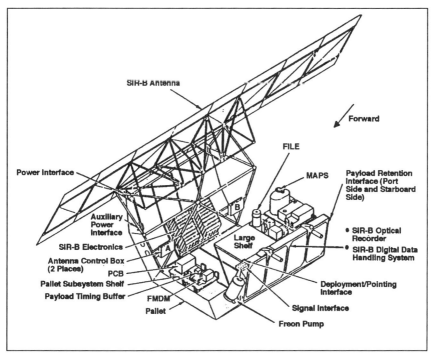

Figure 2–8. OSTA-3 Payload Configuration With FILE, MAPS, and SIR-B

The Large Format Camera was mounted on an MPESS, such as the one used on OSTA-2. It used orbital photography for cartographic mapping and land-use studies at scales of 1:50,000. It obtained 2,289 photographic frames.

The MAPS experiment determined the distribution of carbon monoxide in Earth's lower atmosphere on a global basis, developed an improved understanding of the sources and sinks of atmospheric carbon monoxide, and monitored long-term changes in the total abundance of carbon monoxide within Earth's atmosphere. The data sets of atmospheric carbon monoxide concentration it collected at the start and conclusion of the mission provided the first opportunity to study *in situ* temporal variations in carbon monoxide distribution.

FILE evaluated the utility of multispectral measurements obtained in two spectral channels for classifying surface features or clouds. It was part of an effort to develop advanced sensor systems that in the future could be preprogrammed to acquire imagery of specific types of natural terrain in an automatic fashion. The experiment acquired 240 images over a wide range of environments and successfully classified these scenes.

SIR-B was to use radar imagery acquired under different surface-viewing conditions for various types of surface observations, determine the extent to which subsurface radar penetration occurred in arid environments, and develop improved models of radar backscatter from vegetated terrain and marine areas. The plan was to obtain forty-two hours of digital data that would be analyzed by a science team of forty-three investigators, and eight hours of optical data could be collected as backup. SIR-B actually acquired only seven and a half hours of digital data and eight hours of optical data. Three problems affected the amount of data collected:

1. The Ku-band antenna gimbal failed. It could transmit only prerecorded tape data through the Tracking and Data Relay Satellite System (TDRSS) with special orbiter attitudes. This resulted in acquiring only 20 percent of the planned science data. Therefore, only fifteen investigators received sufficient data (50 to 75 percent) to meet their objectives, twenty-three investigators received a limited amount of data (10 to 50 percent), and six investigators received only a token amount of data.
2. The TDRSS link was lost for twelve hours, forty-two minutes during the mission.
3. Anomalies in the radio frequency feed system to the SIR-B antenna reduced transmitter power and, therefore, degraded the data.

Environmental Observations Program

NASA's Environmental Observations program focused on obtaining and interpreting processes in the magnetosphere, atmosphere, and oceans and extending the capability to predict long- and short-term environmental phenomena and their interaction with human activities. NASA launched two satellite missions in this area—the Stratospheric Aerosol and Gas Experiment (SAGE) and the Earth Radiation Budget Satellite (ERBS)—and worked toward a 1991 launch of the Upper Atmospheric Research Satellite (UARS). In addition, NASA participated in the development and launch of a series of meteorological satellites with NOAA: the NOAA polar-orbiting satellites and the Geostationary Operational Environmental Satellites (GOES).

Stratospheric Aerosol and Gas Experiment

SAGE was part of NASA's Applications Explorer Mission. It represented the first global aerosol data set ever obtained. The experiment complemented two other aerosol satellite experiments—the Stratospheric Aerosol Measurement, flown on Apollo during the Apollo-Soyuz Test Project in 1975, and the Stratospheric Aerosol Measurement II, flown on Nimbus 7, which was launched in 1978 and gathered data at the same time as SAGE. SAGE obtained and used global data on stratospheric aerosols and ozone in various studies concerning Earth's climate and environmental quality. It mapped vertical profiles in the stratosphere of ozone, aerosol, nitrogen dioxide, and molecular extinction in a wide band around the globe. The ozone data extended from approximately nine to forty-six kilometers, the aerosol data ranged from the cloud tops to thirty-five kilometers, the nitrogen dioxide went from about twenty-five to forty kilometers, and the molecular extinction was from about fifteen to forty kilometers. The mission obtained data from tropical to high latitudes for more than three years.

SAGE obtained its information by means of a photometric device. The photometer "looked" at the Sun through the stratosphere's gases and aerosols each time the satellite entered and left Earth's shadow. The device observed approximately fifteen sunrises and fifteen sunsets each twenty-four-hour day—a total of more than 13,000 sunrises and sunsets during its lifetime. The photometer recorded the light in four color bands each time the light faded and brightened. This information was converted to define concentrations of the atmospheric constituents in terms of vertical profiles.

The spacecraft was a small, versatile, low-cost spacecraft that used three-axis stabilization for its viewing instruments. The structure consisted of two major components: a base module, which contained the necessary attitude control, data handling, communications, command, and power subsystems for the instrument module, and an instrument module. The instrument module consisted of optical and electronic subassemblies mounted side by side. The optical assembly consisted of a flat scanning mirror, Cassegrain optics, and a detector package. Table 2–57 contains the instrument module's characteristics. Two solar panels for converting sunlight to electricity extended from the structure. Figure 2–9 shows the SAGE orbit configuration.

SAGE detected and tracked five volcanic eruption plumes that penetrated the stratosphere. It determined the amount of new material each volcano added to the stratosphere. (Mount St. Helens, for example,

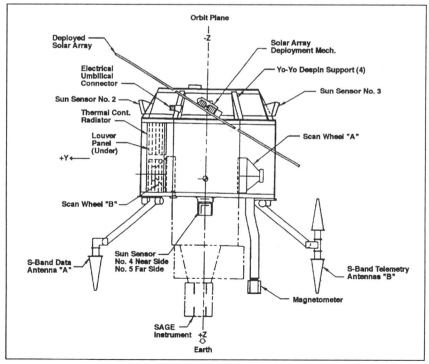

Figure 2–9. SAGE Orbit Configuration

contributed about 0.5×10^6 metric tons for a 100-percent enhancement in background stratospheric aerosol mass.) The characteristics of SAGE are listed in Table 2–58.

Earth Radiation Budget Satellite

ERBS was part of NASA's three-satellite Earth Radiation Budget Experiment (ERBE), which investigated how energy from the Sun is absorbed and re-emitted, or reradiated, by Earth. This process of absorption and reradiation, or reflectance, is one of the principal drivers of Earth's weather patterns. The absorbed solar radiation is converted to heat energy, which increases Earth's temperature and heat content. Earth's heat energy is continuously emitted into space, thereby cooling Earth. The relationship among incident solar energy, reflected solar energy, and Earth-emitted energy is Earth's radiation or energy budget (Figure 2–10). Although observations had been made of incident and reflected solar energy and of Earth-emitted energy, data that existed prior to the ERBE program were not sufficiently accurate to provide an understanding of climate and weather phenomena and to validate climate and long-range weather prediction models. The ERBE program provided observations with increased accuracy, which added to the knowledge of climate and weather phenomena.

Investigators also used observations from ERBS to determine the effects of human activities, such as burning fossil fuels and the use of chlorofluorocarbons, and natural occurrences, such as volcanic eruptions on Earth's radiation balance. The other instruments of the ERBE program were flown on NOAA 9 and NOAA 10.

ERBS was one of the first users of the TDRSS. It was also one of the first NASA spacecraft designed specifically for Space Shuttle deploy-

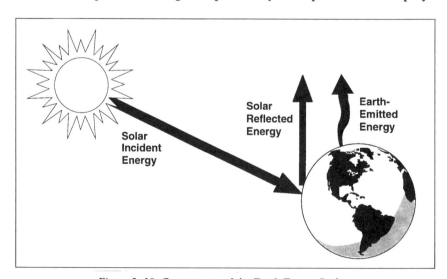

Figure 2–10. Components of the Earth Energy Budget

ment; it was deployed using the Shuttle's Remote Manipulator System. The satellite was equipped with three scientific instruments: SAGE II, the ERBE Non-Scanner, and the ERBE Scanner. Each instrument had one or more contamination doors that protected the instrument's sensitive detectors and optics from accumulating outgassing products from the ERBS spacecraft. Table 2–59 lists the instrument's characteristics.

ERBS provided scientists with the first-ever long-term global monitoring of stratospheric aerosols, including critical ozone data. Investigators used the data to study atmospheric dynamics, ozone chemistry, and ozone depletion. The characteristics of ERBS are in Table 2–60.

Upper Atmospheric Research Satellite

The UARS program continued NASA's investigations of the upper atmosphere carried out by the SAGE and ERBE programs. The national mandate for UARS dates to 1976, when Congress, responding to the identification of new causes of ozone depletion, amended the Space Act and directed NASA to undertake a comprehensive program of research into the upper atmosphere. In 1977, Congress directed NASA to carry out such research "for the purpose of understanding the physics and chemistry of the stratosphere and for the early detection of potentially harmful changes in the ozone in the stratosphere."

NASA stated that the purpose of the mission was to better understand Earth's upper atmosphere, specifically the response of the ozone layer to changes and the role of the upper atmosphere in climate and climate variability. The mission would focus on comprehensive investigations of Earth's stratosphere, mesosphere, and lower thermosphere to understand Earth's upper atmosphere. The major areas to be studied would include energy flowing into and from the upper atmosphere, how sunlight drives chemical reactions in the upper atmosphere, and how gases moved within and between layers of the atmosphere.

NASA's Goddard Space Flight Center would provide the design and definition work with contractor support from the General Electric Space Division. The contractor would be responsible for integrating the instrument module with the bus and flight instruments, conducting environmental testing of the observatory, integrating the observatory into the Space Shuttle, and providing post-launch checkout support. The Goddard Space Flight Center would furnish the Multimission Modular Spacecraft (MMS) bus and flight instruments and design the UARS ground station and data handling facility. Goddard would award a contract for the Central Data Handling Facility, remote analysis computers, and the development of software to perform UARS-unique systems functions.

NASA released its Announcement of Opportunity for the mission in 1978, and the agency selected sixteen experiments and ten theoretical investigations from seventy-five proposals for definition studies in April 1980. In November 1981, NASA narrowed this down to nine instrument experiments, two instruments flown on "flights of opportunity," and ten theoretical investigations. (One "instrument of opportunity," the solar backscattered ultraviolet sensor for ozone, was deleted from the payload in 1984 because an identical instrument was designated to be flown on an operational NOAA satellite during the same timeframe.)

Congress funded the experiments in its fiscal year 1984 budget and approved funding for UARS mission development in its fiscal year 1985 budget. NASA awarded the major observatory contract to General Electric in March 1985 and initiated the execution phase in October 1985. Following the *Challenger* accident, safety concerns led to a redesign of one of the instruments and rebaselining of the mission timeline, with launch rescheduled for the fall of 1991.

Initially, the program concept involved two satellite missions, each with a nominal lifetime of eighteen months and launched one year apart. It was reduced to a single satellite mission in 1982.

In its final configuration, the mission would use the MMS to place a set of nine instruments in Earth orbit to measure the state of the stratosphere and provide data about Earth's upper atmosphere in spatial and temporal dimensions. The remote atmospheric sensors on UARS would make comprehensive measurements of wind, temperature, pressure, and gas species concentrations in the altitude ranges of approximately nine to 120 kilometers. In addition, a tenth instrument, not technically a part of the UARS mission, would use its flight opportunity to study the Sun's energy output. Table 2–61 describes the instruments carried on aboard UARS, what they measured, and their principal investigators. The spacecraft and its instruments were considerably larger than other remote-sensing spacecraft flown up to that time. Figure 2–11 compares the size of UARS with two earlier missions, Nimbus 7 and Landsat-D; Figure 2–12 shows the instrument placement and the MMS.

A chronology of events prior to the September 1991 launch is presented in Table 2–62. It is notable that even with the redesign of one instrument and a rebaselining of the mission timeline because of the *Challenger* accident, NASA launched UARS approximately $30 million below its final budget estimate of $669.5 million and with no schedule delays.

Meteorological Satellites

NASA and NOAA launched and operated two series of meteorological satellites: the NOAA polar-orbiting satellites and the Geostationary Operational Environmental Satellites (GOES)—a group of geosynchronous satellites. A NASA-Department of Commerce agreement dated July 2, 1973, governed both satellite systems and defined each agency's responsibilities. NOAA had responsibility for establishing the observational requirements and for operating the system. NASA was responsible for procuring and developing the spacecraft, instruments, and associated ground stations, for launching the spacecraft, and for conducting an on-orbit checkout of the spacecraft.

SPACE APPLICATIONS

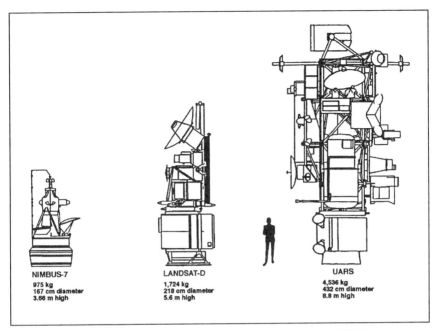

*Figure 2–11. Relative Sizes of Nimbus-7, Landsat D, and UARS
(The width of UARS was essentially that of the Shuttle bay.)*

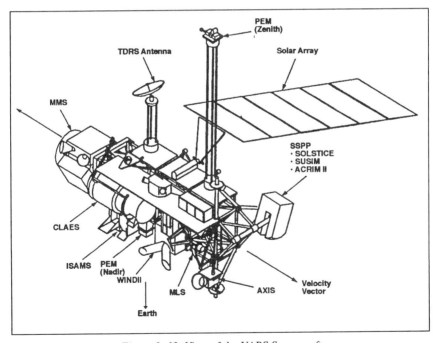

*Figure 2–12. View of the UARS Spacecraft
(From the anti-Sun side, this shows instrument placement, solar array, and the
Multimission Modular Spacecraft. The Halogen Occultation Experiment (HALOE)
and High Resolution Doppler Images instruments cannot be seen from this view.)*

NOAA Polar-Orbiting Satellites

The series of polar-orbiting meteorological satellites that operated during the late 1970s and into the 1990s began with TIROS-N, launched in October 1978. TIROS-N was the operational prototype for the third generation of low-Earth orbiting weather satellites designed and developed by NASA to satisfy the increasing needs of the operational system. The satellites in this TIROS-N series were Sun synchronous, near polar-orbiting spacecraft, and operated in pairs, with one crossing the equator near 7:30 a.m. local time and the second crossing the equator at approximately 1:40 p.m. local time. Operating as a pair, these satellites ensured that nonvisible data for any region of Earth was no more than six hours old.

The NOAA series of satellites was a cooperative effort of the United States (NOAA and NASA), the United Kingdom, and France. NASA funded the development and launch of the first flight satellite (TIROS-N); subsequent satellites were procured and launched by NASA using NOAA funds. The operational ground facilities, including the command and data acquisition stations, the Satellite Control Center, and the data processing facilities (with the exception of the Data Collection System processing facility), were funded and operated by NOAA. The United Kingdom, through its Meteorological Office, Ministry of Defense, provided a stratospheric sounding unit, one of three sounding instruments for each satellite. The Centre Nationale d'Études Spatiales (CNES) of France provided the Data Collection System instrument for each satellite and the facilities needed to process and make the data obtained from this system available to users. CNES also provided facilities for the receipt of sounder data during the blind orbit periods. Details of the TIROS-N satellite can be found in Volume III of the *NASA Historical Data Book*.[4] The satellites launched from 1979 through 1988 are described below.

Instruments on these satellites measured the temperature and humidity of Earth's atmosphere, surface temperature, surface and cloud cover, water-ice-moisture boundaries, and proton and electron flux near Earth. They took atmospheric soundings, measurements in vertical "slices" of the atmosphere showing temperature profiles, water vapor amounts, and the total ozone content from Earth's surface to the top of the atmosphere. Sounding data were especially important in producing global weather analyses and forecasts at the Weather Service's National Meteorological Center. Table 2–63 summarizes the orbit and instrument complement of the NOAA satellites.

The TIROS-N satellites also collected environmental observations from remote data platforms—readings such as wave heights on the oceans, water levels in mountainous steams, and tidal activity. The spacecraft also monitored solar particle radiation in space used, in part, to warn

---

[4]Linda Neuman Ezell, *NASA Historical Data Book, Volume III: Programs and Projects, 1969–1978* (Washington, DC: NASA SP-4012, 1988).

Space Shuttle missions and high-altitude commercial aircraft flights of potentially hazardous solar radiation activity. The NOAA 6 and NOAA 7 satellites were almost identical to the 1978 TIROS-N. The NOAA 8, 9, 10, and 11 satellites were modified versions of TIROS-N and were called Advanced TIROS-N.

The Advanced TIROS-N generation of satellites included a new complement of instruments that emphasized the acquisition of quantitative data of the global atmosphere for use in numerical models to extend and improve long-range (three- to fourteen-day) forecasting ability. In addition, the instruments on these satellites could be used for global search and rescue missions, and they could map ozone and monitor the radiation gains and losses to and from Earth.

NOAA 6. This was the second of eight third-generation operational meteorological polar-orbiting spacecraft. It was the first NOAA-funded operational spacecraft of the TIROS-N series. The satellite greatly exceeded its anticipated two-year lifetime and was deactivated on March 31, 1987. Identical to TIROS-N, NOAA 6 adapted applicable parts of the Defense Meteorological Satellite Program Block 5D spacecraft, built by RCA Corporation and first launched in 1976.

NOAA 6 filled in data-void areas, especially over the oceans, by crossing the equator six hours after TIROS-N, in effect doubling the amount of data made available to the National Meteorological Center in Suitland, Maryland. TIROS-N and NOAA 6, each viewing every part of the globe twice in one twenty-four-hour period, were especially important in providing information from remote locations where more traditional weather-gathering methods could not be used conveniently. Table 2–64 lists the characteristics of NOAA 6.

NOAA B. This satellite went into a highly elliptical rather than the planned circular orbit of 756 kilometers. This was because of one of the Atlas F booster engines developing only 75 percent thrust. The satellite could not operate effectively. It was to have been the second NOAA-sponsored TIROS-N satellite. Its characteristics are in Table 2–65.

NOAA 7. With the successful launch of NOAA 7, designed to replace TIROS-N and join NOAA 6, meteorologists had two polar-orbiting satellites in orbit returning weather and environmental information to NOAA's National Earth Satellite Service. Together, NOAA 6 and NOAA 7 could view virtually all of Earth's surface at least twice every twenty-four hours.

In addition to the data transmitted by earlier NOAA satellites, NOAA 7 provided improved sea-surface temperature information that was of special value to the fishing and marine transportation industries and weather forecasters. Its scanning radiometer, the Advanced Very High Resolution Radiometer (AVHRR), used an additional fifth spectral channel to gather visual and infrared imagery and measurements. Table 2–66 lists the characteristics of each channel. The satellite also carried a joint Air Force-NASA contamination monitor that assessed possible environmental contamination in the immediate vicinity of the spacecraft resulting from its propulsion systems.

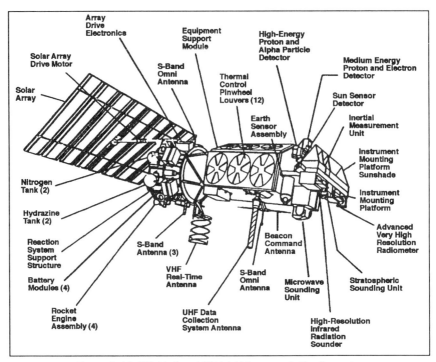

Figure 2–13. NOAA 6 and NOAA 7 Spacecraft Configuration

NOAA 6 and NOAA 7 also served a communications function and could distribute unprocessed sensor data to Earth stations in more than 120 countries in real time as the spacecraft passed overhead. Figure 2–13 shows the NOAA 6 and 7 spacecraft configuration.

NOAA 7 was put in standby mode when its sounder failed and its power system degraded. It was deactivated in June 1986 when the power system failed. Its characteristics are listed in Table 2–67.

NOAA 8. This was the fourth NOAA-funded operational spacecraft of the TIROS-N series to be launched. It was a "stretched" version of the earlier NOAA TIROS-N spacecraft (although not larger in size) and was the first advanced TIROS-N spacecraft with expanded capabilities for new measurement payloads. Because of the need to get an early flight of the Search and Rescue (SAR) mission, NOAA 8 was launched prior to NOAA-D, which did not have a SAR capability.

The satellite experienced problems beginning in June 1984, about 14 months after launch, when it experienced a "clock interrupt" that caused the gyros to desynchronize. Continued clock disturbances interfered with the meteorological instruments, preventing investigators from obtaining good data. In July 1984, NASA and NOAA announced that the satellite appeared to have lost its latitude control system and was tumbling in orbit and unable to relay its signal effectively to Earth. Engineers were able to stabilize the satellite in May 1985, when the defective oscillator gave out and scientists could activate a backup oscillator and reprogram the satellite remotely. It resumed transmission of data and was declared operational in July 1985. It tumbled again on October 30, 1985, and was recovered and reactivated on December 5. Use of the satellite was finally lost on December 29, 1985, following clock and power system failures. Table 2–68 lists NOAA 8's characteristics.

NOAA 9. This was the fifth NOAA funded operational spacecraft of the TIROS-N series and the second in the Advanced TIROS-N spacecraft series. It carried two new instruments, as well as a complement of instruments on previous NOAA satellites. The Solar Backscatter Ultraviolet (SBUV)/2 spectral radiometer acquired data to determine atmospheric ozone content and distribution. It was the successor to the SBUV/1, which flew on Nimbus 7. The Earth Radiation Budget Experiment (ERBE) provided data complementing the Earth Radiation Budget Satellite (ERBS) that NASA launched in October 1984. It made highly accurate measurements of incident solar radiation, Earth-reflected solar radiation, and Earth-emitted longwave radiation at spatial scales ranging from global to 250 kilometers and at temporal scales sufficient to generate accurate monthly averages. Figure 2–14 shows the NOAA 9 spacecraft configuration.

This satellite also carried SAR instrumentation provided by Canada and France under a joint cooperative agreement. It joined similarly equipped COSPAS satellites launched by the Soviet Union. The spacecraft replaced NOAA 7 as the afternoon satellite in NOAA's two polar satellite system. Its characteristics are in Table 2–69.

NOAA 10. This spacecraft circled the globe fourteen times each day, observing a different position on Earth's surface on each revolution as Earth turned beneath the spacecraft's orbit (Figure 2–15). It replaced NOAA 6 as the morning satellite in NOAA's two polar orbit satellite system and restored NOAA's ability to provide full day and night environmental data, including weather reports, and detect aircraft and ships in distress after one of the two TIROS-N satellites shut down in December 1985. (NOAA 6 had been reactivated when NOAA 8 failed.) It was the third of the Advanced TIROS-N spacecraft. The spacecraft was launched from a twenty-five-year-old refurbished Atlas E booster, a launch that had been delayed sixteen times during the previous year because of a series of administrative changes and technical difficulties.

To continue initial support for SAR using the 121.5/243 megahertz (MHz) system and to begin the process for making the system operational for the 406-MHz system, NOAA 10 carried special instrumentation for evaluating a satellite-aided SAR system that would lead to the establishment of a fully operational capability. Less than twenty-four hours after being put into operation on NOAA 10, SARSAT (Search and Rescue Satellite-Aided Tracking) equipment on board picked up the first distress signals of four Canadians who had crashed in a remote area of Ontario. NOAA's characteristics are in Table 2–70.

NOAA 11. This satellite replaced NOAA 9 as the afternoon satellite in NOAA's two polar satellite system. The satellite carried improved instrumentation that allowed for better monitoring of Earth's ozone layer. The launch of NOAA 11 had originally been scheduled for October 1987,

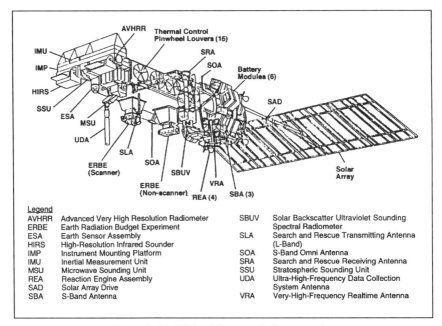

Figure 2–14. NOAA 9 Spacecraft Configuration

but it had been postponed eight times because of management and technical delays.

The Advanced TIROS-N system of satellites normally operated with four gyroscopes—three for directional control and one backup. One gyro on NOAA 11 failed in August 1989, and the backup was put into service. A second gyro failed in 1990, but NASA had developed and transmitted to the satellite software instructions that permitted the satellite to operate fully on two gyros. The characteristics of NOAA 11 are in Table 2–71.

Geosynchronous Operational Environmental Satellites

The impressive imagery of cloud cover produced by the GOES series, as viewed from geostationary (or geosynchronous) orbit, has become a highlight of television weather forecasts. The GOES program has been a joint development effort of NASA and NOAA. NASA provided launch support and also had the responsibility to design, engineer, and procure the satellites. Once a satellite was launched and checked out, it was turned over to NOAA for its operations.

The GOES program has provided systematic, continuous observations of weather patterns since 1974. The pilot Synchronous Meteorological Satellite, SMS-A, was launched in 1974, followed by a second prototype, SMS-B, and an operational spacecraft, SMS-C/GOES-A. Subsequently, GOES-B was successfully launched in 1977, with GOES-C launched in 1978. The GOES spacecraft obtained both day and night information on Earth's weather through a scanner that formed images of Earth's surface and cloud cover for transmission to regional

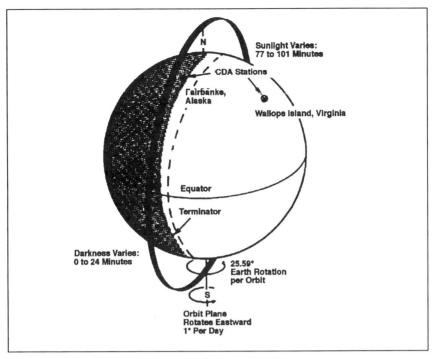

Figure 2–15. NOAA 10 Orbit

data-user stations for use in weather prediction and forecasting.

The GOES satellites during this period (GOES 4 through 7) had similar configurations (Figure 2–16). Beginning with the launch of GOES 4 in 1980 and continuing throughout the series, the instrument complement included an improved Visible/Infrared Spin Scan Radiometer (VISSR) (Figure 2–17). The new VISSR, called the VISSR Atmospheric Sounder, could receive the standard operational VISSR data and also sound the atmosphere in twelve infrared bands, enabling meteorologists to acquire temperature and moisture profiles of the atmosphere (Table 2–72).

Normally, two GOES satellites operated concurrently. GOES-East satellites were stationed at seventy-five degrees west longitude, and GOES-West satellites were located at 135 degrees west longitude. GOES-East observed North and South America and the Atlantic Ocean. GOES-West observed North America and the Pacific Ocean to the west of Hawaii. Together, these satellites provided coverage for the central and eastern Pacific Ocean, North, Central, and South America, and the central and western Atlantic Ocean.

GOES 4. This was the sixth satellite in the GOES series. It provided continuous cloud cover observations from geosynchronous orbit. Initially located at ninety-eight degrees west longitude, it was moved into a geostationary orbit located at 135 degrees west longitude in February 1981 to replace the failing GOES 3 (also known as GOES-C) as the operational GOES-West satellite. GOES 4 was the first geosynchronous satellite capable of obtaining atmospheric temperature and

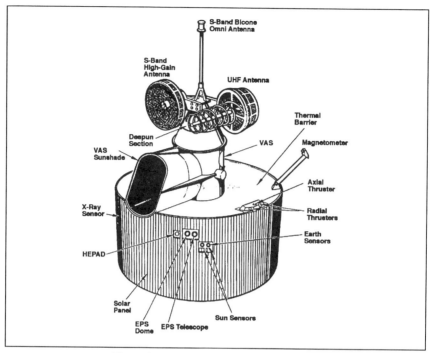

Figure 2–16. GOES Satellite Configuration

water vapor soundings as a function of altitude in the atmosphere. The data were extremely important in forecasting and monitoring the strength and course of highly localized severe storms. It also had the same imaging capability as previous GOES spacecraft.

GOES 4 experienced several anomalies while in orbit. The most serious occurred on November 25, 1982, when the VISSR Atmospheric Sounder's scan mirror stopped during retrace after exhibiting excessively high torque. Efforts to restore either the visible or infrared capability were unsuccessful. The characteristics of GOES 4 are in Table 2–73.

GOES 5. This satellite was placed into a geostationary orbit located seventy-five degrees west longitude and became the operational GOES-East satellite. The satellite failed on July 29, 1984, and GOES 6 (launched in April 1983) was moved into a central location over the continental United States. Table 2–74 lists the characteristics of GOES 5.

GOES 6. This was placed into geostationary orbit located at 135 degrees west longitude and acted as the operational GOES-West satellite. It was moved to ninety-eight degrees west longitude to provide coverage after GOES 5 failed. After the successful launch and checkout of GOES 7 in 1987, it was returned to its original location. GOES 6 failed in January 1989. The satellite's characteristics are in Table 2–75.

GOES G. This satellite, which was planned to become the eastern operational GOES satellite designated as GOES 7, did not reach operational orbit because of a failure in the Delta launch vehicle. NASA attributed this failure to an electrical shortage that shut down the engines on the

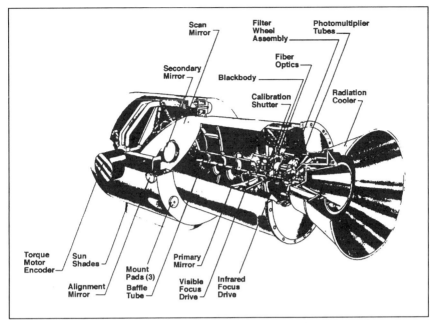

Figure 2–17. VISSR Atmospheric Sounder

launch vehicle. GOES G had the same configuration and instrument complement as earlier GOES spacecraft; its characteristics are in Table 2–76.

GOES 7. The GOES 7 spacecraft was placed into a geostationary orbit located at seventy-five degrees west longitude and acted as the operational GOES-East satellite beginning on March 25, 1987. Its placement allowed GOES 6 to return to its normal position of 135 degrees west longitude from its location at ninety-eight degrees west longitude. GOES 7 was equipped with two encoders: one with two of the same type of tungsten-filament lamps as in the previous GOES spacecraft and the other with light-emitting diodes, which had a longer life expectancy than the original lamps.

The spacecraft was moved to ninety-eight degrees west longitude in July 1989 following the January 1989 failure of GOES 6. It was moved back to 108 degrees west in November 1989. It underwent several more relocations during its more than eight-year lifetime. It was finally shut down in January 1996. The characteristics of GOES 7 are in Table 2–77.

Resource Observations Program

The goals of the Resource Observations program was to assist in solving Earth resources problems of national and global concern through the development and application of space technology and techniques and to conduct research and observations to improve our understanding of the dynamic characteristics of Earth. The program focused on developing and transferring remote-sensing techniques to federal agencies, state, regional, and local governments, private industry, and the scientific community, where these techniques would enhance or supplant existing capabilities or provide a new capability. From 1979 to 1988, NASA launched three resource observations satellites: two Landsat satellites and Magsat.

Landsat Satellites

The Landsat program began in the late 1960s. NASA launched Landsat 1 in July 1972, followed by the launch of Landsat 2 in January 1975 and Landsat 3 in March 1978. These three satellites successfully used the Multispectral Scanner (MSS) to collect and measure the energy reflected or emitted in discrete intervals of the electromagnetic spectrum. The MSS surveyed both renewable and nonrenewable Earth resources. It monitored the reflected solar energy in the green, red, and near-infrared parts of the spectrum and added to the ability to monitor and understand the dynamics and character of the various features and materials on and below the surface of Earth.

The data acquired by Landsat were used worldwide by government agencies, research institutions, and other organizations and individuals seeking information to assist in oil and mineral exploration; agriculture, forestry, and water management; map making; industrial plant site identification and location; and general land-use planning. When Landsat 4 launched, eleven nations could receive and process data directly from the satellite. In addition, more than 100 nations used Landsat data for resource development and management.

NASA was responsible for operating the Landsats through the early 1980s. In January 1983, operations of the Landsat system were transferred to NOAA. In October 1985, the Landsat system was commercialized, and NOAA selected the Earth Observation Satellite Company (EOSAT) to operate the system under a ten-year contract. Under the agreement, EOSAT would operate Landsats 4 and 5, build two new spacecraft (Landsats 6 and 7), have exclusive rights to market Landsat data collected prior to the date of the contract (September 27, 1985) until its expiration date of July 16, 1994, have exclusive right to market data collected after September 27, 1985, for ten years from date of acquisition, and receive all foreign ground station fees.

Landsat 4. This was fourth in a series of near-polar-orbiting spacecraft. In addition to the MSS flown on the earlier Landsat missions, Landsat 4 introduced the Thematic Mapper (TM), whose configuration is shown in Figure 2–18. The TM extended the data set of observations provided by the MSS. It provided data in seven spectral bands, with significantly improved spectral, spatial, and radiometric resolution. Table 2–78 compares the major characteristics of the two instruments.

Both Landsat 4 instruments imaged the same 185-kilometer swath of Earth's surface every sixteen days. The two instruments covered all of Earth, except for an area around the poles, every sixteen days. Image data were transmitted in real time via the Tracking and Data Relay Satellite (TDRS) to its ground terminal at White Sands, New Mexico, beginning August 12, 1983. Prior to that time, the downlink communications mode for MSS data was through the Landsat 4 direct-access S-band link. TM data were transmitted directly to the ground through the X-band.

Landsat 4 consisted of NASA's standard Multimission Modular Spacecraft and the Landsat instrument module (Figure 2–19). The TM was located between the instrument module and the Multimission Modular Spacecraft modular bus, and the MSS was located at the forward end of the instrument module.

NASA launched and checked out the spacecraft, established the precise orbit, and demonstrated that the system was fully operational before transferring management to NOAA. NOAA was responsible for controlling the spacecraft, scheduling the sensors, processing and distributing data from the MSS, and reproducing and distributing public domain data from the TM. NOAA assumed operational responsibility for Landsat 4 on January 31, 1983. The TM remained an experimental development project under direct NASA management.

On February 15, 1983, the X-band transmitter on the spacecraft, which sent data from the TM to ground stations, failed to operate. No further data from the TM would be provided until the TDRS began transmitting TM data in August 1983. The less detailed pictures, which were transmitted from the Multimission Modular Spacecraft on the S-band, continued to be sent. Another problem occurred in 1983 when two solar panels failed. The system was able to continue operating with only two solar panels, but preparations were made to move the spacecraft into a lower orbit, and Landsat D' (D "prime," to become Landsat 5) was readied for a March 1984 launch. However, it was decided to allow Landsat 4 to continue operating, which it did into the 1990s. The satellite's characteristics are in Table 2–79.

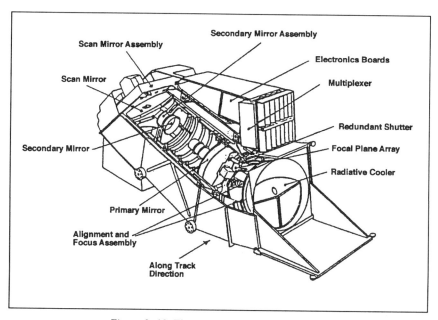

Figure 2–18. Thematic Mapper Configuration

Landsat 5. NASA developed Landsat 5 as Landsat D'. It was intended first to back up and then to replace Landsat 4 when it seemed that Landsat 4's operational days were numbered. However, Landsat 4 continued operating, and Landsat 5 was able to double the amount of remote-sensing data that the system transmitted by providing eight-day rather than sixteen-day repeat coverage. It was virtually identical to Landsat 4, but was modified to prevent the failures experienced on Landsat 4.

Image data were transmitted in real time through the Ku-band via the TDRS to its ground terminal at White Sands, New Mexico. Image data could also be transmitted directly to ground stations through the X-band in addition to or in lieu of transmission via the TDRS. A separate S-band direct link compatible with Landsats 1 through 4 was also provided to transmit MSS data to those stations equipped for receiving only S-band transmissions.

Landsat 5 was turned over to NOAA for management and operations on April 6, 1984. It continued to transmit data into the 1990s. Table 2–80 lists its characteristics.

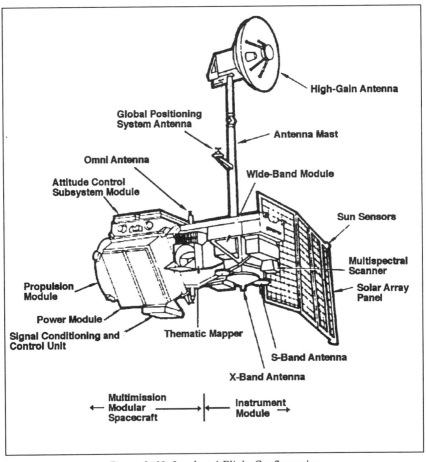

Figure 2–19. Landsat 4 Flight Configuration

Magsat (AEM-C)

Magsat (Magnetic Field Satellite) was the third spacecraft in the Applications Explorer Mission series. From its launch on October 30, 1979, until its reentry on June 11, 1980, its instruments continually measured the near-Earth magnetic field. Magsat was the first spacecraft in near-Earth orbit to carry and use a vector magnetometer to resolve ambiguities in field modeling and magnetic anomaly mapping. The anomalies measured reflected important geologic features, such as the composition and temperature of rock formation, remnant magnetism, and geologic structure on a regional scale. Magsat provided information on the broad structure of Earth's crust with near-global coverage.

Prior to the satellite era, magnetic data from many geographic regions were nonexistent or sparse. The Polar Orbiting Geophysical Observatory (POGO) and the Orbiting Geophysical Observatories 2, 4, and 6 satellites made global measurements of the scalar field from October 1965 through June 1971, and several geomagnetic field models based on POGO data were published. Their magnetometers provided measurements of the scalar field magnitude approximately every half second over an altitude range of about 400 to 1,500 kilometers.

These satellite geomagnetic field measurements mapped the main geomagnetic field originating in Earth's core, determined the long-term temporal, or secular, variations in that field, and investigated short-term field perturbations caused by ionospheric currents. Early in the POGO era, it was thought to be impossible to map crustal anomalies from space. However, while analyzing data from POGO, investigators discovered that the lower altitude data contained separable fields because of anomalies in Earth's crust, thus allowing for the development of a new class of investigations. Magsat data enhanced POGO data in two areas:

1. Vector measurements were used to determine the directional characteristics of anomaly regions and resolved ambiguities in their interpretation.
2. Lower altitude data provided increased signal strength and resolution for detailed studies of crustal anomalies.

Magsat was made of two modules. The base module housed the electrical power supply system, the telemetry system, the attitude control system, and the command and data handling system. The instrument module comprised the optical bench, star cameras, attitude transfer system, magnetometer boom and gimbal systems, scalar and vector magnetometers, and precision Sun sensor. Figure 2–20 shows the orbital configuration.

Magsat's lifetime exceeded its planned minimal lifetime by nearly three months, and it met or exceeded all the accuracy requirements of the scalar and vector magnetometers as well as attitude and position determination. The program was a cooperative effort between NASA and the

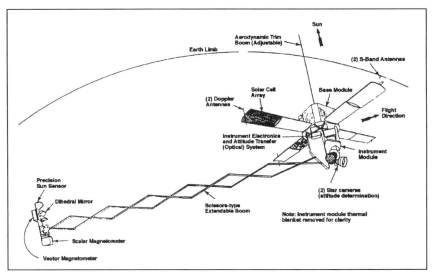

Figure 2–20. Magsat Orbital Configuration

U.S. Geological Survey, which used the Magsat observations and models to update the regional and global magnetic charts and maps that it published. Table 2–81 lists Magsat's characteristics, and Table 2–82 contains the satellite's investigations.

Communications Program

Advanced Communications Technology

NASA's participation in communications satellite programs had been severely curtailed in 1973 because of budget constraints. Not until late 1979, when it became apparent that current communications capabilities would be inadequate to meet the rising demand foreseen for the 1990s, did NASA decide to renew its programs directed at advanced communications satellite research and technology. It gave the Lewis Research Center the lead responsibility for a program that NASA hoped would culminate in the development and launch of a sophisticated communications satellite in 1985 or 1986. NASA concluded that emphasis needed to be placed on developing technology that would open the thirty/twenty-gigahertz (GHz) frequency band (Ka-band) for commercial use. The major advantage of the thirty/twenty-GHz band was the broad frequency range allocated to communications satellite use—five times the band allocated at the C-band and Ku-band that were presently in use.

Although both NASA and Congress agreed on the necessity for such a program, they debated for the next few years over whether the effort should be funded primarily by the government or by industry. Funding for ground-based research, already in the budget, would continue, but money for a flight demonstration, which NASA and industry were convinced would soon be necessary, was removed from both the initial fiscal year 1982 and fiscal year 1983 budget requests. Congress contended that industry should bear more of the cost, but industry representatives responded that, while they were willing to contribute, the cost of a flight demonstration was beyond their means. In hearings before the House Space Subcommittee in July 1981, NASA Associate Administrator Dr. Anthony Calio stated that the United States was already behind Japan and Europe when it came to developing the thirty-twenty-GHz technology. He also agreed that, given the small profit share awarded to satellite builders, industry could not justify funding the demonstration itself.

The initiative was popular with some members of Congress, however, in spite of the Reagan administration's statement that flight testing was not in NASA's mandate. In April 1982, experts in the communications field testified that unless NASA was allowed to continue the program, foreign competitors were likely to gain significantly in the communications market. In May 1982, the Senate Committee on Appropriations earmarked $15.4 million of NASA's fiscal year 1982 budget for work on a thirty/twenty-GHz test satellite by adding to the Urgent Supplemental Bill.

In January 1983, funding for a new Advanced Communications Technology Satellite (ACTS) was placed in the fiscal year 1984 budget. In March, the Lewis Research Center released a request for proposal for the design, development, building, and launch of ACTS, which was then scheduled for a 1988 launch by the Space Shuttle. In August 1984, NASA awarded an industry team headed by RCA's Astro-Electronics Division a $260.3 million contract for the design, development, and fabrication of ACTS. Other major participants were TRW Electronics System Group, Communications Satellite Corporation (Comsat), Motorola Inc., Hughes Aircraft Company, and Electromagnetic Sciences Inc. The ACTS program was to develop advanced satellite communications technologies, including satellite switching and processing techniques and multibeam satellite antennas, using the thirty/twenty-GHz bands. The program would make the ACTS spacecraft and ground systems capabilities for experimentation available to corporations, universities, and government agencies.

The program still did not progress smoothly, however, as funding levels fluctuated during the next few years (see Tables 2–45 and 2–51). NASA more than once reduced its funding request in response to the Reagan administration's attempt to terminate the program. Congress directed NASA to continue the program as planned and restored its funding. These disputes took their toll, and ACTS was not launched until September 1993.

Search and Rescue

NASA's other major communications initiative was in the area of search and rescue. In the Search and Rescue Satellite-Aided Tracking (SARSAT) System, survivors on the ground or on water send up an Emergency Position Indicating Radio Beacon (EPIRB). Distressed planes use the Emergency Locator Transmitter (ELT) to the SARSAT satellite.

A satellite equipped with SARSAT equipment receives the message from the EPIRB or ELT unit and relays it to the Local User Terminal (LUT). The LUT then relays the message to a mission control center, which alerts the Rescue Coordination Center. The Rescue Coordination Center team radios a search-and-rescue unit to look for the missing or distressed persons or vehicles.

The instruments on COSPAS/SARSAT satellites (COSPAS satellites were the Soviet search-and-rescue satellites) were designed to receive 121.5/243- and 406-MHz distress signals from Earth. Signals sent on the 121.5/243-MHz frequencies allowed for location determination within twenty kilometers of the transmission site. These signals were received by the search-and-rescue repeater and transmitted in real time over 1,544.5 MHz to the LUT on the ground.

The instruments could determine the frequency of the distress signal "Doppler shift" caused by the motion of the spacecraft in relation to the beacon. This shift provided a measurement for computation of the emergency location. The distress location alerts were then relayed from the spacecraft to the LUTs on the ground and from there to the mission control centers. With four operational satellites in orbit (NOAA and Soviet satellites), the time until contact between an individual in an emergency situation and a satellite varied from a few minutes to a few hours. Figure 2–21 shows the basic concept of satellite-aided search and rescue.

The use of meteorological satellites for search-and-rescue operations was first envisioned in the late 1950s. NASA began to experiment with "random-access Doppler tracking" on the Nimbus satellite series in the 1970s. In these experiments, instruments located and verified transmissions from remote terrestrial sensors (weather stations, buoys, drifting balloons, and other platforms). The first operational random-access Doppler system was the French ARGOS on the NOAA TIROS satellite series. The 406-MHz search-and-rescue system evolved from this ARGOS system.

The COSPAS/SARSAT program became an international effort in 1976, with the United States, Canada, and France discussing the possibilities of satellite-aided search and rescue. Joint SARSAT testing agreements in 1979 stated that the United States would supply the satellites, Canada would supply the spaceborne repeaters for all frequencies, and France would supply the spaceborne processors for the 406-MHz frequency. The Soviet Union joined the program in 1980, with the Ministry of Merchant Marine agreeing to equip their COSMOS satellites with COSPAS repeaters and processors. Norway joined the program in 1981, also representing Sweden.

COSPAS/SARSAT experimental operations began in 1982. The first COSPAS launch took place on June 30, 1982, and the operations of four North American ground stations began following a period of joint checkout by the United States, the Soviet Union, Canada, and France. The first satellite-aided rescue occurred not long after the launch. The United Kingdom also joined the program. The first SARSAT satellite, NOAA 8, was launched in 1983.

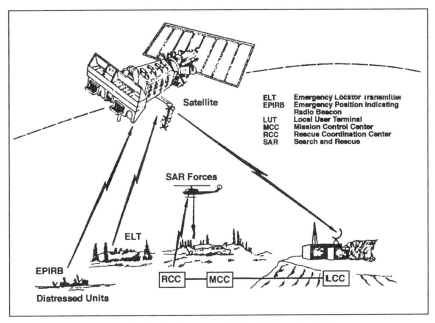

Figure 2–21. Basic Concept of Satellite Search and Rescue

By 1984, the system constellation consisted of two COSPAS and two SARSAT satellites. Bulgaria and Finland also joined the program in 1984. A second SARSAT-Soviet agreement was signed that year, which extended cooperation to 1990. In 1984, NASA turned over the U.S. SARSAT leadership to NOAA, but the space agency continued its role in the areas of research and development.

The full use of the 406-MHz system, designed for global coverage by satellite, was initiated in 1985. Signals sent on the 406-MHz frequency allowed for location determination within five kilometers of the transmission site. In addition, on-board memories stored the 406-MHz data for later transmission in case the signals that were sent in real time were not within range of a ground station. This resulted in global coverage.

The search-and-rescue mission objectives were to:

1. Continue the initial operational use of a spaceborne system to acquire, track, and locate the existing ELTs and EPIRBs that were in the field operating on 121.5 MHz and 243 MHz

2. Demonstrate and provide for operational use of the improved capability for detecting and locating distress incidents utilizing new ELT/EPIRBs operating on 406 MHz (This new capability would provide higher probability of detection and location, greater location accuracy, and coded user information and allow for the necessary growth of an increased population of users. In addition, this capability would allow for global coverage by providing spaceborne processing and storage of the 406-MHz data.)

Operational Communications Satellites

NASA's role in the many operational communications satellites that were launched from 1979 to 1988 was generally limited to providing launch services, with NASA being paid for providing those services. The satellite systems were developed, owned, and operated by commercial enterprises, government agencies from other countries, various commercial or commercial-government consortiums, or the U.S. military. The following sections describe these communications satellites.

ASC Satellites. The American Satellite Company (ASC) began operations in 1974. It was a partnership between Fairchild Industries and Continental Telecom, Inc. Its satellites supplied voice, data, facsimile, and videoconferencing communications services to U.S. businesses and government agencies. Service was provided through an ownership position in the Westar Satellite System and a network of more than 170 Earth stations located in the continental United States, Hawaii, Guam, and Puerto Rico.

Because of the increased demand for ASC's services, in 1981, the company filed an application with the Federal Communications Commission to operate two wholly owned commercial communications satellites. In March 1983, a contract was awarded to RCA Astro Electronics in Princeton, New Jersey, for construction of two ASC spacecraft and the components for a third spacecraft to serve as a ground spare. NASA launched ASC 1 from the Space Shuttle in August 1985 (Table 2–83). ASC 1 operated in both the six/four-GHz (C-band) and fourteen/twelve-GHz (Ku-band) frequencies.

AT&T Satellite System. The American Telephone and Telegraph (AT&T) satellite system consisted of the Comstar satellites and the Telstar satellites. The system began operations in 1976 using the Comstar satellites. The development of the Telstar 3 satellites began in 1980, with the first launch in 1983. Traffic was transferred from the older Comstars to the Telstars, with AT&T maintaining a four-satellite constellation composed of three Telstars and one Comstar. AT&T used its satellites for long-distance high-capacity voice links, television service, and high-speed data and videoconferencing.

Comstar Satellites. Comstar D-4, the only Comstar launched during the 1979–1988 period, was the last in a series of four Comstar satellites that NASA launched for Comsat General Corporation (Table 2–84). Fully leased to AT&T, the satellite had twelve transponders (channels), each capable of relaying 1,500 two-way voice circuits, giving it an overall communications capability of 18,000 simultaneous high-quality, two-way telephone transmissions. Comstar used the same platform as the earlier Intelsat IV series of satellites—the Hughes HS 351.

Telstar 3 Satellites. The Telstar 3 satellites were the second generation of satellites in the AT&T system. AT&T procured them directly rather than through the lease arrangement used for the Comstars. The satellites had the same configuration as the Anik C and SBS satellites and could be launched from a Delta launch vehicle or the Space Shuttle (Tables 2–85, 2–86, and 2–87).

Galaxy Satellites. NASA launched Galaxy 1, 2, and 3 during the early 1980s. The satellites formed the initial elements of the Hughes Communications system of commercial satellites. These vehicles provided C-band television services as well as audio and business telecommunications services. Hughes added to the system in 1988, when it acquired the orbiting Westar 4 and Westar 5 satellites.

The Galaxy spacecraft used the Hughes HS 376 spacecraft. Similar satellites were used for the SBS system, the Telesat satellite system, the Indonesian Palapa satellites, AT&T's Telstar satellites, and the Western Union satellites. Figure 2–22 shows the basic Galaxy spacecraft design.

Each Galaxy satellite had twenty-four transponders and operated in the six/four-GHz C-band. Hughes sold the transponders on Galaxy 1 and Galaxy 3 to private programming owners for the life of each satellite. Galaxy 2 transponders were offered for sale or lease. Galaxy 1 was devoted entirely to the distribution of cable television programming and relayed video signals throughout the contiguous United States, Alaska, and Hawaii (Table 2–88). Galaxy 2 and Galaxy 3 relayed video, voice, data, and facsimile communications in the contiguous United States (Tables 2–89 and 2–90).

RCA Satcom Satellites. RCA American Communications (RCA Americom) launched eight RCA Satcom satellites during the 1979–1988 period. The C-band satellites were Satcom 3, 3R, 4, 5, 6, and 7. The Ku-band satellites were Satcom K-1 and Satcom K-2.

The RCA Satcom satellites formed a series of large, twenty-four-transponder communications satellites. They consisted of a fixed, four-reflector antenna assembly and a lightweight transponder of high-efficiency traveling wavetube amplifiers and low-density microwave filters. The twenty-four input and output multiplex filters and the waveguide sections and antenna feeds were composed of graphite-fiber epoxy composite. Figure 2–23 shows the major physical features of the RCA Satcom satellites.

RCA Americom of Princeton, New Jersey, managed the RCA Satcom program, including the acquisition of the spacecraft and the associated tracking, telemetry, command systems, and launch vehicle support. Spacecraft development and production were the responsibility of RCA's Astro Electronics Division. The Delta Project Office at NASA's Goddard Space Flight Center in Greenbelt, Maryland, was responsible to NASA's Office of Space Transportation Operations for overall project management of the launch vehicle. The Cargo Operations Office at NASA's Kennedy Space Center in Florida was responsible to Goddard for launch operations management. All launch costs incurred by NASA, including the vehicle hardware and launch services, were reimbursed by RCA Americom. The Payload Assist Module (PAM) was procured by RCA directly from the manufacturer, McDonnell Douglas Corporation.

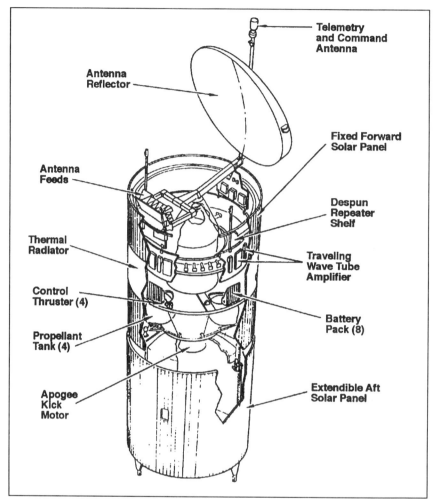

Figure 2–22. Galaxy Components

Satcom 3 was designed for launch by the Delta 3914 (Table 2–91). Beginning with Satcom 3-R, the satellites were designed to be launched either by the Delta 3910/PAM-D or by the Space Shuttle (Table 2–92). (See Table 2–93 for information on Satcom 4.) Satcom 5 was the first RCA satellite to use the Delta 3924 launch vehicle configuration, which used the extended long tank Thor booster, nine Castor IV strap-on motors, and the new Aerojet AJ-118 second stage, but it used the Thiokol TE-364-4 third stage rather than the McDonnell Douglas PAM-D stage (Table 2–94). See Tables 2–95 and 2–96 for information on Satcom 6 and Satcom 7, respectively.) Satcom K-1 and Satcom K-2 (launched in reverse sequence) were heavier spacecraft that were launched by the Space Shuttle, with assistance from a PAM-DII upper stage (Tables 2–97 and 2–98).

SBS Satellites. Satellite Business Systems (SBS) was created on December 15, 1975, by IBM, Comsat, and Aetna Life and Casualty, Inc.

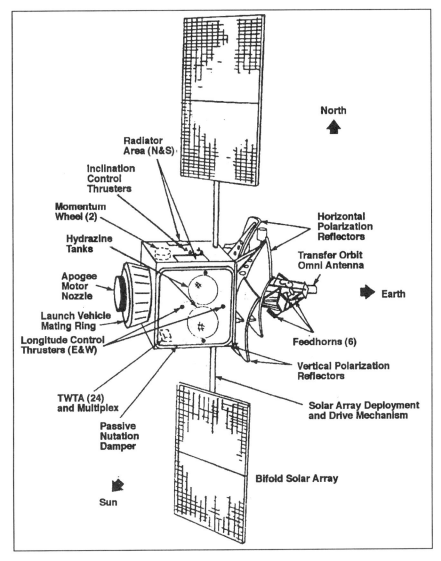

*Figure 2–23. RCA Satcom 3, 3R, and 4
(Satcom 5, 6, and 7 were similar, but the solar panels were in three sections.)*

It was the first private professional satellite digital communications network and the first domestic system to use the twelve- and fourteen-GHz frequencies. In July 1984, Comsat left the consortium and sold its shares to the other two partners. Four satellites were then in orbit. In 1985, IBM and Aetna sold SBS to MCI Communications Corporation. Aetna received cash, and IBM received MCI stock plus ownership of SBS 4, 5, and 6, which it transferred to its subsidiary IBM Satellite Transponder Leasing Corporation. (SBS 5 and SBS 6 had not yet been launched.) The subsidiary and its three satellites were sold to Hughes Communications in 1989.

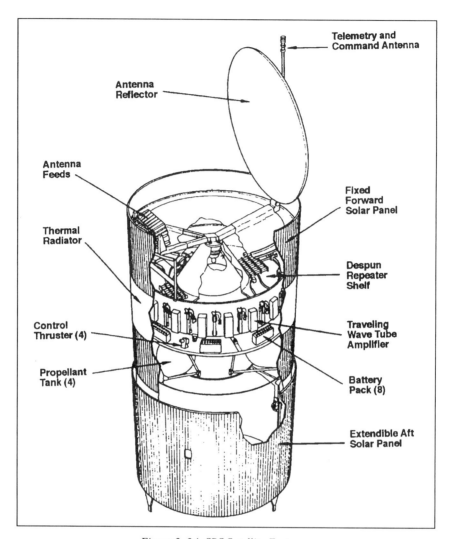

Figure 2–24. SBS Satellite Features

SBS 1 through SBS 5 were very similar in design to the Anik C and several other domestic satellites. (Figure 2–24 illustrates the satellite features.) During launch, the satellite was a compact cylinder. In orbit, the satellite unfolded from one end, and a cylindrical solar array was deployed axially at the other end. When the solar array was deployed, it revealed the main cylindrical body of the satellite, which was also covered with solar cells, except for a mirrored band that served as a thermal radiator. The satellites had ten channels and a capacity for 1,250 two-way telephone conversations per channel, ten simultaneous color television transmissions, or a combination of both. SBS 1 through SBS 4 were launched from NASA vehicles (Tables 2–99, 2–100, 2–101, and 2–102). SBS 5 was launched from an Ariane in September 1988 and is not addressed here.

Westar Satellites. Originally established by Western Union, the Westar satellite system was the first U.S. domestic satellite system. The system relayed data, voice, video, and fax transmissions throughout the continental United States, Hawaii, Puerto Rico, Alaska, and the Virgin Islands. Western Union ended its role as a satellite service provider when it sold the Westar satellites to Hughes Communications in 1988. At the time of the sale, the Westar 3, 4, and 5 satellites were operational. Westar 1 and Westar 2 had already been retired from service (Westar 1 in April 1983 and Westar 2 in 1984). Westar 6 failed to achieve geostationary orbit following its deployment from STS 41-B in February 1984. NASA provided the launch services for the satellites.

Westar 6 was captured and retrieved by an astronaut crew on STS 51-A in February 1984 and returned to Earth for refurbishment. Following its return, the satellite's insurers resold the spacecraft to the Pan Am Pacific Satellite Corporation, which in turn resold it to Asia Satellite, who renamed it AsiaSat 1. The satellite was relaunched in April 1990 aboard a Long March rocket.[5]

The Westar 6S satellite, procured by Western Union as a replacement for Westar 6, was still under development when Western Union was bought out by Hughes. The vehicle was subsequently renamed Galaxy 6.

Westar 1, 2, and 3 were nearly identical to the Canadian Anik A satellites (discussed in Volume III of the *NASA Historical Data Book*). The satellites were spin-stabilized, and the body and all equipment within it spun; only the antenna was despun. The antennas were one and a half meters in diameter and were fed by an array of three horns that produced a pattern optimized for the continental United States. A fourth horn provided a lower-level beam for Hawaii. The communications subsystems had twelve channels with a bandwidth of thirty-six MHz each. Each of twelve spacecraft transponders could relay 1,200 voice channels, one color television transmission with program audio, or data at fifty megabytes per second.

Westar 4, 5, and 6 were larger and had more capacity than the earlier satellites, with twenty-four available channels. Except for communications subsystem details, the satellites were the same as the SBS satellites (addressed above). They had a cylindrical body that was covered with solar cells, except for a band that was a thermal radiator (Figure 2–25). A cylindrical array that surrounded the main body during launch and was deployed in orbit generated additional power. The antenna and the communications equipment were mounted on a platform that was despun during satellite operations. Table 2–103 compares the features of the first generation and the second generation Westar satellites. The characteristics of the Westar 3, 4, 5, and 6 satellites are in Tables 2–104, 2–105, 2–106, and 2–107, respectively.

---

[5]Donald H. Martin, *Communication Satellites, 1958–1992* (El Segundo, CA: The Aerospace Corporation, December 31, 1991), pp. 150–51.

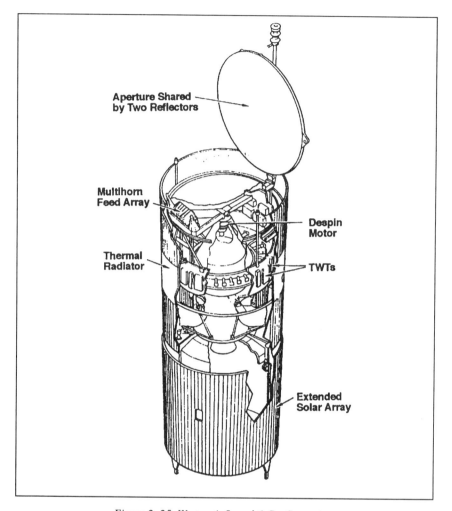

Figure 2–25. Westar 4, 5, and 6 Configuration

Intelsat Satellites. Intelsat (the International Telecommunications Satellite Organization) is an extremely reliable (more than 99 percent) global network of satellites that has provided nearly universal communications coverage except in the polar regions. Intelsat began developing satellites for international public use as soon as the early experimental communications satellite technology had been proven. Starting from a single satellite in 1964, the system grew to a global network using many satellites. Six generations of satellites have been brought into service.

All nations may join Intelsat, and the organization has more than 100 member nations (see Table 2–108). Ownership percentages reflect national investments in Intelsat and are adjusted to reflect each country's use of the system. When Intelsat began, the U.S. ownership was more than 60 percent. As more nations began using the system, this percentage dropped and has been 22 to 27 percent since the late 1970s. Australia, Canada, France, Germany, Italy, Japan, South Korea, and the United Kingdom are the other large owners, with percentages between 2 and 14 percent.[6]

Intelsat was created through the adoption of interim agreements signed by eleven countries that established a global commercial communications satellite system. Since February 12, 1973, Intelsat has operated under definitive agreements, with an organizational structure consisting of an Assembly of Parties (governments that are parties to the Intelsat agreement), a Meeting of Signatories (governments or their designated telecommunications entities that have signed the Operating Agreement), a Board of Governors (responsible for decisions relating to the design, development, construction, establishment, operations, and maintenance of the Intelsat space segment), and an Executive Organ headed by a Director General. The members of the Board of Governors represent countries or groups of countries with relatively large ownership percentages and geographic regions where countries do not have large ownership percentages.

The Intelsat communications system includes the satellites themselves, a large number of ground terminals, and a control center. Intelsat owns the satellites, but each member owns its own terminals. The system has Atlantic, Pacific, and Indian Ocean regions.[7] The number of ground terminals has increased yearly since the system became operational in 1965. Intelsat handles telephone, telegraph, data, and television traffic. Telephone has been the major portion of the traffic. In the early years, almost all Intelsat traffic was voice, but with the growth of television transmissions and, more recently, the surge in nonvoice digital services, revenue. Television accounted for about 10 percent of the revenues, except in months with events of worldwide interest, such as the Olympic Games. The Atlantic region has always had the majority of all Intelsat traffic, almost 70 percent in the early years and decreasing later to about 60 percent. The Pacific region began earlier than the Indian Ocean region because of earlier satellite availability. However, Indian Ocean traffic surpassed Pacific traffic when considerable Hawaiian and Alaskan traffic was transferred to U.S. domestic systems. Pacific traffic, however, has

---

[6]*Ibid.*, p. 83.

[7]Intelsat has four service regions. The Atlantic Ocean Region serves the Americas, the Caribbean, Europe, the Middle East, India, and Africa and generally covers locations from 307 degrees east to 359 degrees east longitude. The Indian Ocean Region serves Europe, Africa, Asia, the Middle East, India, and Australia and covers 327 degrees east to 66 degrees east. The Asia Pacific Region serves Europe, Africa, Asia, the Middle East, India, and Australia and covers 72 degrees east to 157 degrees east. The Pacific Ocean Region serves Asia, Australia, the Pacific, and the western part of North America from 174 degrees east to 183 degrees east. In most discussions, the Asia Pacific Region and the Pacific Ocean Region are treated as a single region.

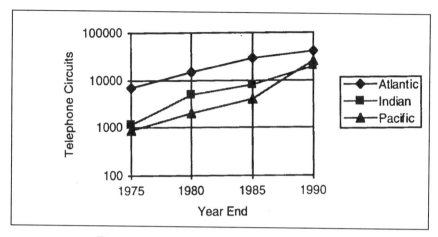

Figure 2–26. Growth of Intelsat Traffic (1975–1990)

continued to grow, as many small nations have begun to use the system.[8] Figure 2–26 shows the growth of Intelsat traffic from 1975 to 1990.

NASA's Lewis Research Center (now Glenn Research Center) managed the Atlas-Centaur launches. Comsat was responsible for firing the apogee kick motor that placed the satellites into near geosynchronous orbit.

Intelsat V. The Intelsat IV-A satellites that were first used in 1975 had a capacity of 6,000 voice circuits and two television channels. They provided a moderate capacity increase over previous satellites without requiring significant ground terminal changes. However, further capacity increases were not practical with a simple stretching of the Intelsat IV/IV-A design, so the development of a new satellite began in 1976. The new series of satellites (Intelsat V) had a capacity of 12,000 voice circuits and two television channels. It has been used in all the Intelsat regions.

The Intelsat V satellites incorporated several new features. These were:

- Frequency reuse through both spatial isolation and dual polarization isolation
- Multiband communications—both fourteen/eleven GHz and six/four GHz
- A contiguous band output multiplexer
- Maritime communications subsystem
- Use of nickel hydrogen batteries in later spacecraft

Two of the new design features required significant ground terminal changes. The use of dual-polarization uplinks and downlinks in the four- and six-GHz bands required improvements at all ground terminals to ensure isolation between the two polarizations. The dual-polarization uplinks and downlinks tripled the satellite capacity in the four- and six-

---

[8]Martin, *Communication Satellites*, pp. 83–85.

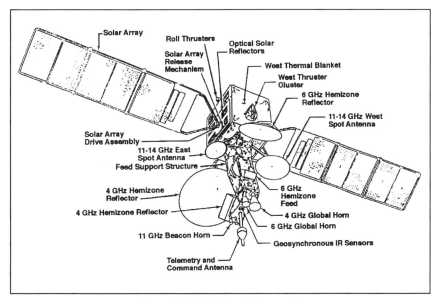

Figure 2–27. Intelsat 5 Spacecraft

GHz bands, compared with the Intelsat IV design. Also, the nations with the largest traffic volumes used the new eleven- and fourteen-GHz bands and two independent beams and needed to construct new terminals for them.[9]

The Intelsat V satellites had a rectangular body of more than one and a half meters across. The Sun-tracking solar arrays, composed of three panels each, were deployed in orbit. An antenna tower on the Earth-viewing face of the body held both the communications and telemetry, tracking, and command antennas and the feed networks for the large reflectors. The tower was fixed to the satellite body, but the three largest reflectors deployed in orbit. The tower was more than four and a half meters tall and was constructed almost entirely of graphite fiber/epoxy materials for strength, light weight, and thermal stability. The entire satellite weighed about 1,928 kilograms at launch and 998 kilograms in orbit and spanned about fifteen and a half meters across the solar array (Figure 2–27).

The initial Intelsat V contract that was awarded to Ford Aerospace and Communications Corporation of the United States called for seven satellites; later an eighth and a ninth were added to the contract. An international team of manufacturers served as subcontractors. Members of the international manufacturing team and their areas of concentration are listed in Table 2–109.

The first Intelsat V launch was in December 1980; the last, the only failure, was in 1984. The eight satellites successfully launched were still in use at the end of 1990. The Intelsat V characteristics are summarized in Tables 2–110 through 2–116.

---

[9]*Ibid.*, pp. 56–57.

Intelsat V-A Series. Intelsat V-A F-10 was the first in the Intelsat V-A series of satellites. Intelsat V-A was a modified Intelsat V design. Its development started in late 1979. As with previous changes to Intelsat satellites, the primary goal was to increase satellite capacity to keep ahead of traffic growth in the Atlantic region. Intelsat V-A satellites had a capacity of 13,500 two-way voice circuits, plus two television channels.

Externally, the satellite was almost identical to Intelsat V. Internally, several changes were made to improve performance, reliability, and communications capacity. Several weight-saving measures compensated for the additional communications hardware. The internal arrangement of the communications hardware was modified for thermal balance. Intelsat V-A satellites did not have the maritime communication subsystem, which was added to Intelsat V-5 launched in September 1982 and Intelsat V-6 through Intelsat V-9. (Intelsat V-7 and V-8 were not launched by NASA.)

The first Intelsat V-A was launched in March 1985 (Table 2–117). Two others were launched later in 1985 (Tables 2–118 and 2–119). A fourth was lost in a launch vehicle failure in 1986. The last two were launched in 1988 and 1989. Only the three 1985 satellites were NASA launches.

Fltsatcom Satellites. The Fltsatcom system (Fleet Satellite Communications) provided worldwide, high-priority, ultrahigh frequency (UHF) communications among naval aircraft, ships, submarines, and ground stations and between the Strategic Air Command and the national command authority network. It supplied military communications capability for the U.S. Air Force with narrowband and wideband channels and the U.S. Navy for fleet relay and fleet broadcast channels. The satellites provided two-way communication, in the 240- to 400-MHz frequency band, between any points on Earth visible from their orbital locations. Between 1979 and 1988, NASA furnished launch services for six

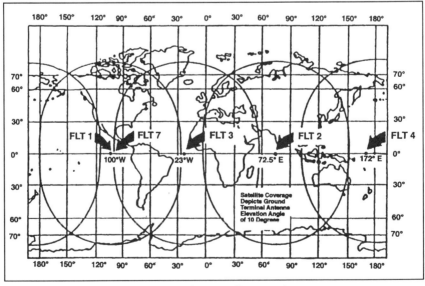

Figure 2–28. Fltsatcom Coverage Areas

Fltsatcom satellites for the U.S. Department of Defense, Fltsatcom 2 through Fltsatcom 7 (Tables 2–120 through 2–125).

Fltsatcom and the Air Force Satellite Communications System shared a set of four Fltsatcom satellites in synchronous equatorial orbits. Figure 2–28 shows the coverage areas of the five operational Fltsatcom satellites.

The Fltsatcom satellites had an hexagonal body with two modules— a spacecraft module and a payload module (Figure 2–29). Fltsatcom 7 had a third module for the extremely high frequency (EHF) communications package that it carried. The spacecraft module contained the attitude control, power, and tracking, telemetry, and command subsystems, as well as the apogee motor. The two solar arrays were mounted on booms attached to this module. The satellite was three-axis stabilized by means of redundant reaction wheels and hydrazine thrusters. This arrangement allowed the antennas to face Earth continuously while being directly attached to the satellite body. The payload module contained the communications subsystem. The transponders on board each satellite carried twenty-three UHF communications channels and one superhigh frequency uplink channel. The Navy used ten of the channels for communications among its land forces, ships, and aircraft. The Air Force used twelve of the channels as part of its satellite communications system for command and control of nuclear forces. One channel was reserved for U.S. national command authorities.

Leasat Satellites. The Leasat satellites (also known by the name Syncom) were leased by the Department of Defense from Hughes Communications Services to replace older Fltsatcom spacecraft for

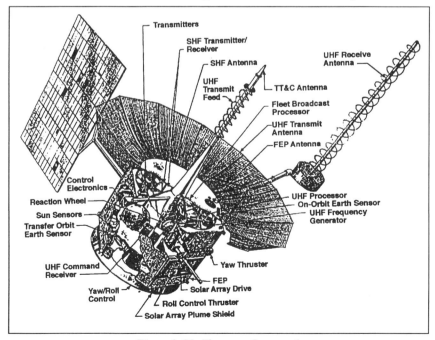

Figure 2–29. Fltsatcom Spacecraft

worldwide UHF communications among ships, planes, and fixed facilities. The spacecraft were designed expressly for launch from the Space Shuttle and used the "frisbee" or rollout method of deployment.

A cradle structure helped install the spacecraft in the orbiter payload bay. This cradle permitted the spacecraft to be installed lying on its side, with its retracted antennas pointing toward the nose of the orbiter and its propulsion system pointing toward the back. Mounting the antennas on deployable structures allowed them to be stowed for launch.

The Leasat satellites did not require a separately purchased upper stage. They contained their own unique upper stage to transfer them from the Shuttle deploy orbit to a geosynchronous circular orbit over the equator.

The satellites used the Hughes HS 381 bus. They were spin-stabilized, with the spun portion containing the solar array and the Sun and Earth sensors for attitude determination and Earth pointing reference, three nickel cadmium batteries for eclipse operation, and all the propulsion and attitude control hardware. The despun platform contained two large helical UHF Earth-pointing communications antennas, twelve UHF communications repeaters, and the majority of the telemetry, tracking, and command equipment.

The contract for Leasat development was awarded in September 1978 to Ford Aerospace and Communications Corporation. The first launch was scheduled for 1982. However, delays in the Shuttle program postponed the launch dates and resulted in a two-year suspension of work on the satellites. Work resumed in 1983, and NASA launched the first two satellites in 1984. NASA launched the third Leasat in April 1985, but the satellite failed to turn on. The Shuttle crew carried out a rescue attempt but was unsuccessful. NASA launched the fourth Leasat in August 1985. The same Shuttle mission then rendezvoused with Leasat 3 and carried out a successful repair, allowing ground controllers to turn the satellite on and orient it. After ensuring that the propellants were warm, Leasat 3 was placed into geosynchronous orbit in November 1985 and began operations in December. Unfortunately, Leasat 4 failed shortly after arriving in geosynchronous orbit, and the wideband channel on Leasat 2 failed in October 1985. The characteristics of these four satellites are in Tables 2–126 through 2–129. NASA launched the fifth and last Leasat in January 1990.[10]

NATO IIID. NATO IIID was the fourth and final NATO III satellite placed in orbit by NASA for the U.S. Air Force and its Space Division acting as agents for NATO. The satellite was spin-stabilized with a cylindrical body and a despun antenna platform on one end. All equipment was mounted within the body, and a three-channel rotary joint connected the communications subsystem with the antennas (Figure 2–30). The spacecraft transmitted voice, data, facsimile, and telex messages among military ground stations.

---

[10]*Ibid.,* p. 115.

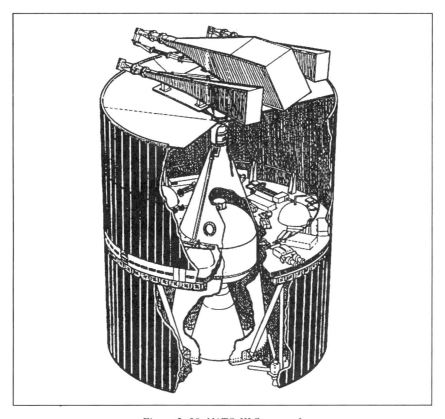

Figure 2–30. NATO III Spacecraft

The NATO communications satellite program began in 1967. The first NATO satellite was launched in 1970. A second satellite was launched in 1971. The NATO III satellites were larger and had significantly greater capabilities than the earlier NATO satellites. NASA launched NATO IIIA in April 1976. NATO IIIB was launched in January 1977 as an orbiting spare. NATO loaned it to the United States to fill the east Pacific operating location of the Defense Space Communications System (DSCS) until at least four DSCS II satellites were available, which occurred in December 1978 with the launch of DSCS II. The United States removed DSCS traffic from NATO IIIB and returned the satellite to its station over the Atlantic Ocean. NATO traffic was switched to NATO IIIB in December 1982, and NATO IIIA was used for ground terminal testing. The flight qualification model was reworked into the third flight model and launched in November 1978; it was put into a dormant state known as orbital storage. NATO IIIC was reactivated and became the primary NATO spacecraft in December 1986, and NATO IIIB became a test vehicle. In 1980, a follow-on contract was issued for a fourth satellite, which NASA launched in November 1984 as NATO IIID (Table 2–130).[11]

---

[11]*Ibid.*, pp. 105–07.

Anik Satellites. Telesat Canada Corporation operated the series of Anik satellites and reimbursed NASA for the cost of its launch services. (Anik means "little brother" in Inuit.) The system began operations in Canada at the beginning of 1973. The first three satellites were designated the Anik A series. Anik A-1 was the world's first geostationary communications satellite launched into orbit for a commercial company. The satellites provided all types of communications services throughout Canada. A single Anik B satellite supplemented the A series and provided additional experimental channels.

The Anik D series replaced the A satellites. The Anik C satellites operated at the same time as Anik D but had a different function. They added to terrestrial communications on high-traffic-density paths and used the twelve- and fourteen-GHz frequencies for service to terminals in urban areas. The four- and six-GHz bands that were used by Anik D were unacceptable because of interference from other users of the band.[12]

The Anik satellites were designed for launch from either a Delta launch vehicle or the Space Shuttle. The characteristics of the Anik C satellites are in Tables 2–132, 2–133, and 2–135, while those of the Anik D satellites are in Tables 2–131 and 2–134; these satellite descriptions are in order of launch date.

Anik C Satellites. Anik C was a spin-stabilized satellite. When in orbit, the antenna was deployed from one end of the satellite, and a cylindrical solar panel was extended from the opposite end. The communications subsystem had sixteen repeaters and used the twelve- and fourteen-GHz bands. Figure 2–31 shows the Anik C configuration.

The Anik C satellites covered only the southern half of Canada because they were designed to connect Canada's urban centers. The use of the twelve- and fourteen-GHz bands allowed the ground terminals to be placed inside cities without interference between the satellite system and terrestrial microwave facilities. Anik C complemented the Anik A and Anik D satellites, which covered all of Canada and were best suited to the distribution of national television or message services that required nationwide access.

The development of Anik C began in April 1978. The first launch (Anik C-3) took place from STS-5 in November 1982. Anik C-3 was the first C series satellite launched because the other C satellites were not as readily accessible; they had been put into ground storage awaiting launch vehicle availability. The second C satellite was launched in June 1983, and the third in April 1985. Traffic did not grow as much as expected when the C series was planned, and Anik C-1 was put into orbital storage and offered for sale. A purchase agreement was made in 1986 by a group that planned to use it for transpacific services, but the agreement was canceled in 1987. By 1989, Telesat began to use the satellite in a limited way, and in 1990, additional traffic was transferred to it in preparation for the introduction of Anik E-1.[13]

---

[12]*Ibid.*, p. 131.
[13]*Ibid.*, p. 136.

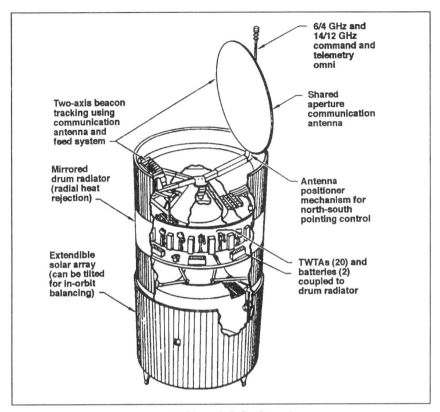

Figure 2–31. Anik C Configuration

Anik D Satellites. The Anik D satellites replaced the Anik A satellites. The satellites were also spin-stabilized, and the structure, support subsystems, thermal radiator, and deployable solar array were almost identical to those of Anik C.

The major difference between the two satellites was in the communications subsystem. Anik D had twenty-four repeaters in the four- and six-GHz bands as compared to the sixteen repeaters in the twelve- and fourteen-GHz bands on Anik C. Figure 2–32 shows the typical geographical coverage of the Anik D satellites from an approximate location of 104 degrees west longitude.

Arabsat Satellite. NASA launched Arabsat-1B from the Space Shuttle in June 1985 (Table 2–136). It was the second in a series of satellites owned by the Arabsat Satellite Communications Organization (or Arabsat). (Arabsat-1A was launched from an Ariane in February 1984.) It was a communications satellite with a coverage area that included the Arab-speaking countries of North Africa and the Middle East (Figure 2–33).

Arabsat was formed in 1976. Saudi Arabia had the largest investment share. The objective of the system was to promote economic, social, and cultural development in the Arab world by providing reliable communications links among the Arab states and in rural areas, developing Arab

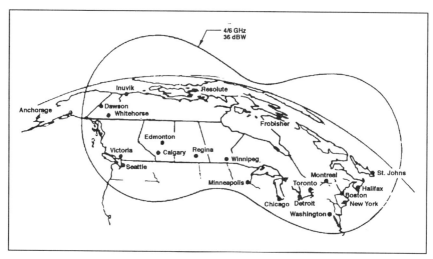

Figure 2–32. Anik D Geographical Coverage at 104 Degrees West Longitude

industrial capabilities in space-related technologies, and introducing new communications services to the area.

The Arabsat Organization purchased the satellites, launch services, and major ground facilities but developed some of the ground equipment within the member nations. The organization awarded a contract to Aerospatiale in May 1981 for three satellites. The satellites included equipment used for other satellites, particularly the Intelsat V series and Telecom 1. It was a three-axis-stabilized design with solar arrays and antennas. The solar arrays were partially deployed in the transfer orbit; the antennas were deployed in synchronous orbit. The satellites contained twenty-five C-band transponders and one television (C/S-band) transponder.[14]

Aussat Satellites. Australia first considered a domestic satellite system in 1966. In 1969, the country began routing some transcontinental telephone circuits through the Intelsat system. During 1970, experiments were conducted using ATS 1 to gather data that would be useful in planning a domestic satellite system.

Studies continued throughout the 1970s. In mid-1979, the government decided to institute a satellite system. In the fall of 1979, the Canadian Hermes satellite (actually CTS) was used for demonstrations of television broadcasting to small terminals at numerous locations. The distribution of television to fifty isolated communities began in 1980 using an Intelsat satellite. Between mid-1979 and April 1982, satellite specifications were developed, a government-owned operating company, Aussat Proprietary, Ltd., was formed, and a satellite contract was signed with Hughes Communications International to develop Australia's first satellite program. Under the contract, Hughes Space and Communications Group built three satellites and two telemetry, tracking, command, and

---

[14]*Ibid.*, p. 268.

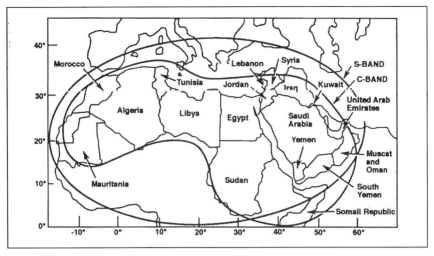

Figure 2–33. Arabsat Coverage Area

monitoring stations. The contract also provided for launch and operational services and ground support.

Aussat provided a wide range of domestic services to the entire continent, its offshore islands, and Papua, New Guinea. This included direct-television broadcast to homesteads and remote communities, high-quality television relays among cities, digital data transmission for both telecommunications and business use, voice applications for urban and remote areas, centralized air traffic control services used as a very high-frequency (VHF) repeater station, and maritime radio coverage.

NASA launched two Aussat satellites for Aussat Proprietary. The system used the Hughes HS 376 spacecraft, the same spacecraft used by Anik, Telstar, Galaxy, and Palapa. Aussat 1 and Aussat 2 were located at geosynchronous orbits at the equator just north of Papua, New Guinea, at 156 degrees east and 164 degrees east longitude (Table 2–137 and 2–138). The satellites were designed to be launched from the Space Shuttle, a Delta, or an Ariane. The Aussat satellites carried fifteen channels, each forty-five MHz wide.

Insat Satellites. NASA launched the Insat satellites for the India Department of Space. The satellites were multipurpose telecommunications/meteorology spacecraft with the capability for nationwide direct broadcasting to community television receivers in rural areas. The spacecraft were built by Ford Aerospace and Communications Corporation under a joint venture of the Department of Space, the Posts and Telegraphs Department of the Ministry of Communications, the India Meteorological Department of the Ministry of Tourism and Civil Aviation, and the Doordarshan of the Ministry of Information and Broadcasting.

The satellites included twelve transponders operating at 5,935–6,424 MHz (Earth-to-satellite) and 3,710–4,200 MHz (satellite-to-Earth) for thick route, thin route, and remote area communications and television program distribution. They also had two transponders operating at 5,855–5,935 MHz

(Earth-to-satellite) and 2,555–2,635 MHz (satellite-to-Earth) for direct-television broadcasting to augmented low-cost community television receivers in rural areas for which direct-television broadcast coverage has been identified as more economical, radio program distribution, national television networking, and disaster warning. The telecommunications component could provide more than 8,000 two-way long-distance telephone circuits potentially accessible from any part of India.

NASA launched Insat 1A from a Delta launch vehicle in 1982 (Table 2–139). The space agency also launched Insat 1B from the Space Shuttle in 1983 (Table 2–140). Insat 1C was launched from an Ariane in 1988 and is not addressed here.

Morelos Satellites. Mexico started domestic use of satellite communications in 1980 by leasing Intelsat capacity on a satellite that was moved to 53 degrees west longitude to provide domestic services for the Western Hemisphere. Mexico also owned one transponder on a U.S. domestic satellite that was used for transmission of television to the United States. In the spring of 1983, Mexico awarded a contract for the construction of a Mexican domestic communications satellite to Hughes Communications. The satellite and the satellite system were called Morelos in honor of a notable person in Mexican history.

The satellite system provided advanced telecommunications to the most remote parts of Mexico, including educational television, commercial programs over the national television network, telephone and facsimile services, and data and business transmissions. The system used eighteen channels at C-band and four channels at Ku-band. The satellites used the popular Hughes HS 376 design.

NASA launched two satellites for the Secretariat of Communications and Transportation, Mexico. Morelos 1 was launched in June 1985, and all traffic from the Intelsat satellite was transferred to it (Table 2–141). NASA launched Morelos 2 in November 1985 (Table 2–142). It was put into a drifting storage orbit just above synchronous altitude. In 1986, it was stabilized at 116 degrees longitude in an orbit with a few degrees inclination. That orbit was phased so that the inclination decreased to zero by 1990 from natural forces. This allowed the satellite to use its scheduled launch date yet not use fuel for stationkeeping until its communications services were required.

Palapa Satellites. The Palapa satellites form Indonesia's domestic satellite system. Meaning "fruits of labor," Palapa satellites provided regional communications among the country's more than 6,000 inhabited islands. The system was operated by a government-owned company, Perumetel until 1993, when a private Indonesian company took over system management.

NASA launched Palapa A1 on July 8, 1976. Operational service began the following month. Palapa A1 and Palapa A2 were removed from service in July 1985 and January 1988, respectively, following the introduction of the Palapa B series, which increased coverage to include the Philippines, Malaysia, and Singapore. Palapa B-1 was launched on STS-7 in 1983

(Table 2–143). Palapa B-2, originally launched by NASA from STS 41-B in February 1984, did not successfully reach orbit and was subsequently retrieved by STS 51-A in November 1984 (Table 2–144). Following the failure of Palapa B-2, Perumetel ordered an identical replacement satellite, Palapa B-2P, which NASA launched in March 1987 on a Delta launch vehicle (Table 2–145). The satellite was sold by its insurers to Sattel Technologies; it was refurbished, relaunched in April 1990, and then resold to Perumetel, with which it was known as Palapa B-2R.[15]

The Palapa B satellites were four times more powerful and twice the size of the Palapa A series. They were based on the frequently used Hughes HS 376 design. Each carried twenty-four C-band transponders and six spares.

UoSAT Satellites. The UoSAT satellites were part of the Oscar program of HAM radio satellites. (Oscar stood for Orbiting Satellite Carrying Amateur Radio.) The satellites were carried as secondary payloads on missions that had excess payload space. NASA launched UoSAT 1 with the Solar Mesospheric Explorer (Table 2–146) and UoSAT 2 with Landsat 5 (Table 2–147).

The UoSATs emphasized microelectronics technology and involved direct contact with the satellites from simple ground terminals located at schools of all levels. UoSAT 1 was the ninth Oscar launch and the first satellite built by the University of Surrey in England. The goal of UoSAT 1 and the UoSAT program was to demonstrate the development of low-cost sophisticated satellites and to use these satellites to promote space science and engineering in education. The satellite was the first satellite designed to transmit data, including pictures of Earth's surface, in a form that could be readily displayed on a domestic television set. It carried a voice synthesizer for "speaking" (in English) information on telemetry, experimental data, and spacecraft operations. The synthesizer had a vocabulary of approximately 150 words, and most standard amateur VHF receivers could listen in with a simple fixed antenna. It carried a series of radio beacons transmitting at different frequencies, two particle counters that provided information on solar activity and auroral events, a magnetometer for measuring the Earth's magnetic field, and an Earth-point camera that covered an area of 500 square kilometers and transmitted images that could be received and stored by simple receivers and displayed on home television sets.

UoSAT 2 was the eleventh Oscar launch. It carried a particle wave experiment, a store-and-forward digital communications experiment, a solid-state slow-scan imaging experiment, VHF/UHF and superhigh frequency (SHF) data downlinks, a multichannel command decoder, a microprocessor-based housekeeping system and data collection facility, digital Sun sensors, horizon sensors, a navigation magnetometer, three axis magnetorquers, a gravity-gradient stabilization system, and an experimental telemetry system.

---

[15]*Ibid.,* p. 256.

Navigational Satellites

NASA launched two series of navigational satellites from 1979 to 1988: the NOVA satellites and the SOOS satellites. Both series were launched for the Navy Transit System from Scout launch vehicles, and both used Oscar spacecraft. The Transit Program was an operational navigation system used by the U.S. Navy and other vessels for worldwide ocean navigation.

Transit was developed at Johns Hopkins University's Applied Physics Laboratory from 1958 to 1962 to provide precision periodic position fixes for U.S. Navy submarines. Subsequently, several commercial companies were contracted to build production models of the spacecraft, which were kept in controlled storage until needed, as well as signal receiver and position computer equipment.

The constellation consisted of two types of spacecraft designated as Oscar and NOVA. The satellites were launched into a polar orbit with a nominal 1,112-kilometer altitude. The last Transit satellite launch was SOOS-3 in August 1988. The program was terminated on December 31, 1996.

NASA and DOD entered into agreements in June 1962 that established the basis for joint utilization of the Scout launch vehicle. These initial agreements were reflected in a memorandum of understanding between NASA and the Air Force Systems Command, dated April 19, 1977. Under this agreement, NASA maintained the Scout launch vehicle system, and DOD used the system capabilities for appropriate missions. DOD requested that NASA provide Scout launches for the Navy Transit and NOVA programs. The Navy reimbursed NASA for the cost of the Scout launch vehicles, Western Strategic Missile Command launch services and mission support requirements, and supporting services, as required.

NOVA Satellites. NOVA was the new-generation, improved Transit navigation satellite. RCA Astro-Electronics performed the initial hardware work under contract to the U.S. Navy's Strategic Systems Project Office, but because of contractual changes, the satellites were returned to the Applied Physics Laboratory at Johns Hopkins University for completion and processing for launch. NASA launched the satellites on a four-stage Scout vehicle into an initial orbit of 342.6 kilometers by 740.8 kilometers. A multiple-burn hydrazine motor then raised and circularized the orbit.

The NOVA spacecraft was an improved Oscar. Improvements included electronics hardened against the effects of radiation, a disturbance compensation system designed to provide stationkeeping capability and remove atmospheric drag and radiation pressure effects, and greater data storage capacity that permitted retention of a long-arc, eight-day navigation message. The NOVA transmitting system consisted of dual five-MHz oscillators, phase modulators, and transmitters operating at 400 MHz and 150 MHz. Dual incremental phase shifters were used to control oscillator offset. The characteristics of the three NOVA satellites are in Tables 2–148, 2–149, and 2–150.

SOOS Satellites. SOOS stands for "Stacked Oscars on Scout." The Navy SOOS mission configuration consisted of two Transit satellites in a stacked configuration. The stacked launch of two satellites and a separation technique placed the two Oscars in virtually the same orbit plane. To make the piggyback launch possible, the lower Oscar spacecraft was modified with a permanently attached graphite epoxy cradle that supported the upper spacecraft in the launch configuration. The characteristics of the four SOOS satellites are in Tables 2–151 through 2–154.

Table 2–1. Applications Satellites (1979–1988)

Launch Date	Satellite	Type of Mission	Owner/ Sponsor	Launch Vehicle
Feb. 18, 1979	SAGE (AEM-2)	Explorer	NASA	Scout
May 4, 1979	Fltsatcom 2	Communications	Dept. of Defense	Atlas Centaur
June 27, 1979	NOAA 6	Meteorological	NOAA	Atlas-F
Aug. 9, 1979	Westar 3	Communications	Western Union	Delta
Oct. 30, 1979	Magsat (AEM-C)	Explorer	NASA	Scout
Dec. 6, 1979	RCA Satcom 3*	Communications	RCA Corp.	Delta
Jan. 17, 1980	Fltsatcom 3	Communications	Dept. of Defense	Atlas Centaur
May 29, 1980	NOAA B*	Meteorological	NOAA	Atlas-F
Sept. 9, 1980	GOES 4	Meteorological	NOAA	Delta
Oct. 30, 1980	Fltsatcom 4	Communications	Dept. of Defense	Atlas Centaur
Nov. 15, 1980	SBS 1	Communications	Satellite Business Systems	Delta
Dec. 6, 1980	Intelsat V F-2	Communications	Intelsat	Atlas Centaur
Feb. 21, 1981	Comstar D-4	Communications	AT&T Corp.	Atlas Centaur
May 15, 1981	NOVA 1	Navigational	U.S. Navy	Scout
May 21, 1981	GOES 5	Meteorological	NOAA	Delta
May 23, 1981	Intelsat V F-1	Communications	Intelsat	Atlas Centaur
June 23, 1981	NOAA 7	Meteorological	NOAA	Atlas-F
Aug. 6, 1981	Fltsatcom 5	Communications	Dept. of Defense	Atlas Centaur
Sept. 24, 1981	SBS 2	Communications	Satellite Business Systems	Delta
Nov. 19, 1981	RCA Satcom 3R	Communications	RCA Corp.	Delta
Dec. 15, 1981	Intelsat V F-3	Communications	Intelsat	Atlas Centaur
Jan. 16, 1982	RCA Satcom 4	Communications	RCA Corp.	Delta
Feb. 25, 1982	Westar 4	Communications	Western Union	Delta
March 4, 1982	Intelsat V F-4	Communications	Intelsat	Atlas Centaur
April 10, 1982	Insat 1A	Communications	India	Delta
June 8, 1982	Westar 5	Communications	Western Union	Delta
July 16, 1982	Landsat 4	Remote Sensing	NOAA	Delta
Aug. 25, 1982	Anik D-1	Communications	Canada	Delta
Sept. 28, 1982	Intelsat V F-5	Communications	Intelsat	Atlas Centaur
Oct. 27, 1982	RCA Satcom 5	Communications	RCA Corp.	Delta
Nov. 11, 1982	SBS 3	Communications	Satellite Business Systems	STS-5
Nov. 12, 1982	Anik C-3	Communications	Canada	STS-5
March 28, 1983	NOAA 8	Meteorological	NOAA	Atlas-E
April 11, 1983	RCA Satcom 6	Communications	RCA Corp.	Delta

Table 2–1 continued

Launch Date	Satellite	Type of Mission	Owner/ Sponsor	Launch Vehicle
April 28, 1983	GOES 6	Meteorological	NOAA	Delta
May 19, 1983	Intelsat V F-6	Communications	Intelsat	Atlas Centaur
June 18, 1983	Anik C-2	Communications	Canada	STS-7
June 18, 1983	Palapa B-1	Communications	Indonesia	STS-7
June 28, 1983	Galaxy 1	Communications	Hughes Communications	Delta
July 28, 1983	Telstar 3-A	Communications	AT&T Corp.	Delta
Aug. 31, 1983	Insat 1B	Communications	India	STS-8
Sept. 8, 1983	RCA Satcom 7	Communications	RCA Corp.	Delta
Sept. 22, 1983	Galaxy 2	Communications	Hughes Communications	Delta
Feb. 3, 1984	Westar 6*	Communications	Western Union	STS 41-B
Feb. 6, 1984	Palapa B-2	Communications	Indonesia	STS 41-B
March 1, 1984	Landsat 5	Remote Sensing	NOAA	Delta
March 1, 1984	UoSAT 2	Communications	University of Surrey	Delta
June 9, 1984	Intelsat V F-9*	Communications	Intelsat	Atlas Centaur
Aug. 31, 1984	SBS 4	Communications	Satellite Business Systems	STS 41-D
Aug. 31, 1984	Leasat 2 (Syncom IV-2)	Communications	Hughes (leased by Dept. of Defense)	STS 41-D
Sept. 1, 1984	Telstar 3-C	Communications	AT&T Corp.	STS 41-D
Sept. 21, 1984	Galaxy 3	Communications	Hughes Communications	Delta
Oct. 5, 1984	Earth Radiation Budget Satellite (ERBS)	Environmental Observations	NASA	STS 41-G
Oct. 12, 1984	NOVA 3	Navigational	U.S. Navy	Scout
Nov. 9, 1984	Anik D-2	Communications	Canada	STS 51-A
Nov. 10, 1984	Leasat 1 (Syncom IV-1)	Communications	Hughes (leased by Dept. of Defense)	STS 51-A
Nov. 13, 1984	NATO IIID	Communications	NATO	Delta
Dec. 12, 1984	NOAA 9	Environmental Observations	NOAA	Atlas-E
Jan. 24, 1985	DOD	n/a	Dept. of Defense	STS 51-C
March 22, 1985	Intelsat V-A F-10	Communications	Intelsat	Atlas Centaur
April 12, 1985	Anik C-1	Communications	Canada	STS 51-D
April 13, 1985	Leasat 3 (Syncom IV-3)	Communications	Hughes (leased by Dept. of Defense)	STS 51-D
June 17, 1985	Morelos 1	Communications	Mexico	STS 51-G
June 18, 1985	Arabsat-1B	Communications	Saudi Arabia	STS 51-G
June 19, 1985	Telstar 3-D	Communications	AT&T Corp.	STS 51-G
June 29, 1985	Intelsat V-A F-11	Communications	Intelsat	Atlas Centaur

Table 2–1 continued

Launch Date	Satellite	Type of Mission	Owner/ Sponsor	Launch Vehicle
Aug. 3, 1985	SOOS-I (Oscar 24 and 30)	Navigational	U.S. Navy	Scout
Aug. 27, 1985	ASC 1	Communications	American Satellite Corp.	STS 51-I
Aug. 27, 1985	Aussat 1	Communications	Australia	STS 51-I
Aug. 29, 1985	Leasat 4 (Syncom IV-4)	Communications	Hughes (leased by Dept. of Defense)	STS 51-I
Sept. 28, 1985	Intelsat V-A F-12	Communications	Intelsat	Atlas Centaur
Oct. 3, 1985	DOD	n/a	Dept. of Defense	STS 51-J
Nov. 26, 1985	Morelos 2	Communications	Mexico	STS 61-B
Nov. 27, 1985	Aussat 2	Communications	Australia	STS 62-B
Nov. 28, 1985	RCA Satcom K-2	Communications	RCA Corp.	STS 61-B
Dec. 12, 1985	AF-16	n/a	Dept. of Defense	Scout
Jan. 12, 1986	RCA Satcom K-1	Communications	RCA Corp.	STS 61-C
May 5, 1986	GOES G*	Meteorological	NOAA	Delta
Sept. 5, 1986	DOD (SDI)	n/a	Dept. of Defense	Delta
Sept. 17, 1986	NOAA 10	Meteorological	NOAA	Atlas-E
Dec. 4, 1986	Fltsatcom F-7	Communications	Dept. of Defense	Atlas Centaur
Feb. 26, 1987	GOES 7	Meteorological	NOAA	Delta
March 20, 1987	Palapa B-2P	Communications	Indonesia	Delta
March 26, 1987	Fltsatcom F-6*	Communications	Dept. of Defense	Atlas Centaur
Sept. 16, 1987	SOOS-2	Navigational	U.S. Navy	Scout
April 25, 1988	SOOS-3	Navigational	U.S. Navy	Scout
June 16, 1988	NOVA 2	Navigational	U.S. Navy	Scout
Aug. 25, 1988	SOOS-4	Navigational	U.S. Navy	Scout
Sept. 24, 1988	NOAA 11	Meteorological	NOAA	Atlas-E
Sept. 29, 1988	DOD	n/a	Dept. of Defense	STS-27

*Mission failed

*Table 2–2. Science and Applications Missions
Conducted on the Space Shuttle*

Date	Payload	STS Mission
Nov. 12, 1981	OSTA-1	STS-2
March 22, 1982	OSS-1 (primarily science payload with some applications experiments)	STS-3
June 18, 1983	OSTA-2	STS-7
Nov. 28, 1983	Spacelab 1 (international mission with ESA)	STS-9
Aug. 30, 1984	OAST-1 (sponsored by the Office of Aeronautics and Space Technology with some experiments contributed by OSTA)	STS 41-D
Oct. 5, 1984	OSTA-3	STS 41-G
April 29, 1985	Spacelab 3 (international mission with ESA)	STS 51-B
July 29, 1985	Spacelab 2 (international mission with ESA)	STS 51-F
Oct. 30, 1985	Spacelab D-1 (German Spacelab with NASA oversight)	STS 61-A

Note: OAST-1 is addressed in Chapter 3, "Aeronautics and Space Research and Technology." OSS-1 and the Spacelab missions are addressed in Chapter 4, "Space Science," in Volume V of the *NASA Historical Data Book*.

*Table 2–3. Total Space Applications Funding History
(in thousands of dollars)*

Year	Request	Authorization	Appropriation	Programmed (Actual)
1979	274,300	280,300	a	274,800 b
1980	332,300	338,300	c	331,620 d
1981	381,700	372,400	331,550 e	331,550
1982	372,900 f	398,600	328,200 g	324,267 h
1983	316,300 i	336,300	341,300	347,700
1984	289,000	313,000	293,000	314,000
1985	344,100	390,100	384,100	374,100
1986	551,800	537,800	519,800 j	487,500
1987	491,100 k	552,600	578,100	562,600
1988	559,300	651,400	641,300	567,500 l

a Undistributed. Total R&D amount = $3,477,200,000.
b Included Resource Observations, Environmental Observations, Applications Systems, Technology Transfer, Materials Processing in Space, and Space Communications funding categories.
c Undistributed. Total R&D amount = $4,091,086,000.
d Communications funding category renamed Communications and Information Systems.
e Reflected recission.
f Amended submission. Original FY 1982 budget submission = $472,900,000.
g Reflected general supplemental appropriation approved September 10, 1982.
h Programmed funding for FY 1982 included Solid Earth Observations, Environmental Observations, Materials Processing in Space, Communications, and Information Systems funding categories. Reflects merger of OSS and OSTA.
i The Offices of Space Science and Space and Terrestrial Applications merged to form the Office of Space Science and Applications. Budget amounts reflected only items that were considered applications. Remaining OSSA budget items (science) can be found in Chapter 4.
j Reflected general reduction of $5,000,000 as well as other cuts made by Appropriations Committee.
k Revised submission. Original FY 1987 budget submission = $526,600,000.
l New Earth Science and Applications funding category incorporated Solid Earth Observations and Environmental Observations.

*Table 2–4. Programmed Budget by Major Budget Category
(in thousands of dollars)*

Budget Category/Fiscal Year	1979	1980	1981	1982	1983
Space Applications	274,800	331,620	331,550	324,267	347,700
Earth Observations	139,400	150,953	151,350	149,400	128,900
Environmental Observations	67,750	105,990	104,100	133,023	156,900
Applications Systems	13,950	24,567	18,100		
Technology Transfer	10,700	10,087	8,100		
Materials Processing in Space	20,400	19,768	18,700	16,244	22,000
Communications	22,600	20,255	31,200	21,300	32,400
Information Systems					7,500

Budget Category/Fiscal Year	1984	1985	1986	1987	1987
Space Applications	314,000	374,100	487,500	562,600	567,500
Solid Earth Observations	76,400	57,600	70,900	72,400	a
Environmental Observations	162,000	212,700	271,600	318,300	b
Earth Science and Applications					389,200
Materials Processing in Space	25,600	27,000	31,000	47,300	62,700
Communications	41,100	60,600	96,400	103,400	94,800
Information Systems	8,900	16,200	17,600	21,200	20,800

a Combined with Environmental Observations to form new Earth Science and Applications funding category.

b Combined with Solid Earth Observations to form new Earth Science and Applications funding category.

Table 2–5. Resource Observations/Solid Earth Observations Funding History (in thousands of dollars) a

Year (Fiscal)	Submission	Authorization	Appropriation	Programmed (Actual)
1979	139,150l b	c	d	139,400 e
1980	141,400	143,400	f	150,953 g
1981	170,300 h	182,600	151,350 i	151,350
1982	165,400 j	165,400	165,400	149,400 k
1983	132,200	132,200	132,200	128,900
1984	74,400	83,400 l	75,400 m	76,400 n
1985	63,600	63,600	63,600	57,600 o
1986	74,900	74,900	74,900	70,900
1987	74,100	74,100	74,100	72,400
1988	76,900	80,800	76,800	p

a   Renamed Solid Earth Observations beginning with FY 1982 programmed funding.
b   Source of data is the NASA Budget Office's *FY 1980 Budget Estimate*. The *Chronological History* for the FY 1979 budget did not include submission or authorization data for the Resource Observations funding category.
c   See note b above. FY 1979 authorization categories and amounts as stated in the *Chronological History FY 1979 Budget Estimates* were: Earth Resources Detection and Monitoring—$157,500,000; Earth Dynamics Monitoring and Forecasting—$8,600,000; Ocean Condition Monitoring and Forecasting—$12,400,000; Environmental Quality Monitoring—$20,200,000; Weather Observation and Forecasting—$22,800,000; Climate Research Program—$12,200,000; and Applications Explorer Missions—$4,200,000.
d   Undistributed. Total FY 1979 R&D appropriation = $3,477,200,000.
e   Included Landsat D, Operational Land Observing System, Magnetic Field Satellite, Shuttle/Spacelab Payload Development, Extended Mission Operations, Geodynamics, Applied Research and Data Analysis, AgRISTARS, Landsat 3, and Heat Capacity Mapping Mission.
f   Undistributed. Total R&D appropriation = $4,091,086,000.
g   Removed Landsat 3 and Heat Capacity Mapping Mission from total Resource Observations funding.
h   Amended submission. Original budget submission = $162,300,000.
i   Reflected recission.
j   Amended submission. Original budget submission = $187,200,000.
k   Removed Payload Development from Solid Earth Observations program funding category. Magsat now included in Extended Operations funding category.
l   House Authorization Committee added $4,000,000 for Research and Analysis to support applications studies related to spaceborne radars and the Global Resource Information System, $2,000,000 to partially restore the OMB reduction of NASA's request for AgRISTARS, and $3,000,000 for Technology Transfer activities, specifically for tests to verify and demonstrate the validity and usefulness of space applications systems. The Senate Authorization Committee added $5,000,000 more to Research and Analysis funding, $1,000,000 to AgRISTARS, and no additional funds to Technology Transfer. The Conference Committee modified this to allow $4,000,000 for Research and Analysis, $2,000,000 for AgRISTARS, and $3,000,000 for Technology Transfer.
m   The Senate Appropriations Committee added $1,000,000 for the multispectral linear array and eliminated all other additional funding.
n   Removed Extended Missions Operations and AgRISTARS from Solid Earth Observations program funding category
o   Removed Landsat 4 from Solid Earth Observations program funding category
p   Programmed amount (calculated in FY 1989) included under new program category: Earth Science and Applications. See Table 2–13.

Table 2–6. Landsat D/Landsat 4 Funding History
(in thousands of dollars)

Year (Fiscal)	Submission	Programmed (Actual)
1979	97,500	97,500
1980	98,663	104,413
1981	88,500	88,500
1982	83,900	81,900
1983	61,700	58,400
1984	16,800	16,800

Table 2–7. Magnetic Field Satellite Funding History
(in thousands of dollars) a

Year (Fiscal)	Submission	Programmed (Actual)
1979	3,900	3,900
1980	1,600	1,600
1981	500	500

a   Included under Extended Mission Operations beginning with FY 1982.

Table 2–8. Shuttle/Spacelab Payload Development Funding History
(in thousands of dollars)

Year (Fiscal)	Submission	Programmed (Actual)
1979	6,000	6,200
1980	1,850	2,031
1981	2,000	2,000
1982	3,300	12,300
1983	13,800	14,500
1984	16,000	17,000
1985	12,100	12,100
1986	23,100	21,800
1987	21,600	21,400 a
1988	20,800 b	27,700 c

a   Renamed Payload and Instrument Development.
b   Submission did not reflect integration of Solid Earth Observations and Environmental Observations into new Earth Sciences Payload and Instrument Development funding category.
c   This amount reflected new Earth Science and Applications funding category. There was now one Earth Science Payload and Instrument Development category that encompassed both the former Solid Earth Observations and Environmental Observations Payload and Instrument Development.

*Table 2–9. Extended Mission Operations Funding History
(in thousands of dollars)*

Year (Fiscal)	Submission	Programmed (Actual)
1979	350	358
1980	1,582	1,904
1981	2,700	2,700
1982	2,800	2,800 a
1983	1,800	1,100

a   Included funding for the operation of Magsat.

Table 2–10. Geodynamics Funding History (in thousands of dollars)

Year (Fiscal)	Submission	Programmed (Actual)
1979	8,200	8,200
1980	12,600	12,600
1981	23,400	23,400
1982	22,900	22,900
1983	26,200	28,100
1984	28,000	28,000
1985	29,900	29,900
1986	31,700	30,000
1987	32,100	31,600
1988	32,400	32,300 a

a   Included under Earth Science and Applications Program funding category.

Table 2–11. Geodynamics Research and Data Analysis Funding History (in thousands of dollars) a

Year (Fiscal)	Submission	Programmed (Actual)
1979	22,200	22,242
1980	12,908	12,405
1981	12,800	12,800
1982	19,500	15,500
1983	13,700	11,800
1984	14,600	14,600
1985	15,600	15,600
1986	20,100	19,100
1987	21,900	19,400
1988	21,100	21,400 b

a   Beginning in FY 1982, all applied research and data analysis funding categories were renamed Research and Analysis.

b   Renamed Land Processes Research and Analysis and included in Earth Science and Applications Program funding.

Table 2–12. AgRISTARS Funding History (in thousands of dollars)

Year (Fiscal)	Submission	Programmed (Actual)
1980	16,000	16,000
1981	31,400	21,450
1982	14,000	14,000
1983	15,000	15,000

Table 2–13. *Environmental Observations Funding History (in thousands of dollars)*

Year (Fiscal)	Submission	Authorization	Appropriation	Programmed (Actual)
1979	67,900 a	b	c	67,750 d
1980	117,200	121,200	e	105,990 f
1981	109,600 g	112,600	104,100 h	104,100 i
1982	135,300 j	145,300	k	133,023 l
1983	128,900	128,900	128,900	156,900 m
1984	163,000	170,000 n	164,000 o	162,000
1985	220,700	220,700	220,700	212,700 p
1986	317,500	311,500	290,500	271,600
1987	336,900 q, r	313,900	346,900	318,300 s
1988	393,800	393,800	378,000	389,200 t

a   Source of data is the NASA Budget Office's *FY 1980 Budget Estimate.* The *Chronological History* for the FY 1979 budget does not include submission and authorization data for the Environmental Observations funding category.

b   See note a above. FY 1979 authorization categories and amounts as stated in the *Chronological History FY 1979 Budget Estimates* were: Earth Resources Detection and Monitoring—$157,500,000; Earth Dynamics Monitoring and Forecasting—$8,600,000; Ocean Condition Monitoring and Forecasting—$12,400,000; Environmental Quality Monitoring—$20,200,000; Weather Observation and Forecasting—$22,800,000; Climate Research Program—$12,200,000; and Applications Explorer Missions—$4,200,000

c   Undistributed. Total FY 1979 R&D appropriation = $3,477,200,000.

d   Included Upper Atmosphere Research Program, Applied Research and Data Analysis, Shuttle/Spacelab Payload Development, Operational Satellite Improvement Program, ERBE, Halogen Occultation Experiment, Extended Mission Operations, National Oceanic Satellite System (NOSS), TIROS N, Nimbus 7, and Seasat.

e   Undistributed. Total R&D appropriation = $4,091,086,000.

f   Removed TIROS N and Seasat from Environmental Observations funding total.

g   Amended submission. Original budget submission = $137,600,000.

h   Reflected recission.

i   Removed Nimbus 7 from Environmental Observations funding total and added NOSS.

j   Amended submission. Original budget submission = $194,600,000.

k   Undistributed. Total FY 1982 R&D appropriation = $4,740,900,000.

l   Removed Applied Research and Data Analysis from Environmental Observations funding category. Added Upper Atmosphere Research Satellite (UARS) Experiments and Mission Definition to Environmental Observations funding category.

m   Removed Halogen Occultation Experiment from Environmental Observations funding history.

n   The House Authorization Committee added $2,000,000 for Technology Development and $1,000,000 for the Sun-Earth Interaction Study to the NASA submission. The Senate Authorization Committee added $2,000,000 for Space Physics/Technology Development, specifically for university research teams conducting experiments on the origin of plasmas in the Earth's neighborhood (OPEN), $4,000,000 for UARS Experiments, $2,000,000 for Atmospheric Dynamics, and $2,000,000 for Oceanic Research and Analysis to the NASA submission. The Conference Committee modified this authorization to allow $2,000,000 for OPEN and $5,000,000 for UARS Experiments and Atmospheric and Ocean Sensors.

Table 2–13 continued

o   The Senate Appropriations Committee added $2,000,000 to the NASA submission for UARS/OPEN Definition Studies. The Conference Committee reduced this by $1,000,000.

p   Added Payload and Instrument Development, Interdisciplinary Research and Analysis, Tethered Satellite System, and Scatterometer to Environmental Observations program funding category. Removed Operational Satellite Improvement Program from Environmental Observations program funding category.

q   Revised submission. Original FY 1987 budget submission = $367,900,000.

r   Submission, authorization, and appropriation data did not reflect new program budget category: Earth Science and Applications. See Table 2–5.

s   Removed ERBE and added Ocean Topography Experiment and Airborne Science and Applications funding categories.

t   Renamed Earth Science and Applications Program. New funding category incorporated Geodynamics from former Solid Earth Observations category and combined Payload and Instrument Development from both Solid Earth Observations and Environmental Observations funding categories.

Table 2–14. Upper Atmospheric Research Program Funding History (in thousands of dollars)

Year (Fiscal)	Submission	Programmed (Actual)
1979	(14,500)	(14,500) a
1980	12,500	12,400
1981	13,500	13,500
1982	13,000	20,500 b
1983	27,700	27,700
1984	28,500	28,435
1985	31,000	31,000
1986	33,000	31,100
1987	33,400	32,700
1988	32,700	32,700

a   Program was transferred from Space Science to Space Applications in January 1979; FY 1979 funding was not included in total.
b   Renamed Upper Atmosphere Research and Analysis with FY 1984 budget submission and FY 1982 actuals.

Table 2–15. Upper Atmospheric Research and Data Analysis Funding History (in thousands of dollars)

Year (Fiscal)	Submission	Programmed (Actual)
1979	33,876	33,726
1980	48,670	48,750
1981	48,100	48,100
1982	47,000	a

a   Programmed amounts found under new funding categories: Atmospheric Dynamics and Radiation Research and Analysis (Table 2–25) and Oceanic Processes Research and Development (Table 2–26)

Table 2–16. Interdisciplinary Research and Analysis Funding History (in thousands of dollars)

Year (Fiscal)	Submission	Programmed (Actual)
1985	1,000	1,000
1986	1,000	1,000
1987	1,100	1,100
1988	1,100	1,100

Table 2–17. Shuttle/Spacelab Resource Observations Payload Development Funding History (in thousands of dollars)

Year (Fiscal)	Submission	Programmed (Actual)
1979	7,750	7,750
1980	9,600	9,600
1981	1,700	1,700
1982	4,100	4,100
1983	3,700	3,700
1984	7,600	7,600
1985	7,800	7,800 a
1986	5,600	5,300
1987	12,000	9,700
1988	4,100 b	27,700 c

a Renamed Payload and Instrument Development.
b Payload and Instrument Development funding category was only for Environmental Observations Program and did not reflect new funding category of Earth Science and Applications Program.
c Incorporated amounts from both Solid Earth Observations and Environmental Observations Payload and Instrument Development funding categories.

Table 2–18. Operational Satellite Improvement Program Funding History (in thousands of dollars)

Year (Fiscal)	Submission	Programmed (Actual)
1979	6,100	6,100
1980	7,400	7,400
1981	9,200	7,200
1982	6,000	6,000
1983	6,000	6,000
1984	600	600

Table 2–19. Earth Radiation Budget Experiment Funding History (in thousands of dollars)

Year (Fiscal)	Submission	Programmed (Actual)
1979	7,000	7,000
1980	17,000	13,720
1981	20,300	20,300
1982	24,000	24,000
1983	24,000	24,000
1984	15,500	15,500
1985	8,100	8,100
1986	—	1,900

Table 2–20. Halogen Occultation Experiment Funding History (in thousands of dollars)

Year (Fiscal)	Submission	Programmed (Actual)
1979	3,600	3,600
1980	8,000	8,000
1981	4,500	4,500
1982	5,000	5,000

Table 2–21. Halogen Occultation Extended Mission Operations Funding History (in thousands of dollars)

Year (Fiscal)	Submission	Programmed (Actual)
1979	1,250	1,250
1980	5,800	5,800
1981	8,000	8,000
1982	11,400	16,100
1983	22,800	22,800
1984	27,400	27,400
1985	29,500	29,500
1986	37,000	35,000
1987	33,600	33,600 a
1988	14,800	14,700

a Renamed Mission Operations and Data Analysis.

Table 2–22. National Oceanic Satellite System Funding History (in thousands of dollars)

Year (Fiscal)	Submission	Programmed (Actual)
1981	5,800	800

Table 2–23. Nimbus 7 Funding History (in thousands of dollars)

Year (Fiscal)	Submission	Programmed (Actual)
1979	3,624	3,624
1980	500	500

Table 2–24. Upper Atmospheric Research Satellite Experiments and Mission Definition Funding History (in thousands of dollars)

Year (Fiscal)	Submission	Programmed (Actual)
1982	6,000	6,000
1983	14,000	14,000 a
1984	20,000	20,000
1985	55,700	55,700
1986	124,000	114,000
1987	114,200	113,800
1988	89,600	89,200

a Renamed Upper Atmosphere Research Satellite Mission.

Table 2–25. Atmospheric Dynamics and Radiation Research and Analysis Funding History (in thousands of dollars)

Year (Fiscal)	Submission	Programmed (Actual)
1982	a	22,300
1983	26,500	26,500
1984	27,500	27,465
1985	28,500	28,500
1986	30,300	28,700
1987	31,900	31,300
1988	31,400	31,400

a Included under Applied Research and Data Analysis (see Table 2–15).

Table 2–26. Oceanic Processes Research and Development Funding History (in thousands of dollars)

Year (Fiscal)	Submission	Programmed (Actual)
1982	a	16,900
1983	17,000	17,000
1984	18,200	18,200
1985	19,400	19,400
1986	20,600	17,400
1987	20,800	18,000
1988	20,200	20,100

a Included under Applied Research and Data Analysis (see Table 2–15).

Table 2–27. Space Physics/Research and Analysis Funding History (in thousands of dollars)

Year (Fiscal)	Submission	Programmed (Actual)
1982	No submission	12,123
1983	15,200	15,200
1984	16,700	16,800
1985	16,700	16,700
1986	17,800	16,800
1987	21,000	20,800

Table 2–28. Tethered Satellite System Funding History (in thousands of dollars)

Year (Fiscal)	Submission	Programmed (Actual)
1985	3,000	3,000
1986	4,500	6,400
1987	1,000	5,500 a

a Renamed Tethered Satellite Payloads.

Table 2–29. Scattermometer Funding History (in thousands of dollars)

Year (Fiscal)	Submission	Programmed (Actual)
1985	12,000	12,000
1986	14,000	14,000
1987	32,900	32,900
1988	22,700	22,600

Table 2–30. Ocean Topography Experiment Funding History (in thousands of dollars)

Year (Fiscal)	Submission	Programmed (Actual)
1987	19,000	18,900
1988	75,000	74,500

Table 2–31. Airborne Science and Applications Funding History (in thousands of dollars)

Year (Fiscal)	Submission	Programmed (Actual)
1987	No submission	(27,600) a
1988	21,900	21,800

a Previously funded under Physics and Astronomy Suborbital Program funding category.

SPACE APPLICATIONS

Table 2–32. Applications Systems Funding History (in thousands of dollars)

Year (Fiscal)	Submission	Authorization	Appropriation	Programmed (Actual)
1979	15,700	a		13,950 b
1980	24,200	24,200	c	24,567
1981	18,100	18,100	18,100 d	18,100 e
1982	13,200 f	13,200	13,200	g
1983	11,700	11,700	11,700	h

a   Applications Systems funding category did not appear in *Chronological History* of FY 1979 budget.
b   Included Airborne Instrumentation Research Program, Shuttle/Spacelab Mission Design and Integration, and NASA Integrated Payload Planning.
c   Undistributed. Total FY 1980 R&D appropriation = $4,091,086,000.
d   Reflected recission.
e   Included only Airborne Instrumentation Research Program.
f   Amended submission. Original budget submission = $14,400,000.
g   Programmed amounts for Applications Systems appropriation included with Suborbital Program in Physics and Astronomy funding category (Space Science funding).
h   Applications System Airborne Instrumentation Research Program efforts continued under Suborbital Program (Space Science funding). Program budget category eliminated in FY 1982.

Table 2–33. Airborne Instrumentation Research Program Funding History (in thousands of dollars)

Year (Fiscal)	Submission	Programmed (Actual)
1979	5,800	6,530
1980	15,547	15,567
1981	18,100	18,100
1982	13,200	a

a   Programmed amounts for Applications Systems appropriation included with Suborbital Program in Physics and Astronomy funding category (Space Science funding).

Table 2–34. Shuttle/Spacelab Mission Design and Integration Funding History (in thousands of dollars) a

Year (Fiscal)	Submission	Programmed (Actual)
1979	6,400	6,260
1980	7,300	7,300

a   Funding responsibility for FY 1981 and subsequent years transferred from Space Applications to Space Science.

*Table 2-35. NASA Integrated Payload Planning Funding History
(in thousands of dollars) a*

Year (Fiscal)	Submission	Programmed (Actual)
1979	2,000	1,160
1980	1,700	7,400

a	Funding responsibility for FY 1981 and subsequent years transferred from Space Applications to Space Science.

*Table 2-36. Materials Processing in Space Funding History
(in thousands of dollars)*

Year (Fiscal)	Submission	Authorization	Appropriation	Programmed (Actual)
1979	20,400	20,400	a	20,400 b
1980	19,800	19,800	c	19,768 d
1981	22,200	24,900	18,700 e	18,700 f
1982	27,700 g	31,700	h	16,244 i
1983	23,600	28,600	23,600	22,000
1984	21,600	26,600	23,600	25,600
1985	23,000	28,000	23,000	27,000
1986	34,000	36,000	34,000	31,000 j
1987	39,400 k	43,900	39,400	47,300
1988	45,900	50,000	65,900	62,700

a	Undistributed. Total FY 1979 R&D appropriation = $3,477,200,000.
b	Included Space Processing Applications Rocket (SPAR) project, Applied Research and Data Analysis, and Shuttle/Spacelab Payload Development.
c	Undistributed. Total FY 1980 R&D appropriation = $4,091,086,000.
d	Added Materials Experiment Operations to FY 1980 Materials Processing funding categories.
e	Reflected recission.
f	Removed SPAR project from Materials Processing funding categories.
g	Amended submission. Original FY 1982 budget submission = $32,100,000.
h	Undistributed. Total FY 1982 R&D appropriation = $4,740,900,000.
i	Removed Shuttle/Spacelab Payload Development funding category
j	Added Microgravity Shuttle/Space Station Payloads funding category. Removed Materials Experiment Operations funding category from Materials Processing in Space.
k	Revised submission. Original FY 1987 budget submission = $43,900,000.

Table 2-37. Materials Processing Research and Data Analysis Project Funding History (in thousands of dollars)

Year (Fiscal)	Submission	Programmed (Actual)
1979	4,400	4,850
1980	6,450	7,200
1981	10,950	9,230
1982	12,000	14,000
1983	13,100	13,100
1984	11,000	11,000
1985	11,700	11,700
1986	12,400	12,100
1987	13,900	13,900
1988	12,900	12,900

Table 2-38. Shuttle/Spacelab Materials Processing Payload Development Funding History (in thousands of dollars)

Year (Fiscal)	Submission	Programmed (Actual)
1979	12,400	11,950
1980	11,218	10,468
1981	10,750	8,157
1982	8,800	a

a Activities and funding transferred to the Physics and Astronomy Shuttle Payload Development and Mission Management area.

Table 2-39. Materials Processing Experiment Operations (Microgravity Shuttle/Space Station Payloads) Funding History (in thousands of dollars) a

Year (Fiscal)	Submission	Programmed (Actual)
1980	—	(533) b
1981	(1,900) c	1,310
1982	3,000	4,244
1983	8,900	8.900
1984	12,600	14,600
1985	15,300	15,300
1986	22,600	18,900 d
1987	34,000	33,400
1988	49,800	49,800

a Renamed Microgravity Shuttle/Space Station Payloads in FY 1986.
b Included under Materials Processing Shuttle/Spacelab Payload Development funding category.
c Included under Materials Processing Shuttle/Spacelab Payload Development funding category.
d Funding category was renamed and restructured as Microgravity Shuttle/Space Station Payloads. This category consolidated ongoing activities that provided a range of experimental capabilities for all scientific and commercial participants in the Microgravity Science and Applications program. These included Shuttle mid-deck experiments, the Materials Experiment Assembly, and the Materials Science Laboratory, which was carried in the orbiter bay. Included activities had been included under Materials Experiment Operations.

*Table 2–40. Technology Transfer Funding History
(in thousands of dollars)*

Year (Fiscal)	Submission	Authorization	Appropriation	Programmed (Actual)
1979	10,950 a	n/a	n/a	10,700 b
1980	10,300	10,300	c	10,087 d
1981	7,500 e	11,500	8,100	8,100
1982 f	5,000	—	—	—

a	Source of data is the FY 1979 current estimate found in the FY 1980 budget estimates. No Technology Transfer funding category appears in the *Chronological History* of the FY 1979 budget submissions. Therefore, no authorization or appropriations figures are available.
b	Included Applications Systems Verification and Transfer, Regional Remote Sensing Applications, User Requirements and Supporting Activities, and Civil Systems.
c	Undistributed. Total FY 1980 funding category = $4,091,086,000.
d	Removed Civil Systems from Technology Transfer funding total.
e	Amended submission. Original submission = $12,500,000.
f	Technology Transfer program funding eliminated beginning with FY 1982.

*Table 2–41. Applications Systems Verification and Transfer Funding
History (in thousands of dollars)*

Year (Fiscal)	Submission	Programmed (Actual)
1979	1,150	900
1980	1,700	1,700
1981	1,400	700

*Table 2–42. Regional Remote Sensing Applications Funding History
(in thousands of dollars)*

Year (Fiscal)	Submission	Programmed (Actual)
1979	3,500	3,500
1980	3,657	3,655
1981	2,700	2,400
1982	2,000	a

a Funding eliminated.

Table 2–43. User Requirements and Supporting Activities Funding History (in thousands of dollars)

Year (Fiscal)	Submission	Programmed (Actual)
1979	4,500	4,500
1980	4,730	4,732
1981	6,000	5,000
1982	3,000	a

a Funding eliminated.

Table 2–44. Civil Systems Funding History (in thousands of dollars)

Year (Fiscal)	Submission	Programmed (Actual)
1979	1,800	1,800

Table 2–45. Space Communications Funding History
(in thousands of dollars)

Year (Fiscal)	Submission	Authorization	Appropriation	Programmed (Actual)
1979	22,000	22,000	a	22,600 b
1980	19,400	19,400	c	20,255 d, e
1981	29,000	29,000	31,200 f	31,200 g
1982	20,900 h	34,000	i	21,300 j, k
1983	19,900	34,900 l	39,900 m	32,400
1984	21,100	24,100	21,100	41,100 n
1985	20,600	60,600 o	60,600	60,600
1986	106,200	101,200	101,200	96,400
1987	19,500	99,500 p	96,500	103,400 q
1988	20,500	104,500 r	97,500	94,800 s

a   Undistributed. Total FY 1979 R&D appropriation = $3,477,200,000.
b   Included Search and Rescue Mission, Technical Consultation and Support Studies, Applied Research and Data Analysis, Follow-On Data Analysis and Operations, Applications Data Service Definition, Data Management, and Adaptive Multibeam Phased Array (AMPA) System.
c   Undistributed. Total FY 1980 R&D appropriation = $4,091,086,000.
d   Referred to as Communications and Information Systems in FY 1980 programmed budget data material and NASA FY 1982 budget estimate.
e   Removed Follow-On Data Analysis, Applications Data Service Definition, and Data Management from FY 1980 Communications and Information Systems funding total.
f   Reflected recission.
g   Added Experiment Coordination and Operations Support and Information Systems funding categories.
h   Final revised submission. Original FY 1982 budget submission (January 1981) = $35,600,000. Amended submission (March 1981) = $30,300,000.
i   Undistributed. Total FY 1982 R&D appropriation = $4,740,900,000.
j   Added Experiment Coordination and Operations Support to Communications funding category.
k   Budget category referred to as Communications Program in FY 1984 NASA budget estimate (FY 1982 actual cost data).
l   The House Authorization Committee added $5,000,000 for 30/20-GHz test and evaluation flights. The Senate Authorization Committee added $15,000,000 to allow for a large proof-of-concept of communications operations in the 30/20-GHz frequency range. The final authorization added a total of $15,000,000 to NASA's budget submission.
m   The Appropriations Committee restored the entire $20,000,000 addition to NASA's budget submission. See Table 2–51.
n   Large difference between programmed and appropriated amounts reflected an increase in funding to the ACTS program. See Table 2–51.
o   Increase reflected Authorization Committee disagreement with NASA's restructuring of ACTS flight program. The Committee directed NASA "to proceed with the flight program and make the necessary future requests for budget authority as required." See Table 2–51.
p   The Authorization Committee directed NASA to continue the ACTS program in spite of the Reagan administration's attempts to terminate it.
q   Technical Consultation and Support Studies renamed Radio Science and Support Studies. Research and Analysis renamed Advanced Communications Research.
r   The Authorization Committee once again restored funds for the ACTS program that were removed by the Reagan administration. See Table 2–51.
s   Added Communications Data Analysis funding category.

Table 2–46. Space Communications Search and Rescue Mission Funding History (in thousands of dollars)

Year (Fiscal)	Submission	Programmed (Actual)
1979	8,000	8,000
1980	5,000	2,530
1981	4,800	4,800
1982	2,300	2,300
1983	3,700	3,700
1984	3,800	3,800
1985	2,400	2,400
1986	1,300	1,100
1987	1,000	1,385
1988	1,300	1,300

Table 2–47. Space Communications Technical Consultation and Support Studies Funding History (in thousands of dollars)

Year (Fiscal)	Submission	Programmed (Actual)
1979	3,100	3,100
1980	2,982	3,182
1981	3,100	3,145
1982	2,600	2,600
1983	2,600	2,600
1984	2,700	2,700
1985	2,900	2,900
1986	2,600	2,518
1987	3,200	3,050 a
1988	2,542	2,586

a   Renamed Radio Science and Support Studies.

Table 2–48. Space Communications Research and Data Analysis Funding History (in thousands of dollars)

Year (Fiscal)	Submission	Programmed (Actual)
1979	3,900	3,900
1980	6,200	6,200
1981	16,600	16,600
1982	10,000	15,400
1983	5,100	5,100
1984	8,500	8,500
1985	9,100	9,100
1986	10,400	9,770
1987	13,000	13,384 a
1988	14,136	13,992

a   Renamed Advanced Communications Research.

Table 2–49. Communications Data Analysis Funding History (in thousands of dollars)

Year (Fiscal)	Submission	Programmed (Actual)
1988	1,322	1,322

Table 2–50. Applications Data Service Definition Funding History (in thousands of dollars)

Year (Fiscal)	Submission	Programmed (Actual)
1979	No category listed	100
1980	2,400	2,245
1981	—	a

a Funding category not broken out separately.

Table 2–51. Advanced Communications Technology Satellite Funding History (in thousands of dollars)

Year (Fiscal)	Submission	Programmed (Actual)
1983	20,000	20,000
1984	5,000 a	25,000
1985	45,000	45,000
1986	85,000	81,900
1987	85,000	84,600
1988	75,600	75,600

a Reflected NASA's restructuring of the program to encompass only an experimental ground test program. Congress disagreed with the restructuring and directed NASA to continue with the program as originally planned. See Table 2–45.

Table 2–52. Information Systems Program Funding History (in thousands of dollars)

Year (Fiscal)	Submission	Authorization	Appropriation	Programmed (Actual)
1981	Included in Communications and Information Systems figures (see Table 2–45)			
1982	Included in Communications and Information Systems figures (see Table 2–45)			
1983 a	7,500 b	Included in Communications and Information Systems figures (see Table 2–45)		7,500
1984 c	8,900	8,900	8,900	8,900 d
1985	16,200	16,200	16,200	16,200
1986	19,200	19,200	19,200	17,600
1987	21,200	21,200	21,200	21,200
1988	22,300	22,300	22,300	20,800

a   Included only Data Systems funding category.
b   New program-level funding category.
c   FY 1984 was the first year that the Information Systems Program appeared as a separate appropriation in the *Chronological History* budget submissions. Previous programmed amounts were a subcategory under the Communications and Information Systems appropriation category.
d   Information Systems Program included Data Systems and Information Systems funding categories.

Table 2–53. Data Systems Funding History (in thousands of dollars)

Year (Fiscal)	Submission	Programmed (Actual)
1980	4,500	10,600
1981	No category	
1982	No submission	4,300
1983	7,500 a	7,500
1984	7,900	7,900
1985	8,400	8,400
1986	9,000	8,500
1987	9,400	10,000
1988	9,700	9,600

a   Included in Information Systems funding category.

Table 2–54. Information Systems Funding History (in thousands of dollars) a

Year (Fiscal)	Submission	Programmed (Actual)
1984	1,000	1,000
1985	7,800	7,800
1986		9,100
1987		11,200
1988		11,200

a   Information Systems funding category was a subcategory under the Information Systems program.

Table 2–55. OSTA-1 Payload

Principal Investigator	Institution	Experiment
Charles Elachi	Jet Propulsion Laboratory, Pasadena, California	Shuttle Imaging Radar-A (SIR-A) evaluated using spaceborne imaging radar for geologic exploration, with emphasis on mineral and petroleum exploration and fault mapping. A secondary goal was to determine the capability to combine SIR-A data with Landsat data and improve the usefulness of both (Figure 2–4).
Alexander F.H. Goetz	Jet Propulsion Laboratory, Pasadena, California	Shuttle Multispectral Infrared Radiometer obtained radiometric data in 10 spectral bands from a statistically significant number of geological areas around the world.
Roger T. Schappell	Martin Marietta Aerospace, Denver, Colorado	Feature Identification and Location Experiment developed video techniques to provide methods for identifying, spectrally classifying, and physically locating surface features or clouds.
Henry G. Reichle, Jr.	NASA Langley Research Center, Hampton, Virginia	Measurement of Air Pollution From Satellites measured the distribution of carbon monoxide in the middle and upper troposphere and traced its movement between the Northern and Southern Hemispheres.
Hongsuk H. Kim	NASA Goddard Space Flight Center, Greenbelt, Maryland	Ocean Color Experiment evaluated a passive ocean color sensing technique for mapping the concentration of chlorophyll-producing phytoplankton in the open ocean.
Bernard Vonnegut	State University of New York at Albany	Night-Day Optical Survey of Lightning studied the convective circulation in storms and the relationship to lightning discharges using a motion picture camera to film the lightning flashes of nighttime thunderstorms.
Allan H. Brown	University of Pennsylvania	Heflex Bioengineering Test determined the effect of near weightlessness and soil moisture content on dwarf sunflower growth.

Table 2–56. OSTA-2 Experiments

Investigation	Principal Investigator	Institution
MEA Experiments		
Liquid Phase Miscibility Gap Materials	Stanley H. Gelles	S.H. Gelles Associates, Columbus, Ohio
Vapor Growth of Alloy-Type Semiconductor Crystals	Herbert Wiedemeier	Rensselaer Polytechnic Institute, Troy, New York
Containerless Processing of Glass Forming Melts	Delbert E. Day	University of Missouri–Rolla
MAUS Experiments		
Stability of Metallic Dispersions	Guenther H. Otto	German Aerospace Research Establishment (DVFLR), Federal Republic of Germany
Particles at a Solid/Liquid Interface	Hermann Klein	German Aerospace Research Establishment (DVFLR), Federal Republic of Germany

Table 2–57. OSTA-2 Instrument Module Characteristics

Detector wavelength	0.385, 0.45, 0.6, 1.0 microns
Field of view	0.15 milliradians (0.5 km)
Altitude range	10 km to 100 km above Earth horizon
Altitude resolution	1 km
Detector operating temperature	19 degrees to 30 degrees C
Scan rate	15 km/sec
Sampling rate	64 samples/sec
Information bandwidth	1 Hz/km/channel
Radiometer resolution	3,000:1
Signal-to-noise ratio (1.0 micron channel)	1.5×15^5 at peak

Table 2–58. SAGE (AEM-2) Characteristics

Launch Date	February 18, 1979
Launch Vehicle	Scout
Range	Wallops Flight Center
Lead NASA Center	Goddard Space Flight Center/Langley Research Center
Owner	NASA
NASA Mission Objectives	Determine a global database for stratospheric aerosols and ozone and use these data sets for a better understanding of Earth's environmental quality and radiation budget; specifically: • Develop a satellite-based remote-sensing technique for measuring stratospheric aerosols and ozone • Map vertical extinction profiles of stratospheric aerosols and ozone from 78 degrees south to 78 degrees north latitude • Investigate the impact of natural phenomena, such as volcanoes and tropical upwellings, on the stratosphere • Investigate the sources and sinks of stratospheric ozone and aerosols
Orbit Characteristics	
Apogee (km)	661
Perigee (km)	548
Inclination (deg.)	54.9
Period (min.)	96.7
Weight (kg)	147
Dimensions	Base module: 65 cm; overall height including antenna: 161.85 cm; six-sided prism
Power Source	Solar paddles and batteries
Instruments	Four-spectral channel radiometer
Contractor	Ball Aerospace Systems Division, Ball Corp.; Boeing Aerospace Company
Remarks	The satellite was turned off April 15, 1982, after the spacecraft's battery failed. It decayed in April 1989.

Table 2–59. ERBS Instrument Characteristics

Sensor	Measured Quantities	No. of Channels/ Frequencies	Spectral Range/Frequency Range	Resolution
ERBE Non-Scanner	Total energy of Sun's radiant heat and light	1 4	0.2–3.5 μm 1.2–50.0 μm 0.2–50.0 μm	100 km across swath Full solar disk
ERBE Scanner	Reflected solar radiation, Earth-emitted radiation			
SAGE II	Stratospheric aerosols, O₃, NO₂, water vapor	7	0.385–1.02 μm	0.5 km

Table 2–60. Earth Radiation Budget Satellite Characteristics

Launch Date	October 5, 1984
Launch Vehicle	STS 41-G (*Challenger*)
Range	Kennedy Space Center
Lead NASA Center	Goddard Space Flight Center; Langley Research Center
Owner	NASA
NASA Mission Objectives	Increase knowledge of Earth's climate and weather systems, particularly how climate is affected by radiation from the Sun by measuring the distribution of aerosols and gases in the atmosphere
Orbit Characteristics	
Apogee (km)	603
Perigee (km)	602
Inclination (deg.)	57.0
Period (min.)	96.8
Weight (kg)	2,307 at launch
Dimensions	4.6 m x 3.8 m x 1.6 m
Power Source	Solar panels and batteries
Instruments	ERBE Non-Scanner had five sensors: two wide field-of-view sensors viewed the entire disc of Earth from limb to limb; two medium field-of-view sensors viewed a 10-degree region; and the fifth sensor measured the total output of radiant heat and light from the Sun. ERBE Scanner instrument was a scanning radiometer that measured reflected solar radiation and Earth-emitted radiation. Stratospheric Aerosol and Gas Experiment (SAGE II) was a Sun-scanning radiometer that measured solar radiation attenuation caused by the constituents in the atmosphere.
Contractor	TRW Defense and Space Systems; Ball Brothers
Remarks	It was still operating as of October 1994.

Table 2–61. UARS Instruments and Investigators

Instrument	Description and Primary Measurements	Principal Investigator	Institution
UARS Species and Temperature Measurements			
CLAES (Cryogenic Limb Array Etalon Spectrometer)	Neon and CO_2 cooled interferometer sensing atmospheric infrared emissions; T, CF_2, Cl_2, $CFCl_3$, $ClONO_2$, CH_4, O_3, NO_2, N_2O, HNO_3, and H_2O	A.E. Roche	Lockheed Palo Alto Research Laboratory, Palo Alto, California
ISAMS (Improved Stratospheric and Mesospheric Sounder)	Mechanically cooled radiometer sensing atmospheric infrared emissions; T, O_3, NO, NO_2, N_2O, HNO_3, H_2O, CH_4, and CO	F.W. Taylor	Oxford University, Oxford, United Kingdom
MLS (Microwave Limb Sounder)	Microwave radiometer sensing atmospheric emissions; ClO and H_2O_2	J.W. Waters	Jet Propulsion Laboratory, Pasadena, California
HALOE (Halogen Occultation Experiment)	Gas filter/radiometer sensing sunlight occulted by the atmosphere; HF and HCl	J.M. Russell, III	NASA Langley Research Center, Hampton, Virginia
UARS Wind Measurements			
HRDI (High Resolution Doppler Imager)	Fabry-Perot spectrometer sensing atmospheric emission and scattering; two-component wind: 10–110 km	P.B. Hays	University of Michigan, Ann Arbor, Michigan
WINDII (Wind Imaging Interferometer)	Michelson interferometer sensing atmospheric emission and scattering; two-component wind: 80–110 km	G.G. Shepherd	York University, York, Canada

SPACE APPLICATIONS

Table 2–61 continued

Instrument	Description and Primary Measurements	Principal Investigator	Institution
UARS Energy Input Measurements			
SUSIM (Solar Ultraviolet Spectral Irradiance Monitor)	Full disk solar irradiance spectrometer incorporating on-board calibration; solar spectral irradiance: 120–400 nm	G.E. Brueckner	Naval Research Laboratory, Washington, D.C.
SOLSTICE (Solar Stellar Irradiance Comparison Experiment)	Full disk solar irradiance spectrometer incorporating stellar comparison; solar spectral irradiance: 115–440 nm	G.J. Rottman	University of Colorado, Boulder, Colorado
PEM (Particle Environment Monitor)	X-ray proton and electron spectrometers; *in situ* energetic electrons and protons; remote sensing of electron energy deposition	J.D. Winningham	Southwest Research Institute, San Antonio, Texas
Instrument of Opportunity			
ACRIM II (Active Cavity Irradiance Monitor II)	Full disk solar irradiance radiometer; continuation of solar constant measurements	R.C. Willson	Jet Propulsion Laboratory, Pasadena, California

Table 2–62. UARS Development Chronology

Date	Event
1978	The UARS project concept is developed. The objective of UARS, as stated by OSSA, is to provide the global database necessary for understanding the coupled chemistry and dynamics of the stratosphere and mesosphere, the role of solar radiation in driving the chemistry and dynamics, and the susceptibility of the upper atmosphere to long-term changes in the concentration and distribution of key atmospheric constituents, particularly ozone. OSSA defines the project as a crucial element of NASA's long-term program in upper atmospheric research—a program initiated in response to concerns about ozone depletion.
July 1978	UARS Science Working Group final report is published.
Sept. 1978	UARS Announcement of Opportunity is released.
April 25, 1980	NASA selects 26 investigations to be studied for possible inclusion on the UARS mission proposed for the late 1980s. Of the 26 investigations, 23 are from the United States, 2 are from the United Kingdom, and 1 is from France. Each country is responsible for funding its investigation. The initial study phase cost to the United States, including its investigations, is estimated to be $5 million over the next 2 years. The mission is to have two satellites launched 1 year apart from the Space Shuttle.
Feb. 18, 1981	The current cost of UARS is estimated at $400–$500 million.
May 12, 1981	Because of a $110 million cutback in space applications funding, the development of instruments for UARS is delayed.
Nov. 1981	NASA selects nine experimental and ten theoretical teams for UARS. The experimental teams are to develop instruments to make direct measurements of upper atmospheric winds, solar ultraviolet irradiance, energetic particle interactions with the upper atmosphere, and densities of critical chemical species as a function of altitude. The theoretical teams are to develop and apply models of the upper atmosphere, which, when combined with the new data to be acquired, should increase understanding of the upper atmospheric chemistry and dynamics and improve the capability to assess the impact of human activities on the delicate chemical processes in the stratosphere.
Dec. 24, 1981	UARS instrument developers are selected.
Jan. 26, 1982	NASA reprograms FY 1982 funds so that the UARS experiment budget is increased from $5 million to $6 million to enhance the long lead development work on selected payloads.
Aug. 1982	The mission is reduced from two to one spacecraft. The project now calls for 11 instruments. Instruments (including one each from Britain and France) enter Phase C/D development (Design and Development or Execution). Run-out cost for instruments through projected 1988 launch is estimated at $200 million. Total estimated mission cost of $500 million includes procurement of the MMS (at $200 million).
Aug. 4, 1982	Goddard Space Flight Center director states hope that UARS will receive FY 1984 new start funding. The UARS would use the MMS. The spacecraft was planned to be launched in 1988.

Table 2–62 continued

Date	Event
Aug. 31, 1982	NASA officials state that UARS could be helpful in understanding the cloud of volcanic dust currently covering the lower latitudes of the globe and that UARS will provide insight on how this volcanic cloud affects climate.
Feb. 3, 1983	NASA declares that it does not need the fifth orbiter for UARS. UARS mission is not included by OMB (Office of Management and Budget) in FY 1984 budget. NASA proceeds with instrument development and now expects to seek UARS as a FY 1985 new start.
Feb. 10, 1983	OMB wants NASA to find a way to reduce the price of the design for UARS. Because funding for instruments was previously approved, eventual project approval is not in question.
Feb. 17, 1983	Goddard investigates modifying the command and data handling module of the MMS so it will be compatible with UARS.
Sept. 9, 1983	Goddard announces plans to issue a preliminary RFP, for industry comment, for system design of the UARS observatory and design and fabrication of an instrument module compatible with the MMS bus.
Aug. 19, 1983	NASA announces plans to build UARS on a spare MMS bus. It will also include a refurbished attitude control system from the Solar Maximum Mission. The mission now includes nine instruments. The launch date has been delayed until the fall of 1989 because UARS is not included in the FY 1984 budget.
Dec. 1983	Objectives state that UARS will study energy flowing into and from the upper atmosphere, chemical reactions in the upper atmosphere, and how gases are moved within and between layers of the atmosphere. UARS will be located 600 kilometers high. The current estimated costs are $570–$670 million. NASA currently has $27.7 million for upper atmosphere research and $14 million for UARS experiments and definition. By using the MMS design, NASA hopes to save $30–$36 million.
Jan. 31, 1984	NASA requests FY 1985 funding for UARS.
March 1984	RFP is issued for system design of UARS observatory and design and fabrication of instrument module compatible with MMS bus.
April 9, 1984	Lockheed Missile and Space Co. begins building the CLAES sensor, which will be used on UARS. The instrument is designed to measure concentrations of nitrogen oxides, ozone, chlorine compounds, carbon dioxide, and methane, among other atmospheric constituents, and to record temperatures.
May 31, 1984	NASA states that using the MMS attitude control module will save 75 percent of the costs over building a new attitude control module.
July 1984	UARS Execution Phase Project Plan is approved.
July 1984	NASA proposes that the WINDII instrument replaces the French WINTERS instrument.
Nov. 7, 1984	Goddard announces plans to award a sole source contract to Fairchild Space Co. to build the MMS for UARS.
Feb. 4, 1985	NASA's FY 1985 budget includes UARS.
March 6, 1985	NASA awards a $145.8 million contract to General Electric Co.'s Valley Forge Space Center in Philadelphia for UARS observatory. The GE Space Center will be responsible for the design of the observatory system and the design and fabrication of a module compatible with the MMS. The launch is scheduled for October 89.

Table 2-62 continued

Date	Event
April 18, 1985	The estimated cost for UARS is currently at $630–$700 million.
June 25, 1985	A review of UARS Support Instrumentation Requirements Document is requested. The document requests a review of the deep space network as a backup to TDRSS for emergency support.
Aug. 27, 1985	Goddard awards a $16.3 million contract to the Fairchild Space Co. in Germantown, Maryland. Fairchild will be responsible for providing the MMS for UARS. Under the contract, Fairchild will fabricate the structure and harness for the spacecraft, refurbish a spare Communications and Data Handling Mode, and integrate and test the assembled spacecraft. Fairchild will also be responsible for the refurbishing of the thermal louvers on the Solar Max module.
Oct. 1985	Observatory work plan review is complete; execution phase is initiated.
Nov. 1985	The central data handling facility contract is awarded to Computer Sciences Corporation.
Jan. 1986	The WINDII contract is awarded.
June 1986	The central data handling facility hardware contract is awarded to Science Systems and Applications, Inc.
March 1987	The CLAES cryogen redesign is to comply with recommendations arising from the *Challenger* investigation.
April 1987	Observatory Preliminary Design Review is conducted.
Fall 1987	Rebaseline results from the *Challenger* accident; launch is rescheduled for the fall of 1991.
March 1988	Observatory Critical Design Review is conducted.
Jan. 1989	Technicians at Goddard make final adjustments to the MMS being fitted for the UARS spacecraft. UARS is scheduled for deployment from the Space Shuttle *Discovery* in September 1991.
July 6, 1989	ACRIM II is delivered.
July 21, 1989	SOLSTICE is delivered.
July 27, 1990	The United States and the Soviet Union announce that they will share the information they have regarding the hole in the ozone layer over Antarctica. The Soviet Union has been acquiring information about the hole in the ozone layer through its Meteor-3; the United States has been collecting information through NASA's TOMS (Total Ozone Mapping Spectrometer).
Aug. 22, 1989	SUSIM is delivered.
Sept. 13, 1989	HALOE is delivered.
Dec. 19, 1990	NASA announces the crew members for UARS, which is scheduled for launch in November 1991.
March 21, 1991	The projected launch date for UARS is October 1991. The Tracking Data Relay Satellite mission originally scheduled to launch in July has been pushed to August. The Defense support mission has been moved from August to December. These changes were made to preserve the NASA's capability to fly *Discovery* with the UARS payload during its required science window.
Sept. 12, 1991	UARS is launched from STS-48 (*Discovery*).

Table 2-63. *NOAA Satellite Instruments (1978-1988)*

Satellite	Orbit	AVHRR a	HIRS/2	MSU	SSU	ERBE	SBUV/2	SEM	DCS	SAR
TIROS N	PM	1	X	X	X			X	X	
NOAA 6	AM	1	X	X	X			X	X	
NOAA 7	PM	2	X	X	X			X	X	
NOAA 8	AM	1	X	X	X			X	X	X
NOAA 9	PM	2	X	X	X	X	X	X	X	X
NOAA 10	AM	1	X	X	X	X		X	X	X
NOAA 11	PM	2	X	X	X		X	X	X	X

Legend: AVHRR—Advanced Very High Resolution Radiometer; DCS—Data Collection and Location System; ERBE—Earth Radiation Budget Experiment; HIRS—High-Resolution Infrared Radiation Sounder; MSU—Microwave Sounding Unit; SAR—Search and Rescue; SBUV—Solar Backscatter Ultraviolet Spectral Radiometer; SEM—Space Environment Monitor; and SSU—Stratospheric Sounding Unit.

a   Two versions of the AVHRR were flown. The AVHRR/1 had four channels, and the AVHRR/2 had five channels, resulting in different response functions.

Table 2–64. NOAA 6 Characteristics

Launch Date	June 27, 1979
Launch Vehicle	Atlas F
Range	Western Test Range
Lead NASA Center	Goddard Space Flight Center
Owner	National Oceanic and Atmospheric Administration
NASA Mission Objectives	Launch the spacecraft into a Sun-synchronous orbit of sufficient accuracy to enable it to accomplish its operational mission requirements and conduct an in-orbit evaluation and checkout of the spacecraft
NOAA Objectives	Collect and send data of Earth's atmosphere and sea surface as part of the National Operational Environmental Satellite System (NOESS) to improve forecasting ability
Orbit Characteristics	
Apogee (km)	801
Perigee (km)	786
Inclination (deg.)	98
Period (min.)	100.7
Weight (kg)	1,405
Dimensions	3.71 m high and 1.88 m diameter unstowed; 4.91 m high and 2.37 m diameter with solar arrays extended
Power Source	Solar array and two 30 AH nickel cadmium batteries
Instruments	1. Advanced Very High Resolution Radiometer (AVHRR) provided digital data for each of four spectral intervals.
	2. Data Collection and Location System (DCS) was a random-access system that located and/or collected data from remote fixed and free-floating terrestrial and atmospheric platforms.
	3. High Energy Proton-Alpha Detector (HEPAD) sensed protons and alphas from a few hundred MeV up through relativistic particles above 850 NeV
	4. Medium Energy Proton Electron Detector (MEPED) sensed protons, electrons, and ions with energies from 30 keV to several tens of MeV.
	5. Space Environment Monitor (SEM) was a multichannel charged-particle spectrometer that provided measurements on the population of Earth's radiation belts and on particle precipitation phenomena resulting from solar activity.
	6. Total Energy Detector (TED) used a programmed swept electrostatic curved-plate analyzer to select particle type/energy and a channeltron detector to sense/qualify the intensity of the sequentially selected energy bands.

Table 2–64 continued

	7. TIROS Operational Vertical Sounder (TOVS) determined radiances needed to calculate temperature and humidity profiles of the atmosphere from the planetary surface into the stratosphere. It consisted of three instruments: HIRS/2, SSU, and MSU. – High Resolution Infrared Sounder (HIRS/2) measured incident radiation in 20 spectral regions of the infrared spectrum, including long and short wave regions. – Stratospheric Sounding Unit (SSU) used a selective absorption technique to make temperature measurements in three channels. – Microwave Sounding Unit (MSU) provided four channels for the TOVS in the 60-GHz oxygen absorption region. These were accurate in the presence of clouds. The passive microwave measurements could be converted into temperature profiles of the atmosphere from Earth's surface to 20 km.
Contractor	RCA Astro Electronics

Table 2–65. NOAA B Characteristics

Launch Date	May 29, 1980
Launch Vehicle	Atlas F
Range	Western Space and Missile Center
Lead NASA Center	Goddard Space and Flight Center
Owner	National Oceanic and Atmospheric Administration
NASA Mission Objectives	Launch the spacecraft into a Sun-synchronous orbit of sufficient accuracy to enable it to accomplish its operational mission requirements and to conduct an in-orbit evaluation and checkout of the spacecraft
NOAA Objectives	Collect and send data of Earth's atmosphere and sea surface as part of the NOESS to improve forecasting ability
Orbit Characteristics	Did not reach proper orbit
Weight (kg)	1,405
Dimensions	3.71 m high and 1.88 m diameter unstowed; 4.91 m high and 2.37 m diameter with solar arrays extended
Power Source	Solar array and two 30 AH nickel cadmium batteries
Instruments	Same as NOAA 6
Contractor	RCA Astro Electronics

Table 2–66. Advanced Very High Resolution Radiometer Characteristics

Characteristics	Channels				
	1	2	3	4	5
Spectral range (micrometers)	0.58 to 0.68	0.725 to 1.0	3.55 to 3.93	10.3 to 11.3	11.4 to 12.4
Detector	Silicon	Silicon	InSb	(HgCd)Te	(HgCd)Te
Resolution (km at nadir)	1.1	1.1	1.1	1.1	1.1
Instantaneous field of view (milliradians)	1.3 sq.	1.3 sq.	1.3 sq.	1.3 sq.	1.3 sq.
Signal-to-noise ratio at 0.5 albedo	>3:1	>3:1	—	—	—
Noise-equivalent temperature difference at (NEΔT) 300 degrees K	—	—	<0.12 K	<0.12 K	<0.12 K
Scan angle (degrees)	±55	±55	±55	±55	±55

Optics—8-inch diameter afocal Cassegrain telescope
Scanner—360-rpm hysteresis synchronous motor with beryllium scan mirror
Cooler—Two-stage radiant cooler, infrared detectors controlled at 105 or 107 degrees K
Data output—10-bit binary, simultaneous sampling at 40-kHz rate

Table 2–67. NOAA 7 Characteristics

Launch Date	June 23, 1981
Launch Vehicle	Atlas F
Range	Western Space and Missile Center
Lead NASA Center	Goddard Space Flight Center
Owner	National Oceanic and Atmospheric Administration
NASA Mission Objectives	Launch the spacecraft into a Sun-synchronous orbit of sufficient accuracy to enable it to accomplish its operational mission requirements and to conduct an in-orbit evaluation and checkout of the spacecraft
NOAA Objectives	Collect and send data of Earth's atmosphere and sea surface as part of the NOESS to improve forecasting ability
Orbit Characteristics	
Apogee (km)	847
Perigee (km)	829
Inclination (deg.)	98.9
Period (min.)	101.7
Weight (kg)	1,405
Dimensions	3.71 m high and 1.88 m diameter unstowed; 4.91 m high and 2.37 m diameter with solar arrays extended
Power Source	Solar array and two 30 AH nickel cadmium batteries
Instruments	Same as NOAA 6 with the exception of the AVHRR, which had five channels rather than four. In addition, the U.S. Air Force provided a contamination monitor to assess contamination sources, levels, and effects for consideration on future spacecraft. This instrument flew for the first time on this mission.
Contractor	RCA Astro Electronics

Table 2–68. NOAA 8 Characteristics

Launch Date	March 28, 1983
Launch Vehicle	Atlas E
Range	Western Space and Missile Center
Lead NASA Center	Goddard Space and Flight Center
Owner	National Oceanic and Atmospheric Administration
NASA Mission Objectives	Launch the spacecraft into a Sun-synchronous orbit of sufficient accuracy to enable it to accomplish its operational mission requirements and to conduct an in-orbit evaluation and checkout of the spacecraft
NOAA Mission Objectives	To collect and send data of Earth's atmosphere and sea surface as part of the NOESS to improve forecasting ability
Orbit Characteristics	
Apogee (km)	825.5
Perigee (km)	805
Inclination (deg.)	98.6
Period (min.)	101.2
Weight (kg)	1,712
Dimensions	3.71 m high and 1.88 m diameter unstowed; 4.91 m high and 2.37 m diameter with solar arrays extended
Power Source	Solar array and two 30AH nickel cadmium batteries
Instruments	Same as NOAA 6 instruments with the addition of the Search and Rescue (SAR) system. The SAR on NOAA 8 could detect and locate existing emergency transmitters operating at 121.5 MHz and 245 MHz, as well as experimental transmitters operating at 406 MHz (see "Communications Program" section in this chapter).
Contractor	RCA Astro Electronics

Table 2–69. NOAA 9 Characteristics

Launch Date	December 12, 1984
Launch Vehicle	Atlas E
Range	Vandenberg Air Force Base
Lead NASA Center	Goddard Space Flight Center
Owner	National Oceanic and Atmospheric Administration
NASA Mission Objectives	• Launch the spacecraft into a Sun-synchronous orbit of sufficient accuracy to enable it to accomplish its operational mission requirements, conduct an in-orbit evaluation and checkout of the spacecraft, and, upon completion of this evaluation, turn the operational control of the spacecraft over to the NOAA National Environmental Satellite Data and Information Service (NESDIS)
	• Successfully acquire data from the Earth Radiation Budget Experiment (ERBE) instruments for application in scientific investigations aimed at improving our understanding of the processes that influence climate and climate changes
	• Acquire data from the Solar Backscatter Ultraviolet (SBUV/2) instrument to determine stratospheric ozone concentrations on a global basis
NOAA Mission Objectives	Collect and send data of Earth's atmosphere and sea surface as part of the NOESS in acquiring daily global weather information for the short- and long-term forecasting needs of the National Weather Service
Orbit Characteristics	
Apogee (km)	863
Perigee (km)	839
Inclination (deg.)	99.1
Period (min.)	102.2
Weight (kg)	1,712
Dimensions	4.91 m high; 1.88 m diameter with solar array extended
Power Source	Solar array and two 30AH nickel cadmium batteries

Table 2–69 continued

Instruments	Same instruments as NOAA 8 with the addition of SBUV/2 and ERBE: • ERBE consisted of a medium and wide field-of-view nonscanning radiometer and a narrow field-of-view scanning radiometer. The radiometers would measure Earth radiation energy budget components at satellite altitude; make measurements from which monthly average Earth radiation energy budget components can be derived at the top of the atmosphere on regional, zonal, and global scales; and provide an experimental prototype for an operational ERBE instrument for future long-range monitoring programs. • SBUV/2 consisted of two instruments: the Monochrometer and the Cloud Cover Radiometer. The Monochrometer was a spectral scanning ultraviolet radiometer that could measure solar irradiance and scene radiance (back-scattered solar energy) over a spectral range of 160 to 400 nanometers. The Cloud Cover Radiometer detected clouds that would contaminate the signal. Experiment objectives were to make measurements from which total ozone concentration in the atmosphere could be determined to an accuracy of 1 percent, make measurements from which the vertical distribution of atmospheric ozone could be determined to an accuracy of 5 percent, and measure the solar spectral irradiance from 160 to 400 nanometers.
Contractor	RCA Astro Electronics

Table 2–70. NOAA 10 Characteristics

Launch Date	September 17, 1986
Launch Vehicle	Atlas E
Range	Vandenberg Air Force Base
Lead NASA Center	Goddard Space Flight Center
Owner	National Oceanic and Atmospheric Administration
NASA Mission Objectives	Launch the spacecraft into a Sun-synchronous orbit of sufficient accuracy to enable it to accomplish its operational mission requirements, conduct an in-orbit evaluation and checkout of the spacecraft, and, upon completion of this evaluation, turn the operational control of the spacecraft over to the NOAA NESDIS
NOAA Mission Objectives	Collect and send data of Earth's atmosphere and sea surface as part of the NOESS to improve forecasting ability
Orbit Characteristics	
Apogee (km)	823
Perigee (km)	804
Inclination (deg.)	98.7
Period (min.)	101.2
Weight (kg)	1,712
Dimensions	4.91 m high; 1.88 m diameter with solar panels expanded
Power Source	Solar array and two 30 AH nickel cadmium batteries
Instruments	Same as NOAA 9 instruments, including NASA's ERBE, but with a "dummy" SBUV and a "dummy" SSU. The SSU, which was provided by the United Kingdom through its Meteorological Office, Ministry of Defense, was flown only on "afternoon" satellites beginning with NOAA 9.
Contractor	RCA Astro Electronics

Table 2-71. NOAA 11 Characteristics

Launch Date	September 24, 1988
Launch Vehicle	Atlas E
Range	Western Space and Missile Center
Lead NASA Center	Goddard Space Flight Center
Owner	National Oceanic and Atmospheric Administration
NASA Mission Objectives	Launch the spacecraft into a Sun-synchronous orbit of sufficient accuracy to enable it to accomplish its operational mission requirements, to conduct an in-orbit evaluation and checkout of the spacecraft, and upon, completion of this evaluation, to turn the operational control of the spacecraft over to the NOAA NESDIS
NOAA Mission Objectives	Collect and send data of Earth's atmosphere and sea surface as part of the NOESS to improve forecasting ability
Orbit Characteristics	
Apogee (km)	865
Perigee (km)	849
Inclination (deg.)	98.9
Period (min.)	102.1
Weight (kg)	1,712
Dimensions	4.91 m high; 1.88 m diameter
Power Source	Solar array and two 30 AH nickel cadmium batteries
Instruments	Same instruments as NOAA 9 with the exception of ERBE
Contractor	RCA Astro Electronics

Table 2-72. VISSR Atmospheric Sounder Infrared Spectral Bands

Spectral Band	Central Wavelength (mm)	Spatial Resolution (km)	Weighting Function Peak (mb)	Absorbing Constituent
1	14.73	13.8	70	CO_2
2	14.48	13.8	125	CO_2
3	14.25	6.9 and 13.8	200	CO_2
4	14.01	6.9 and 13.8	500	CO_2
5	13.33	6.9 and 13.8	920	CO_2
6	4.525	13.8	850	CO_2
7	12.66	6.9 and 13.8	Surf.	H_2O
8	11.17	6.9 and 13.8	Surf.	Window
9	11.17	6.9 and 13.8	600	H_2O
10	6.725	6.9 and 13.8	400	H_2O
11	4.444	13.8	300	CO_2
12	3.945	13.8	Surf.	Window

Table 2–73. GOES 4 Characteristics

Launch date	September 9, 1980
Launch vehicle	Delta 3914
Range	Eastern Test Range
Lead NASA Center	Goddard Space Flight Center
Owner	National Oceanic and Atmospheric Administration
NASA Mission Objectives	Launch the satellite into a synchronous orbit of sufficient accuracy to enable the spacecraft to provide the capability for continuous observations of the atmosphere on an operational basis, flight-test the satellite in orbit and, when checked out, turn the spacecraft over to NOAA for operational use, and demonstrate, validate, and assess the temperature and moisture soundings from the VISSR Atmospheric Sounder
NOAA Mission Objectives	Collect and relay weather data to enable forecasters and other scientists to study severe storms and storm-spawned phenomena, such as hail, flash floods, and tornadoes, by monitoring weather over Canada, the United States, and Central and South America
Orbit Characteristics	
Apogee (km)	35,795
Perigee (km)	35,780
Inclination (deg.)	4.1
Period (min.)	1,436.2
Weight (kg)	444 (in orbit)
Dimensions	4.43 m high from the S-band omni antenna rod to the apogee boost motor nozzle aperture; 2.15 m diameter spin-stabilized drum
Power Source	Solar panels and two nickel cadmium batteries

Table 2–73 continued

Instruments	1. VISSR Atmospheric Sounder was capable of simultaneous imaging in the visible portion of the spectrum with a resolution of 0.9 km and the infrared portion of the spectrum with a resolution of 6.9 km, multispectral imaging simultaneously in five spectral bands (one visible and four selectable from the 12 infrared bands), and a dwell sounding mode from which moisture, temperature, and vertical structure of the atmosphere may be determined. 2. Space Environmental Monitor (SEM) provided direct quantitative measurements of the major effects of solar activity for use in real-time solar forecasting and subsequent research, detected unusual solar flares with high levels of radiation, measured the strength of solar winds, and measured the strength and direction of Earth's magnetic field. 3. Data Collection and Location System (DCS) provided communications relay from data collection platforms on land, at sea, and in the air to the Command and Data Acquisition Station (CDA), as well as the interrogation of platforms from the CDA via the satellite. 4. Telemetry, Tracking, and Command used S-band frequencies for transmission of wideband visual data to the CDA, for relay of "stretched" data from the CDA via the spacecraft to facilities operated by NOAA, and for transmission of weather facsimile data to local ground stations equipped to receive S-band automatic picture transmission data; UHF for transmissions from data collection platforms to the spacecraft and then to the CDA on the S-band; and VHF and S-band for commanding the spacecraft, for telemetry, and for transmitting the space environment monitoring data.
Contractors	Hughes Aircraft, Ball Aerospace, Panametrics, Ford Aerospace and Communications Corp.

Table 2–74. GOES 5 Characteristics

Launch date	May 22, 1981
Launch vehicle	Delta 3914
Range	Eastern Test Range
Lead NASA Center	Goddard Space Flight Center
Owner	National Oceanic and Atmospheric Administration
NASA Mission Objectives	Launch the satellite into a synchronous orbit of sufficient accuracy to enable the spacecraft to provide the capability for continuous observations of the atmosphere on an operational basis, flight test the satellite in orbit and, when checked out, turn the spacecraft over to NOAA for operational use, and continue the demonstration and validation of the temperature and moisture soundings from the VISSR Atmospheric Sounder
NOAA Mission Objectives	Collect and relay weather data to enable forecasters and other scientists to study severe storms and storm-spawned phenomena such as hail, flash floods, and tornadoes, by monitoring weather over Canada, the United States, and Central and South America
Orbit Characteristics	
Apogee (km)	35,792
Perigee (km)	35,782
Inclination (deg.)	1.2
Period (min.)	1,435.9
Weight (kg)	444 (in orbit)
Dimensions	4.43 m high from the S-band omni antenna rod to the apogee boost motor nozzle aperture; 2.15 m diameter spin-stabilized drum
Power Source	Solar panels and two nickel cadmium batteries
Instruments	Same as GOES 4
Contractors	Hughes Aircraft, Ball Aerospace, Panametrics, Ford Aerospace and Communications Corp.

Table 2–75. GOES 6 Characteristics

Launch date	April 28, 1983
Launch vehicle	Delta 3914
Range	Eastern Space and Missile Center
Lead NASA Center	Goddard Space Flight Center
Owner	National Oceanic and Atmospheric Administration
NASA Mission Objectives	Launch the satellite into a synchronous orbit of sufficient accuracy to enable the spacecraft to provide the capability for continuous observations of the atmosphere on an operational basis and flight-test the satellite in orbit and, when checked out, turn the spacecraft over to NOAA for operational use
NOAA Mission Objectives	Collect and relay weather data to enable forecasters and other scientists to study severe storms and storm-spawned phenomena such as hail, flash floods, and tornadoes, by monitoring weather over Canada, the United States, and Central and South America
Orbit Characteristics	
Apogee (km)	35,891
Perigee (km)	35,776
Inclination (deg.)	0.1
Period (min.)	1,436.4
Weight (kg)	444 in orbit
Dimensions	4.43 m high from the S-band omni antenna rod to the apogee boost motor nozzle aperture; 2.15 m diameter spin-stabilized drum
Power Source	Solar panels and two nickel cadmium batteries
Instruments	Same as GOES 4
Contractors	Hughes Aircraft, Ball Aerospace, Panametrics, Ford Aerospace and Communications Corp.

Table 2–76. GOES G Characteristics

Launch date	May 3, 1986
Launch vehicle	Delta 3914
Range	Cape Canaveral Air Force Station
Lead NASA Center	Goddard Space Flight Center
Owner	National Oceanic and Atmospheric Administration
NASA Mission Objectives	Launch the satellite into a synchronous orbit of sufficient accuracy to enable the spacecraft to provide the capability for continuous observations of the atmosphere on an operational basis, flight-test the satellite in orbit and, when checked out, turn the spacecraft over to NOAA for operational use, and determine usefulness of instant alert capabilities of geosynchronous search and rescue systems and to develop and test processing techniques for geosynchronous search and rescue data
NOAA Mission Objectives	Collect and relay weather data to enable forecasters and other scientists to study severe storms and storm-spawned phenomena, such as hail, flash floods, and tornadoes
Orbit Characteristics	Did not achieve orbit
Weight (kg)	1,712 at launch
Dimensions	4.43 m high from the S-band omni antenna rod to the apogee boost motor nozzle aperture; 2.15 m diameter spin-stabilized drum
Power Source	Solar panels and two nickel cadmium batteries
Instruments	Same as GOES 4
Contractors	Hughes Aircraft, Ball Aerospace, Panametrics, Ford Aerospace and Communications Corp.

Table 2–77. GOES 7 Characteristics

Launch date	February 26, 1987
Launch vehicle	Delta 3924
Range	Cape Canaveral Air Force Station
Lead NASA Center	Goddard Space Flight Center
Owner	National Oceanic and Atmospheric Administration
NASA Mission Objectives	Launch the satellite into a geosynchronous orbit of sufficient accuracy to enable the spacecraft to provide the capability for continuous observations of the atmosphere on an operational basis, flight-test the satellite in orbit and, when checked out, turn the spacecraft over to NOAA for operational use, determine the usefulness of instant alert capabilities of geosynchronous search and rescue systems, and develop and test processing techniques for geosynchronous search and rescue data
NOAA Mission Objectives	Transmit cloud cover images from a geosynchronous orbit and atmospheric temperature profiles, collect space environmental data, and conduct an experiment for detecting emergency distress signals on the ground from geosynchronous orbit
Orbit Characteristics	
Apogee (km)	35,796
Perigee (km)	35,783
Inclination (deg.)	4.3
Period (min.)	1,436.2
Weight (kg)	456 in orbit
Dimensions	4.43 m high from the S-band omni antenna rod to the apogee boost motor nozzle aperture; 2.15 m diameter spin-stabilized drum
Power Source	Solar array and two nickel cadmium batteries
Instruments	Same as GOES 4
Contractors	Hughes Aircraft, Ball Aerospace, Panametrics, Ford Aerospace and Communications Corp.

Table 2–78. Landsat 4 Instrument Characteristics

	Thematic Mapper		Multispectral Scanner	
Spectral Band	Micrometers	Radiometric Sensitivity (NEΔP) %	Micrometers	Radiometric Sensitivity (NEΔP) %
1	0.45–0.52	0.8	0.5–0.6	0.57
2	0.52–0.60	0.5	0.6–0.7	0.57
3	0.63–0.69	0.5	0.7–0.8	0.65
4	0.76–0.9	0.5	0.8–1.1	0.70
5	1.55–1.75	1.0		
6	2.08–2.35	2.4		
7	10.40–12.50	0.5K (NEΔT)		
Ground IFOV	30M (bands 1–6)		83M (bands 1–4)	
Data Rate	85 Mb/s		15 Mb/s	
Quantization Levels	256		64	
Weight (kilograms)	246		58	
Size (meters)	1.1 x 0.7 x 2.0		0.35 x 0.4 x 0.9	
Power (watts)	345		81	

Table 2–79. Landsat 4 Characteristics

Launch Date	July 16, 1982
Launch Vehicle	Delta 3920
Range	Vandenberg Air Force Base
Lead NASA Center	Goddard Space Flight Center
Customer/Sponsor	National Oceanic and Atmospheric Administration
Mission Objectives	• Acquire multispectral, high-spatial resolution images of solar radiation reflected from Earth's surface and, for the Thematic Mapper, the emitted radiation in the thermal infrared region of the electromagnetic spectrum
	• Provide continuing Earth remote-sensing information and to encourage continued national and international participation in land remote-sensing programs
	• Assess the capabilities of the new Thematic Mapper sensing system and to exploit new areas of the infrared and visible light spectrum at higher resolution
	• Establish a technical and operational proficiency that can be used to help define the characteristics necessary for potential future operational land remote-sensing systems
Orbit Characteristics	
Apogee (km)	700
Perigee (km)	699
Inclination (deg.)	98.2
Period (min.)	98.8
Weight (kg)	1,941
Dimensions	4 m long; 2 m wide (deployed)
Power Source	Solar array and batteries
Instruments	1. Multispectral Scanner (MSS) scanned cross-track swaths of 185 km imaging six scan lines across in each of the four spectral bands simultaneously, focusing the scanned Earth image on a set of detectors. The instantaneous field of view of each detector subtended an Earth area square of 83 cm.
	2. Thematic Mapper (TM) was a seven-band multispectral high-resolution scanner that collected, filtered, and detected radiation from Earth in a swath 185 km wide.
Contractor	General Electric (Landsat 4 spacecraft), Hughes Aircraft (TM and MSS), Fairchild Industries (Multimission Modular Spacecraft)

Table 2–80. Landsat 5 Characteristics

Launch Date	March 1, 1984
Launch Vehicle	Delta 3920
Range	Western Test Range
Lead NASA Center	Goddard Space Flight Center
Customer/Sponsor	National Oceanic and Atmospheric Administration
Mission Objectives	• Acquire multispectral, high-spatial resolution images of solar radiation reflected from Earth's surface and, for the TM, the emitted radiation in the thermal infrared region of the electromagnetic spectrum
	• Launch the spacecraft into a polar orbit of sufficient accuracy to enable the spacecraft to provide the capability of acquiring MSS and TM scenes on a global basis for a period of 1 year
	• Flight-test the spacecraft in orbit and, when checked out, turn the spacecraft and MSS over to NOAA for operational use
	• Demonstrate the capability to process up to 50 TM scenes per day to produce tapes and film masters and complete the transfer of TM operations and data processing to NOAA as agreed to by NASA and NOAA
	• Perform evaluations of TM and MSS data quantifying some of the observational advantages of TM versus MSS imagery
Orbit Characteristics	
Apogee (km)	700
Perigee (km)	699
Inclination (deg.)	98.2
Period (min.)	98.8
Weight (kg)	1,941
Dimensions	4 m long, 2 m wide (deployed)
Power Source	Solar array and batteries
Instruments	Same as Landsat 4
Contractor	General Electric (spacecraft), Hughes Aircraft (TM and MSS), Fairchild Industries (MMS)

Table 2–81. Magsat Characteristics

Launch Date	October 30, 1979
Launch Vehicle	Scout
Date of Reentry	June 11, 1980
Range	Western Test Range
Customer/Sponsor	NASA Office of Space and Terrestrial Applications and U.S. Geological Survey
Lead NASA Center	Goddard Space Flight Center
Mission Objectives	Develop a worldwide vector magnetic field model suitable for the U.S. Geological Survey update and refinement of world and regional magnetic charts, compile crustal magnetic anomaly maps with spatial resolution of 350 km or better, interpret anomalies in conjunction with correlative data in terms of geologic/geophysical models of Earth's crust, and increase understanding of the origin and nature of the geomagnetic field and its temporal variations
Orbit Characteristics	
Apogee (km)	551
Perigee (km)	350
Inclination (deg.)	96.8
Period (min.)	93.6
Weight (kg)	183
Dimensions	Instrument module: height—874 cm with trim boom extended, diameter—77 cm with solar panels and magnetometer boom extended, width—340 cm tip to tip with solar array deployed, length—722 cm along flight path with magnetometer boom and solar array deployed Base module: diameter—66 cm, height—61 cm
Power Source	Solar panels
Instruments	1. Scalar Magnetometer was a dual lamp cesium vapor magnetometer that measured the magnitude of Earth's crustal magnetic field. 2. Vector Magnetometer was a three-axis fluxgate magnetometer that measured magnetic field direction as well as magnitude.
Experiments	Thirty-two investigations were selected in response to an Announcement of Opportunity issued September 1, 1978. They included 13 foreign investigations from Australia, Brazil, Canada, France, India, Italy, Japan, and the United Kingdom, as well as investigations from the United States. The general resources categories were: geophysics, geology, field modeling, marine studies, magnetosphere/ionosphere, and core/mantle studies. Data distribution was through the National Space Science Data Center. Table 2–82 lists the investigations.
Contractor	Applied Physics Laboratory, Johns Hopkins University

Table 2–82. Magsat Investigations

Principal Investigator	Organization	Research Area
Geophysics		
R.L. Coles	The Geomagnetic Service of Canada	Reduction, Verification, and Interpretation of Magsat Data Over Canada
B.N. Bhargava	Indian Institute of Geomagnetism	Magnetic Anomaly and Magnetic Field Map Over India
W.J. Hinze	Purdue University	Processing and Interpretation of Magnetic Anomaly Data Over South America
G.R. Keller	University of Texas, El Paso	Synthesis of Data for Crustal Modeling of South America
P. Gasparini	University of Naples, Italy	Crustal Structures Under the Active Volcanic Areas of the Mediterranean
N. Fukushima	University of Tokyo	Proposal From Japanese National Team for Magsat Project
C.R. Bentley	University of Wisconsin	Investigation of Antarctic Crust and Upper Mantle
M.A. Mayhew	Business and Technology Systems, Inc., Seabrook, Maryland	Magsat Anomaly Field Inversion and Interpretation for the United States
J.L. leMouel	Institut de Physique du Globe, Toulouse, France	Data Reduction, Studies of Europe, Central Africa, and Secular Variation
J.C. Dooley	Bureau of Mineral Resources, Canberra, Australia	The Regional Field and Crustal Structure of Australia and Antarctica
B.D. Johnson	Macquarie University, Australia	Crustal Properties of Australia and Surrounding Regions
Geology		
R.S. Carmichael	University of Iowa	Crustal Structure and Mineral Resources in the U.S. Midcontinent
D.H. Hall	University of Manitoba, Canada	Lithostratigraphic and Structural Elements in the Canadian Shield
I. Gill Pacca	Universidade de Sao Paulo, Brazil	Structure, Composition, and Thermal State of the Crust in Brazil
D.A. Hastings	Michigan Technological University	Precambrian Shields and Adjacent Areas of West Africa and South America
D.W. Strangeway	University of Toronto, Canada	Analysis of Anomaly Maps Over Portions of the Canadian and Other Shields

Table 2–82 continued

Principal Investigator	Organization	Research Area
I.J. Won	North Carolina State University, Raleigh, North Carolina	Compatibility Study of the Magsat Data and Aeromagnetic Data in the Eastern Piedmont, United States
S.E. Haggerty	University of Massachusetts, Amherst, Massachusetts	The Mineralogy of Global Magnetic Anomalies
M.R. Godiver	ORSTROM, Paris, France	Magnetic Anomaly of Bangui
Field Modeling		
D.R. Baraclough	Institute of Geological Sciences, Edinburgh, UK	Spherical Harmonic Representation of the Main Geomagnetic Field
D.P. Stern	NASA/Goddard Space Flight Center	Study of Enhanced Errors and of Secular Variation
M.A. Mayhew	Business and Technology Systems, Inc., Seabrook, Maryland	Equivalent Source Modeling of the Main Field
B.P. Gibbs	Business and Technology Systems, Inc., Seabrook, Maryland	Field Modeling by Optimal Recursive Filtering
Marine Studies		
C.G.A. Harrison	University of Miami, Florida	Investigations of Medium Wavelength Anomalies in the Eastern Pacific
J.L. LaBrecque	Lamont-Doherty Geological Observatory, Palisades, New York	Analysis of Intermediate Wavelength Anomalies Over the Oceans
R.F. Brammer	The Analytical Sciences, Corp., Reading, Massachusetts	Satellite Magnetic and Gravity Investigation of the Eastern Indian Ocean
Magnetosphere/Ionosphere		
D.M. Klumpar	University of Texas, Richardson, Texas	Effects of External Current Systems on Magsat Data Utilizing Grid Cell Modeling
J.R. Burrows	National Research Council of Canada	Studies of High Latitude Current Systems Using Magsat Vector Data
T.A. Potemra	Johns Hopkins University	Corrective Information on High-Latitude External Fields
R.D. Regan	Phoenix Corporation, McLean, Virginia	Improved Definition of Crustal Magnetic Anomalies in Magsat Data
Core/Mantle Studies		
E.R. Benton	University of Colorado, Boulder, Colorado	Field Forecasting and Fluid Dynamics of the Core
J.F. Hermance	Brown University, Providence, Rhode Island	Electromagnetic Deep-Probing of the Earth's Interior: Crustal Resource

Table 2–83. ASC-1 Characteristics

Launch Date	August 27, 1985
Launch Vehicle	STS 51-I (*Discovery*)/PAM-D
Range	Kennedy Space Center
Mission Objectives	Launch the satellite with sufficient accuracy to allow the PAM-D and spacecraft propulsion system to place the spacecraft into stationary geosynchronous orbit while retaining sufficient stationkeeping propulsion to meet the mission lifetime requirements
Owner	American Satellite Company
Orbit Characteristics	
Apogee (km)	35,796
Perigee (km)	35,777
Inclination (deg.)	0.1
Period (min.)	1,436.1
Weight (kg)	665 (in orbit)
Dimensions	Main body: 1.625 m x 1.320 m x 1.320 m Spans: 14 m with solar array extended
Shape	Cube
Power Source	Solar array panels and two nickel cadmium batteries
Contractor	RCA Astro Electronics
Remarks	ASC-1 was the first satellite to have encrypted command links, a security feature that prevented unauthorized access to the satellite command system. It was in a geosynchronous orbit at approximately 128 degrees west longitude.

Table 2–84. Comstar D-4 Characteristics

Launch Date	February 21, 1981
Launch Vehicle	Atlas Centaur
Range	Eastern Space and Missile Center
Mission Objectives	Launch the satellite into a transfer orbit which that would enable the spacecraft apogee motor to inject the spacecraft into a synchronous orbit
Owner	American Telephone and Telegraph Co. (AT&T)
Orbit Characteristics	
Apogee (km)	35,794
Perigee (km)	35,784
Inclination (deg.)	1.9
Period (min.)	1,436.2
Weight (kg)	1,484 (before launch)
Dimensions	6.1 m high; 2.44 m diameter
Shape	Cylindrical
Power Source	Solar array and batteries
Contractor	Hughes Aircraft
Remarks	Comstar D-4 became operational on May 5, 1981. It was located at approximately 127 degrees west longitude.

SPACE APPLICATIONS

Table 2–85. Telstar 3-A Characteristics

Launch Date	July 28, 1983
Launch Vehicle	Delta 3920/PAM-D
Range	Eastern Space and Missile Center
Mission Objectives	Launch the satellite on a two-stage Delta 3920 with sufficient accuracy to allow the MDAC PAM-D and spacecraft propulsion system to place the spacecraft into stationary geosynchronous orbit while retaining sufficient station-keeping propulsion to meet the mission lifetime requirements
Owner	AT&T
Orbit Characteristics	
Apogee (km)	35,796
Perigee (km)	35,778
Inclination (deg.)	0
Period (min.)	1,436.1
Weight (kg)	653 (in orbit)
Dimensions	6.48 m high (deployed); 2.74 m diameter
Shape	Cylindrical
Power Source	Solar cells and nickel cadmium batteries
Contractor	Hughes Aircraft
Remarks	Also called Telstar 301, the spacecraft was placed in a geosynchronous orbit at approximately 96 degrees west longitude above the equator.

Table 2–86. Telstar 3-C Characteristics

Launch Date	September 1, 1984
Launch Vehicle	STS 41-D (*Discovery*)/PAM-D
Range	Kennedy Space Center
Mission Objectives	Launch the satellite with sufficient accuracy to allow the MDAC PAM-D and spacecraft propulsion system to place the spacecraft into stationary geosynchronous orbit while retaining sufficient stationkeeping propulsion to meet the mission lifetime requirements
Owner	AT&T
Orbit Characteristics	
Apogee (km)	35,791
Perigee (km)	35,782
Inclination (deg.)	0
Period (min.)	1,436.1
Weight (kg)	653 (in orbit)
Dimensions	6.48 m high (deployed); 2.74 m diameter
Shape	Cylindrical
Power Source	Solar cells and nickel cadmium batteries
Contractor	Hughes Aircraft
Remarks	Telstar 3-C was placed into a geosynchronous orbit at approximately 85 degrees west longitude. It was also called Telstar 302.

Table 2–87. Telstar 3-D Characteristics

Launch Date	June 19, 1985
Launch Vehicle	STS-51 G (*Discovery*)/PAM-D
Range	Kennedy Space Center
Mission Objectives	Launch the satellite with sufficient accuracy to allow the MDAC PAM-D and spacecraft propulsion system to place the spacecraft onto stationary geosynchronous orbit while retaining sufficient stationkeeping propulsion to meet the mission lifetime requirements
Owner	AT&T
Orbit Characteristics	
Apogee (km)	35,804
Perigee (km)	35,770
Inclination (deg.)	0
Period (min.)	1,436.1
Weight (kg)	653 (in orbit)
Dimensions	6.48 m high (deployed); 2.74 m diameter
Shape	Cylindrical
Power Source	Solar cells and nickel cadmium batteries
Contractor	Hughes Aircraft
Remarks	Telstar 3-D was placed in a geostationary orbit at approximately 125 degrees west longitude. It was also called Telstar 303.

Table 2–88. Galaxy 1 Characteristics

Launch Date	June 28, 1983
Launch Vehicle	Delta 3920/PAM-D
Range	Eastern Space and Missile Center
Mission Objectives	Launch the satellite on a two-stage Delta 3920 launch vehicle with sufficient accuracy to allow the MDAC PAM-D and the spacecraft propulsion system to place the satellite into a stationary geosynchronous orbit while retaining sufficient stationkeeping propulsion to meet the mission lifetime requirements
Owner	Hughes Communications Inc.
Orbit Characteristics	
Apogee (km)	35,797
Perigee (km)	35,780
Inclination (deg.)	0
Period (min.)	1,436.2
Weight (kg)	519 at beginning of life
Dimensions	2.16 m diameter; 2.8 m long (stowed); 6.8 m long (with solar panel and antenna reflector deployed)
Shape	Cylinder
Power Source	K-7 solar cells and two nickel cadmium batteries
Contractor	Hughes Communications
Remarks	Galaxy 1 was devoted entirely to distributing cable television programming. It had a geostationary orbit at approximately 133 degrees west longitude. It operated until April 1994.

Table 2–89. Galaxy 2 Characteristics

Launch Date	September 22, 1983
Launch Vehicle	Delta 3920/PAM-D
Range	Eastern Space and Missile Center
Mission Objectives	Launch the satellite on a two-stage Delta 3920 with sufficient accuracy to allow the MDAC PAM-D and the satellite propulsion system to place the satellite into a stationary geosynchronous orbit while retaining sufficient stationkeeping propulsion to meet the mission lifetime requirements
Owner	Hughes Communications Inc.
Orbit Characteristics	
Apogee (km)	35,799
Perigee (km)	35,782
Inclination (deg.)	0
Period (min.)	1,436.2
Weight (kg)	519 at beginning of life
Dimensions	2.16 m diameter; 2.8 m long (stowed); 6.8 m long (with solar panel and antenna reflector deployed)
Shape	Cylinder
Power Source	K-7 solar cells and two nickel cadmium batteries
Contractor	Hughes Communications
Remarks	Galaxy 2 had a geostationary orbit above the equator at approximately 74 degrees west longitude. It operated until May 1994.

Table 2–90. Galaxy 3 Characteristics

Launch Date	September 21, 1984
Launch Vehicle	Delta 3920/PAM-D
Range	Eastern Space and Missile Center
Mission Objectives	Launch the satellite on a two-stage Delta 3920 launch vehicle with sufficient accuracy to allow the MDAC PAM-D and the satellite propulsion system to place the satellite into a stationary geosynchronous orbit while retaining sufficient stationkeeping propulsion to meet the mission lifetime requirements
Owner	Hughes Communications Inc.
Orbit Characteristics	
Apogee (km)	35,792
Perigee (km)	35,783
Inclination (deg.)	0
Period (min.)	1,436.2
Weight (kg)	519 at beginning of life
Dimensions	2.16 m diameter; 2.8 m long (stowed); 6.8 m long (with solar panel and antenna reflector deployed)
Shape	Cylinder
Power Source	K-7 solar cells and two nickel cadmium batteries
Contractor	Hughes Communications
Remarks	Galaxy 3 was placed in a geosynchronous orbit at approximately 93.5 degrees west longitude. It operated until September 30, 1995.

Table 2-91. Satcom 3 Characteristics

Launch Date	December 6, 1979
Launch Vehicle	Delta 3914
Range	Eastern Space and Missile Center
Mission Objectives	Place the RCA satellite into a synchronous transfer orbit of sufficient accuracy to allow the spacecraft propulsion systems to place the spacecraft into a stationary synchronous orbit while retaining sufficient stationkeeping propulsion to meet the mission lifetime requirements
System Objectives	Provide communications coverage for all 50 states, be capable of operating all 24 transponder channels at specified power throughout the minimum 8-year life, and be compatible with the Delta 3914 launch vehicle
Owner	RCA American Communications (RCA Americom)
Orbit Characteristics	Transfer orbit—did not achieve final orbit
Apogee (km)	35,798
Perigee (km)	162
Inclination (deg.)	23.9
Period (min.)	630
Weight (kg)	895
Dimensions	Base plate: 119 cm x 163 cm; main body height: 117 cm
Shape	Rectangular with two solar panels extended on booms from opposite sides and an antenna and reflector mounted on one end
Power Source	Solar cells and nickel cadmium batteries
Contractor	RCA Americom Astro-Electronics Division
Remarks	The satellite was destroyed during the firing of the apogee kick motor on December 10, 1979. This was the third RCA satellite launched by NASA.

Table 2-92. Satcom 3-R Characteristics

Launch Date	November 19, 1981
Launch Vehicle	Delta 3910
Range	Eastern Space and Missile Center
Mission Objectives	Launch the RCA satellite along a suborbital trajectory on a two-stage Delta 3910 launch vehicle with sufficient accuracy to allow the payload propulsion system to place the spacecraft into a stationary synchronous orbit while retaining sufficient stationkeeping propulsion to meet the mission lifetime requirements
System Objectives	Provide communications coverage for Alaska, Hawaii, and the contiguous 48 states, be capable of operating all 24 transponder channels at specified power throughout the minimum 10-year life, including eclipse periods, and be compatible with the Delta 3910 launch vehicle
Owner	RCA Americom
Orbit characteristics	
Apogee (km)	35,794
Perigee (km)	35,779
Inclination (deg.)	0.1
Period (min.)	1,436.1
Weight (kg)	1,082 (at launch)
Dimensions	Baseplate: 119 cm x 163 cm; main body height: 117 cm
Shape	Rectangular with two solar panels extended on booms from opposite sides and an antenna and reflector mounted on one end
Power Source	Solar cells and nickel cadmium batteries
Contractor	RCA Americom Astro-Electronics Division
Remarks	RCA Satcom 3-R was placed into geosynchronous orbit at approximately 132 degrees west longitude above the equator. This spacecraft and future RCA spacecraft were designed for launch by the Space Shuttle or by the Delta 3910/PAM-D launch vehicle.

Table 2–93. Satcom 4 Characteristics

Launch Date	January 19, 1982
Launch Vehicle	Delta 3910
Range	Eastern Space and Missile Center
Mission Objectives	Launch the RCA satellite along a suborbital trajectory on a two-stage Delta 3910 launch vehicle with sufficient accuracy to allow the payload propulsion system to place the spacecraft into a stationary synchronous orbit while retaining sufficient stationkeeping propulsion to meet the mission lifetime requirements
System Objectives	Provide communications coverage for Alaska, Hawaii, and the contiguous 48 states, be capable of operating all 24 transponder channels at specified power throughout the minimum 10-year life, including eclipse periods, and be compatible with the Delta 3910 launch vehicle
Owner	RCA Americom
Orbit characteristics	
Apogee (km)	35,795
Perigee (km)	35,781
Inclination (deg.)	0
Period (min.)	1,436.2
Weight (kg)	1,082 at launch; 598 in orbit
Dimensions	Baseplate: 119 cm x 163 cm; main body height: 117 cm
Shape	Rectangular with two solar panels extended on booms from opposite sides and an antenna and reflector mounted on one end
Power Source	Solar cells and nickel cadmium batteries
Contractor	RCA Americom Astro-Electronics Division
Remarks	RCA Satcom 4 was placed into geosynchronous orbit located at approximately 83 degrees west longitude.

Table 2-94. Satcom 5 Characteristics

Launch Date	October 27, 1982
Launch Vehicle	Delta 3924
Range	Eastern Space and Missile Center
Mission Objectives	Launch the RCA spacecraft into a synchronous transfer orbit on a three-stage Delta 3924 launch vehicle with sufficient accuracy to allow the spacecraft apogee kick motor to place the spacecraft into a stationary synchronous orbit while retaining sufficient stationkeeping propulsion to meet the mission lifetime requirements
System Objectives	Increase traffic capacity per satellite, assure longer satellite life with improved reliability, and make the satellite compatible with existing terrestrial and space facilities
Owner	RCA Americom
Orbit characteristics	
Apogee (km)	35,792
Perigee (km)	35,783
Inclination (deg.)	0
Period (min.)	1,436.2
Weight (kg)	1,116 at launch; 598.6 in orbit
Dimensions	Main body: 142 cm x 163 cm x175 cm
Shape	Rectangular with two solar panels extended on booms from opposite sides and an antenna and reflector mounted on one end
Power Source	Solar cells and nickel cadmium batteries
Contractor	RCA Americom Astro-Electronics Division
Remarks	RCA Satcom 5 (also called Aurora) was the first in a new series of high-traffic-capacity, 24-transponder communications satellites. It was the first RCA satellite to be launched from the Delta 3924 launch vehicle. The spacecraft was placed into a geosynchronous orbit located at approximately 128 degrees west longitude.

Table 2-95. Satcom 6 Characteristics

Launch Date	April 11, 1983
Launch Vehicle	Delta 3924
Range	Eastern Space and Missile Center
Mission Objectives	Launch the RCA satellite into synchronous transfer orbit on a three-stage Delta 3924 launch vehicle with sufficient accuracy to allow the spacecraft apogee kick motor to place the spacecraft into a stationary synchronous orbit while retaining sufficient stationkeeping propulsion to meet the mission lifetime requirements
System Objectives	Serve the commercial, government, video/audio, and Alaskan domestic communication traffic markets: • Government: provide voice/video and high-speed data to federal agencies via RCA-owned Earth stations located on various government installations • Video/audio services: provide point-to-point and point-to-multipoint distribution of TV, radio, and news services to broadcasters, cable TV operators, and publishers • Alascom services: provide Alascom, Inc., the long-distance common carrier for Alaska, the satellite capacity for interstate and intrastate message and video transmission
Owner	RCA Americom
Orbit Characteristics	
Apogee (km)	35,794
Perigee (km)	35,779
Inclination (deg.)	0
Period (min.)	1,436.1
Weight (kg)	1,116 at launch, 598.6 in orbit
Dimensions	Main body: 142 cm x 163 cm x 175 cm
Shape	Rectangular with two solar panels extended on booms from opposite sides and an antenna and reflector mounted on one end
Power Source	Solar panels and nickel cadmium batteries
Contractor	RCA Americom Astro-Electronics Division
Remarks	RCA Satcom 6 (also called Satcom IR) was the second of a new series of high-traffic-capacity, 24-transponder communications satellites. It replaced the RCA Satcom 1, which was launched in 1975. It was placed in a geosynchronous orbit at approximately 128 degrees west longitude.

Table 2–96. Satcom 7 Characteristics

Launch Date	September 8, 1983
Launch Vehicle	Delta 3924
Range	Eastern Space and Missile Center
Mission Objectives	Launch the RCA spacecraft into a synchronous orbit on a three-stage Delta 3924 launch vehicle with sufficient accuracy to allow the spacecraft apogee kick motor to place the spacecraft into a stationary synchronous orbit while retaining sufficient stationkeeping propulsion to meet the mission lifetime requirements
System Objectives	Serve the commercial, government, video/audio, and Alaskan domestic communication traffic markets: • Government: provide voice/video and high-speed data to federal agencies via RCA-owned Earth stations located on various government installations • Video/audio services: provide point-to-point and point-to-multipoint distribution of TV, radio, and news services to broadcasters, cable TV operators, and publishers • Alascom services: provide Alascom, Inc., the long-distance common carrier for Alaska, the satellite capacity for interstate and intrastate message and video transmission
Owner	RCA Americom
Orbit Characteristics	
Apogee (km)	35,794
Perigee (km)	35,779
Inclination(deg.)	0
Period (min.)	1,436.1
Weight (kg)	1,116 at launch; 598.6 in orbit
Dimensions	Main body: 142 cm x 163 cm x175 cm
Shape	Rectangular with two solar panels extended on booms from opposite sides and an antenna and reflector mounted on one end
Power Source	Solar panels and nickel cadmium batteries
Contractor	RCA Americom Astro-Electronics Division
Remarks	RCA Satcom 7 (also called Satcom 2R) replaced the RCA Satcom 2 that was launched in 1976. It was placed in geosynchronous orbit at approximately 72 degrees west longitude.

Table 2–97. Satcom K-2 Characteristics

Launch Date	November 28, 1985
Launch Vehicle	STS-61B (*Atlantis*)/PAM-DII
Range	Kennedy Space Center
Mission Objectives	Launch communications satellite successfully
System Objectives	Provide communications coverage for the 48 continental U.S. states or either the eastern half or western half
Owner	RCA Americom
Orbit Characteristics	
Apogee (km)	35,801
Perigee (km)	35,774
Inclination (deg.)	0.1
Period (min.)	1,436.2
Weight (kg)	7,225.3 (includes PAM-DII)
Dimensions	Main structure: 170 cm x 213 cm x 152 cm
Shape	Three-axis stabilized rectangular box and two deployable arms
Power Source	Solar array and three-battery system back-up
Contractor	RCA Americom Astro-Electronics Division
Remarks	RCA Satcom K-2 was the first in a series of communications satellites operating in the Ku-band part of the spectrum. The PAM-DII was used for the satellite's upper stage because of the satellite's heavy weight. The satellite was placed into a geosynchronous orbit at approximately 81 degrees west longitude.

Table 2–98. Satcom K-1 Characteristics

Launch Date	January 12, 1986
Launch Vehicle	STS 61-C (*Columbia*)/PAM-DII
Range	Kennedy Space Center
Mission Objectives	Launch communications satellite successfully
System Objectives	Provide communications coverage for the 48 continental states or either the eastern or the western half of the country
Owner	RCA Americom
Orbit Characteristics	
Apogee (km)	35,795
Perigee (km)	35,780
Inclination (deg.)	0
Period (min.)	1,436.2
Weight (kg)	7225.3 (includes PAM DII)
Dimensions	Main structure: 170 cm x 213 cm x 152 cm
Shape	Three-axis stabilized rectangular box and two deployable arms
Power Source	Solar array and three-battery system back-up
Contractor	RCA Americom Astro-Electronics Division
Remarks	Satcom K-1 was the second in a series of three planned communications satellites operating in the Ku-band part of the spectrum. It was placed into an orbital position at approximately 85 degrees west longitude.

Table 2-99. SBS-1 Characteristics

Launch Date	November 15, 1980
Launch Vehicle	Delta 3910/PAM-D
Range	Eastern Space and Missile Center
Mission Objectives	Launch the satellite along a suborbital trajectory on a two-stage Delta 3910 vehicle with sufficient accuracy to allow the spacecraft propulsion systems to place the spacecraft into a stationary synchronous orbit while retaining sufficient stationkeeping propulsion to meet the mission lifetime requirements
Owner	Satellite Business Systems: IBM, Comsat General, Aetna Insurance
Orbit Characteristics	
Apogee (km)	35,797
Perigee (km)	35,777
Inclination (deg.)	0.7
Period (min.)	1,436.1
Weight (kg)	555 on orbit
Dimensions	6.6 m high (deployed); 2.16 m diameter
Shape	Cylindrical
Power Source	Solar cells and two nickel cadmium batteries
Contractor	Hughes Aircraft
Remarks	This launch marked the first use of the Payload Assist Module (PAM-D) in place of a conventional third stage. SBS-1 was the first satellite capable of transmitting point-to-point data, voice, facsimile, and telex messages within the continental United States as routine commercial service in the 12/14 GHz (K-) band; prior K-band service on ATS-6, CTS, and Telesat-D was experimental. SBS-1 was placed into geosynchronous orbit at approximately 106 degrees west longitude.

Table 2–100. SBS-2 Characteristics

Launch Date	September 24, 1981
Launch Vehicle	Delta 3910/PAM-D
Range	Eastern Space and Missile Center
Mission Objectives	Launch the satellite along a suborbital trajectory on a two-stage Delta 3910 vehicle with sufficient accuracy to allow the spacecraft propulsion system to place the spacecraft into a stationary synchronous orbit while retaining sufficient stationkeeping propulsion to meet the mission lifetime requirements
Owner	Satellite Business Systems: IBM, Comsat General, Aetna Insurance
Orbit Characteristics	
Apogee (km)	35,789
Perigee (km)	35,785
Inclination (deg.)	0
Period (min.)	1,436.1
Weight (kg)	555 on orbit
Dimensions	6.6 m high (deployed); 2.16 m diameter
Shape	Cylindrical
Power Source	Solar cells and two nickel cadmium batteries
Contractor	Hughes Aircraft
Remarks	SBS-2 was placed in geostationary orbit at approximately 97 degrees west longitude

Table 2–101. SBS-3 Characteristics

Launch Date	November 11, 1982
Launch Vehicle	STS-5 (*Columbia*)/PAM-D
Range	Kennedy Space Center
Mission Objectives	Launch the satellite with sufficient accuracy to allow the spacecraft propulsion system to place the spacecraft into a stationary synchronous orbit while retaining sufficient stationkeeping propulsion to meet the mission lifetime requirements
Owner	Satellite Business Systems: IBM, Comsat General, Aetna Insurance
Orbit Characteristics	
Apogee (km)	35,788
Perigee (km)	35,786
Inclination (deg.)	0
Period (min.)	1,436.1
Weight (kg)	555 on orbit
Dimensions	6.6 m high (deployed); 2.16 m diameter
Shape	Cylindrical
Power Source	Solar cells and two nickel cadmium batteries
Contractor	Hughes Aircraft
Remarks	This was the first launch from the Shuttle cargo bay. SBS-3 was placed in geostationary orbit at approximately 95 degrees west longitude

Table 2–102. SBS-4 Characteristics

Launch Date	August 31, 1984
Launch Vehicle	STS 41-D (*Discovery*)
Range	Kennedy Space Center
Mission Objectives	Launch the satellite with sufficient accuracy to allow the spacecraft propulsion system to place the spacecraft into a stationary synchronous orbit while retaining sufficient stationkeeping propulsion to meet the mission lifetime requirements
Owner	Satellite Business Systems: IBM, Comsat General, Aetna Insurance
Orbit Characteristics	
Apogee (km)	35,793
Perigee (km)	35,781
Inclination (deg.)	0
Period (min.)	1,436.1
Weight (kg)	555 on orbit
Dimensions	6.6 m high (deployed); 2.16 m diameter
Shape	Cylindrical
Power Source	Solar cells and two nickel cadmium batteries
Contractor	Hughes Aircraft
Remarks	SBS-4 was placed in geostationary orbit at approximately 91 degrees west longitude

Table 2–103. Westar Satellite Comparison

Feature	First Generation Westar 1, 2, and 3	Second Generation Westar 4, 5, and 6
Launch Vehicle	Delta 2914	Delta 3910
Weight, Beginning of Life (kg)	306	584
Service, GHz Channels	6/412	6/424
Dimensions (cm) Height Diameter	 345 190	 659 (deployed) 279 (stowed) 216
Power Capability, Watts Beginning of Life End of Life	 307 822	 262 684
Traveling Wave Tube (TWT) Output Power, Watts	5.0	7.5
Design Life, Years	7	10
Performance EIRP, dBW	 33.0 (CONUS) 24.5 (Alaska, Hawaii)	 34.0 (CONUS) 31.0 (Alaska) 28.3 (Hawaii) 27.2 (Puerto Rico)
G/T, dB/°K	-7.4 (CONUS) -14.4 (Alaska, Hawaii)	-6.0 (CONUS) 31.0 (Alaska) -10.9 (Hawaii) -10.9 (Puerto Rico)

Table 2–104. Westar 3 Characteristics

Launch Date	August 9, 1979
Launch Vehicle	Delta 2914
Range	Eastern Test Range
Mission Objectives	Place the satellite into a synchronous transfer orbit of sufficient accuracy to allow the spacecraft propulsion system to place the spacecraft into stationary synchronous orbit while retaining sufficient stationkeeping propulsion to meet the mission lifetime requirements
Owner	Western Union
Orbit Characteristics	
Apogee (km)	35,794
Perigee (km)	35,780
Inclination (deg.)	0
Period (min.)	1,436.2
Weight (kg)	572 in transfer orbit
Dimensions	1.56 m high; 1.85 m diameter
Shape	Cylindrical (drum)
Power Source	Solar cells and battery system
Contractor	Hughes Aircraft
Remarks	Because Westar 1 and Westar 2 were still operating at the time Westar 3 was launched, it was placed into a storage geosynchronous orbit over the equator at approximately 91 degrees west longitude until Westar 1 was removed from service. Westar 3 was in use until it was turned off in January 1990.

Table 2–105. Westar 4 Characteristics

Launch Date	February 25, 1982
Launch Vehicle	Delta 3910
Range	Eastern Space and Missile Center
Mission Objectives	Launch the satellite along a suborbital trajectory on a two-stage Delta 3910 launch vehicle with sufficient accuracy to allow the payload propulsion system to place the spacecraft into a stationary synchronous orbit while retaining sufficient stationkeeping propulsion to meet the mission lifetime requirements
Owner	Western Union
Orbit Characteristics	
Apogee (km)	35,796
Perigee (km)	35,778
Inclination (deg.)	0.1
Period (min.)	1,436.1
Weight (kg)	585 (after apogee motor was fired)
Dimensions	6.84 m high (deployed); 2.16 m diameter
Shape	Cylindrical (drum)
Power Source	Solar cells and battery system
Contractor	Hughes Aircraft
Remarks	The satellite was positioned at approximately 99 degrees west longitude above the equator. It operated until November 1991.

Table 2–106. Westar 5 Characteristics

Launch Date	June 8, 1982
Launch Vehicle	Delta 3910
Range	Eastern Space and Missile Center
Mission Objectives	Launch the satellite along a suborbital trajectory on a two-stage Delta 3910 launch vehicle with sufficient accuracy to allow the payload propulsion system to place the spacecraft into a stationary synchronous orbit while retaining sufficient stationkeeping propulsion to meet the mission lifetime requirements.
Owner	Western Union
Orbit Characteristics	
Apogee (km)	35,796
Perigee (km)	35,783
Inclination (deg.)	0
Period (min.)	1,436.3
Weight (kg)	585 (after apogee motor was fired)
Dimensions	6.84 m high (deployed); 2.16 m diameter
Shape	Cylindrical (drum)
Power Source	Solar cells and battery system
Contractor	Hughes Aircraft
Remarks	Westar 5 was placed in a geostationary position at approximately 123 degrees west longitude. It replaced Westar 2. It operated until May 1992.

Table 2–107. Westar 6 Characteristics

Launch Date	February 3, 1984
Launch Vehicle	STS 41-B (*Challenger*)/PAM-D
Range	Kennedy Space Center
Mission Objectives	Launch the satellite along a suborbital trajectory on a two-stage Delta 3910 launch vehicle or on the Space Shuttle with sufficient accuracy to allow the payload propulsion system to place the spacecraft into a stationary synchronous orbit while retaining sufficient stationkeeping propulsion to meet the mission lifetime requirements
Owner	Western Union
Orbit Characteristics	Did not reach proper orbit
Weight (kg)	607.8 (after apogee motor was fired)
Dimensions	6.84 m high (deployed); 2.16 m diameter
Shape	Cylindrical (drum)
Power Source	Solar cells and battery system
Contractor	Hughes Aircraft
Remarks	Westar 6 failed to reach its intended geostationary orbit because of a failure of the PAM-D. It was retrieved by the STS 51-A mission in November 1984 and returned to Earth for refurbishment.

Table 2–108. Intelsat Participants

Intelsat Member Countries (as of 1985)

Afghanistan	Guinea, People's Revolutionary Republic of	Norway
Algeria		Oman
Angola		Pakistan
Argentina	Haiti	Panama
Australia	Honduras	Paraguay
Austria	Iceland	Peru
Bangladesh	India	Philippines
Barbados	Indonesia	Portugal
Belgium	Iran, Islamic Republic of	Qatar
Bolivia	Iraq	Saudi Arabia
Brazil	Ireland	Senegal
Cameroon	Israel	Singapore
Canada	Italy	South Africa
Central African Republic	Ivory Coast	Spain
Chad	Jamaica	Sri Lanka
China, People's Republic of	Japan	Sudan
Chile	Jordan	Sweden
Columbia	Kenya	Switzerland
Congo	Korea, Republic of	Syria
Costa Rica	Kuwait	Tanzania
Cyprus	Lebanon	Thailand
Denmark	Libya	Trinidad and Tobago
Dominican Republic	Liechtenstein	Tunisia
Ecuador	Luxembourg	Turkey
Egypt	Madagascar	Uganda
El Salvador	Malaysia	United Arab Emirates
Ethiopia	Mali	United Kingdom
Fiji	Mauritania	United States
Finland	Mexico	Upper Volta
France	Monaco	Vatican City State
Gabon	Morocco	Venezuela
Germany, Federal Republic of	Netherlands	Viet Nam
	New Zealand	Yemen Arab Republic
Ghana	Nicaragua	Yugoslavia
Greece	Niger	Zaire
Guatemala	Nigeria	Zambia

Table 2–108 continued

Intelsat Non-Signatory		
Users	Hungary	Romania
Bahrain	Kiribati	Seychelles
Botswana	Liberia	Sierra Leone
Brunei	Malawi	Solomon Islands
Burma	Maldives	Somalia
Cook Islands	Mauritius	Surinam
Cuba	Mozambique	Togo
Czechoslovakia	Nauru, Republic of	Tonga
Djibouti	New Guinea	U.S.S.R.
Gambia	Papua	Western Samoa
Guyana	Poland	

Other Territory Users		
American Samoa	French Guiana	Netherlands Antilles
Ascension Island	French Polynesia	New Caledonia
Azores	French West Indies	Van Uatu
Belize	Gibraltar	
Bermuda	Guam	
Cayman Islands	Hong Kong	

Table 2–109. International Contributors to Intelsat

Manufacturer (Country)	Contribution
Aerospatiale (France)	Initiated the structural design that formed the main member of the spacecraft modular design construction; supplied the main body structure thermal analysis and control
GEC-Marconi (United Kingdom)	Produced the 11-GHz beacon transmitter used for Earth station antenna tracking
Messerschmitt-Bolkow-Blohm (Federal Republic of Germany)	Designed and produced the satellites' control subsystem and the solar array
Mitsubishi Electric Corporation (Japan)	Designed and produced the 6-GHz and the 4-GHz Earth coverage antennas; also manufactured the power control electronics and, from an FACC design, the telemetry and command digital units
Senia (Italy)	Designed and built the six telemetry, command, and ranging antennas, two 11-GHz beacon antennas and two 14/11-GHz spot beam antennas; also built the command receiver and telemetry transmitter, which combined to form a ranging transponder for determining the spacecraft position in transfer orbit
Thomson-CSF (France)	Built the 10W, 11-GHz traveling wave tubes (10 per spacecraft)

Table 2–110. Intelsat V F-2 Characteristics

Launch Date	December 6, 1980
Launch Vehicle	Atlas-Centaur
Range	Eastern Space and Missile Center
Mission Objectives	Launch the satellite into a transfer orbit that enables the spacecraft apogee motor to inject the spacecraft into a synchronous orbit
Comsat Objectives	Fire the apogee motor, position the satellite into its planned geostationary position, and operate and manage the system for Intelsat
Owner	International Telecommunications Satellite Consortium
Orbit Characteristics	
Apogee (km)	35,801
Perigee (km)	35,774
Inclination (deg.)	0
Period (min.)	1,436.2
Weight (kg)	1,928 at launch
Dimensions	Main body: 1.66m x 2 m x 1.77 m; height: 6.4 m; solar array span: 15.5 m
Shape	Box
Power Source	Solar arrays and rechargeable batteries
Contractor	Ford Aerospace and Communication
Remarks	Intelsat V F-2 (also designated 502) was the first of the Intelsat V series. It was positioned in an orbit at approximately 22 degrees west longitude in the Atlantic region.

Table 2–111. Intelsat V F-1 Characteristics

Launch Date	May 23, 1981
Launch Vehicle	Atlas-Centaur
Range	Cape Kennedy
Mission Objectives	Launch the satellite into a transfer orbit that enables the spacecraft apogee motor to inject the spacecraft into a synchronous orbit
Comsat Objectives	Fire the apogee motor, position the satellite into its planned geostationary position, and operate and manage the system
Owner	Intelsat
Orbit Characteristics	
Apogee (km)	35,800
Perigee (km)	35,778
Inclination (deg.)	0
Period (min.)	1,436.2
Weight (kg)	1,928 at launch
Dimensions	Main body: 1.66m x 2 m x 1.77 m; height: 6.4 m; solar array span: 15.5 m
Shape	Box
Power Source	Solar array and rechargeable batteries
Contractor	Ford Aerospace and Communication
Remarks	Also designated Intelsat 501, it was positioned in the Atlantic region and later moved to the Pacific region.

Table 2–112. Intelsat V F-3 Characteristics

Launch Date	December 15, 1981
Launch Vehicle	Atlas-Centaur
Range	Cape Canaveral
Mission Objectives	Launch the satellite into a transfer orbit that enables the spacecraft apogee motor to inject the spacecraft into a synchronous orbit
Comsat Objectives	Fire the apogee motor, position the satellite into its planned geostationary position, and operate and manage the system
Owner	Intelsat
Orbit Characteristics	
Apogee (km)	35,801
Perigee (km)	35,772
Inclination (deg.)	0
Period (min.)	1,436.1
Weight (kg)	1,928 at launch
Dimensions	Main body: 1.66m x 2 m x 1.77 m; height: 6.4 m; solar array span: 15.5 m
Shape	Box
Power Source	Solar array and rechargeable batteries
Contractor	Ford Aerospace and Communication
Remarks	Also designated Intelsat 503, it was positioned in the Atlantic region and later moved into the Pacific region.

Table 2–113. Intelsat V F-4 Characteristics

Launch Date	March 3, 1982
Launch Vehicle	Atlas-Centaur
Range	Cape Canaveral
Mission Objectives	Launch the satellite into a transfer orbit that enables the spacecraft apogee motor to inject the spacecraft into a synchronous orbit
Comsat Objectives	Fire the apogee motor, position the satellite into its planned geostationary position, and operate and manage the system
Owner	Intelsat
Orbit Characteristics	
Apogee (km)	35,808
Perigee (km)	35,767
Inclination (deg.)	0.1
Period (min.)	1,436.2
Weight (kg)	1,928 at launch
Dimensions	Main body: 1.66m x 2 m x 1.77 m; height: 6.4 m; solar array span: 15.5 m
Shape	Box
Power Source	Solar array and rechargeable batteries
Contractor	Ford Aerospace and Communication
Remarks	Also designated Intelsat 504, it was positioned in the Indian Ocean region.

Table 2–114. Intelsat V F-5 Characteristics

Launch Date	September 28, 1982
Launch Vehicle	Atlas-Centaur
Range	Cape Canaveral
Mission Objectives	Launch the satellite into a transfer orbit that enables the spacecraft apogee motor to inject the spacecraft into a synchronous orbit
Comsat Objectives	Fire the apogee motor, position the satellite into its planned geostationary position, and operate and manage the system
Owner	Intelsat
Orbit Characteristics	
Apogee (km)	35,805
Perigee (km)	35,769
Inclination (deg.)	0.1
Period (min.)	1,436.2
Weight (kg)	1,928 at launch
Dimensions	Main body: 1.66m x 2 m x 1.77 m; height: 6.4 m; solar array span: 15.5 m
Shape	Box
Power Source	Solar array and rechargeable batteries
Contractor	Ford Aerospace and Communication
Remarks	Also designated Intelsat 505, it was positioned in the Indian Ocean region. This flight carried a Maritime Communications Services package for the first time for the Maritime Satellite Organization (Inmarsat) to provide ship/shore/ship communications.

SPACE APPLICATIONS

Table 2–115. Intelsat V F-6 Characteristics

Launch Date	May 19, 1983
Launch Vehicle	Atlas-Centaur
Range	Cape Canaveral
Mission Objectives	Launch the satellite into a transfer orbit that enables the spacecraft apogee motor to inject the spacecraft into a synchronous orbit
Comsat Objectives	Fire the apogee motor, position the satellite into its planned geostationary position, and operate and manage the system
Owner	Intelsat
Orbit Characteristics	
Apogee (km)	35,810
Perigee (km)	35,765
Inclination (deg.)	0
Period (min.)	1,436.2
Weight (kg)	1,996 at launch
Dimensions	Main body: 1.66m x 2 m x 1.77 m; height: 6.4 m; solar array span: 15.5 m
Shape	Box
Power Source	Solar array and rechargeable batteries
Contractor	Ford Aerospace and Communication
Remarks	Also designated Intelsat 506, it was positioned in the Atlantic region. It carried the Marine Communications Services package for Inmarsat.

Table 2–116. Intelsat V F-9 Characteristics a

Launch Date	June 9, 1984
Launch Vehicle	Atlas-Centaur
Range	Cape Canaveral
Mission Objectives	Launch the satellite into a transfer orbit that enables the spacecraft apogee motor to inject the spacecraft into a synchronous orbit
Comsat Objectives	Fire the apogee motor, position the satellite into its planned geostationary position, and operate and manage the system
Owner	Intelsat
Orbit Characteristics	Did not reach useful orbit
Weight (kg)	1,928 at launch
Dimensions	Main body: 1.66m x 2 m x 1.77 m; height: 6.4 m; solar array span: 15.5 m solar array span
Shape	Box
Power Source	Solar array and rechargeable batteries
Contractor	Ford Aerospace and Communication
Remarks	The satellite did not reach useful orbit. A leak in the Centaur liquid oxygen tank at the time of Atlas and Centaur separation and the accompanying loss of liquid oxygen through the tank opening precipitated events that compromised vehicle performance and resulted in loss of the mission. This was the first launch of the new lengthened Atlas Centaur rocket.

a Intelsat F-7 and F-8 were launched by an Ariane and are not addressed here.

Table 2–117. Intelsat V-A F-10 Characteristics

Launch Date	March 22, 1985
Launch Vehicle	Atlas-Centaur
Range	Eastern Space and Missile Center
NASA Objectives	Launch the satellite into a transfer orbit, orient it, and spin it at 2 rpm about its longitudinal axis, enabling the spacecraft apogee motor to inject the spacecraft into a synchronous orbit
Intelsat Objectives	Fire the apogee motor, position the satellite into its planned geostationary position, and operate and manage the system
Owner	Intelsat
Orbit Characteristics	
Apogee (km)	35,807
Perigee (km)	35,768
Inclination (deg.)	0
Period (min.)	1,436.1
Weight (kg)	1,996 at launch
Dimensions	Main body: 1.66m x 2 m x 1.77 m; height: 6.4 m; solar array span: 15.5 m
Shape	Box
Power Source	Solar array with rechargeable batteries
Contractor	Ford Aerospace and Communications
Remarks	The first in a series of improved commercial communication satellites, the satellite was positioned in the Pacific Ocean region.

Table 2–118. Intelsat V-A F-11 Characteristics

Launch Date	June 30, 1985
Launch Vehicle	Atlas-Centaurr
Range	Eastern Space and Missile Center
NASA Objectives	Launch the satellite into a transfer orbit, orient it, and spin it a 2 rpm about its longitudinal axis, enabling the spacecraft apogee motor to inject the spacecraft into a synchronous orbit
Intelsat Objectives	Fire the apogee motor, position the satellite into its planned geostationary position, and operate and manage the system
Owner	Intelsat
Orbit Characteristics	
Apogee (km)	35,802
Perigee (km)	35,772
Inclination (deg.)	0
Period (min.)	1,436.1
Weight (kg)	1,996 at launch
Dimensions	Main body: 1.66m x 2 m x 1.77 m; height: 6.4 m; solar array span: 15.5 m
Shape	Box
Power Source	Solar panels and rechargeable batteries
Contractor	Ford Aerospace and Communications
Remarks	The satellite was placed into a geostationary final orbit at 332.5 degrees east longitude.

Table 2-119. Intelsat V-A F-12 Characteristics

Launch Date	September 29, 1985
Launch Vehicle	Atlas-Centaur
Range	Eastern Space and Missile Center
NASA Objectives	Launch the satellite into a transfer orbit, orient it, and spin it at 2 rpm about its longitudinal axis, enabling the spacecraft apogee motor to inject the spacecraft into a synchronous orbit
Intelsat Objectives	Fire the apogee motor, position the satellite into its planned geostationary position, and operate and manage the system
Owner	Intelsat
Orbit Characteristics	
Apogee (km)	35,802
Perigee (km)	35,772
Inclination (deg.)	0
Period (min.)	1,436.1
Weight (kg)	1,996 at launch
Dimensions	Main body: 1.66m x 2 m x 1.77 m; height: 6.4 m; solar array span: 15.5 m
Shape	Box
Power Source	Solar panels and rechargeable batteries
Contractor	Ford Aerospace and Communications
Remarks	The satellite was positioned in the Atlantic Ocean region. This was the last commercial mission for the Atlas Centaur rocket. Future Intelsat missions were planned to be launched from the Space Shuttle or the Ariane.

Table 2–120. Fltsatcom 2 Characteristics

Launch Date	May 4, 1979
Launch Vehicle	Atlas-Centaur
Range	Eastern Space and Missile Center
Mission Objectives	Launch the satellite into a transfer orbit that enables the spacecraft apogee motor to inject the satellite into a synchronous orbit
Owner	U.S. Department of Defense
Orbit Characteristics	
Apogee (km)	35,837
Perigee (km)	35,736
Inclination (deg.)	4.7
Period (min.)	1,436.1
Weight (kg)	1,005 (in orbit)
Dimensions	Main body: 2.5 m diameter x 1.3 m high; height including antenna: 6.7 m
Shape	Hexagonal spacecraft module and attached payload module
Power Source	Solar arrays and batteries
Contractor	TRW Systems
Remarks	Fltsatcom 2 was initially placed into a geostationary orbit at approximately 23 degrees west longitude after Fltsatcom 3 was deployed, Fltsatcom 2 was moved to a position at approximately 72.5 degrees east longitude to carry Indian Ocean traffic. This marked the 50th Atlas Centaur launch.

Table 2–121. Fltsatcom 3 Characteristics

Launch Date	January 17, 1980
Launch Vehicle	Atlas-Centaur
Range	Eastern Test Range
Mission Objectives	Launch the satellite into a transfer orbit that enables the spacecraft apogee motor to inject the spacecraft into a synchronous orbit
Owner	Department of Defense
Orbit Characteristics	
Apogee (km)	35,804
Perigee (km)	35,767
Inclination (deg.)	4.3
Period (min.)	1,436.1
Weight (kg)	1,005 (in orbit)
Dimensions	Main body: 2.5 m diameter x 1.3 m high; height including antenna: 6.7 m
Shape	Hexagonal spacecraft module and attached payload module
Power Source	Solar arrays and batteries
Contractor	Defense and Space Systems Group, TRW, Inc.
Remarks	Fltsatcom 3 was placed in geostationary orbit at approximately 23 degrees west longitude.

SPACE APPLICATIONS

Table 2–122. Fltsatcom 4 Characteristics

Launch Date	October 30, 1980
Launch Vehicle	Atlas-Centaur
Range	Eastern Test Range
Mission Objectives	Launch the satellite into a transfer orbit that enables the spacecraft apogee motor to inject the satellite into synchronous orbit
Owner	Department of Defense
Orbit Characteristics	
Apogee (km)	35,811
Perigee (km)	35,765
Inclination (deg.)	4.0
Period (min.)	1,436.2
Weight (kg)	1,005 (in orbit)
Dimensions	Main body: 2.5 m diameter x 1.3 m high; height including antenna: 6.7 m
Shape	Hexagonal spacecraft module and attached payload module
Power Source	Solar arrays and batteries
Contractor	Defense and Space Systems Group, TRW, Inc.
Remarks	Fltsatcom 4 was placed into a geostationary orbit at approximately 172 degrees east longitude above the equator.

Table 2–123. Fltsatcom 5 Characteristics

Launch Date	August 6, 1981
Launch Vehicle	Atlas-Centaur
Range	Eastern Space and Missile Center
Mission Objectives	Launch the satellite into a transfer orbit that enables the spacecraft apogee motor to inject the satellite into synchronous orbit
Owner	U.S. Department of Defense
Orbit Characteristics	
Apogee (km)	36,284
Perigee (km)	36,222
Inclination (deg.)	4.6
Period (min.)	1,460.0
Weight (kg)	1,039 (in orbit)
Dimensions	Main body: 2.5 m diameter x 1.3 m high; height including antenna: 6.7 m
Shape	Hexagonal spacecraft module and attached payload module
Power Source	Solar arrays and batteries
Contractor	Defense and Space Systems Group, TRW, Inc.
Remarks	The satellite reached geostationary orbit, but an imploding payload shroud destroyed the primary antenna, rendering the satellite useless.

Table 2–124. Fltsatcom 7 Characteristics

Launch Date	December 4, 1986
Launch Vehicle	Atlas-Centaur
Range	Cape Canaveral
Mission Objectives	Launch the satellite into an inclined transfer orbit and orient the spacecraft in its desired transfer orbit attitude
Owner	Department of Defense
Orbit Characteristics	
Apogee (km)	35,875
Perigee (km)	35,703
Inclination (deg.)	4.3
Period (min.)	1,436.2
Weight (kg)	1,128.5
Dimensions	Main body: 2.5 m diameter x 1.3 m high; height including antenna: 6.7 m
Shape	Hexagonal spacecraft module and attached payload module
Power Source	Solar arrays and batteries
Contractor	Defense and Space Systems Group, TRW, Inc.
Remarks	Fltsatcom 7 carried an experimental EHF package in addition to the equipment carried on previous missions. The satellite was placed into a geosynchronous orbit at approximately 100 degrees west longitude.

Table 2–125. Fltsatcom 6 Characteristics

Launch Date	March 26, 1987
Launch Vehicle	Atlas-Centaur
Range	Eastern Space and Missile Center
Mission Objectives	Launch the satellite into an inclined transfer orbit and orient the spacecraft in its desired transfer orbit attitude
Owner	Department of Defense
Orbit Characteristics	Did not achieve orbit
Weight (kg)	1,048 (after firing of apogee boost motors)
Dimensions	Main body: 2.5 m diameter x 1.3 m high; height including antenna: 6.7 m
Shape	Hexagonal spacecraft module and attached payload module
Power Source	Solar arrays and batteries
Contractor	Defense and Space Systems Group, TRW, Inc.
Remarks	Fltsatcom 6 did not achieve proper orbit because of a lightning strike.

Table 2–126. Leasat 2 Characteristics

Launch Date	September 1, 1984
Launch Vehicle	STS-41D (*Discovery*)
Range	Kennedy Space Center
Mission Objectives	Launch the satellite into successful transfer orbit
Owner	Leased from Hughes Communications Inc. by U.S. Department of Defense
Orbit Characteristics	
Apogee (km)	35,788
Perigee (km)	35,782
Inclination (deg.)	0.7
Period (min.)	1,436.2
Weight (kg)	1,315 on orbit
Dimensions	6 m long (deployed); 4.26 m diameter
Shape	Cylinder
Power Source	Solar array and nickel cadmium batteries
Contractor	Hughes Communications
Remarks	The launch of Leasat 2 was postponed from June 1984 because the Shuttle launch was delayed. The satellite occupied a geostationary position located at approximately 177 degrees west longitude

Table 2–127. Leasat 1 Characteristics

Launch Date	November 10, 1984
Launch Vehicle	STS 51A (*Discovery*)
Range	Kennedy Space Center
Mission Objectives	Launch the satellite into successful transfer orbit
Owner	Hughes Communications (leased to Department of Defense)
Orbit Characteristics	
Apogee (km)	35,890
Perigee (km)	35,783
Inclination (deg.)	0.9
Period (min.)	1,436.0
Weight (kg)	1,315 on orbit
Dimensions	6 m long (deployed); 4.26 m diameter
Shape	Cylinder
Power Source	Solar array and nickel cadmium batteries
Contractor	Hughes Communications
Remarks	Leasat 1 was positioned in geostationary orbit at approximately 16 degrees west longitude.

Table 2–128. Leasat 3 Characteristics

Launch Date	April 13, 1985
Launch Vehicle	STS-51-D (*Discovery*)
Range	Kennedy Space Center
Mission Objectives	Launch the satellite into successful transfer orbit
Owner	Hughes Communications (leased to Department of Defense)
Orbit Characteristics	
Apogee (km)	35,809
Perigee (km)	35,768
Inclination (deg.)	1.4
Period (min.)	1,436.2
Weight (kg)	1,315 on orbit
Dimensions	6 m long (deployed); 4.26 m diameter
Shape	Cylinder
Power Source	Solar array and nickel cadmium batteries
Contractor	Hughes Communications
Remarks	The Leasat 3 sequencer failed to start despite attempts by the crew to activate it. The satellite remained inoperable until it was repaired in orbit by the crew of STS 51-I in August 1985. It was placed in geosynchronous orbit in November 1985 and began operations in December.

Table 2–129. Leasat 4 Characteristics

Launch Date	August 29, 1985
Launch Vehicle	STS-51-I (*Discovery*)
Range	Kennedy Space Center
Mission Objectives	Launch the satellite into successful transfer orbit
Owner	Hughes Communications (leased to Department of Defense)
Orbit Characteristics	
Apogee (km)	36,493
Perigee (km)	35,079
Inclination (deg.)	1.4
Period (min.)	1,436.1
Weight (kg)	1,315 on orbit
Dimensions	6 m long (deployed); 4.26 m diameter
Shape	Cylinder
Power Source	Solar array and nickel cadmium batteries
Contractor	Hughes Communications
Remarks	Leasat 4 was placed into geosynchronous orbit on September 3, 1985. It functioned normally for about 2 days, at which time the communications payload failed. Efforts to restore the satellite were unsuccessful.

SPACE APPLICATIONS

Table 2–130. NATO IIID Characteristics

Launch Date	November 14, 1984
Launch Vehicle	Delta 3914
Range	Eastern Space and Missile Center
Mission Objectives	Place the satellite into synchronous transfer orbit of sufficient accuracy to allow the spacecraft propulsion system to place the satellite into a stationary synchronous orbit while retaining sufficient stationkeeping propulsion to meet the mission lifetime requirements
Owner	North Atlantic Treaty Organization
Orbit Characteristics	
Apogee (km)	35,788
Perigee (km)	35,783
Inclination (deg.)	3.2
Period (min.)	1,436.1
Weight (kg)	388 (after apogee motor fired)
Dimensions	3.1 m long including antennas; 2.18 m diameter
Shape	Cylindrical
Power Source	Solar array and battery charge control array
Contractor	Ford Aerospace and Communications
Remarks	NATO-IIID was positioned at approximately 21 degrees west longitude.

Table 2–131. Anik D-1 Characteristics

Launch Date	August 26, 1982
Launch Vehicle	Delta 3920/PAM-D
Range	Eastern Space and Missile Center
Mission Objectives	Launch the satellite on a two-stage Delta 3920 vehicle with sufficient accuracy to allow the MDAC PAM-D and the spacecraft propulsion system to place the spacecraft into a stationary synchronous orbit while retaining sufficient stationkeeping propulsion to meet the mission lifetime requirements
Owner	Telesat Canada Corporation
Orbit Characteristics	
Apogee (km)	35,796
Perigee (km)	35,776
Inclination (deg.)	0
Period (min.)	1,436.0
Weight (kg)	730 in orbit
Dimensions	6.7 m high with solar panel and antenna deployed; 2.16 m diameter
Shape	Cylindrical
Power Source	Solar panels and nickel cadmium batteries
Contractor	Hughes Aircraft
Remarks	Anik D-1 was the first of two satellites built for Telesat/Canada to replace the Anik A series. It was located in geostationary orbit at approximately 104.5 degrees west longitude. It remained in service until February 1995.

Table 2–132. Anik C-3 Characteristics

Launch Date	November 12, 1982
Launch Vehicle	STS-5 (*Columbia*)/PAM-D
Range	Kennedy Space Center
Mission Objectives	Launch the satellite into transfer orbit, permitting the spacecraft propulsion system to place it in stationary synchronous orbit for communications coverage over Canada
Owner	Telesat Canada Corporation
Orbit Characteristics	
Apogee (km)	35,794
Perigee (km)	35,779
Inclination (deg.)	0
Period (min.)	1,436.1
Weight (kg)	567 in orbit
Dimensions	2 m high; 1.5 m diameter
Shape	Cylindrical
Power Source	Solar panels and nickel cadmium batteries
Contractor	Hughes Aircraft
Remarks	Anik C-3 was placed in geostationary orbit at approximately 114.9 degrees west longitude.

Table 2–133. Anik C-2 Characteristics

Launch Date	June 18, 1983
Launch Vehicle	STS-7 (*Challenger*)
Range	Kennedy Space Center
Mission Objectives	Launch the satellite into transfer orbit, permitting the spacecraft propulsion system to place it in stationary synchronous orbit for communications coverage over Canada
Owner	Telesat Canada Corporation
Orbit Characteristics	
Apogee (km)	35,791
Perigee (km)	35,782
Inclination (deg.)	0
Period (min.)	1,436.2
Weight (kg)	567 in orbit
Dimensions	2 m high; 1.5 m diameter
Shape	Cylindrical
Power Source	Solar panels and nickel cadmium batteries
Contractor	Hughes Aircraft
Remarks	Anik C-2 was placed in geostationary orbit at approximately 110 degrees west longitude

Table 2–134. Anik D-2 Characteristics

Launch Date	November 9, 1984
Launch Vehicle	STS 51-A (*Discovery*)/PAM-D
Range	Kennedy Space Center
Mission Objectives	Launch the satellite with sufficient accuracy to allow the MDAC PAM-D and the spacecraft propulsion system to place the spacecraft into a stationary synchronous orbit while retaining sufficient stationkeeping propulsion to meet the mission lifetime requirements
Owner	Telesat Canada Corporation
Orbit Characteristics	
Apogee (km)	35,890
Perigee (km)	35,679
Inclination (deg.)	0.9
Period (min.)	1,436.0
Weight (kg)	730 in orbit
Dimensions	6.7 m high with solar panel and antenna deployed; 2.16 m diameter
Shape	Cylindrical
Power Source	Solar panels and nickel cadmium batteries
Contractor	Hughes Aircraft
Remarks	Anik D-2 was placed in geostationary orbit at approximately 110 degrees west longitude. It was removed from service in March 1995.

Table 2–135. Anik C-1 Characteristics

Launch Date	April 13, 1985
Launch Vehicle	STS-51D (*Discovery*)/PAM-D
Range	Kennedy Space Center
Mission Objectives	Launch the satellite into transfer orbit, permitting the spacecraft propulsion system to place it in stationary synchronous orbit for communications coverage over Canada
Owner	Telesat Canada Corporation
Orbit Characteristics	
Apogee (km)	35,796
Perigee (km)	35,777
Inclination (deg.)	0.1
Period (min.)	1,436.0
Weight (kg)	567 in orbit
Dimensions	2 m high; 1.5 m diameter
Shape	Cylindrical
Power Source	Solar panels and nickel cadmium batteries
Contractor	Hughes Aircraft
Remarks	Anik C-1 was placed in geostationary orbit at approximately 107 degrees west longitude.

Table 2–136. Arabsat-1B Characteristics

Launch Date	June 18, 1985
Launch Vehicle	STS 51-G (*Discovery*)/PAM-D
Range	Kennedy Space Center
Mission Objectives	Launch satellite into transfer orbit of sufficient accuracy to allow the spacecraft propulsion systems to place it in stationary synchronous orbit for communications coverage
Owner	Saudi Arabia
Orbit Characteristics	
Apogee (km)	35,807
Perigee (km)	35,768
Inclination (deg.)	0
Period (min.)	1,436.2
Weight (kg)	700 kg in orbit
Dimensions	2.26 m x 1.64 m x 1.49 m with a two-panel solar array 20.7 m wide
Shape	Cube
Power Source	Solar array and batteries
Contractor	Aerospatiale
Remarks	The satellite was placed in geosynchronous orbit at approximately 26 degrees east longitude. It began drifting east in October 1992 and went out of service in early 1993.

Table 2–137. Aussat 1 Characteristics

Launch Date	August 27, 1985
Launch Vehicle	STS 51-I (*Discovery*)/PAM-D
Range	Kennedy Space Center
Mission Objectives	Successfully launch the satellite into transfer orbit
Owner	Australia
Orbit Characteristics	
Apogee (km)	35,794
Perigee (km)	35,781
Inclination (deg.)	0
Period (min.)	1,436.2
Weight (kg)	655 in orbit
Dimensions	2.8 m long stowed; 6.6 m deployed; 2.16 m diameter
Shape	Cylindrical
Power Source	Solar cells and nickel cadmium batteries
Contractor	Hughes Communications
Remarks	Aussat 1 (also called Optus A1) was placed in geosynchronous orbit at approximately 160 degrees east longitude.

Table 2–138. Aussat 2 Characteristics

Launch Date	November 27, 1985
Launch Vehicle	STS 61-B (*Atlantis*)/PAM-D
Range	Kennedy Space Center
Mission Objectives	Successfully launch the satellite into transfer orbit
Owner	Australia
Orbit Characteristics	
Apogee (km)	35,794
Perigee (km)	35,780
Inclination (deg.)	0
Period (min.)	1,436.2
Weight (kg)	655 in orbit
Dimensions	2.8 m long stowed; 6.6 m deployed; 2.16 m diameter
Shape	Cylindrical
Power Source	Solar cells and nickel cadmium batteries
Contractor	Hughes Communications
Remarks	Aussat 2 (also called Optus A2) was placed in geosynchronous orbit at approximately 156 degrees east longitude.

Table 2-139. Insat 1A Characteristics

Launch Date	April 10, 1982
Launch Vehicle	Delta 3910
Range	Eastern Space and Missile Center
Mission Objectives	Launch the satellite along a suborbital trajectory on a two-stage Delta 3910 launch vehicle with sufficient accuracy to allow the payload propulsion system to place the spacecraft into a stationary synchronous orbit while retaining sufficient stationkeeping propulsion to meet the mission lifetime requirements
Owner	Department of Space for India
Orbit Characteristics	
Apogee (km)	35,936
Perigee (km)	35,562
Inclination (deg.)	0.1
Period (min.)	1,434.2
Weight (kg)	650 in orbit
Dimensions	1.6 m x 1.4 m x 2.2 m
Shape	Cube
Power Source	Solar arrays and nickel cadmium batteries
Contractor	Ford Aerospace and Communications
Remarks	The initial attempt to open the C-band uplink antenna was unsuccessful. Deployment was finally accomplished by blasting the antenna with reaction control jets beneath it. The S-band downlink antenna was successfully deployed, but the accompanying release of the solar sail did not occur. This resulted in the Moon being in the field of view of the active Earth sensor. The unpredicted Moon interference caused the satellite attitude reference to be lost. The command link was broken as the satellite attitude changed. As a result, safing commands could not be received, all fuel was consumed, and the satellite was lost in September 1982.

Table 2–140. Insat 1B Characteristics

Launch Date	August 31, 1983
Launch Vehicle	STS-8 (*Challenger*)
Range	Kennedy Space Center
Mission Objectives	Launch the satellite along a suborbital trajectory with sufficient accuracy to allow the payload propulsion system to place the spacecraft into a stationary synchronous orbit while retaining sufficient stationkeeping propulsion to meet the mission lifetime requirements
Owner	Department of Space for India
Orbit Characteristics	
Apogee (km)	35,819
Perigee (km)	35,755
Inclination (deg.)	0.1
Period (min.)	1,436.2
Weight (kg)	650 in orbit
Dimensions	1.6 m x 1.4 m x 2.2 m
Shape	Cube
Power Source	Solar arrays and nickel cadmium batteries
Contractor	Ford Aerospace and Communications
Remarks	Insat 1B was placed in geosynchronous orbit at approximately 74 degrees east longitude.

Table 2–141. Morelos 1 Characteristics

Launch Date	June 17, 1985
Launch Vehicle	STS 51-G (*Discovery*)/PAM-D
Range	Kennedy Space Center
Mission Objectives	Launch the satellite into transfer orbit, permitting the spacecraft propulsion system to place it in stationary synchronous orbit for communications coverage
Owner	Mexico
Orbit Characteristics	
Apogee (km)	35,794
Perigee (km)	35,780
Inclination (deg.)	1.1
Period (min.)	1,436.1
Weight (kg)	645 in orbit
Dimensions	6.6 m long (deployed); 2.16 m diameter
Shape	Cylindrical
Power Source	Solar cells and nickel cadmium batteries
Contractor	Hughes Communications
Remarks	The satellite was positioned at approximately 113.5 degrees west longitude.

Table 2–142. Morelos 2 Characteristics

Launch Date	November 27, 1985
Launch Vehicle	STS 61-B (*Atlantis*)/PAM-D
Range	Kennedt Space Center
Mission Objectives	Launch the satellite into transfer orbit, permitting the spacecraft propulsion system to place it in stationary synchronous orbit for communications coverage
Owner	Mexico
Orbit Characteristics	
Apogee (km)	35,794
Perigee (km)	35,780
Inclination (deg.)	1.1
Period (min.)	1,436.1
Weight (kg)	645 in orbit
Dimensions	6.6 m long (deployed); 2.16 m diameter
Shape	Cylindrical
Power Source	Solar cells and nickel cadmium batteries
Contractor	Hughes Communications
Remarks	Morelos 2 was not activated once it achieved its geosynchronous storage orbit. It was allowed to drift to its operational orbit at approximately 116.8 degrees west longitude. It began operations in March 1989.

Table 2–143. Palapa B-1 Characteristics

Launch Date	June 18, 1983
Launch Vehicle	STS-7 (*Challenger*)
Range	Kennedy Space Center
Mission Objectives	Launch the satellite into a transfer orbit that permits the spacecraft propulsion system to place it in stationary geosynchronous orbit for communications
Owner	Indonesia
Orbit Characteristics	
Apogee (km)	35,788
Perigee (km)	35,783
Inclination (deg.)	0
Period (min.)	1,436.1
Weight (kg)	630 at beginning of life in orbit
Dimensions	2 m high; 1.5 m diameter
Shape	Cylindrical
Power Source	Solar panels and nickel cadmium batteries
Contractor	Hughes Communications
Remarks	This satellite replaced Palapa A-1 in geosynchronous orbit at approximately 83 degrees east longitude.

Table 2–144. Palapa B-2 Characteristics

Launch Date	February 6, 1984
Launch Vehicle	STS 41-B (*Challenger*)/PAM-D
Range	Kennedy Space Center
Mission Objectives	Launch the satellite into a circular orbit with sufficient accuracy to allow the PAM-D stage and the spacecraft apogee kick motor to place the spacecraft into a stationary geosynchronous orbit while retaining sufficient station-keeping propulsion to meet the mission lifetime requirements
Owner	Indonesia
Orbit Characteristics	Did not achieve proper orbit
Apogee (km)	1,190
Perigee (km)	280
Inclination (deg.)	28.2
Period (min.)	99.5
Weight (kg)	630 in orbit
Dimensions	2 m high; 1.5 m diameter
Shape	Cylindrical
Power Source	Solar panels and nickel cadmium batteries
Contractor	Hughes Communications
Remarks	Palapa B-2 was to be placed into geostationary orbit, but it did not reach its location because the PAM failed. The spacecraft was retrieved by STS 51-A and returned to Earth for refurbishment. The satellite was relaunched as Palapa B-2R in April 1990.

Table 2–145. Palapa B-2P Characteristics

Launch Date	March 20, 1987
Launch Vehicle	Delta 3920
Range	Eastern Space and Missile Center
Mission Objectives	Launch the satellite into a circular orbit on a two-stage Delta 3920 launch vehicle with sufficient accuracy to allow the PAM-D stage and the spacecraft apogee kick motor to place the spacecraft into a stationary geosynchronous orbit while retaining sufficient stationkeeping propulsion to meet the mission lifetime requirements
Owner	Indonesia
Orbit Characteristics	
Apogee (km)	35,788
Perigee (km)	35,788
Inclination (deg.)	0
Period (min.)	1,436.2
Weight (kg)	630 in orbit
Dimensions	2 m high; 1.5 m diameter
Shape	Cylindrical
Power Source	Solar panels and nickel cadmium batteries
Contractor	Hughes Communications
Remarks	The satellite was positioned in geosynchronous orbit at approximately 113 degrees east longitude.

Table 2–146. UoSAT 1 Characteristics

Launch Date	October 6, 1981
Launch Vehicle	Delta 2310
Range	Western Test Range
Mission Objectives	Provide radio amateurs and educational institutions with an operational satellite that could be used with minimal ground stations for studying ionosphere and radio propagation conditions
Owner	University of Surrey, United Kingdom
Orbit Characteristics	
Apogee (km)	470
Perigee (km)	469
Inclination (deg.)	97.6
Period (min.)	94
Weight (kg)	52
Dimensions	42.5 cm square, 83.5 cm high
Shape	Rectangular
Power Source	Batteries
Contractor	University of Surrey
Remarks	UoSAT 1 was a piggyback payload with the Solar Mesospheric Explorer. It had some initial difficulty with transmitting data because of interference from a 145-MHz telemetry transmitter that was overcome by shifting to a redundant 435-MHz command system.

SPACE APPLICATIONS

Table 2–147. UoSAT 2 Characteristics

Launch Date	March 1, 1984
Launch Vehicle	Delta 3920
Range	Western Space and Missile Center
Mission Objectives	Stimulate interest in space science and engineering among radio amateurs, school children, students, colleges, and universities; provide professional and amateur scientists with a low-Earth-orbit reference for magnetospheric studies to be carried out concurrently with AMPTE and Viking missions, while supporting ground-based studies of the ionosphere; and advance further developments in cost-effective spacecraft engineering with a view to establishing a low-cost spacecraft system design for use in future STS Get-Away Special launches and other secondary payload opportunities
Owner	University of Surrey, United Kingdom
Orbit Characteristics	
Apogee (km)	692
Perigee (km)	674
Inclination (deg.)	98.1
Period (min.)	98.4
Weight (kg)	60
Dimensions	35 cm x 35 cm x 65 cm
Shape	Cube
Power Source	Batteries
Contractor	University of Surrey
Remarks	UoSAT 2 was a piggyback payload with Landsat 5.

Table 2–148. NOVA 1 Characteristics

Launch Date	May 15, 1981
Launch Vehicle	Scout
Range	Western Space and Missile Center
Mission Objectives	Place the Navy satellite in a transfer orbit to enable the successful achievement of Navy objectives
Owner	Department of Defense (Navy)
Orbit Characteristics	
Apogee (km)	1,182
Perigee (km)	1,164
Inclination (deg.)	90
Period (min.)	109.2
Weight (kg)	166.7
Dimensions	Body: 52.07 cm diameter; attitude control section: 26.7 cm diameter, 76.2 cm length
Shape	Octagonal body topped by a cylindrical attitude control section
Power Source	Solar cells and nickel cadmium batteries
Contractor	RCA Astro Electronics and Applied Physics Laboratory
Remarks	NOVA 1 was the first in a series of advanced navigational satellites built for the Navy. The satellite failed in March 1991.

Table 2–149. NOVA 3 Characteristics

Launch Date	October 12, 1984
Launch Vehicle	Scout
Range	Western Space and Missile Center
Mission Objectives	Place the satellite in a transfer orbit to enable the successful achievement of Navy objectives
Owner	Department of Defense (Navy)
Orbit Characteristics	
Apogee (km)	1,200
Perigee (km)	1,149
Inclination (deg.)	90
Period (min.)	108.9
Weight (kg)	166.7
Dimensions	Body: 52.07 cm diameter; attitude control section: 26.7 cm diameter, 76.2 cm length
Shape	Octagonal body topped by a cylindrical attitude control section
Power Source	Solar cells and nickel cadmium batteries
Contractor	RCA Astro Electronics and Applied Physics Laboratory
Remarks	NOVA 3 was the second in the series of improved transit navigation satellites. The satellite failed in December 1993.

Table 2–150. NOVA 2 Characteristics

Launch Date	June 16, 1988
Launch Vehicle	Scout
Range	Western Space and Missile Center
Mission Objectives	Place the satellite in a transfer orbit to enable the successful achievement of Navy objectives
Owner	Department of Defense (Navy)
Orbit Characteristics	
Apogee (km)	1,199
Perigee (km)	1,149
Inclination (deg.)	89.9
Period (min.)	108.9
Weight (kg)	166.7
Dimensions	Body: 52.07 cm diameter; attitude control section: 26.7 cm diameter, 76.2 cm length
Shape	Octagonal body topped by a cylindrical attitude control section
Power Source	Solar cells and nickel cadmium batteries
Contractor	RCA Astro Electronics and Applied Physics Laboratory
Remarks	Third in a series of improved transit navigation satellites launched by NASA for the U.S. Navy, the satellite failed in June 1996.

Table 2–151. SOOS-1 (Oscar 24/Oscar 30) Characteristics

Launch Date	August 3, 1985
Launch Vehicle	Scout
Range	Western Space and Missile Center
Mission Objectives	Place the Navy SOOS-1 mission into an orbit that will enable the successful achievement of Navy objectives
Owner	Department of Defense (Navy)
Orbit Characteristics	
Apogee (km)	Oscar 24: 1,257; Oscar 30: 1,258
Perigee (km)	1,002
Inclination (deg.)	89.9
Period (min.)	107.9
Weight (kg)	128 (both Oscars and interface cradle)
Dimensions	25 cm long; 46 cm diameter
Shape	Octagonal prism
Power Source	Four solar panels
Contractor	RCA Americom Astro-Electronics Division
Remarks	Oscar 24 and Oscar 30 were part of U.S. Navy Transit (Navy Navigation Satellite System). The satellites were launched into polar orbit at the same time.

Table 2–152. SOOS-2 (Oscar 27/Oscar 29) Characteristics

Launch Date	June 16, 1987
Launch Vehicle	Scout
Range	Western Space and Missile Center
Mission Objectives	Place the Navy SOOS-2 mission into an orbit that will enable the successful achievement of Navy objectives
Owner	Department of Defense (Navy)
Orbit Characteristics	
Apogee (km)	1,175 and 1,181
Perigee (km)	1,017 and 1,181
Inclination (deg.)	90.3
Period (min.)	107.2
Weight (kg)	128 (both Oscars and interface cradle)
Dimensions	25 cm long; 46 cm diameter
Shape	Octagonal prism
Power Source	Four solar panels
Contractor	RCA Americom Astro-Electronics Division
Remarks	This was in use through 1996.

Table 2–153. SOOS-3 (Oscar 23/Oscar 30) Characteristics

Launch Date	April 25, 1988
Launch Vehicle	Scout
Range	Western Space and Missile Center
Mission Objectives	Place the Navy SOOS-3 mission into an orbit that will enable the successful achievement of Navy objectives
Owner	Department of Defense (Navy)
Orbit Characteristics	
Apogee (km)	1,302 and 1,316
Perigee (km)	1,017 and 1,018
Inclination (deg.)	129.6
Period (min.)	108.6 and 108.7
Weight (kg)	128 (both Oscars and interface cradle)
Dimensions	25 cm long; 46 cm diameter
Shape	Octagonal prism
Power Source	Four solar panels
Contractor	RCA Americom Astro-Electronics Division
Remarks	This had an improved downlink antenna and a frequency synthesizer that gave the capability of selecting other downlink frequencies. This allowed monitoring of stored-in-orbit spacecraft on a frequency offset that did not interfere with satellites broadcasting on "operational" frequency. It was operational through 1996.

Table 2–154. SOOS-4 (Oscar 25 and Oscar 31) Characteristics

Launch Date	August 24, 1988
Launch Vehicle	Scout
Range	Western Space and Missile Center
Mission Objectives	Place the Navy SOOS-4 mission into an orbit that will enable the successful achievement of Navy objectives
Owner	Department of Defense (Navy)
Orbit Characteristics	
Apogee (km)	1,176 and 1,178
Perigee (km)	1,032 (both)
Inclination (deg.)	90.0
Period (min.)	107.4
Weight (kg)	128 (both Oscars and interface cradle)
Dimensions	25 cm long; 46 cm diameter
Shape	Octagonal prism
Power Source	Four solar panels
Contractor	RCA Americom Astro-Electronics Division
Remarks	This had an improved downlink antenna and a frequency synthesizer that gave the capability of selecting other downlink frequencies. This allowed monitoring of stored-in-orbit spacecraft on a frequency offset that did not interfere with satellites broadcasting on "operational" frequency. It was operational through 1996.

CHAPTER THREE
AERONAUTICS AND SPACE RESEARCH AND TECHNOLOGY

CHAPTER THREE

AERONAUTICS AND SPACE RESEARCH AND TECHNOLOGY

Introduction

The federal government's involvement with aeronautics preceded NASA's establishment by many years. In 1915, Congress mandated that the National Advisory Committee for Aeronautics (NACA) "supervise and direct the scientific study of the problems of flight, with a view to their practical solution." In the National Aeronautics and Space Act of 1958 that established NASA, Congress stated that NASA would be involved in "aeronautical and space activities" using "aeronautical and space vehicles." The law defined aeronautical and space activities as:

(a) research into, and the solution of, problems of flight within and outside the Earth's atmosphere; (b) the development, construction, testing, and operation for research purposes of aeronautical and space vehicles; (c) the operation of a space transportation system including the Space Shuttle, upper stages, space platforms, and related equipment; and (d) such other activities as may be required for the exploration of space.[1]

It also defined aeronautical and space vehicles as "aircraft, missiles, satellites, and other space vehicles, manned and unmanned, together with related equipment, devices, components, and parts."[2] It can safely be said that NASA Office of Aeronautics and Space Technology (OAST) activities have covered all these areas.

OAST's aeronautics research and technology program from 1979 to 1988 was derived from several technological disciplines and spanned the flight spectrum from hovering to hypersonic aircraft. OAST provided technology results well in advance of specific applications needs and conducted long-term independent research without

---

[1]"Aeronautics: The NASA Perspective," NASA Fact Sheet, February 10, 1981.
[2]U.S. Congress, *NASA Aeronautics and Space Act of 1958 (as Amended)*, sec. 103 (Washington, DC: U.S. Government Printing Office, 1958).

the payoff of known immediate mission applications. The disciplinary research applied to all classes of vehicles and related to capabilities that were yet undefined. In addition, OAST's technology research enhanced the capabilities of specific classes of vehicles, such as subsonic transport, rotorcraft, high-performance military aircraft, and supersonic and hypersonic vehicles.

Space research and technology took both a disciplinary approach and a vehicle-specific approach. Disciplines represented in the program included propulsion, space energy, aerothermodynamics, materials and structures, controls and guidance, automation and robotics, space human factors, computer science, sensors, data and communications systems, and spaceflight systems. The space research and technology program developed and improved technologies and components for the Space Shuttle and for the future Space Station and also participated in missions and experiments launched from and conducted on the Shuttle.

OAST's fundamental involvement with other agencies and with industry differed from other NASA organizations. In the area of general aviation, OAST worked with the Federal Aviation Administration (FAA), the Department of Transportation, and aircraft manufacturers to improve aircraft and aviation safety and to lessen any harmful impact of flight on the environment. In the area of high-performance aircraft, OAST research supported the needs of the military, and NASA continually participated in joint projects with the Department of Defense (DOD) and sometimes shared the financial costs of these projects.

OAST's activities have benefited the U.S. economy. Congress regularly, in its deliberations on NASA's budget, noted that aeronautics was one area in which the United States had a positive balance of trade and also contributed to creating a large number of jobs. Congress generally deemed NASA's aeronautics deserving of steady support. For instance, in the Conference Report that accompanied the FY 1982 budget authorization, the committee expressed its concern

with a recent trend toward lower levels of Federal support for aeronautical research and technology development. NASA's research and technology for decades has been the wellspring for U.S. aviation development from which the nation's military, commercial, and general aviation leadership has evolved. This has meant millions of jobs for Americans with a wide range of trade and professional skills in every region of the country. It has meant billions in favorable balance of trade over the years.... It has meant billions of dollars returned to the Federal treasury in tax revenues.[3]

---

[3]"Authorizing Appropriations to the National Aeronautics and Space Administration," Conference Report, November 21, 1982, *Chronological History FY 1982 Budget Submission,* prepared by the NASA Comptroller, Budget Operations Division, p. 51.

This recognition of the benefits of NASA's aeronautics activities helped OAST secure a reasonably steady level of funding.

From 1979 to 1988, three policy statements issued by the Executive Office of the President's Office of Science and Technology Policy (OSTP) helped define OAST's focus. The first policy statement resulted from a 1982 multi-agency review of national aeronautical research and technology policy. The group, chaired by Victor Reis, assistant director of OSTP, addressed two questions:

1. Was aeronautics a mature technology, and was continued investment justified by potential benefits?
2. What were the proper government roles in aeronautical research and technology, and did the present institutional framework satisfy these roles or should it be changed?

The group stated that the aerospace industry "has evolved into a major U.S. enterprise that provided about 1.2 million jobs in the United States in 1981." It concluded that "the present institutional framework allowed implementation of the U.S. government role in developing aeronautical research and technology." It recommended that the government meet the following national aeronautics goals:

Aeronautics
1. Maintain a superior military aeronautical capability
2. Provide for the safe and efficient use of the national airspace system, vehicles operated within the system, and facilities required for those operations
3. Maintain an environment in which civil aviation services and manufacturing can flourish
4. Ensure that the U.S. aeronautical industry has access to and is able to compete fairly in domestic and international markets consistent with U.S. export policy

Aeronautical Research and Technology
1. Ensure the timely provision of a proven technology base to support future development of superior U.S. aircraft
2. Ensure the timely provision of a proven technology base for a safe, efficient, and environmentally compatible air transportation system[4]

OSTP, chaired by G.A. Keyworth II, science advisor to the President, issued the second policy statement in March 1985. It spelled out specific

---

[4] "Aeronautical Research and Technology Policy," Vol. I: Summary Report, Executive Office of the President, Office of Science and Technology Policy, November 1982, pp. 14, 21–23.

goals in subsonics, supersonics, and transatmospherics. These goals were the basis for NASA's future aeronautics program planning.[5]

The subsonic goal aimed to provide technology for an entirely new generation of fuel-efficient, affordable U.S. aircraft operating in a modernized national airspace system. The supersonic goal focused on developing "pacing technologies" for sustained supersonic cruise capability for efficient long-distance flight. The transatmospheric goal encompassed the pursuit of research toward a capability for extremely fast passenger transportation between points on Earth, as well as for a vehicle that could provide routine cruise and maneuvers into and out of the atmosphere with takeoffs and landings from conventional runways.

The third policy statement, issued in February 1987, stated that although the United States had made progress in reaching the 1985 goals, "greater achievement" was necessary. The committee, chaired by the science advisor to the President, William R. Graham, presented an eight-point action plan to achieve the national goals and "remain a viable competitor in the world aviation marketplace."[6] The action plan summary was as follows:

1. Increase innovative industry research and development efforts given the certainty of intensifying global competition and the importance of new technology for U.S. competitiveness
2. Aggressively pursue the National Aerospace Plane program, assuring maturation of critical technologies leading to an experimental airplane
3. Develop a fundamental technology, design, and business foundation for a long-range supersonic transport in preparation for a potential U.S. industry initiative
4. Expand domestic research and development collaboration by creating an environment that reflects the new era of global competition
5. Encourage government aeronautical research in long-term emerging technology areas that provide high payoffs
6. Strengthen American universities for basic research and science education through enhanced government and aerospace industry support and cooperation
7. Improve the development and integration of advanced design, processing, and computer-integrated manufacturing technologies to transform emerging research and development results into affordable U.S. products

---

[5]"National Aeronautical R&D Goals: Technology for America's Future," Executive Office of the President, Office of Science and Technology Policy, March 1985.

[6]"National Aeronautical R&D Goals: Agenda for Achievement," Executive Office of the President, Office of Science and Technology Policy, February 1987.

8. Enhance the safety and capacity of the National Airspace System through advanced automation and electronics technology and new vehicle concepts, including vertical and short takeoff and landing aircraft

The Last Decade Reviewed (1969–1978)

From 1969 to 1978, NASA carried out aeronautics and space research and technology activities in two organizations: the Office of Advanced Research and Technology (OART) until 1972 and OAST beginning in 1972. The goals were to build a research and technology base, conduct systems and design studies, and carry out systems and experimental programs. Work included the broad categories of air transportation system improvement, spacecraft subsystem improvement, support to the military, and the application of technology to nonaerospace systems.

Research

Until 1970, NASA included basic research as one of its major divisions. The results of basic research added to the pool of knowledge and did not apply to any ongoing project. This program was divided into four sections: fluid dynamics, electrophysics, materials, and applied mathematics.

Space Vehicle Systems

This division dealt with problems vehicles might encounter during launch, ascent through the atmosphere, spaceflight, and atmospheric entry. NASA conducted research in the areas of lifting-body research and planetary entry research.

Guidance, Control, and Information Technology

From 1969 to 1978, NASA worked at improving the operational electronics systems, while reducing their size, weight, cost, and power requirements. Several NASA centers directed a variety of projects with this goal in mind.

Human Factor Systems

This directorate was responsible for the human factors systems program, which held that humans were a critical component of the spacecraft system or part of a human-machine system. Investigators were concerned with the interaction between the pilot/astronaut and the vehicle that affected health, comfort, survival, and decision-making skills. NASA conducted research into the various systems that were found on aircraft and that would be found on the Space Shuttle. Researchers also investigated long-term exposure to the space environment.

Space Power and Propulsion Systems

Researchers during the 1970s investigated lighter, more efficient propulsion systems than the chemical propulsion systems of the 1960s. Both electric and nuclear propulsion received much attention. Efforts in chemical propulsion were devoted to solving the Shuttle's main engine design problems. NASA also carried out joint research into nuclear propulsion with the Atomic Energy Commission. In addition, NASA tested various methods of generating power using chemical, electric, and nuclear sources.

Aeronautics

NASA reorganized OART in 1970 to emphasize improving aeronautical research, which NASA had been accused of neglecting. Both staff and budget levels were increased to provide additional resources. NASA abolished basic research divisions and carried out aeronautics activities in three offices—aeronautical operating systems, aeronautical research, and aeronautical propulsion—and had special offices devoted to short takeoff and landing (STOL) aircraft and experimental transport aircraft. It also added an office for the Military Aircraft Support Program. The Aeronautics Division conducted projects in the areas of general aviation, environmental factors, vertical/STOL aircraft, supersonic/hypersonic aircraft, and military support.

Aeronautical and Space Research and Technology (1979–1988)

OAST focused on aeronautical research and technology and on space research and technology. Within these two major areas, work took place in two prime fields: research and systems. Research was generally disciplinary in nature and focused on aerodynamics, materials and structures, propulsion, aerothermodynamics, energy conversion, controls and human factors, computer science, and information sciences. Systems-focused work was often multidisciplinary and had more immediate application. NASA's systems activities supported existing NASA projects such as the Space Shuttle, developed enabling technology for future projects such as the Space Station, and provided support to the military.

In addition, in the early part of the 1980s, OAST supported national energy needs through its Energy Technology Program. The Department of Energy and other federal agencies sponsored NASA's work in this area, which encompassed a variety of projects. These included solar cell power systems, automotive power systems, industrial gas turbine development, solar heating and cooling, wind turbine generators, solar thermal electric conversion, energy storage, and advanced coal extraction and processing.

Aeronautics

Many OAST efforts focused on improving flight efficiency. Beginning in the 1970s and continuing into the 1980s, the Aircraft Energy Efficiency program spanned several disciplines and focused on developing solutions that could be applied to existing vehicles, to their spinoffs expected within a few years, and to new classes of aircraft designed specifically to be fuel efficient. New advances in turboprop research promised considerable fuel savings while maintaining the performance and cabin environment of modern turbofan aircraft. New composite materials were also being developed that would result in reduced cost and weight. In addition, OAST analyzed ways to increase lift and reduce drag in wings, shaping them to meet the needs of the new generation of aircraft. Aircraft drag reduction research emphasized techniques for maintaining laminar boundary-layer airflow over larger segments of aircraft wings and other surfaces. In addition, the oblique wing was extended to the requirements of supersonic flight and showed increased fuel economy. Other wing configurations also added to flight efficiency.

Another research target was the large-capacity STOL aircraft. STOL and vertical takeoff and landing (VTOL) aircraft were planned for use at airports located close to populated areas. These locations demanded aircraft with a low noise level. NASA, along with industry, developed an experimental engine that produced a significant reduction in generated noise. The quiet, clean, short-haul experimental engine had a goal of providing power for a four-engine, 150-passenger STOL transport that generated relatively low noise. This engine began test runs at Lewis Research Center in the late 1970s.The Quiet Short-haul Research Aircraft (QSRA), evaluated by Ames Research Center, was another "quiet" aircraft that incorporated the propulsive lift system.

Related developments in propulsion system thrust-to-weight ratios, propulsive lift control, and understanding low-speed aerodynamics provided advances in vertical and short takeoff and landing (V/STOL) and short takeoff and vertical landing (STOVL) technology. In 1986, the United States and the United Kingdom signed a joint research agreement to develop advanced STOVL (ASTOVL) technologies and to reduce the risks associated with developing this type of aircraft. Also, NASA and Canada agreed to test a full-scale STOVL model designated as the E-7.

Rotary wing aircraft was another primary focus of NASA's aeronautics activities during this decade. Capable of STOL and VTOL performance, rotorcraft had both civilian and military applications. Two flight vehicles formed the cornerstones of NASA's rotorcraft research. The Tilt-Rotor Research Aircraft had twin rotors and power plants mounted at the ends of a high wing. The rotors could be tilted from horizontal, permitting vertical flight, to vertical, permitting horizontal flight. The Sikorsky Rotor Systems Research Aircraft (RSRA) used helicopter rotor heads as the basic lifting system but were designed to be able to test a wide variety of rotor systems. The RSRA could be flown as a conventional helicopter, or as a compound helicopter, with fixed wings installed to "unload" the rotor by assuming some of the lift. Both aircraft flew out of Ames Research Center and at other locations.

Aviation safety has traditionally been a focus of NASA aeronautical research. During this decade, NASA carried out research on wind shear, icing, heavy rain, lightning, and combustible materials on aircraft. NASA conducted many of these activities cooperatively with the FAA.

One of NASA's aeronautics missions was to provide support to the military. The Highly Maneuverable Aircraft Technology (HiMAT) program worked with the military to resolve the problems associated with combining high maneuverability, high speed, and a human pilot. The project had two basic tasks: to study the interrelated problems of all aspects of the flight of a typical advanced fighter configuration and to contribute to the design of future fighter types by furnishing fundamental aerodynamic and structural loads data to assist designers. The HiMAT remotely piloted research vehicle made its first flight on July 27, 1979, from Dryden Flight Research Center.

In 1985, the X-29A flight research program began. The unique forward-swept wing of the aircraft was made of composite materials that reduced the wing's weight up to 20 percent, compared to the weight of conventional aft-swept wings. The forward-mounted "canards" were computer-adjusted forty times a second to improve flight efficiency and aircraft agility. [7]

A new convertible gas turbine engine and other propulsion systems were developed and demonstrated at Lewis Research Center. The gas turbine engine allowed the engine's output to take the form of either shaft power or fan power. This type of propulsion system was required for advanced high-speed rotorcraft concepts such as the X-wing, in which rotor blades operating in a spinning mode for takeoff and landing were stopped and locked in place as an X-shaped fixed wing for high-speed flight. A propfan propulsion system, also developed and tested at Lewis, received the 1987 Robert J. Collier Trophy for developing the technology for and testing of advanced fuel-efficient subsonic aircraft propulsion systems.

In his 1986 State of the Union address, President Ronald Reagan announced the initiation of the joint NASA-DOD National Aerospace Plane research program that was planned to lead to an entirely new family of aerospace vehicles. Reagan stated that "we are going forward with research on a new Orient Express that could, by the end of the next decade, take off from Dulles Airport [located near Washington, D.C.], accelerate up to 25 times the speed of sound, attaining low Earth orbit or flying to Tokyo within two hours."[8] The goal of the program was to develop hypersonic and transatmospheric technologies for a new class of aero-

---

[7]Canards are horizontal stabilizers used to control pitch.

[8]"State of the Union Address," in *Presidential Papers of Ronald Reagan* (Washington, DC: U.S. Government Printing Office, February 1986).

space vehicles characterized by horizontal takeoff and landing, single-stage operation to orbital speeds, and sustained hypersonic cruise within the atmosphere using airbreathing rather than rocket propulsion. These technologies, it was hoped, would lead to a new flight research vehicle (the X-30). Between 1986 and 1989, the main goal of the technology development phase was to develop and test an integrated airframe/propulsion system that could operate efficiently from takeoff to orbit. A series of developmental contracts awarded during 1986 and 1987 focused on propulsion systems and certain aircraft components.

Several new NASA facilities supported OAST's research programs. In 1985, a new National Transonic Facility opened at Langley Research Center that permitted engineers to test models in a pressurized tunnel in which air was replaced by the flow of supercooled nitrogen. As the nitrogen vaporized into gas in the tunnel, it provided a medium more dense and viscous than air, offsetting scaling inaccuracies of smaller models tested in the tunnel. In 1987, the Numerical Aerodynamic Simulation Facility, located at Ames, became operational. It could make 1 billion calculations per second. For the first time, aircraft designers could routinely simulate the three-dimensional airflow patterns around an aircraft and its propulsion system. Also at Ames, a complement to the existing 12.2-meter by 24.4-meter closed-circuit tunnel became operational at the end of 1987. It had a test section 24.4 meters high and 36.6 meters wide, three times as large in cross-section as the parent tunnel. The original tunnel's fans were also replaced, raising its speed from 370 to 555 kilometers per hour.

Space

NASA's space research and technology program provided advanced technology to ensure continued U.S. leadership in civil space programs. The program focused on technology to develop more capable and less costly space transportation systems, large space systems with growth potential such as the Space Station, geosynchronous communications platforms, lunar bases, crewed planetary missions, and advanced scientific, Earth observation, and planetary exploration spacecraft. All NASA centers were involved, along with significant industry and university participation.

Many of NASA's space technology programs from 1979 to 1988 were concerned with the problems of providing power, controls and structures, and assembly of large space structures. Other research areas included spacesuit studies, research for more efficient reentry from space, advanced power systems for future lunar and Mars bases, and lighter weight tanks for cryogenic fuels. Still other investigations concentrated on control systems for future large lightweight spacecraft and the assembly of large space structures with teleoperated manipulators, as well as a program to allow free-flying telerobots to grapple and dock with gyrating satellites to stabilize and repair the spacecraft.

OAST and the NASA centers were heavily involved with the Space Shuttle. They developed and demonstrated the Shuttle's thermal protection system and continued to improve the composition and durability of the materials. They developed the experiments for the Orbiter Experiments Program, which flew on the first Space Shuttle mission. This program evaluated the aerodynamic, aerothermodynamic, acoustic, and other stress phenomena involved in spaceflight, particularly during the orbiter's return to the atmosphere at hypersonic velocity. OAST participated in Shuttle payloads, developing the Long Duration Exposure Facility and many of its experiments that flew on STS 41-C in 1984, as well as the OAST-1 mission, which flew on STS 41-D, also in 1984.

Many of OAST's technology activities were applicable to Space Station development. Researchers at Langley Research Center and Marshall Space Flight Center developed a mobile work station concept from which astronauts in spacesuits could assemble large structures in space. Methods were developed for the "toolless" assembly of large structures in the weightless environment of space using lightweight composite columns and unique specialized joints. Automation and robotics were important OAST discipline areas. The ACCESS and EASE experiments on STS 61-B tested assembling erectable structures in space.

A major effort went into improving the batteries, solar cells, and solar arrays used on spacecraft. Solar cell and solar array technology improved conversion efficiency, reduced mass and cost, and increased the operating life of these essential components.

The space research and technology program also provided development support for the planetary program and the Earth-orbiting spacecraft programs. OAST developed a computer program that used artificial intelligence techniques to perform automatic spacecraft operations for use on the Voyager project during the Uranus encounter in 1986. In a different area, one of the barriers to planetary return missions was the cost and complexity of return propulsion. Planetary return missions would be less costly if propellant could be produced at or on the planet. OAST researchers successfully demonstrated methods for producing liquid oxygen from a simulated Martian atmosphere by using electrolytic techniques.

In 1987, NASA's Civilian Space Technology Initiative began. Its objective was to advance the state of technology in key areas in which capabilities had eroded and stagnated over the previous decade. This program had a short-term perspective—it was designed to address high-priority national and agency needs of the 1990s.[9] The program included research in technologies to enable efficient, reliable access to and operations in Earth orbit and to support science missions. The program had three technology thrusts: space transportation, space science, and space operations. NASA also encouraged academic sector participation through programs such as the University Space Design program and the University Space Engineering Research program.

---

[9]*Aeronautics and Space Report of the President, 1988 Activities* (Washington, DC: U.S. Government Printing Office, 1990), p. 59.

Management of NASA's Aeronautics and Space Technology Program

As mentioned in the previous section, OAST had two primary focuses: aeronautics research and technology and space research and technology. In addition, for a period of time, the office also managed a program for energy technology. Until its August 1984 reorganization, OAST had a division devoted to aerospace research that managed the various discipline areas, an aeronautical systems division, and a space systems division (Figure 3–1). The research division was called the Research and Technology Division until mid-1982 and then the Aerospace Research Division until August 1984. The Aeronautical Systems and Space Systems Divisions managed vehicle-specific and system-specific activities. The disciplinary areas changed slightly between 1979 and 1984. Under the Research and Technology Division were offices for Electronics and Human Factors, Aerodynamics, Materials and Structures, Propulsion, and Space Power and Propulsion. Under the Aerospace Research Division were offices for Controls and Human Factors, Computer Science and Electronics, Fluid and Thermal Physics, Materials and Structures, and Space Energy Conversion.

After the August 1984 reorganization, OAST had two divisions that managed system-related and vehicle-specific work, the Aeronautics Division and the Space Division, as well as a number of disciplinary divisions. The disciplinary divisions interacted with both the Aeronautics and Space Divisions. When the National Aerospace Plane program was established, it became a separate program office (Figure 3–2).

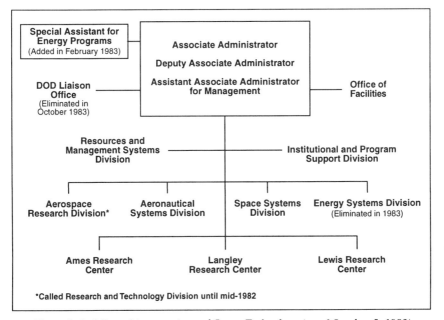

Figure 3–1. Office of Aeronautics and Space Technology (as of October 5, 1983)

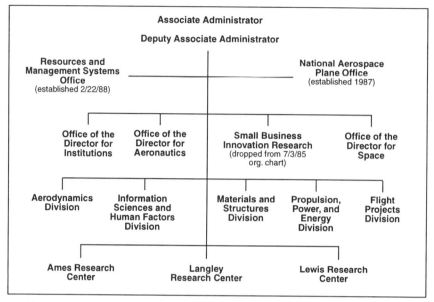

Figure 3–2. Office of Aeronautics and Space Technology (as of August 14, 1984)

OAST also had management responsibility for Ames Research Center at Moffett Field, California, Langley Research Center in Hampton, Virginia, and Lewis Research Center in Cleveland, Ohio. These centers conducted almost all of NASA's aeronautics technology research as well as a considerable amount of space technology research. Dryden Flight Research Facility in Edwards, California, which had been an independent NASA center, became a directorate of Ames Research Center on October 1, 1981. Aeronautical and research activities at the two locations were integrated and staff functions combined. This arrangement continued until 1994, when Dryden again became an autonomous NASA center.

Phase I (Pre-1984 Reorganization)

James Kramer served as associate administrator from October 23, 1977, until his retirement on September 30, 1979. He was replaced by Walter B. Olstad, who served as acting associate administrator until Dr. Jack Kerrebrock became the next associate administrator in June 1981. Kerrebrock remained at that post until July 1983. Dr. Raymond S. Colladay followed and served briefly as acting associate administrator until John J. Martin took the reins at the start of 1984. He remained in place through the 1984 reorganization until April 1985.

Division directors in place at the beginning of 1979 were:

- Donald A. Beattie, director of the Energy Systems Division
- William S. Aiken, Jr., acting director of the Aeronautical Systems Division

- Dell P. Williams, III, acting director of the Space Systems Division
- George C. Deutsch, director of the Research and Technology Division

Beattie remained in place until the Energy Systems Division was disestablished at the end of 1982. Aiken moved from acting division director to division director of the Aeronautical Systems Division in 1981 and remained at that post through the 1984 reorganization. Williams remained with the Space Systems Division until 1983, when Henry O. Slone became acting director, followed by Dr. Leonard A. Harris, who was acting director until he was appointed division director in 1984. Deutsch led the Research and Technology Division until 1981, when Frederick Povinelli succeeded him as acting director. Raymond Colladay became director in July 1981 and remained until the formation of the Aerospace Research Division in 1982. Dr. Leonard A. Harris led the Aerospace Research Division until he was replaced by Cecil C. Rosen III, who served as acting director until the 1984 reorganization.

Phase II (Post-1984 Reorganization)

John Martin remained as associate administrator until April 1985 when Raymond Colladay became acting associate administrator. Colladay was appointed associate administrator effective June 14, 1985. He remained at that post until February 1988, when Dr. William F. Ballhaus, Jr., became acting associate administrator, followed by Dr. Robert Rosen as acting associate administrator later in 1988.

The Aeronautical Systems Division became the Aeronautics Division with the 1984 reorganization. William Aiken, Jr., served as director for Aeronautics until Cecil Rosen III became acting director in mid-1985. He was appointed as division director later that year.

The Space Systems Division was renamed the Space Division, with Leonard Harris serving as director. He remained at that post until mid-1987, when James T. Rose briefly took the position. Frederick Povinelli later became director in October 1987.

Individual disciplinary divisions replaced the Aerospace Research Division in the August 1984 reorganization. Division directors at the time of the reorganization were:

- Information Sciences and Human Factors—Lee B. Holcomb
- Aerodynamics—Gerald G. Kayten
- Materials and Structures—Samuel L. Venneri
- Propulsion, Power, and Energy—Linwood C. Wright (acting)
- Flight Projects—Jack Levine

Kayten remained as Aerodynamics Division director until Dr. Randolph A. Graves assumed the post first as acting director in late 1986 and then as division director in 1987. He left in 1988, and Paul Kutler became acting division director for a brief period.

Wright remained as acting director of the Propulsion, Power, and Energy Division until Robert Rosen became division director in mid-1985. Rosen was followed by Edward A. Gabris in 1986. He remained until Gregory Reck assumed the post as acting director in late 1987 and then as director early in 1988.

Duncan E. McIver was the first director of the National Aerospace Plane Office, which was formed in 1987.

Money for Aeronautics and Space Research and Technology

NASA's funding for aeronautics and space research and technology activities grew in real dollars during the decade but decreased as a percentage of NASA's total budget. Taking into account the rate of inflation during the 1980s, the buying power of the additional dollars may have been negligible. Tables 3–1 through 3–61 show the funding for these activities.

Aeronautics and Space Research and Technology Programs

In FY 1987, total programmed funding jumped more than $13.6 million as the new Civil Space Technology Initiative became part of the budget. This program included focused systems technology programs supporting transportation, operations, and science consistent with the goals of the U.S. space program. In all, from 1979 to 1988, the OAST budget comprised between 6.8 and 10.8 percent of the total NASA Research and Development (R&D) budget (including Space Flight Control and Data Communications, or SFC&DC) (Figure 3–3).

Programmed funding for aeronautics and space research programs often differed from the amounts that Congress authorized or appropriated, even when Congress specified funding for a particular activity. NASA's aeronautics and space research activities were more intertwined with both military priorities and the activities of industry than the agency's programs in other areas. Thus, whether one of NASA's programs continued or was cancelled depended somewhat on whether

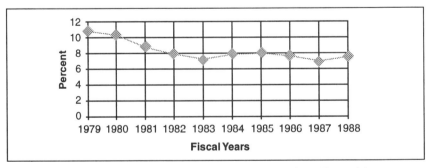

Figure 3–3. Percentage of R&D and SFC&DC Budget Allocated for Office of Aeronautics and Space Technology Activities

the military or industry shared some of the funding burden. The reduction or elimination of funding from non-NASA sources part way through a program could have forced temporary or permanent cessation of the activity.

Budget data in this section (request or submission, authorization, and appropriation) for the major budget categories come from the annual Budget Chronological Histories. Request or submission data for the more detailed budget items comes from the annual budget estimates produced by NASA's Budget Office. No corresponding authorization or appropriations data for these activities were available. All programmed (actual) figures come from NASA's Budget Office budget estimates. (Note that these "budget estimate" volumes contain estimates for a future period as well as actual amounts for a period that has ended.)

During FY 1979, a significant reordering of budget categories took place. Thus, the budget categories that were used for the budget requests often no longer existed by the time the programmed or actual budget was known. This was especially true in the Spacecraft Research and Technology area. Sometimes it is obvious from the item description that only a name change took place or two categories merged into one. These are noted. However, in other cases, no obviously equivalent budget category was substituted. These are also noted.

Aircraft Energy Efficiency Program

The Aircraft Energy Efficiency Program was a joint effort funded by NASA and industry. It included activities that fell in the Transport Aircraft Systems Technology and the Advanced Propulsion Systems areas of the Systems Technology Program. The budget categories listed in Table 3–21 were funded prior to FY 1979. Engine Component Improvement terminated in FY 1981, Advanced Turboprops received funding through FY 1985, and Composite Primary Aircraft Structures received funding through FY 1983. The other budget categories received funding through FY 1982.

Aeronautics Research and Technology Programs

NASA's aeronautics programs have been noteworthy in the extent that they have been cooperative efforts with other government agencies—particularly DOD through its Defense Advanced Research Projects Agency (DARPA) and the FAA—and with industry. Many NASA aeronautics research efforts began at the suggestion of industry or other agencies or because industry or the military identified a need and a large proportion of NASA-developed technologies saw their practical demonstration in the commercial or military sector. The following sections address aeronautics programs in the areas of flight efficiency, high-performance aircraft, and aircraft safety and operations.

Flight Efficiency

Aircraft Energy Efficiency Program

The oil crisis of the 1970s focused attention on the cost and efficient use of fuel and led Congress to push NASA to develop ways to increase fuel efficiency. NASA's Aircraft Energy Efficiency (ACEE) program, a NASA-industry effort, examined the problem of fuel efficiency and worked to develop solutions that could be applied to existing transports, to derivative vehicles expected within the next few years, and to new classes of aircraft designed specifically to be fuel efficient.

The basic goal of the ACEE program was to learn how to use fuel energy more efficiently for propulsion and lift. The program worked on improving aircraft engines, reducing drag of aircraft as they traveled through the atmosphere, decreasing the weight of the materials that comprised airframes, and finding more efficient ways of propelling aircraft through the atmosphere. Researchers believed that reductions in drag, as well as other improvements in the ratio of lift to drag, would improve the range capabilities of aircraft and reduce the operating cost of aircraft. In addition, new materials and structural concepts, combined with the use of active controls, could produce lighter and smaller airframes.

NASA initiated the program in FY 1976 following the oil embargo of the Organization of the Petroleum Exporting Countries (OPEC). Although the program was originally motivated by fuel conservation concerns, it was soon apparent that in addition to fuel savings, improving technology for aircraft and engine efficiency could also improve the competitive position of the United States in the worldwide multibillion dollar air transport marketplace. Industry shared the ACEE costs with NASA.

The cross-disciplinary program included six subsonic technologies or programs: Engine Component Improvement, Energy Efficiency Engine, Advanced Turboprop Program, Energy Efficient Transport, Composite Primary Aircraft Structures, and Laminar Flow Control. Several NASA centers participated in the program. Langley Research Center was responsible for technology programs in aerodynamics and in materials and structures. Langley and Ames Research Center shared wind tunnel testing. Dryden Flight Research Center conducted flight research. Lewis Research Center carried out propulsion research.[10]

The program demonstrated the benefits of NASA-industry cooperation in developing and validating advanced technology for use in civil applications. In addition, many of the concepts had future applications for both civil and military transport.

Engine Component Improvement. This project involved improving existing engine components by using improved aerodynamics and mate-

---

[10] "Propulsion/ACEE," *NASA Facts*, NF-93/8-81 (Washington, DC: U.S. Government Printing Office, 1981), pp. 1–2.

rials, applying clearance control techniques, and increasing the bypass ratio. It produced engine component technology for significantly better performance and performance retention in engine retrofits and new projects. Elements included fan blade improvements, turbine aerodynamics, blade cooling seals, and active clearance control. It projected a fuel saving of 5 percent. Applications appeared as early as 1978 and were applied to all modern transport engines. General Electric and Pratt & Whitney took part in this project.

Energy Efficient Engine. This program, completed in 1983, made possible a much greater reduction in cruise fuel consumption and accelerated technology readiness for incorporation into a new generation of fuel-efficient engines than was possible with earlier engines. The program's technologies included compressor, fan, and turbine-gas-path improvements; improved blading and clearance control; and structural advances. The program demonstrated or identified design and technology advances that could reduce turbofan engine fuel consumption by an estimated 15 to 20 percent. Technology derived from the program was applied to the CF6-80C, PW2037, and PW4000 engines. General Electric and Pratt & Whitney also participated in this program.

Energy Efficient Transport. This project focused on aerodynamic and control concepts, such as high-aspect-ratio, low-sweep supercritical wing technology, new high-lift devices; propulsion-airframe integration, digital avionics, and active controls. It led to the application of winglets and the wing load alleviation system.

Composite Primary Aircraft Structures. This project was built on previous NASA composite research and cooperative efforts with industry in the development and flight service validation of secondary structural components. Researchers fabricated and successfully flight-tested primary composite empennage structures, but efforts did not progress as planned to the validation of large wing and fuselage structures, nor did the project resolve manufacturing technology or cost problems.

Laminar Flow Control. This research was also part of the ACEE program. It is addressed elsewhere in this chapter.

Advanced Turboprop Project. This project was directed at greater efficiency for future turboprop-powered aircraft cruising at or near jet transport speed (Mach 0.65 to 0.85). The advanced turboprop, or unducted fan, technology was an important option for medium-range transports with fuel savings of 25 percent or more, as compared to equally advanced turbofan engines. The project successfully tested thin, swept-tip, multibladed, high-speed single-rotation, as well as dual-rotation, geared and ungeared versions of turboprop propulsion systems.

This project was the most successful of the ACEE efforts. Lewis Research Center and its industry team received the 1987 Robert J. Collier Trophy for their accomplishments in this area. They were recognized for developing the technology and testing advanced turboprop propulsion systems that provided dramatic reductions in fuel usage and operating costs for subsonic transport aircraft.

Although the rewards were the greatest of all the ACEE projects, the challenges were also plentiful. Because of political opposition, funding was very limited, and additional studies had to be performed to support the value of the advanced turboprop and to identify the most critical technical issues. Areas of technical concern included propeller efficiency at cruise speed, propeller and aircraft interior noise, installation aerodynamics, and maintenance costs.[11]

The development in propulsion technology grew out of the 1973 oil embargo, which had increased fuel costs from about 25 percent of airline direct operating costs to about half. In January 1975, the Senate Aeronautical and Space Science Committee asked NASA to help resolve the fuel crisis. NASA responded with a NASA, Department of Transportation, FAA, and DOD task force that reported on concepts with fuel-saving potential.[12] Among them was an advanced turboprop advocated by NASA's Lewis Research Center and Hamilton Standard Division of United Technologies, the last major propeller manufacturing company in the United States. This design overcame the high-speed compressibility losses of conventional propellers but was controversial because of the perception that using propellers was a return to an outmoded technology.[13] However, the prospect of lower ticket prices was an incentive to accept the new "old" technology.

In 1974, Lewis engineers began evaluating the high-speed turboprop propulsion system to see whether propeller blade redesign might lead to lower fuel consumption. They joined with Hamilton Standard to explore different types of blade shapes intended to allow greater tip speed that would permit propfan-driven aircraft to fly at jetliner speeds while retaining the inherently better fuel consumption found on propeller-driven aircraft.

In 1976, Hamilton Standard performed a series of wind tunnel and other ground tests on an SR-1 (single-rotating) model to investigate how sweep affected propfan performance and noise at speeds of Mach 0.8. The tests resulted in a new type of rotary thruster with extremely thin blades that swept away from the direction of rotation. The researchers conducting the ground tests found that this system could provide jetliner speed at fuel savings of perhaps 30 percent or more if driven by an advanced type of engine.[14] The propfan, as the new concept was called, had eight or more thin, highly swept blades, unlike conventional turboprops, which had up to four straight, large-diameter blades (Figure 3–4).

---

[11]Roy D. Hager and Deborah Vrabel, *Advanced Turboprop Project* (Washington, DC: NASA SP-495, 1988), p. 5.

[12]The full name of the task force was the NASA Inter-Center Aircraft Fuel Conservation Technology Task Force, headed by NASA's James Kramer.

[13]John R. Facey, "Return of the Turboprops," *Aerospace American,* October 1988, p. 16. Facey was the advanced turboprop program manager for OAST at NASA Headquarters.

[14]James J. Haggerty, "Propfan Update," *Aerospace,* Fall/Winter 1986, p. 10.

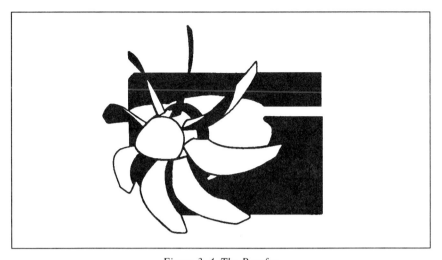

Figure 3–4. The Propfan
(Eight or ten thin, stiff, highly swept turboprop blades allowed speeds comparable to those of subsonic jet transports that were presently in use.)

Between 1976 and 1978, propfan research received only minimal funding. The efforts of a small group of engineers were rewarded, however, in 1978, when the Advanced Turboprop Project formally began, with overall project management at Lewis Research Center. The project's goal was to establish both single- and counter-rotating propfan technologies for Mach 0.65 to 0.85 applications.[15]

The first phase, which lasted through 1980, was called enabling technology. It focused on building subscale propeller models to test and establish the feasibility of the propfan concept. Researchers verified the projected performance, fuel savings, and structural integrity of the different blades under actual operating conditions. They also worked to bring the level of cabin noise and vibration to the point where passenger comfort levels approached those of turbofan-powered airliners. In addition, they verified that propfan-powered aircraft could meet airport and community noise standards as stated in the Federal Air Regulations, Part 36. Hamilton Standard's design studies evaluated the structural characteristics of several large-scale blade configurations. They also conducted preliminary flight research at Dryden to determine propfan source noise using a JetStar aircraft with a powered propeller model mounted above the fuselage and microphones implanted in the airframe.[16]

Other researchers focused on identifying the most suitable configuration for an advanced turboprop aircraft. Two basic installations were tested: the wing-mounted tractor and the aft-mounted pusher (Figure 3–5). The aim was to provide a comprehensive database to assist industry in

---

[15] Hager and Vrabel, *Advanced Turboprop Project*, p. v.

[16] NASA report to the House Science and Technology Committee, as reprinted in *Aerospace Daily,* September 8, 1981, pp. 37–38.

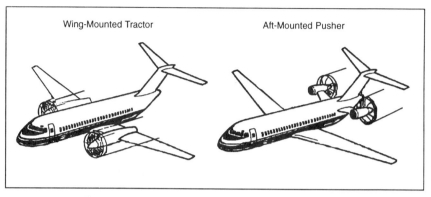

Figure 3–5. Basic Propeller Installation Configurations

choosing a configuration, including whether single- or counter-rotation best suited the application. For an effective installation, the wing and nacelle had to be integrated to avoid drag penalties and aircraft stability and control problems.

The second phase of the Advanced Turboprop Project, called large-scale integration, began in 1981. This phase concentrated on obtaining definitive data on noise at cruise conditions, fuselage noise attenuation, efficient wing mounting, and large-scale blade design. During 1981, NASA and Congress considered accelerating the program in response to strong industry interest. However, this would have required an increase in the total cost of the program, and the goal was abandoned.

In 1981, NASA selected Hamilton Standard to design large-scale, single-rotating propeller assemblies that would be suitable for flight testing. The first phase of the procurement included building a 0.6-meter-diameter aero-elastic propeller model; design and fabrication of a large-scale 2.7-meter-diameter propeller assembly with fixed pitch, ground-adjustable blades; and static and high-speed wind tunnel tests of the large-scale blade assembly. In the second phase of the contract, Hamilton Standard delivered three additional large-scale 2.7-meter-diameter, variable-pitch rotor assemblies that were used for additional static and wind tunnel tests.

During FY 1983 and FY 1984, Congress added $15 million to the amount allocated for the program in anticipation of a contract for the propeller test assembly. In March 1984, NASA selected Lockheed-Georgia Company as the prime contractor, responsible for the overall design of a flight test vehicle and supervision of an industry team that included Hamilton Standard, Allison Gas Turbine Division of General Motors, Rohr Industries, Gulfstream Aerospace Corporation, and Lockheed-California Company. Lewis Research Center was assigned management responsibility for development of the Propfan Test Assessment (PTA) technology effort designed to provide generic data on propfans for dissemination to industry.

In October 1985, before the test assessment program formally began, Lewis tested the new, highly loaded, multi-bladed propellers (called

SR-7A) for use at speeds up to Mach 0.85 and at altitudes compatible with commercial air support system requirements in the Lewis transonic wind tunnel. Using the hardware from an earlier propfan, Hamilton Standard engineers built the first Large-scale Advanced Propfan (LAP) assembly (SR-7L) and tested it at Wright-Patterson Air Force Base in a static propeller test rig they designed and built. Testing of the 2.7-meter-diameter propfan, powered by an electric drive motor in the rig test, began in late August. The propfan assembly completed the test in good mechanical condition. In November 1985, the propeller was shipped to Hamilton Standard to be prepared for the high-speed wind tunnel tests.

High-speed testing was conducted in the S1 wind tunnel at Modane, France. NASA used this tunnel because it was large enough to test the full 2.7-meter-diameter assembly at Mach 0.8 and at 3,658-meter-altitude conditions. The final test series on the SR-7A was performed in the Lewis transonic wind tunnel in early 1987. The tests recorded performance data and completed the high-speed acoustic measurements. The data agreed favorably with predictions and with earlier data.[17]

After static testing was completed under the LAP project, NASA used the SR-7L propfan for further evaluation as part of a complete turboprop propulsion system in the Propfan Test Assessment (PTA) project, under a contract with Lockheed-Georgia. The objectives of this project were to verify the structural integrity of the blading; evaluate the acoustic characteristics of a large-scale propfan at cruise conditions; test the compatibility of the engine, fan, and nacelle; measure propulsion system performance; and acquire data on propulsion system temperatures and stresses.[18]

The PTA project formally began in the summer of 1986 with fifty hours of static testing conducted at a Rohr Industries facility in California. All test objectives were met—the propulsion system functioned according to design, all control systems operated satisfactorily, and the flight instrumentation system operated as planned. Propfan blade stresses and propulsion system temperatures, pressures, and vibrations were within specified limits, and specific fuel consumption was better than expected. The static tests successfully cleared the propulsion system for flight tests.[19]

While NASA was pursuing propfan research in the direction of a single-rotation tractor system, General Electric (GE) Company submitted an unsolicited proposal for a counter-rotation blade concept. In November 1983, Lewis Research Center awarded GE a $7.2 million contract for aircraft propulsion technology research based on modern counter-rotation blade concepts. This approach for a gearless, dual-rotation pusher propulsion system was known as the Unducted Fan, or UDF™. The UDF had two counter-rotating external fans, each with eight sweptback blades

---

[17] Hager and Vrabel, *Advanced Turboprop Project*, p. 54.
[18] Haggerty, "Propfan Update," p. 11.
[19] Hager and Vrabel, *Advanced Turboprop Project*, p. 67.

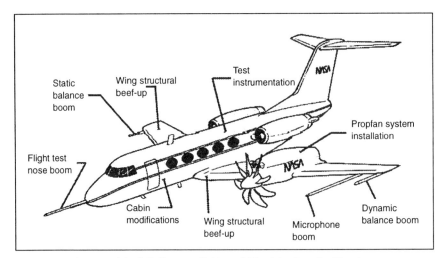

Figure 3–6. Modified Gulfstream II Aircraft Used for Propfan Test Assessment

driven directly by a counter-rotating internal turbine. This gearless design eliminated the weight of a gearbox and its oil cooling system. The UDF had a design rating of 111,200 newtons thrust—a power level intended for commercial transports in the 100- to 160-passenger range.

Model testing began in October 1984 at Lewis and at a Boeing facility. In August 1985, in cooperation with NASA, GE began an extensive ground test program on a full-scale demonstrator engine. The tests, which covered 100 hours and 100 flight cycles, concluded in July 1986. They included successful tests at thrust ratings above the design level and demonstrated a specific fuel consumption rate 20 percent better than for the turbofans then available. Following culmination of the tests on the proof-of-concept engine, GE started assembling a second prototype engine that flew on a McDonnell Douglas MD-80 transport in May 1987. The UDF used 40 to 50 percent less fuel than the engine it replaced. Cabin noise could be kept to less than that of the standard MD-80.[20]

After completing the ground tests, both the LAP and the UDF propulsion systems underwent flight tests. The LAP was tested in a wing-mount installation on a modified Gulfstream II testbed aircraft under a NASA-contracted program with Lockheed (Figure 3–6). Testing took place in May 1987 at Lockheed-Georgia's facility. The UDF was tested as an aft-mounted pusher on a Boeing 727 as part of a GE/Boeing cooperative program. These flight tests began in August 1986 at GE's Mojave, California, test facility. The tests evaluated the structural integrity of the blades and measured the noise both inside and outside the Gulfstream II testbed.

In 1987, three series of flight tests verified the readiness of advanced turboprop propulsion technology for commercial engine systems development. The flight tests included the NASA/GE/Boeing tests of the UDF

---

[20]Facey, "Return of the Turboprops," p. 19.

engine on a B-727 aircraft, the NASA/Lockheed PTA of a single-rotation advanced turboprop on a Gulfstream II aircraft, and GE/McDonnell Douglas flight tests of the UDF on an MD-80 aircraft.

NASA continued to work with Lockheed to prove in flight that large, unducted propellers with a radically swept design were a feasible alternative to higher-cost turbofan propulsion systems. Flight tests held in March 1988, at Lockheed facilities in Georgia, examined ways to reduce interior noise levels. Research data were recorded simultaneously for more than 600 parameters using instrumentation such as microphones and accelerometers, strain gauges, temperature, and pressure-measurement gauges.

The final flights in the PTA project were held during May and June 1988. Instruments measured instantaneous pressure on propfan blade surfaces at several flight speeds with a range of power settings on the eight-bladed propfan. After these tests ended, the aircraft were delivered to Johnson Space Center, where the advanced turboprop system was removed and the aircraft modified to a Shuttle training aircraft. The PTA project ended in June 1989.

Other Flight Efficiency Activities

Supercritical Wing/Mission Adaptive Wing. The supercritical wing was a design concept envisioned by Dr. Richard T. Whitcomb, a research engineer at NASA's Langley Research Center, during the 1960s. Whitcomb developed wing shapes that he theorized would make a transonic aircraft much more fuel efficient, either increasing its speed or range or decreasing the amount of fuel it consumed.[21] During the early 1970s, his concepts were tested on an F-8A Crusader at Dryden Flight Research Center.

When increases in the price of oil refocused research efforts more on efficiency than on speed, Whitcomb modified his supercritical wing design for maximum aerodynamic efficiency. The modified wing was one way of improving the lift-drag ratio. The unusual airfoil section controlled the flow over the wing; it avoided the sudden increase in drag that would occur with conventional airfoils operating in high-speed airflow. In addition, it showed this lower drag feature in spite of an increased thickness of the wing section. Consequently, a properly designed supercritical wing reduced wing drag, increased the internal volume for fuel storage, increased the structural efficiency of the wing, and led to lower weight. It showed the potential for fuel savings of 10 to 15 percent, and the design was incorporated into many transport airplanes.

The military also used a supercritical wing on a General Dynamics F-111 aircraft to see how it might benefit military aircraft in its Transonic

---

[21]Lane E. Wallace, *Flights of Discovery: 50 Years at the NASA Dryden Flight Research Center* (Washington, DC: NASA SP-4309, 1996), p. 90.

Aircraft Technology (TACT) program, which began in 1972. Test results showed that a supercritical wing could improve aircraft performance. The F-111 TACT kept flying through the early 1980s, testing different drag-reducing aerodynamic modifications.

The C-17 transport, as well as other military transports, also used the supercritical wing. This wing design enhanced the range, cruising speed, and fuel efficiency of the aircraft by producing weaker shock waves that created less drag and permitted high efficiency.[22]

The TACT program provided impetus to further wing research under NASA's Advanced Fighter Technology Integration (AFTI) program. The initial AFTI experiment was the Mission Adaptive Wing (MAW), built by Boeing under a $24 million contract from the Air Force Flight Dynamics Laboratory. The MAW was tested on a modified General Dynamics F-111 TACT aircraft at NASA's Dryden Flight Research Facility. The F-111 AFTI flight research program focused on four automatic modes: cruise camber control, maneuver enhancement/gust alleviation, maneuver camber control, and maneuver load control. It ran from 1985 to 1988.

Internal hydraulic actuators in the MAW flexed the composite-covered aircraft wing to adjust the amount of its camber (curvature), depending on flight conditions. It could flex enough to generate the additional lift needed for slow speeds, eliminating the need for lift-producing devices such as slats and flaps. It could then change to a supercritical wing platform for transonic flight and adjust to a near-symmetrical section for supersonic speeds. The smooth, variable camber wing was expected to yield a 25- to 30-percent improvement in aircraft range and more capability for tight evasive maneuvers. It was also expected to result in increased fatigue life, improved handling, and a more stable weapons platform. Tests indicated that the drag reduction from a MAW design would have 25 percent more range for a low-altitude mission and 30 percent more range for a high-altitude mission. Mission load factors could also be 20 to 30 percent better.[23]

Winglets. Winglets are small, nearly vertical fins installed on an airplane's wing tips to help produce a forward thrust in the vortices that typically swirl off the end of the wing, thereby reducing drag. Whitcomb investigated winglet aerodynamics that matured into an applicable technology. He tested several designs in the wind tunnels at Langley Research Center and chose the best configuration for a flight research program. The concept was demonstrated in flight on a corporate Gates Model 28 Longhorn series Learjet and further tested on a large DC-10 aircraft as part of the ACEE program.

NASA installed winglets on a KC-135A tanker, on loan from the Air Force, and flight-tested it at Dryden Flight Research Center in 1979 and

---

[22]"NASA Contributions to the C-17 Globemaster III," *NASA Facts*, FS-1996-05-06-LaRC (Hampton, VA: Langley Research Center, May 1996).

[23]Remarks by Louis Steers, director of NASA's MAW effort, speaking at an industry briefing session on the AFTI/F-111 program, printed in *Antelope Valley Press,* August 4, 1988.

1980. The research showed that the winglets could increase an aircraft's range by as much as 7 percent at cruise speeds. The first industry application of the winglet concept was in general aviation business jets, but winglets were also incorporated into most new commercial and military transport jets.[24]

Laminar Flow Research. One problem for modern civil air transports traveling at about 800 kilometers per hour occurs in the boundary layer, a thin sheet of flowing air that moves along the surfaces of the wing, fuselage, and tail of an airplane. At low speeds, this layer follows the aircraft contours and is smooth—a condition referred to as "laminar." At high speeds, the boundary layer changes from laminar to turbulent, creating friction and drag that wastes fuel. It was estimated that the maintenance of laminar flow over the wing and tail surfaces of long-range transports could reduce fuel consumption by 25 percent or more. Researchers developed three methods for increasing laminar flow and controlling the behavior of laminar/turbulent boundary layers:

1. Natural laminar flow, which reduced skin-friction drag by shaping and passive control
2. Laminar flow control and hybrid laminar flow control, which reduced skin-friction drag by combined shaping and active control
3. The development of low Reynolds-number airfoils, which reduced pressure drag by shaping with and without passive or active control.[25]

NASA conducted natural laminar flow experiments on the variably swept-wing F-111 during the late 1970s. These experiments investigated how changing the sweep of a wing affected the degree of its laminar flow. Research in the early 1980s, using a Navy Grumman F-14 Tomcat, investigated sweep angles greater than those found on the F-111. This research told investigators how much sweep could be incorporated into a subsonic wing before it began to lose its laminar flow properties.[26]

The laminar flow control concept called for maintaining laminar flow by removing the turbulent boundary layer by suction (Figure 3–7). Suction required developing porous or slotted aircraft surfaces and lightweight pumping systems.[27] The concept had been well established,

---

[24]Wallace, *Flights of Discovery,* p. 93.

[25]William D. Harvey, Head, Fluid Dynamics Branch, Transonic Aerodynamics Division, NASA Langley Research Center, "Boundary-Layer Control for Drag Reduction," paper presented at the First International Pacific Air and Space Technology Conference, Melbourne, Australia, November 1987, pp. 2, 9. The Reynolds number is a ratio used to calculate flow characteristics; it is useful in characterizing a flow in a simulated environment, such as a wind tunnel.

[26]Wallace, *Flights of Discovery,* p. 95.

[27]"Aircraft Energy Efficiency Program: Laminar Flow Control Technology," *NASA Facts,* NF-86/8-79 (Washington DC: U.S. Government Printing Office, 1979).

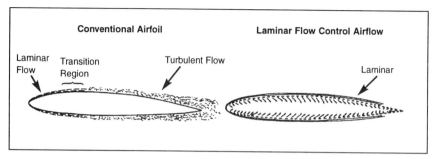

Figure 3–7. Laminar Flow Control Through Suction

verified in wind tunnel tests, and demonstrated in various flight tests, particularly the X-21 flight research program performed by the Northrop Corporation in the 1960s under an Air Force contract. This program demonstrated that under controlled conditions, laminar flow could be established and maintained over essentially the entire wing surface where suction was applied. The Laminar Flow Control Project, which began in 1976, demonstrated that the technology was ready for practical application to commercial transports during the next decades. Figure 3–8 shows some of the concerns regarding the implementation of laminar flow control on a typical aircraft.

The program continued in the early 1980s, and researchers at Langley Research Center predicted that modern construction techniques would allow full-size wings to be built that approached the smoothness of highly accurate wind tunnel scale models and flight test wings. During 1982,

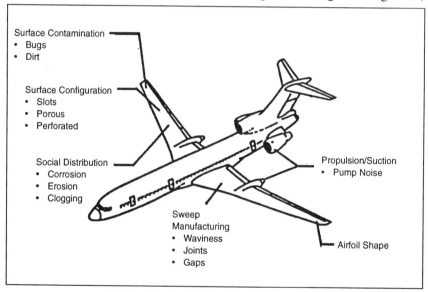

Figure 3–8. Factors Affecting Laminar Flow
(Bugs and dirt can contaminate the surface. Corrosion, erosion, and clogging can affect the airflow through the tiny slots. Manufacturing irregularities can reduce the effectiveness of the system.)

flight tests at Langley substantiated those predictions. The aircraft had either bonded wing skins made of aluminum, integrally stiffened milled skins of aluminum, or skins made of composites. The research also provided data on the effect of other factors on subsonic laminar flow, such as aircraft speed and insects being splattered on aircraft wings. The researchers found that wherever insects hit and stuck to the surface of the aircraft, they interfered with the smoothness of the boundary layer of air.

In 1985, NASA installed two experimental laminar flow control devices on its business-size JetStar aircraft that incorporated techniques to help prevent leading-edge contamination—that is, disturbance of the laminar flow by insects, ice, and other obstructions adhering to the leading edges of an aircraft's wings. The flights took place in widely separated areas of the United States to experience a wide variety of contaminant conditions.

Following this research, in September 1987, NASA selected Boeing to provide data on the aerodynamic and operational effectiveness of a hybrid system to achieve laminar airflow control at flight conditions representative of high subsonic speeds of commercial and military transport airplanes. During the three-year program, designers developed a shield for use while an aircraft was close enough to the ground to encounter insects, ordinarily within 300 meters of the ground. This shield, called a Krueger flap, folded flush against the wing's lower surface when not in use, but when extended forward and upward, it became the leading edge of the wing—the part that encountered the insects. On a Boeing 757 used to demonstrate the flap, the suction system used with the flap not only created laminar flow over the leading part of the wing but also demonstrated that laminar flow continued behind it to cover 65 percent of the distance to the trailing edge.

Beginning in the late 1980s, NASA started examining laminar flow on aircraft traveling at supersonic speeds. NASA acquired two F-16XL aircraft and began research flights at Dryden Flight Research Center in 1991 in a joint activity with Rockwell.

Riblets. Riblets also reduce drag-producing air turbulence and increase fuel efficiency. Investigators at Langley Research Center discovered in 1984 wind tunnel experiments that barely visible grooves, each shaped like a tiny "v," on the surface of an airplane, no more than two-thousandths of an inch deep, would favorably alter the turbulent flow of air that formed over the surface of a moving airplane. The 3M Company used this technology to design and produce test specimens of riblets in tape form with an adhesive backing that would be pressed into place on an aircraft's surface.

Oblique Wing Research. NASA's oblique wing research successfully demonstrated an aircraft wing that could be pivoted obliquely from zero to sixty degrees during flight. The wing was demonstrated on a small subsonic jet-powered research aircraft called the AD-1 in a program conducted between 1979 and 1982. The first sixty-degree angle skew was reached on its twenty-third flight on April 24, 1981.

The oblique wing concept originated in 1945 with Robert T. Jones, at NASA's Ames Research Center, but the idea was not pursued until the late 1960s. Analytical and wind tunnel studies indicated that a transport-size oblique wing flying at 1,600 kilometers per hour might achieve twice the fuel economy of more conventional wings.[28] The studies stated that, at high speeds, pivoting the wing up to sixty degrees would decrease aerodynamic drag, permitting increased speed and longer range with the same fuel expenditure. At lower speeds, during takeoffs and landings, the wing would be perpendicular to the fuselage like a conventional wing to provide maximum lift and control qualities. As the aircraft gained speed, the wing would be pivoted to increase the oblique angle, thereby reducing the drag and decreasing fuel consumption.[29]

NASA demonstrated the concept on the AD-1 aircraft, which was delivered to Dryden Flight Research Center in February 1979. During seventy-nine flights that took place over eighteen months, the wing was pivoted incrementally until the full sixty-degree angle was reached in 1981. The aircraft continued to be flown for another year, obtaining data at various speeds and wing pivot angles until the final flight in August 1982. Although successful, the concept had not been incorporated into any production aircraft at the time this volume went to press.

NASA began a follow-up program to the AD-1 in the early 1980s. The goal of the program was to modify the NASA F-8 digital fly-by-wire research aircraft to a supersonic oblique wing configuration. In 1983, researchers completed a feasibility study of the oblique wing concept operating at supersonic speeds. In November 1984, NASA solicited proposals for the preliminary design phase of a joint NASA-Navy program to design, construct, and evaluate an oblique wing during supersonic flight research conditions. The solicitation marked the second phase of a planned four-phase program that was also to define the aircraft's expected flight performance and determine the potential operational capabilities for Navy applications. Phase 3 was to include detailed design work, fabrication, and ground testing of a composite, aero-elastically tailored oblique wing. The composite wing was to be constructed so that the bending stresses of flight would not degrade the wing's aerodynamic efficiency. Phase 4 was to consist of a flight support contract to the Phase 3 contractor. A twelve-month, approximately forty-flight test program was planned to take place in 1986 and 1987 from Dryden.[30]

Rockwell received the contract for the design work. However, the work did not progress beyond the design stage. The Navy canceled the program near the end of the second phase just before the modifications were set to begin, and the modifications to the F-8 never took place.

---

[28] Wallace, *Flights of Discovery,* p. 94.

[29] "The AD-1," *NASA Facts On-Line,* Dryden Flight Research Center, November 1994.

[30] "NASA Seeks Design for Supersonic Oblique Wing Testbed," *Aviation Week & Space Technology,* December 3, 1984.

Powered Lift Technology. Powered lift enables aircraft to operate from short or reduced-length runways because the aircraft can take off and land vertically or after traveling only a short distance on the ground. Variations of this technology include short takeoff and landing (STOL), vertical takeoff and landing (VTOL), vertical/short takeoff and landing (V/STOL), and short takeoff and vertical landing (STOVL) aircraft. Most of these technologies appeared on some variation of rotorcraft. However, the QSRA and the C-17 Globemaster III also incorporated STOL technology.

The subsonic STOL aircraft's enhanced in-flight capabilities include steep-gradient and curved-flight departures and approaches, high rates of climb, steep final descents, high maneuverability, rapid response for aborted landing, and low landing approach airspeeds. These characteristics allow for aircraft that:

- Require less airspace in the near-terminal area
- Require less ground space at the terminal
- Operate in smaller spaces relatively quietly
- Have improved crashworthiness and survivability because of their low-speed capability at near-level fuselage attitudes
- When equipped with modern avionics, can operate in very low visibility in adverse weather[31]

These aircraft are useful in both civilian and military situations. STOL concepts investigated by NASA included the augmentor wing and the upper-surface-blown flap and research with the four-engine QSRA.

Quiet Short-haul Research Aircraft. The QSRA originated as a proof-of-concept vehicle and a research tool. It was designed to demonstrate new forms of lift that researchers believed might one day be used in commercial and STOL aircraft. It validated the technology of a propulsive lift system that used upper-surface blowing.

The QSRA program began in 1974. NASA obtained an aircraft and several high-bypass-ratio geared engines at no cost for use in the program. Boeing assembled the aircraft, and rollout occurred on March 31, 1978. The initial flight testing for airworthiness took place at Boeing, and the aircraft was then delivered to Ames Research Center in August 1978, where a flight evaluation was conducted.

The high-performance STOL characteristics resulted from its new moderately swept wing, designed and built by Boeing. It incorporated the upper-surface-blowing propulsive-lift technique in its design. Four acoustically treated jet engines were mounted on top of the wing so that the fan air from the engines was directed across the upper

---

[31]W.H. Deckert and J.A. Franklin, Ames Research Center, *Powered-Lift Aircraft Technology* (Washington, DC: NASA Office of Management, Scientific and Technical Information Division, 1989), p. 3.

surface of the wing and flaps to create very high levels of lift as compared to conventional wings. The design gross weight of the aircraft was 22,700 kilograms. Even with four turbofan engines, it could operate at lower noise levels than most current small business jet airplanes—an attractive feature.

In June and July 1980, NASA and the U.S. Navy used the QSRA for more than 500 landings on a simulated aircraft carrier deck in an investigation of the application of propulsive-lift technology to aircraft carriers. This was followed by the initiation of a joint NASA-Navy program that used the QSRA to evaluate the application of advanced propulsive-lift technology to naval aircraft carrier operations. This consisted of thirty-six "touch and go" landings and sixteen full-stop landings and takeoffs. The aircraft demonstrated new technology for quieter jet engine operations while also providing the performance for operations from airports with very short runways. The QSRA also successfully completed a forty-three-flight evaluation program in January 1981, at Ames Research Center, where test pilots made short runway landings with malfunctions in the aircraft that were intentionally created.

The Quiet, Clean, Short-haul Experimental Engine (QCSEE) was a related development. Test runs began at Lewis Research Center in the late 1970s. The goal of the program was to produce a power plant for a four-engine 150-passenger STOL transport with a small and relatively low noise footprint. The STOL technology around which NASA developed the QCSEE used the engine exhaust to produce lift. In one case, the exhaust was blown directly over external flaps to produce the added lift for STOL. In the other, part of the bypass air was ducted to blow over the upper surface of the wing to generate additional lift. Both of these engine types were built and successfully tested.

C-17 Globemaster III. The first C-17 Globemaster III rolled off the assembly line in 1991—the culmination of a lengthy process that began in 1979 when DOD began its Cargo-Experimental program. In 1981, the Air Force selected McDonnell Douglas as the manufacturer of the aircraft. The company used NASA-derived technologies to produce the aircraft.

The aircraft used a powered lift system, or "externally blown flap," that enabled the aircraft to make slow, steep approaches with heavy cargo loads. The steep approach helped pilots make precision landings. This was accomplished by diverting engine exhaust downward, giving the wing more lift. In this system, the engine exhaust from pod-mounted engines impinged directly on conventional slotted flaps and was deflected downward to augment the wing lift. This allowed aircraft with blown flaps to operate at roughly twice the lift coefficient of conventional jet transport aircraft. Researchers studied the concept extensively in wind tunnels at Langley Research Center, including tests of flying models in the nine-meter by eighteen-meter tunnel. The Air Force procurement specification included a STOL capability. Researchers investigated this capability on flight simulators and the Augmentor Wing Research Aircraft at Ames Research Center.[32]

Subsonic V/STOL applied concepts that used a lifting rotor, a tilt-rotor, and the X-wing configuration. The military used subsonic V/STOL technology in its Harrier aircraft. Civil opportunities for subsonic V/STOL aircraft included:

- Ocean resource operations, with "terminals" on oil rigs, ships, and mineral exploration platforms
- Direct city center to city center transportation
- Direct corporate office to factory service
- Transportation for underdeveloped countries
- Transportation for inaccessible communities
- Search and rescue
- Emergency medical services
- Disaster relief

NASA used its National Full-Scale Aerodynamics Complex wind tunnels at Ames Research Center to determine the low- and medium-speed aerodynamic characteristics of high-performance aircraft, rotorcraft, and fixed-wing powered-lift V/STOL aircraft.

Powered Lift Rotorcraft Research. NASA and DOD also developed several rotary-wing-based aircraft that used powered lift technology. These included the XH-59A, advancing blade concept aircraft during the 1970s, the JVX or tilt-rotor aircraft, and the RSRA/X-wing aircraft. These aircraft had the common ability to take off and land vertically like a helicopter, but in flight, they used a variety of technologies to operate as conventional fixed-wing aircraft.

Rotor Systems Research Aircraft/X-Wing Program. Jointly funded by NASA and the U.S. Army, the RSRA aircraft program began in the early 1970s. The program investigated ways to increase rotor aircraft speed, performance, reliability, and safety and to reduce helicopter noise, vibration, and maintenance. There were two aircraft in the program manufactured by Sikorsky Aircraft Division, United Technologies Laboratories, for Langley Research Center. After initial flight testing at Langley, the two aircraft were transferred to Ames Research Center for an extensive flight research program by Ames and the Army.

The RSRA could be configured to fly as a helicopter or as a compound helicopter and could be fitted with a variety of experimental and developmental rotor systems for research purposes. The compound configuration had fixed wings providing a portion of the needed lift and auxiliary jet engines; it could accommodate rotor systems too small to support the aircraft. Table 3–62 compares the helicopter and compound configurations.

---

[32]"NASA Contributions to the C-17 Globemaster III."

A unique rotor vibration isolation system prevented the transmission of main rotor vibrations to the fuselage structure. This allowed for the installation of various rotor systems with a wide range of vibration characteristics without modifying the fuselage. At the same time, the system provided precise measurements and control of rotor forces and of aircraft maneuvering flight parameters over a wide range of operating conditions.[33]

NASA and DARPA initiated a follow-up program to investigate the X-wing concept. Sikorsky was selected in early 1984 to work with NASA and DARPA on converting one of the two RSRAs to a demonstrator aircraft for the X-wing concept. It was envisioned that the four-blade X-wing would operate like a standard helicopter rotor for vertical and low-speed flight, but could be stopped and function as a wing for high-speed forward flight. It was expected that X-wing technology would lead to rotorcraft that could operate at greater speeds and altitudes than existing helicopters.[34]

The modified RSRA airframe could be configured in three flight modes: fixed wing (airplane), helicopter, and compound. In the compound mode, the RSRA could transition between fixed-wing and helicopter configurations. For fixed-wing configuration taxi and flight testing, the tail rotor would remain in place, attached to the rudder pedals for yaw control. The main rotor system would be removed. In the helicopter configuration (X-wing), twin GE T58–GE–5/100 gas turbine engines powered the rotor system.[35]

One of the two RSRA Sikorskys was designated an X-wing demonstration aircraft under the contract with Sikorsky. The second RSRA was based at Dryden for fixed-wing configuration testing, which began on May 8, 1984. This marked the first time the RSRA in the compound configuration was flown in the airplane mode. The tests were conducted with the RSRA equipped with its tail rotor but no main rotor and test speeds limited to less than 463 kilometers per hour and also with the RSRA completely rotorless for higher speed flights. Tests were carried out at altitudes up to 3,000 meters. A total of thirteen tests were conducted.

The modified RSRA with the X-wing system mounted on it was rolled out on August 19, 1986, at Sikorsky's facilities in Connecticut. Although researchers did not foresee replacing conventional fixed-wing or rotorcraft with the X-wing aircraft, they envisioned that X-wing aircraft would provide enhanced capabilities to perform missions that called for the low-speed efficiency and maneuverability of helicopters combined with the high cruise speed of fixed-wing aircraft. The aircraft had a 13.7-meter, variable-incidence conventional wing that could support the full weight of the aircraft in flight. The aircraft

---

[33] "Advanced Research Aircraft," NASA Activities, May 1979, p. 11.

[34] "X-Wing Contract," *Aviation Week & Space Technology,* January 2, 1984, p. 23.

[35] "NASA Nears Fixed-Wing Tests on RSRA Research Aircraft," *Aviation Week & Space Technology,* January 30, 1984, p. 54.

was expected to demonstrate convertibility from rotary to wing-borne flight and to efficiently combine the vertical lift and stable hover characteristics of conventional helicopters with the high cruise speed of fixed-wing aircraft.[36]

The aircraft was twenty-one and a half meters long by five and a half meters high and had a five-blade tail rotor just over three meters in diameter. The design gross weight was 15,093 kilograms. Power for the X-wing/RSRA rotor came from two T58-GE-10 engines. Two TF34-GE-400As provided thrust for forward flight.[37]

The X-wing used a four-bladed helicopter-like rotor system that would rotate for takeoffs, landings, and low-speed flight. The rotor system would be stopped in flight at speeds of approximately 281 to 370 kilometers per hour to act as a fixed x-shaped wing for high-speed flight. In the x-shape, two blades would be swept forward at forty-five-degree angles, and two would be swept to the rear at the same angles. The prime objective of the program was the successful demonstration in flight of conversion of the rotor-wing system from fixed to rotating and back again.

A computer-controlled air-circulation control system would provide lift. It would first be used with the X-wing not rotating and then rotating. As testing proceeded, rotor turning and circulation control development would enable researchers to gradually achieve more lift with the rotor rather than depending on the aircraft fixed wing. Advanced composite materials were used in the four-rotor/wing blades.[38]

Plans called for the RSRA/X-wing aircraft to be flown in the fall of 1986 first as a fixed-wing aircraft without the rotor and then with the X-wing installed in a fixed position. The next phase would include full operation of the X-wing blowing systems, with the rotor stationary in normal horizontal flight. Ground testing of the X-wing in rotary mode would follow, and then conversion test flights between rotary and horizontal flight modes would complete the program. However, the tests were delayed by a series of technical problems linked to design changes and the extensive reassembly required after the aircraft was shipped from Sikorsky in Connecticut to Edwards Air Force Base in California.

The delays and cost overruns led NASA and DARPA to scale down the X-wing flight test program in August 1987 to a low-level research effort that concentrated on basic research objectives and postponed the demonstration of conversion from rotary to fixed-rotor flight modes. The conversion demonstration would have required the development of complex digital computers and software, and developing the flight hardware

---

[36]"NASA/DOD Hybrid Research Aircraft Rolled Out," *NASA News,* Release 86-113, August 19, 1986.

[37]"Sikorsky Rolls Out X-Wing Demonstrator," *Aviation Week & Space Technology,* August 25, 1986, p. 19.

[38]"X-Wing Research Aircraft Set for Delivery to NASA," *NASA News,* Release No. 86-13, September 18, 1986.

for the X-wing concept proved to be far more complex than was first thought.[39]

Initial flight tests were made in November and December 1987 without the X-wing rotor. The flights evaluated the basic stability of the aircraft in the first of three rotor-off configurations. The contract with Sikorsky ended in December 1987. Further flight tests and modification work on the X-wing RSRA were halted in January 1988 while NASA and DARPA assessed the program's future.

JVX/Tilt-Rotor. The JVX/tilt-rotor program was NASA's second primary research effort involving rotorcraft. NASA contributed to the JVX program through the transfer of generic tilt-rotor technology. NASA also provided facilities and expertise to address technology issues specific to the JVX.[40]

Tilt-rotor aircraft operated as helicopters at low speeds and as fixed-wing propeller-driven aircraft at higher speeds. This permitted vertical takeoff and landing, longer cruising range, and speeds up to 640 kilometers per hour (as compared to conventional helicopters, which were limited to less than 320 kilometers per hour).

Concepts for tilt-rotor VTOL aircraft had been first studied in the late 1940s, and related investigations continued into the 1970s. During the early 1970s, the joint NASA-Army XV-15 Tilt Rotor Research Aircraft (TRRA) program began. This aircraft, developed by Bell Helicopter Textron, was a third-generation tilt-rotor V/STOL aircraft. The 12.8-meter long, 5,900-kilogram craft was powered by two 1,120-kilowatt turbine engines located in the wing tip nacelles that rotate with the rotors. The XV-15 was the first research aircraft with rotors that were designed to be tilt rotors. The XV-3 that had been designed earlier had helicopter-designed rotors that could be tilted.

By the early 1980s, tests with the XV-3 and XV-15 research aircraft and other supporting research had proven that the critical design issues could be successfully addressed. The joint NASA-Army TRRA program provided the confidence level necessary for DOD to initiate full-scale development of the JVX. The V-22 Osprey was the designation for the military version of the JVX (Figure 3–9). It was based on the Bell XV-15 tilt-rotor demonstrator.

Scale-model wind tunnel testing was conducted at Langley Research Center to investigate JVX spin characteristics and to establish aeroelastic stability boundaries for the JVX preliminary design. The Vertical Motion Simulator at Ames Research Center was used in two design and development tests to validate the JVX math model and evaluate the flight control system characteristics. Critical performance testing completed at the Ames Outdoor Aerodynamic Research Facility provided new data on hover efficiency and wing download.[41]

---

[39]"Technological Problems, Rising Costs Force X-Wing Program to Scale Down," *Aviation Week & Space Technology,* October 19, 1987, p. 23.

[40]William S. Aiken, Jr., NASA Director for Aeronautics, to Lynn Heninger, memorandum, July 9, 1985.

[41]*Ibid.*

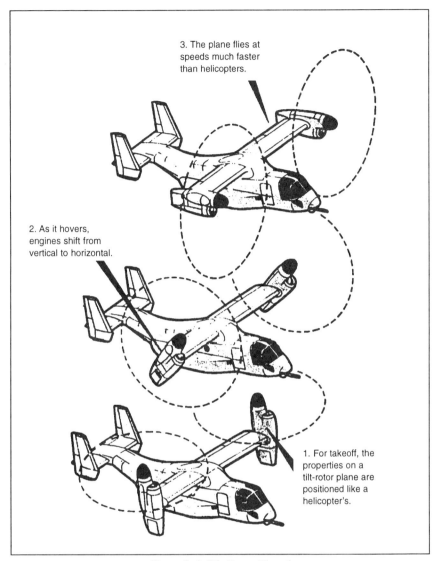

Figure 3–9. Tilt-Rotor Aircraft

NASA also provided in-house expertise, analysis routines, and basic research. Langley researchers provided improved analytical methods to industry and worked closely with the contractors to analyze the JVX wing/rotor aero-elastic coupling characteristics. They gave similar assistance in composite construction, flying qualities, performance, engine inlet design, and rotor dynamics. NASA developed computer programs that continued to be used by Bell and Boeing to make key design choices. Basic research tasks included airfoil design for an advanced technology rotor, crashworthiness concepts, fatigue analysis, cockpit integration, and an XV-15 flight evaluation of side-arm controllers.

The tilt-rotor concept had civil potential because of its VTOL- and STOL-mode capabilities, fuel efficiency, and low noise and vibration levels. An effort was made to fund a civilian version of the JVX that would enable passengers to board at special facilities near city centers and fly into the center of another city, saving commuting time and reducing congestion at major commercial airports. However, the cost was too high. The FAA estimated that a tilt-rotor aircraft built to carry thirty passengers would cost between $15 million and $19 million—two to five times the price of a comparably sized turboprop.

A related research program initiated by NASA included the design, fabrication, and flight evaluation of advanced technology blades (ATB), known as the SV-15/ATB program. Program objectives were to improve the SV-15's VTOL performance, expand the conversion envelope between helicopter and airplane modes of flight, and at least maintain cruise propulsive efficiency. The results from static (hovering) tests of the isolated-full-scale ATB rotor verified theoretical predictions. The first flight of the XV-15/ATB was in late 1987.

The objective of another XV-15 research program was to establish the viability of a three-axis sidearm controller as a primary controller for tilt-rotor aircraft. The first flight with the sidearm controller occurred in June 1985. Ongoing research with the XV-15 included support for the V-22 Osprey tilt-rotor program, flight evaluation of new tilt-rotor steel hubs, and more complete determination of rotor downwash characteristics, documentation of handling qualities, and STOL performance.

Advanced Short Takeoff and Vertical Landing (ASTOVL). The ASTOVL program was a cooperative research effort between the United States and the United Kingdom. NASA, DOD, and the United Kingdom signed a memorandum of understanding in February 1986 to proceed with a research program to investigate various propulsion concepts. The program would assess the relative potential as well as the joint research required for advancement of these technologies to future ASTOVL aircraft. The program aimed to reduce the technological risk associated with potential future ASTOVL combat aircraft. Those aircraft would have the capabilities of an advanced supersonic fighter aircraft with the added advantage of landing vertically when necessary.

NASA awarded contracts to Allison, General Electric, and Pratt & Whitney to evaluate the four propulsion concepts. NASA and DOD also awarded contracts to study airframe design to General Dynamics, Grumman, Lockheed, and McDonnell Douglas. British participants included British Aerospace and Rolls Royce. Reviews of the four concepts were held late in 1987 and in 1988. Remote augmented lift systems and ejector augmentors were selected for further studies early in 1989.

Aircraft Control With Computerized Aircraft Systems. The digital fly-by-wire (DFBW) system replaced conventional mechanical flight controls with an electronic flight control system that was coupled with a digital computer. It allowed the control surfaces of an aircraft to be operated electronically through a computer system. The pilot would move the aircraft's stick, which sent a command to the flight control computer. The computer would calculate the necessary control surface movements and send a command to the actuator to move the control surfaces. The development and early tests of the system occurred during the 1970s.

Draper Laboratory, which had developed an extensive software development process for the Apollo program, developed the flight-critical software for the DFBW program. Dryden engineers, in turn, adapted Draper's methods to develop all the subsequent flight control system software used at the center.[42]

The first DFBW flight occurred in May 1972, using an F-8C research aircraft and a single Apollo digital computer with an analog backup. This phase of the DFBW program validated the fly-by-wire concept and showed that a refined system—especially in large aircraft—would greatly enhance flying qualities by sensing motion changes and applying pilot inputs instantaneously.

Phase II of the program began in 1973. During this phase, developers replaced the Apollo hardware with a triply redundant digital computer system, the IBM AP 101, which would be more like a system that industry would use and which was also selected for the Space Shuttle control system. Computer synchronization, redundancy management, and the demonstration of data bus concepts that reduced the amount of hard-wiring necessary in the control system were also developed during Phase II of the DFBW program.[43]

The F-8 was also used both to get the "bugs" out of the AP-101 computer and to remedy a problem that pilots encountered on the fifth approach and landing test of the unpowered Space Shuttle *Enterprise* in October 1977. Pilot-induced oscillation can occur on computerized control system aircraft because the linkage is no longer direct between the pilot's control stick and the control surfaces. This results in a greater possibility that the pilot's input and the aircraft's response will become unsynchronized. The human tendency is to respond to what is seen, and a pilot's actions can "fight" an aircraft's control system, causing overcontrol and unplanned movement, sometimes at a dangerous level.

When the problem appeared during the approach and landing test, NASA scheduled an additional series of test flights with the F-8 and other aircraft to try to replicate the problem and experiment with solutions. These tests occurred in March and April 1978 and provided needed data to develop a solution, a P10 suppression filter.[44] The Shuttle was launched beginning in 1980, using DFBW for descent, approach, and landing maneuvers and experiencing a perfect safety record in this part

---

[42]Wallace, *Flights of Discovery,* p. 114.

[43]*Ibid.,* p. 115.

[44]James Tomayko, "Digital Fly-by-Wire: A Case of Bidirectional Technology Transfer," *Aerospace Historian,* March 1986, pp. 15–18.

of the flight. In 1978, the F-18 Hornet became the first production DFBW aircraft.[45]

Farther into Phase II, in August 1984, the F-8 aircraft was given resident backup software technology designed to tolerate errors in its digital control system without the use of analog or hardware backup. Early flight tests were successful.

The DFBW program lasted 13 years. The 210th and final flight of the program took place on April 2, 1985. The F-8 program proved the feasibility of DFBW aircraft and gave the technology enough credibility to encourage industry to incorporate computerized flight control systems in new aircraft designs, such as the later models of the F-16 and the Boeing 777.[46]

Throughout the 1980s, researchers continued to improve and use DFBW technology. The X-29 high-performance research aircraft, flown from 1984 through 1992, used DFBW technology in its flight control system to sense flight conditions (including aircraft attitude, speed, and altitude), to process this information, and to continually adjust the control surfaces, transmitting up to fifty commands a second to provide artificial stability for the aircraft, which had an inherently unstable design. The X-29 used a triply redundant three-computer digital system, each with analog backups. If one digital system failed, the remaining two would take over. If two digital computers failed, the flight control system would switch to the analog mode. If one of the analog computers failed, the two remaining analog computers would take over. The risk of failure in the X-29's system was less than the risk of a mechanical failure in a conventional system. The digital system allowed relatively easy software changes to modify the "control calculations" or control laws to suit research needs or changing flight conditions.

Research during the 1970s on the Integrated Propulsion Control System, which used a General Dynamics F-111E, led to flight research with an advanced digitally controlled engine designed by Pratt & Whitney. This engine, with the Digital Electronic Engine Control (DEEC) system, was installed on Dryden's F-15 and flown from 1981 to 1983. The DEEC engines allowed engine stall-free performance throughout the entire F-15 flight envelope, faster throttle response, improved air-start capability, and an increase of 305 meters of altitude in afterburner capability.[47]

A follow-up effort to DEEC research mixed a digital jet engine control system, a mated digital flight control system, an on-board general-purpose computer, and an integrated architecture that allowed all components to communicate with each other. A modified F-15 jet aircraft performed the first flight of the Highly Integrated Digital Electronic

---

[45]Wallace, *Flights of Discovery,* p. 116.
[46]*Ibid.*
[47]*Ibid.*, p. 120.

Control (HIDEC) system on June 25, 1986, from Dryden. It marked the first time such large-scale integration efforts were attempted in aircraft systems. The HIDEC F-15 also had a dual-channel, fail-safe digital flight control system programmed in Pascal. It was linked to the Military Standard 1553B and an H009 data bus that tied all other electronic systems together. The HIDEC technology permitted researchers to adjust the operation of the engines to suit the flight conditions of the aircraft. This extended engine life, increased thrust, and reduced fuel consumption. HIDEC also added flight control information such as altitude, Mach number, angle of attack, and sideslip. The HIDEC system actively adapted to varying flight conditions, allowing the engine to operate closer to its stall boundary to gain additional thrust.

The Advanced Digital Engine Control System (ADECS) also used the F-15. This system traded excess engine stall margin for improved performance that was achieved through the integrated and computerized flight and engine control systems. The engine stall margin—the amount that engine-operating pressures must be reduced to avoid an engine stall—was continually monitored and adjusted by the integrated system, based on the flight profile and real-time performance needs.

Using this information, ADECS freed up engine performance that would otherwise be held in reserve to meet the stall margin requirement. Improved engine performance obtained through ADECS could take the form of increased thrust, reduced fuel flow, or lower engine operating temperatures because peak thrust was not always needed.

The initial ADECS engineering work began in 1983. Research and demonstration flights with ADECS began in 1986. These flights displayed increases in engine thrust of 10.5 percent and up to 15 percent lower fuel flow at constant thrust. The increased engine thrust observed with ADECS improved the aircraft's rate of climb 14 percent at 12,192 meters, and its time to climb from 3,048 meters to 12,192 meters was reduced 13 percent. Increases of 14 percent and 24 percent, respectively, in acceleration were also experienced at intermediate and maximum power settings. No stalls were encountered during even aggressive maneuvering, although intentional stalls were induced to validate ADECS methodology.[48]

High-Performance Aircraft

High-performance aircraft technologies were generally developed to support military objectives. DOD— and particularly its research arm, DARPA—often generated these efforts and usually also contributed at least part of the funds. However, because NASA had a hand in the technology development, the technologies were sometimes also transferred to

---

[48]"F-15 Flight Research Facility," *NASA Facts On-Line*, FS-1994-11-022-DFRC, Dryden Flight Research Center, November 1994.

the civilian sector. An example was the X-29 aircraft. Its technologies were developed and intended for both civilian and military aircraft.

HiMAT

The HiMAT (Highly Maneuverable Aircraft Technology) subscale research vehicles flown from Dryden from mid-1979 to January 1983 demonstrated advanced fighter aircraft technologies that could be used to develop future high-performance military aircraft. Two vehicles were used in the research program that was conducted jointly by NASA and the Air Force Flight Dynamics Laboratory at Wright-Patterson Air Force Base in Ohio. The North American Aircraft Division of Rockwell International built the vehicles.

The two HiMATs were equipped with different instrumentation but had identical fundamental designs. The first aircraft was configured to fly at transonic and supersonic speeds and was equipped with accelerometers. The second vehicle was designed to acquire subsonic performance data and was heavily equipped with strain gauges, accelerometers, and pressure sensor orifices.

The first HiMAT flight took place on July 27, 1979, at Edwards Air Force Base in California. The aircraft flew for twenty-two minutes of stability and control tests before landing on the dry lakebed. The early HiMAT flights involved "gentle" maneuvers. The aircraft gradually increased the complexity of its maneuvers and underwent modifications in preparation for supersonic flight. The initial supersonic flight of the first HiMAT aircraft took place on May 11, 1982, flying at a maximum speed of Mach 1.2 at 12,192 meters altitude and remaining at supersonic speed for 7.5 minutes. The second supersonic flight took place on May 15, 1982, when the aircraft demonstrated a supersonic design point of Mach 1.4 and three g's acceleration at 12,192 meters altitude. It flew for five minutes at supersonic speed and achieved a maximum acceleration of just under four g's at Mach 1.4.

The second HiMAT aircraft made its first research data acquisition flight on May 26, 1982. It collected airspeed data and pressure, loads, and deflection data for aero-elastic tailoring assessment. It sustained a 5-percent negative static margin. On its second research data acquisition flight on June 2, 1982, the aircraft flight test maneuver autopilot acquired high-fidelity flight test data during wind-up turns and pushover, pullup maneuvers. The maximum acceleration attained was eight g's. It achieved its maximum Mach number of 0.9 at 11,582 meters altitude.

The final flight occurred on January 11, 1983. The two vehicles flew a total of twenty-six times during the three-and-a-half-year program.[49]

The program investigated aircraft design concepts, such as relaxed static stability control, that could be incorporated on the fighter aircraft of

---

[49]"HiMAT," *NASA Facts On-Line,* FS-1994-11-025-DFRC, Dryden Flight Research Center, November 1994.

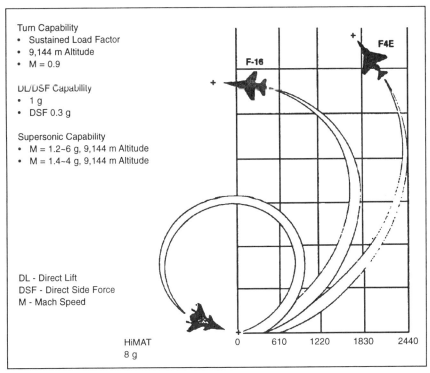

Figure 3–10. Increased Turning Capability of HiMAT Compared With Other Aircraft

the 1990s. Testing concentrated on high-g maneuvers at transonic and supersonic speeds. The vehicles provided data on the use of composites, aero-elastic tailoring, close-coupled canards (the smaller forward set of horizontal stabilizers), winglets (small vertical extensions of the wing tips), and the interaction of these then-new technologies on each other. Throughout the HiMAT test program, static stability—the tendency of an aircraft to return to its original attitude after being disturbed—was gradually reduced by relocating lead ballast from the nose to the tail of the aircraft to shift the center of gravity aft.[50] Turning performance of the canard-configured vehicle was improved by moving the center of gravity aft, although it reduced the aircraft's normal static stability.

The unique shape of HiMAT permitted high-gravity turns at transonic speeds—965 to 1,290 kilometers per hours. The rear-mounted swept wings and a forward controllable canard coupled to the flight control system provided the vehicles with twice the turning capability of military fighters (Figure 3–10).

About 30 percent of the materials used to build each HiMAT were composites. These materials—glass fibers and graphites—gave the structures additional strength for increased maneuverability and the high

---

[50]William B. Scott, "HiMAT Maneuvering Goals Surpassed in Flight Test," *Aviation Week & Space Technology*, June 21, 1982, p. 38.

gravitational loads encountered during their flights. In HiMAT, graphite composites were used for the skin on the fuselage, wings, canards, engine inlet, vertical tails, and the wing and canard spars. Glass fiber composites were used for the leading edges of the outboard wings.

Both sets of airfoils were aero-elastically tailored to twist and bend in flight to the most favorable shape to achieve maximum performance for the particular flight conditions. The vehicle used the increased lift from the combination of the canards and wings to increase maneuverability at both subsonic and supersonic speeds.[51]

About one-half the size of a standard crewed fighter and powered by a small jet engine, the HiMAT vehicles were launched from NASA's B-52 carrier aircraft at an altitude of about 13,716 meters. A NASA research pilot flew them remotely from a ground station with the aid of a television camera mounted in the HiMAT cockpits. When the research portion of a HiMAT flight ended, the pilot landed the vehicle remotely on the dry lakebed adjacent to Dryden. The HiMATs were flown remotely because it was a safe way to test advanced technologies without subjecting a pilot to a high-risk environment. Remotely piloted research vehicles such as HiMAT could also be flown more economically than larger crewed vehicles.[52]

Each HiMAT had a DFBW control system instead of a conventional system. Lightweight wires replaced the heavier hydraulic lines and metal linkages that most aircraft used to transfer control commands to the movable surfaces on the wings and tail. Pilot commands were fed via telemetry to an on-board computer that sent electrical commands to the flight control surfaces. Fly-by-wire flight control systems were lighter in weight, were more versatile in terms of automatic features than conventional systems, and provided basic aircraft stability. This technology also saved weight and increased performance because the size of the normal stabilizing surfaces could be reduced.

The plane also incorporated an integrated propulsion system that used a digital computer to control the aircraft's entire propulsion system, instead of a conventional hydromechanical system. The system integrated control of the jet engine and nozzle, which vectored (tilted) during flight, permitting additional maneuverability without adverse interaction.

The vehicles were seven meters long and had a wingspan of close to about four and a half meters (Figure 3–11). They weighed 1,543 kilograms at launch and were powered by a General Electric J85 turbojet producing 22,240 newtons of thrust. The vehicles had a top speed of Mach 1.4 (Table 3–63).[53]

---

[51]"HiMAT Research Plane to Make First Flight," *NASA News*, Release 79-90, June 28, 1979, p. 2.

[52]"HiMAT," FS-1994-11-025-DFRC.

[53]*Ibid.*

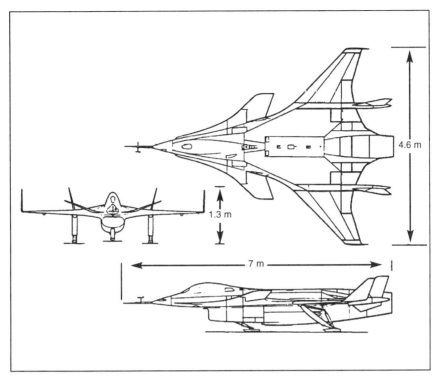

Figure 3–11. HiMAT Dimensions

One technology tested on the HiMAT vehicles that appeared later on other aircraft included the extensive use of composites that became common on military and commercial aircraft. Other technologies from the HiMAT tests appearing on other aircraft were the rear-mounted wing and forward canard configuration used on the X-29 research aircraft flown at Dryden and the winglets that were used on many private and commercial aircraft to lessen wingtip drag, increase stability, and enhance fuel savings.

X-29 Technologies

The X-29 research aircraft demonstrated the forward-swept wing configuration as well as the DFBW technology and flight control system addressed earlier. In December 1981, DARPA and the Air Force Flight Dynamics Laboratory selected Grumman Aircraft Corporation to build two X-29 aircraft, the first new X-series aircraft in more than a decade. The research aircraft were designed to explore the forward-swept wing concept and to validate studies that claimed the aircraft would provide better control and lift qualities in extreme maneuvers, reduce aerodynamic drag, and fly more efficiently at cruise speeds.

DARPA initially funded the X-29 program. NASA managed and conducted the X-29 flight research program at Dryden. The initial flight of the first X-29 took place on December 14, 1984, and the second first flew

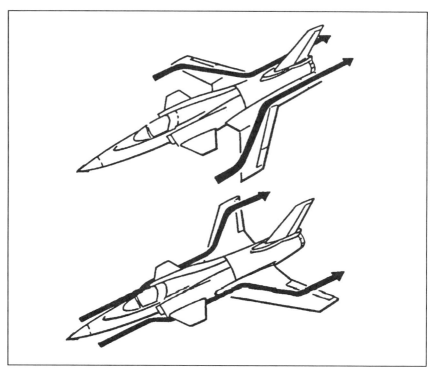

Figure 3–12. Forward-Swept Wing
(This design, shown in the top diagram, directs airflow inward behind the aircraft rather than outward.)

on May 23, 1989. Both flights were from Dryden. Table 3–64 gives the X-29's characteristics.

Forward-Swept Wing. The thirty-degree forward-swept wing configuration on the X-29 was mounted well back on the fuselage, while its canards—horizontal stabilizers to control pitch—were in front of the wings instead of on the tail. The complex geometries of the wings and canards combined to provide exceptional maneuverability, supersonic performance, and a light structure (Figure 3–12). The reverse airflow did not allow the wingtips and their ailerons to stall at higher angles of attack (the direction of the fuselage relative to the airflow).[54] Research results showed that the configuration of forward-swept wings, coupled with movable

---

[54] Angle of attack (alpha) is an aeronautical term that describes the angle of an aircraft's body and wings relative to its actual flight path. During maneuvers, pilots often fly at extreme angles of attack—with the nose pitched up while the aircraft continues in its original direction. This can lead to conditions in which the airflow around the aircraft becomes separated from the airfoils. At high angles of attack, the forces produced by the aerodynamic surfaces, including lift provided by the wings, are reduced. This often results in insufficient lift to maintain altitude or control of the aircraft.

canards, gave pilots excellent control response at up to forty-five degrees angle of attack.[55]

Aero-elastic Tailoring. Germany first attempted to design an aircraft with a forward-swept wing during World War II, but the effort was unsuccessful because the technology and materials did not then exist to construct the wing rigidly enough to overcome bending and twisting forces without making the aircraft too heavy. The introduction of composite materials in the 1970s allowed for the design of airframes and structures that were stronger than those made of conventional materials, yet were lightweight and able to withstand tremendous aerodynamic forces. The use of composites made from carbon, Kevlar, glass, and other fibers embedded in a plastic matrix allowed a wing to be built that could resist the divergent forces encountered at high speeds. This technology, called aero-elastic tailoring, allowed the wing to bend, but it limited twist and eliminated structural divergence during flight.[56]

The X-29 wing had composite wing covers that used 752 crisscrossed tapes comprising 156 layers at their thickest point. The wing covers made up the top and bottom of the wing torsion box, the major structural element of the X-29 wing.

Thin Supercritical Wing. The composite wing also incorporated a thin supercritical wing section that was approximately half as thick as the supercritical wing flown on the F-8 (Figure 3–13). The thin supercritical wing design delayed and softened the onset of shock waves on the upper surface of a wing, deteriorating the smooth flow over the wing and causing a loss of lift and an increase of drag. The design was particularly effective at transonic speeds.

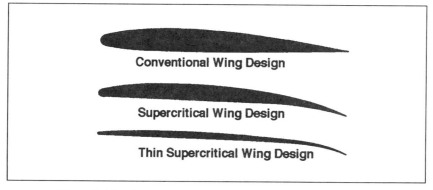

Figure 3–13. Relative Thickness of Conventional, Supercritical, and Thin Supercritical Wing Designs

---

[55]"The X-29," *NASA Facts On-Line*, FS-98-04-008-DFRC, Dryden Flight Research Center, April 1998.

[56]Structural divergence is the deformation or the breaking off of the wing in flight.

Variable Camber. The X-29's flaperons (combination of flaps and ailerons) were composed of two segments. This feature allowed what was, in effect, a change of camber or wing curvature. The segmented flaperon could be "straightened" to adapt the wing to supersonic flight, creating the best combination of lift and drag for that speed range.

Close-Coupled Variable Incidence Canards. The canards, forward of and in line with the X-29's wings, provided the primary pitch control, shared the aerodynamic load with the wing, and added lift. The close-coupled canards channeled the airflow over the inboard wing area to resist wing root stall. Both right and left canards could move independently thirty degrees up or sixty degrees down.[57]

Strake Flaps. The strakes—the horizontal surfaces that extended along the rear fuselage from wing to the exhaust nozzle of the aircraft—were equipped with thirty-inch-long flaps that augmented the canards for pitch control.

Three-Surface Pitch Control. Simultaneous and continuous operation of the canards, flaperons, and strake flaps minimized trim drag and maximized the X-29's responsiveness at the onset of maneuvers. The canards provided primary pitch control; the flaperons provided roll control, high lift, and camber adjustments; and the strake flaps augmented the canards at low speeds, such as rotation for takeoff or recovery from a deep stall.

F-18 High Angle of Attack

NASA used an F-18 Hornet fighter aircraft in its High Angle of Attack Research Vehicle (HARV) program. This program, which began in 1987, attempted to expand what researchers called the "stall barrier"—the tendency of aircraft to stall and become uncontrollable at high angles of attack and slow speeds. This tendency greatly limited an aircraft's performance and maneuverability.[58]

NASA used the HARV to explore the use of thrust vectoring at high angles of attack. The research program produced technical data at high angles of attack to validate computer codes and wind tunnel research. The successful validation of these data could give engineers and aircraft designers a better understanding of aerodynamics, the effectiveness of flight controls, and airflow phenomena at high angles of attack. This could lead to design methods that provided better control and maneuverability in future high-performance aircraft and helped prevent dangerous spins and related crashes. The database would permit more efficient computer-aided design of aircraft and was expected to decrease wind tunnel and flight testing time. Costly postproduction design "fixes" could also be minimized.

The HARV program was a joint effort of NASA's Dryden Flight, Ames, Langley, and Lewis Research Centers. Ames examined aerody-

---

[57]"The X-29," FS-98-04-008-DFRC.
[58]Wallace, *Flights of Discovery,* p. 103.

namic and vortex control concepts. Dryden had responsibility for flight vehicle demonstration and testing. Lewis investigated the thrust vector nozzle and propulsion technologies. Langley made extensive use of its wind tunnel and computer facilities to generate much of data that were being validated.

The first phase of high alpha flights began in mid-1987 using an unmodified aircraft. Investigators conducted visual studies of the airflow over various parts of the aircraft up to fifty-five degrees angle of attack. Special tracer smoke was released through small ports just forward of the leading-edge extensions near the nose and was photographed as it followed airflow patterns around the aircraft. Also photographed in the airflow were short pieces of yarn (tufts) taped on the aircraft, as well as an oil-based dye released onto the aircraft surfaces from 500 small orifices around the vehicle's nose.

The airflow patterns of smoke, dye, and tufts were recorded on film and videotape and compared with computer and wind tunnel predictions. Additional data obtained included air pressures recorded by sensors located in a 360-degree pattern around the nose and at other locations on the aircraft. The first phase lasted two and a half years and consisted of 101 research flights.

In 1987, NASA selected McDonnell Douglas Corporation to equip the research aircraft with a thrust vector control system about the pitch and yaw axes.[59] The system had an easily programmable research flight control system that allowed research into flight control concepts using various blends of aerodynamics and thrust vector control at subsonic and high alpha flight conditions. These thrust-vectoring paddles helped stabilize the aircraft at extremely high angles of attack. The modified Hornet was used for subsequent phases of the program, which was still under way in 1996.

X-31

The development of the X-31, a highly maneuverable fighter-type plane, began in the late 1980s. Funded by DOD and West Germany, the program used NASA's Dryden Flight Research Center for some of its testing.

Hypersonics: The National Aerospace Plane Program

NASA's hypersonic research in the late 1970s and early 1980s was conducted primarily at Langley Research Center under a minimal budget. Researchers at Langley developed subscale versions of the scramjet (supersonic combustion ramjet) and conducted numerous tests in supersonic

---

[59] "McDonnell Douglas Selected for Contract Negotiations," *NASA News,* May 1, 1987.

combustion.[60] The advent of high-speed digital computers and advanced metal-matrix composites increased the rate of progress in this field.

Developments in computational fluid dynamics, principally at NASA's Ames Research Center, paralleled the development in scramjet technology. The advent of supercomputing capabilities allowed for more detailed analyses and simulation of the aerodynamics and thermodynamics associated with sustained hypersonic cruise and exiting and entering Earth's atmosphere at various trajectories. Advanced computational fluid dynamics codes also assisted in understanding the supersonic airflow through scramjet configurations.[61]

In 1982, DARPA initiated an effort at Langley called Copper Canyon, which would be Phase I of the National Aerospace Plane (NASP) program. This phase incorporated recent research in the areas of hypersonic propulsion, advanced materials and structures, and computational fluid dynamics. Technically, the largest challenge was in the field of propulsion technology. The proposed vehicle needed a combination engine that covered a wide range of Mach speeds. In the lower speed range up to Mach 5, turbojet or subsonic ramjet engines were required, but above those speeds, the vehicle required either the scramjet or a combination scramjet and scramrocket. In contrast to ramjets, scramjets do not slow the air to subsonic speed so that the air can be used to burn liquid hydrogen, but rather, they burn the hydrogen in supersonic streams at lower temperatures. This would increase engine efficiency, proponents of the program stated, and could lead to a significant reduction in launch costs to low-Earth orbit.

The program's goal was to develop and demonstrate the technologies needed to fly an aircraft into orbit by using airbreathing propulsion instead of rockets. The eventual intent was to build and fly an actual experimental transatmospheric vehicle that would take off horizontally from a conventional runway, accelerate from 0 to Mach 25, and be capable of leaving Earth's atmosphere, then enter into low-Earth orbit, return to the atmosphere, and land, again horizontally. Its airbreathing engines (scramjet technology) would use oxygen from the environment to burn its fuel rather than carry its own oxygen supply, as rockets do.[62]

---

[60] "NASA 'Hyper-X' Program Established—Flights Will Demonstrate Scramjet Technologies," *NASA Facts On-Line,* FS-1998-07-27-LaRC, Langley Research Center, July 1998. A scramjet is a ramjet engine in which the airflow through the whole engine remains supersonic. A ramjet is an air-breathing engine similar to a turbojet but without mechanical compressor or turbine. Compression is accomplished entirely by ram and is thus sensitive to vehicle forward speed and is nonexistent at rest.

[61] John D. Moteff, "The National Aero-Space Plane Program: A Brief History," *CRS Report for Congress,* 88-146 SPR (Washington, DC: Congressional Research Service, The Library of Congress, February 17, 1988), p. 3.

[62] John D. Moteff, "National Aero-Space Plane," *CRS Report for Congress* (Washington, DC: Congressional Research Service, The Library of Congress, updated January 2, 1991 (archived)), p. 2.

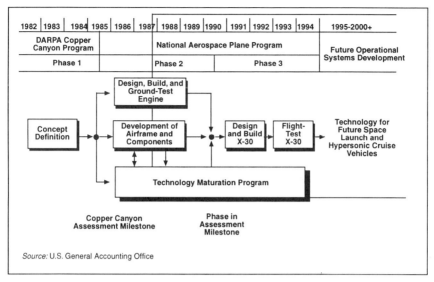

Figure 3–14. National Aerospace Plane Program Schedule and Milestones

It was envisioned that horizontal takeoff and landing would provide flexible basing and reduce reliance on the launch and landing facilities at Kennedy Space Center and Edwards Air Force Base, respectively. They might also reduce operational costs and shorten turnaround time. The aircraft would demonstrate the capability for flying single-stage-to-orbit without carrying large disposable fuel tanks or having stages that separated as the vehicle gained altitude. It would also be unique in that the engines would be integrated with the airframe rather than separate units that were bolted to the wings.[63] Other goals were a powered landing capability and maximum aircraft-like maneuverability.[64] The plane would have both military and civilian applications.

Program Development. The program was to consist of three phases (Figure 3–14). Phase I, Copper Canyon, began at Langley Research Center in 1982. This phase, concept definition, focused on scramjet technology and involved several government agencies and private firms and universities in tests and design studies to determine the feasibility of transatmospheric vehicles. During this phase, researchers investigated a hydrogen-based power aircraft that would be capable of horizontal takeoff and landing and operating at speeds between Mach 12 and 25 at altitudes between 30,480 and 106,680 meters.

In 1985, DARPA and NASA completed the definition of an air-breathing aerospace plane, and NASA stated its conviction that a hypersonic transatmospheric vehicle was technically feasible. NASA's

---

[63]Larry Schweikart, "Hypersonic Hopes: Planning for NASP, 1986–1991," *Air Power History,* Spring 1994, p. 36.

[64]Moteff, "National Aero-Space Plane," p. 3.

Dr. Raymond S. Colladay, the associate administrator of OAST, cited "significant activities" at NASA in support of the hypersonic vehicle. These included:

- A cooperative program with DARPA to develop a database for the required combined cycle airbreathing engine
- Continuing scramjet research
- Identification of airframe/propulsion integration as the key to achieving acceptable performance for a hypersonic cruise airplane
- Space Shuttle experiments to produce data important to hypersonics and transatmospherics
- Planned major facility modifications to permit the full-scale verification of scramjet combustion systems at the high-temperature tunnel at Langley Research Center, testing of combined cycle engine concepts at the propulsion system lab at Lewis Research Center, and flow-field studies at the hypersonic tunnel at Ames Research Center
- A joint program with the Office of Naval Research and the Air Force Office of Scientific Research to initiate a new research program at universities in FY 1986 in hypersonic viscous flows[65]

DOD also expressed optimism. U.S. Air Force Maj. General Donald J. Kutyna stated that DOD had decided to proceed with a $500 million program to design a hypersonic plane that could fly around the globe in less than two hours and in the highest reaches of the atmosphere. He envisioned this vehicle capable of providing a low-cost method for launching satellites and other equipment critical to the Strategic Defense Initiative.

Funding responsibility for the program would be divided between NASA and DOD, with NASA assuming 20 percent of the funding burden and DOD assuming the other 80 percent.[66] During the early research and development activities, NASA would carry a larger portion of the funding burden.

As it advanced in late 1985, the program was a large team effort. In addition to NASA and DOD (represented by DARPA), the U.S. Air Force, U.S. Navy, and Strategic Defense Initiative Organization also participated. DOD was responsible for overall management of the joint program. NASA had lead responsibility for overall technology direction, application studies, and the design, fabrication, and flight testing of experimental flight vehicles. Within DOD, the Air Force was assigned overall responsibility for the program. In the 1986 memorandum of understanding, DARPA was given the lead for early technology develop-

---

[65]"NASA Moving Out on Hypersonic Vehicle Research," *Defense Daily*, August 1, 1985, p. 172.

[66]Brendan M. Greeley, Jr., "U.S. Moves Toward Aerospace Plane Program," *Aviation Week & Space Technology*, December 16, 1985, p. 16.

ment (Phase II), and the Air Force had the lead for Phase III technology development.[67]

Phase II began in 1986, following the formal establishment of the NASP program in 1985. This technology development phase consisted of the accelerated development of key technologies, airframe design, propulsion module development, and ground tests of the propulsion system up to Mach 8—the then-current practical limit of wind tunnels for engine tests.[68] NASA and the Air Force awarded numerous contracts in the spring of 1986. The contracts in the general areas of propulsion and airframe called for research and development in propulsion, aerodynamics, computational fluid mechanics, advanced structures, and high-temperature materials that would lead to the design of a NASP flight research vehicle called the X-30. Potential total contract value was more than $450 million.[69] In November 1986, the NASA administrator approved Duncan E. McIver's appointment as director of the NASP Office.[70]

President Ronald Reagan strongly advocated the program. When he mentioned a hypothetical commercial vehicle in his February 1986 State of the Union address, in his call for research into "a new Orient Express," he was really referring to the NASP program.[71]

Design Concepts. Four design concepts were under consideration (Figure 3–15). The blended body was elliptically shaped and used an engine integrated in the lower surface of the airframe. The design had structural weight and thermal protection advantages, but the baseline concept that was selected offered better low-speed control and efficiency.

The cone body featured an aerodynamically shaped cylindrical airframe ringed by engines. The advantages of that concept included large thrust capabilities and large fuel capacity. Compared with the baseline, the cone body was less aerodynamically efficient and had less vehicle stability and control.

The combination body had a turtle-shaped body with rounded scramjets located on the lower surface of the airframe. Although this design was as efficient as the wing body, the combination body had a higher structural weight and required greater thermal protection.[72]

---

[67]"Memorandum of Understanding Between the Department of Defense and the National Aeronautics and Space Administration for the Conduct of the National Aero-Space Plane Program," June 1986, National Historical Reference Collection, NASA History Office, NASA Headquarters, Washington, DC.

[68]"The National Aerospace Plane Program," *Aerospace,* Spring 1986, p. 2.

[69]"National Aerospace Plane Program Awards Contracts," *NASA News,* April 7, 1986.

[70]"Duncan McIver Appointed Director, National Aero-Space Plane Office," *Headquarters Bulletin,* NASA, January 5, 1987, p. 6.

[71]Moteff, "The National Aero-Space Plane Program," p. 1.

[72]Stanley W. Kandebo, "Researchers Pursue X-30 Spaceplane Technologies for 1990 Evaluation," *Aviation Week & Space Technology,* August 8, 1988, p. 50.

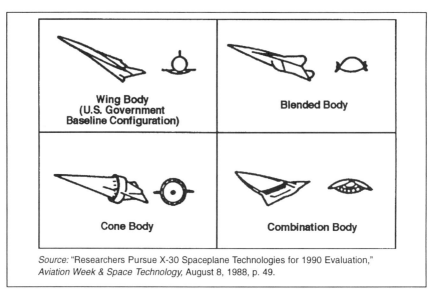

Figure 3–15. Four Generic X-30 Designs With Fully Integrated Engines and Airframes

All of the designs featured an integrated engine and airframe. The vehicle would be about the size of a Boeing 727 transport and use three to five scramjet engines and a single rocket that produced approximately 200,000 to 300,000 newtons of thrust. Its weight at takeoff would be approximately 113,000 to 136,000 kilograms. The X-30 test vehicle would have little payload capacity beyond the ability to carry a crew and test instrumentation and would require about 45,360 kilograms of slush hydrogen (partly liquid and partly frozen) per mission.

The design baseline for the X-30 (as of August 1988) was the wing body configuration (Figure 3–16). The wing body had a rounded fuselage and positioned the engine underneath the airframe. Although the design was aerodynamically efficient, permitted a large fuel tank, and offered good low-speed control, problems existed in integrating the airframe afterbody with the engine exhaust nozzles.

During this period, the participants expressed confidence that the program would progress as planned. Colladay testified before Congress that the NASP program was making good technical progress and said that initial applications of the vehicle would most likely be for the government, either as a launch system or as a strategic military vehicle. Presidential Science Advisor Dr. William R. Graham told the Senate subcommittee on space that only an insurmountable technical barrier could prevent the United States from proceeding with the plane, and no such barrier was presently foreseen.[73] Air Force Colonel Len Vernamonti, chief of the NASP program, agreed that researchers had encountered no obstacles in their theoretical work on the plane.

---

[73]"Graham Sees No Barrier to X-30 Space Plane," *Defense Daily*, March 2, 1987, p. 1.

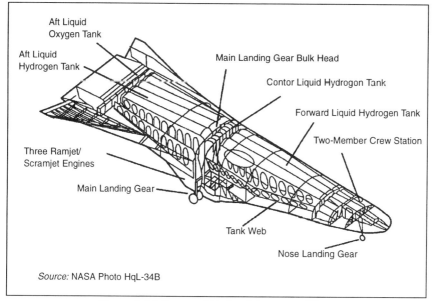

Figure 3–16. Proposed National Aerospace Plane

However, others within DOD expressed concerns. DOD Director of Operational Test and Evaluation John Krings told the Senate subcommittee on industry and technology that existing facilities were "barely adequate to support the experimentation and feasibility demonstration phases for [new] technology programs, let alone the development and operational testing and evaluation that will be required as they matured."[74] A Defense Science Board task force recommended that DOD slow the schedule for producing a NASP experimental vehicle by at least one year because the advanced technology components, such as materials, relied on by engine and airframe designers were not yet available.[75]

When the program outgrew DARPA's traditional R&D functions in 1988, the program moved to the Air Force in preparation for the development of a flight test vehicle. DOD signed the new memorandum of understanding in August 1988, and NASA signed it in September of that year.[76] If times had been different, NASA might have offered to assume program responsibility. But in 1988, NASA was involved with

---

[74]"New Space Systems Test Facilities to Cost $7 Billion," *Aviation Week & Space Technology,* April 27, 1987, p. 85.

[75]"DSB Expected to Propose Slowdown in NASP Program, More Technology Research," *Inside the Pentagon,* September 18, 1987, p. 1.

[76]"Memorandum of Understanding Between the Department of Defense and the National Aeronautics and Space Administration for the Conduct of the National Aero-Space Plane Program," September 1986, NASA Historical Reference Collection.

reinvigorating the Space Shuttle program, planning a new heavy-lift rocket, and working on the Space Station.[77]

The program became more controversial as it progressed, and funding problems developed. The Senate Armed Services Committee reduced the Air Force's FY 1988 request for R&D funds for the NASP program from $236 million to $200 to boost NASA's share of the program costs, which increased 1 percent to 19.2 percent in the FY 1988 budget.[78] The program was threatened with up to a 33-percent budget cut in FY 1988 from congressional actions, which could lead to at least a one-year slip. This would require that private-sector contractors continue to fund the program heavily with their own money, which, in some areas, amounted to four times the government's contribution.[79]

Funding constraints pushed the X-30 about a year behind schedule as of the spring of 1988, with the first flight delayed to 1994 or 1995. Also, although NASA stated that both the NASP program and the Space Station deserved a sufficient level of funding, it found the two programs competing for limited funds. Beginning in 1989, funding levels generally dropped. President George Bush's Secretary of Defense, Richard Cheney, proposed eliminating DOD funding of the NASP program and recommended transferring the program entirely to NASA. The President's budget showed that DOD's share of the program would be transferred to NASA along with program management. Congress restored the program's joint NASA-DOD funding and recommended that DOD retain program management. Congress also recommended that Phase II be extended and that a decision whether to proceed with building the X-30 be postponed until March 1993.

Phase III was to have begun in 1990. This phase called for the selection of one engine contractor and one airframe contractor to design and build two X-30s to explore propulsion performance above Mach 8. Structures and materials needed to fabricate such a vehicle would be developed and tested. It was originally intended that a decision to proceed with this phase would be made in 1988. However, as the events just described show, at the end of 1988, technology development had not yet progressed to a point where a decision could be made.

DOD pulled out of the program in 1993. It survived until FY 1994, when Congress reduced NASA's funding to $80 million.[80] It eliminated all remaining funding in FY 1995.

---

[77]Schweikart, "Hypersonic Hopes," p. 43.

[78]"Defense Digest," *Defense Daily,* May 15, 1987, p. 91.

[79]"Washington Roundup," *Aviation Week & Space Technology,* November 2, 1987, p. 21.

[80]Stanley W. Kandebo, "NASP Cancelled, Program Redirected," *Aviation Week & Space Technology,* June 14, 1993, p. 33.

Safety and Flight Management

Operational and safety problems have been traditional topics for NASA aeronautical research. Flights in bad weather, landings on wet runways, and airport approaches during periods of high-density traffic flow have been studied and improved by NASA programs, often working cooperatively with the FAA. NASA programs were conducted in technological areas, such as materials and structures and guidance and control, and in human factors areas, such as how pilots interact with various cockpit displays or react to unexpected weather conditions.

Transport Systems Research Vehicle

Although not a program, NASA's Transport Systems Research Vehicle (TSRV) deserves special mention. This Boeing 737-100 was the prototype 737, acquired by Langley Research Center in 1974 to conduct research into advanced transport aircraft technologies. In the twenty years that followed, the airplane participated in more than twenty different research projects, particularly focused on improving the efficiency, capacity, and safety of the air transportation system. It played a significant role in developing and gaining acceptance for numerous transport technologies, including "glass cockpits," airborne wind shear detection systems, a data link for air traffic control communications, the microwave landing system, and the satellite-based global positioning system (GPS).

The TSRV's unique research equipment included a complete second cockpit in the cabin (Figure 3–17). The plane had three major subsystems. One subsystem operated the actual flight controls of the airplane. A second subsystem provided computerized navigation functions, which controlled the airplane's flight path. The third subsystem operated the electronic flight displays in the aft cockpit. The on-board computer equipment was regularly upgraded to keep pace with rapid developments in computer technology.

The aircraft served as a focus for joint NASA-industry research efforts as well as joint efforts with other government agencies.[81] The following sections address programs that made use of this unique vehicle from 1979 to 1988. Table 3–65 gives the aircraft's specifications.

Terminal-Configured Vehicle/Advanced Transport Operating System Program

The Terminal-Configured Vehicle (TCV) program, a joint NASA-FAA effort, began in 1973. In June 1982, the name of the TCV program was changed to the Advanced Transport Operating System (ATOPS)

---

[81]Lane E. Wallace, *Airborne Trailblazer: Two Decades With NASA Langley's 737 Flying Laboratory* (Washington, DC: NASA SP-4216, 1994), p. vii.

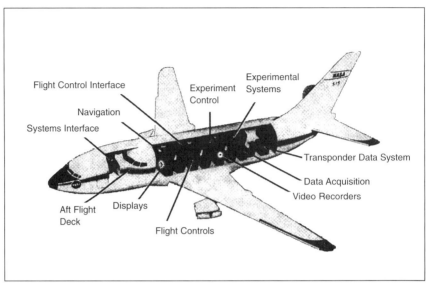

*Figure 3–17. NASA Transport Systems Research Vehicle (TSRV)
(The plane was flown from the research cockpit, located in the forward fuselage.
Safety pilots in the conventional cockpit served as backup to the research pilots and
could fly the airplane if required. Seating behind the research cockpit was for flight
test engineers who monitored and interpreted the video display system.
The TSRV could be reconfigured for various research purposes.)*

program, to reflect its renewed emphasis on commercial air transportation system issues, rather than on individual airplane technologies.[82]

One area of the program addressed the techniques needed to achieve time-controlled descent to an airport. The program used the TSRV to investigate advanced technology for conventional takeoff and landing aircraft. The program examined approach paths for noise abatement and improved airport acceptance rates, cockpit displays of traffic information, and profile and time-based navigation (which would use a computer to calculate an optimum fuel-efficient flight plan to sort out and sequence arriving aircraft in a time-based traffic control system that matched airport demand to its capacity and allowed closer spacing of aircraft).

Research could place aircraft at a point in space, for example, at the start of the descent to the airport within a few seconds. If there were unfavorable winds, that time might increase by as much as ten seconds; however, that compared with perhaps two minutes' accuracy with conventional methods of air traffic control. The descent itself, handled by the "smart" avionics in the TSRV, would be done along a flight path that used minimum fuel, so there would be potential fuel savings by using the system. Other potential payoffs included routine operations in bad weather, pilot participation in the traffic control system loop by using a cockpit

---

[82] *Ibid.*, p. 21.

display of traffic, reduced lateral separation and spacing, and reduced runway occupancy time. All of these factors tended to increase the capacity of an airport in all kinds of weather.

Several research projects in the ATOPS program were geared toward improving the internal systems and operation of transport aircraft. One was the Digital Autonomous Terminal Access Communication (DATAC) project. Boeing had developed the technology for a single, global data bus that would carry the information between the different components of an airplane's systems. NASA expressed an interest in the system for its TSRV, and in 1983, the initiative became the joint DATAC project. Boeing designed and built the data bus and the terminals that provided the interface between the data bus and the computers or components using the system. NASA engineers designed the interface boxes and software that would convert the data from the format needed for transmission on the data bus to a format the TSRV's computers and experimental systems could understand. By 1984, the DATAC system was installed and operating successfully on the TSRV. In 1985, Boeing became interested in using the system on its new airliners and incorporated it in its new transports, the 777s.

The Total Energy Control System (TECS) project attempted to make an autopilot/autothrottle system perform more like an actual pilot by designing a more efficient, integrated system that would make better use of an airplane's stored energy. From 1979 to 1981, NASA contracted with engineers at the Boeing Commercial Airplane Company to develop the control laws the system would require. Engineers at Boeing designed a system that would use the throttle and the elevator to control the energy state of the plane and the distribution of that energy from flight path to normal flight speed.

TECS was first tested successfully in the Boeing 737 simulator at Langley Research Center. NASA engineers then programmed it into TSRV flight computers and conducted twenty hours of flight testing in 1985. The system worked as expected, and the pilots liked the system. Nevertheless, because implementing TECS would require complete redesign of the automatic control system on commercial airliners, it was not incorporated into any of Boeing's commercial planes. It was, however, used on the uncrewed Condor aircraft that was remotely piloted.[83]

Cockpit Technology

As pilots moved from landing aircraft on a straight path that often approached ten miles or more to relying on steep, curved approach paths with final distances as short as one mile, they required a more accurate picture of the airplane's position at all times. They also had to control the airplane's progress precisely and monitor accurately any automatic systems so they could take over if necessary. This degree of monitoring and

---

[83]*Ibid.*, pp. 81–82.

management was virtually impossible with the conventional displays used during the mid-1970s. A new technology used cathode ray tube displays, developed as part of the TCV program, to process the raw aircraft system and flight data into an integrated, easily understood picture of the aircraft.[84]

The TCV experiments with electronic flight displays examined the effectiveness of the displays and how they could be used in a transport cockpit. In addition to validating the benefits of the basic equipment, researchers investigated and evaluated several display concepts to examine whether they would improve pilot awareness and the ability to compensate and correct for flight path errors.

Much of the development work in the early 1980s was conducted in the TSRV simulator at Langley Research Center, which duplicated the aft flight deck on the TSRV. The "all-glass" concept presented information to crew members on eight electronic displays that matched the TSRV aircraft. The crew members used the simulator to investigate new concepts in flight station design that would provide for safer and more efficient system operations by reducing clutter and improving the orderly flow of information controlled by the flight crew. Using the simulator allowed for the evaluation of various displays and also permitted research on improving situational awareness, air traffic control communication, flight management options, traffic awareness, and weather displays.[85] Promising display concepts were then incorporated into the TSRV's aft flight deck for operational testing.

The initial displays were monochrome cathode ray tubes. These were replaced by eight twenty-centimeter-squared electronic color displays representing the technology to become available in commercial transports of the future. The state-of-the-art color displays were driven by new on-board computers and specially developed computer software. These new technologies allowed information to be displayed more clearly than would be the case on existing electromechanical and first-generation electronic displays on current aircraft. The displays gave the pilots integrated, intuitively understandable information that provided a more accurate picture of the airplane's exact situation at all times. Pilots were expected to use this information to monitor and control airplane progress much more effectively and precisely than by using conventional displays.

Later in the 1980s, NASA began investigating the technology necessary to design "error-tolerant" cockpits that included a model of pilot behavior. The system used this model to monitor pilots' activities, such as track pilot actions, infer pilot intent, detect unexpected actions, and alert the crew to potential errors. A related investigation at Ames Research

---

[84]*Ibid.*, pp. 26–27.

[85]Randall D. Grove, ed., *Real-Time Simulation User's Guide; "The Red Book"* (Hampton, VA: Analysis and Simulation Branch, NASA Langley Research Center, January 1993), ch. 3, sec. 3.3.3 [no page numbers].

Center, using the Man-Vehicle Systems Research Facility, examined the human side of the people-machine relationship, including human error, fatigue, stress, and the effects of increasingly automated technologies on flight crew performance.

The advent of computerization and automation in the cockpits of commercial airliners resulted in a variety of benefits. Aircraft could travel on more fuel-efficient flight paths, use more reliable equipment that had greater flexibility for upgrades, and operate with only two pilots, no matter how large the aircraft. However, the new technology led to some unexpected problems. Human factors became an integral part of design analysis, and researchers looked closely at optimum levels of pilot workload and ways to keep pilots involved in the computerized systems. Initially, there was some concern that the pilots' workload would be decreased to the point where their skills would also lessen. However, researchers found that their workload actually increased to too high a level. One of the components of the system, the control and display unit, required so much attention that the pilots would neglect to look out the windows for visual information. Training had to be adjusted so that pilots learned when it was appropriate to use the control and display units and when to hand-fly the aircraft.[86]

Wind Shear

Wind shear refers to any rapidly changing wind current. It is characterized by almost instantaneous reversals of wind speed and direction. Microbursts are local, short-lived severe downdrafts that radiate outward as they rush toward the ground. They can produce extremely strong wind shear. As a downdraft spreads both downward and outward from a cloud, it creates an increasing headwind over the wings of an oncoming aircraft. This headwind causes a sudden leap in airspeed, and the plane lifts.

If the pilot is unaware that wind shear caused the increase in speed, the reaction will be to reduce engine power. However, as the plane passes through the shear, the wind quickly becomes a downdraft and then a tailwind. The speed of air over the wings decreases, and the extra lift and speed rapidly fall to below original levels. Because the plane is then flying on reduced power, it is vulnerable to sudden loss of airspeed and altitude. The pilot may be able to escape the microburst by increasing power to the engines. But if the shear is strong enough, the aircraft may crash.[87] Figure 3–18 illustrates the effects of wind shear on an aircraft.

Wind shear poses the greatest danger to aircraft during takeoff and landing, when the plane is close to the ground and has little extra speed or time or room to maneuver. During landing, the pilot has already reduced engine power and may not have time to increase speed enough to

---

[86]Wallace, *Airborne Trailblazer,* pp. 36–37.

[87]"Making the Skies Safe From Windshear," *NASA Facts,* NF176 (Hampton, VA: Langley Research Center, June 1992).

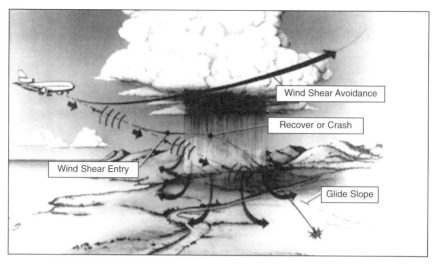

Figure 3–18. Artist's Depiction of the Effect of Wind Shear on an Aircraft (Wind shear is dangerous to aircraft primarily during takeoff and landing.) (NASA Photo 92-HC-423)

escape the downdraft. During takeoff, an aircraft is near stall speed and thus is very vulnerable to wind shear.

Microburst wind shear often occurs during thunderstorms. But it can also arise in the absence of rain near the ground. Some of the sensor systems that Langley Research Center tested worked better in rain, while others performed more successfully during dry conditions.

Beginning in 1976, more than 100 U.S. airports installed the FAA-developed ground-based low-level wind shear alert system, which consisted of an array of wind velocity measuring instruments located at various spots around an airport. The system compared the wind direction and velocity readings from the different sensors and, if significant variations between sensors were detected, transmitted an alert to the air traffic controllers, who then notified pilots in the area. The system, however, could not measure winds above the ground sensors, record vertical wind forces, or predict the approach of wind shears. Although this system was an improvement over existing detection methods, an on-board warning system with the capability to warn pilots of wind shear in time for them to avoid it was still needed.[88]

In 1986, Langley and the FAA signed a memorandum of agreement authorizing the start of a program to develop technology for detecting and avoiding hazardous wind shear. The five-year $24 million research project, the Airborne Windshear Detection and Avoidance Program, came in response to congressional directives and National Transportation Safety Board recommendations that followed three fatal accidents and numerous other nonfatal accidents linked to wind shear. In 1988 the FAA directed

---

[88]Wallace, *Airborne Trailblazer,* p. 58.

that all commercial aircraft must have on-board wind shear detection systems installed by the end of 1993.

The program had three major goals. The first goal was to find a way to characterize the wind shear threat in a way that related to the hazard level it presented for aircraft. The second was to develop airborne remote-sensor technology to provide accurate, forward-looking wind shear detection. The third was to design flight management concepts and systems to transfer that information to pilots so they could respond effectively to a wind shear threat.[89]

The program covered five major technology areas: technology assessment, present position sensor integration, hazard characterization, pilot factors in wind shear, and effects of heavy rain. The effort produced a database on microbursts and detection systems with data gathered from analyses, simulations, laboratory tests, and flight tests that would help the FAA certify predictive wind shear detection systems for installation on all commercial aircraft.

Roland Bowles, manager of the Langley wind shear research program, devised the "F-Factor" as an index to describe the hazard level of the wind shear. The index, which would be displayed in the cockpit, measured the loss in rate-of-climb capability that would result from flying into a wind shear. The higher the F-Factor, the greater the hazard. Information from past wind shear accidents indicated that the wind shear became a serious hazard when the F-Factor reached 0.1. Thus, the cockpit warning would be preset to alert the crew whenever that point was reached.[90] The F-Factor of a wind shear also would indicate how much extra power an airplane needed to fly through it without losing airspeed or altitude.[91]

Experts agree that avoidance is the best approach to take when encountering a wind shear situation. NASA, the FAA, and industry partners developed three systems that would warn pilots of wind shear so that they could avoid it: microwave radar, light detecting and ranging (LIDAR), and infrared (Figure 3–19). These three systems had been discussed in a 1983 report released by the National Academy of Sciences that recommended continued research into airborne wind shear detection systems. The systems gave pilots from ten to forty seconds' advance warning of the approaching wind shear. (Pilots need ten to forty seconds of warning to avoid wind shear; fewer than ten seconds is not enough time to react, while changes in atmospheric conditions can occur if more than forty seconds elapse.) Flight tests of the three systems began in the summer of 1991 in Orlando, Florida, and in Denver, Colorado, once more using the TSRV.

In addition to the sometimes fatal impact that wind shear has had on airplanes, some investigators believe that severe wind shear affected the Space Shuttle *Challenger* in its 1986 accident and may have magnified

---

[89]*Ibid.*, p. 61.

[90]"The Hazard Index: Langley's 'F-Factor,'" *NASA Facts,* NF177 (Hampton, VA: Langley Research Center, June 1992).

[91]Wallace, *Airborne Trailblazer,* p. 63.

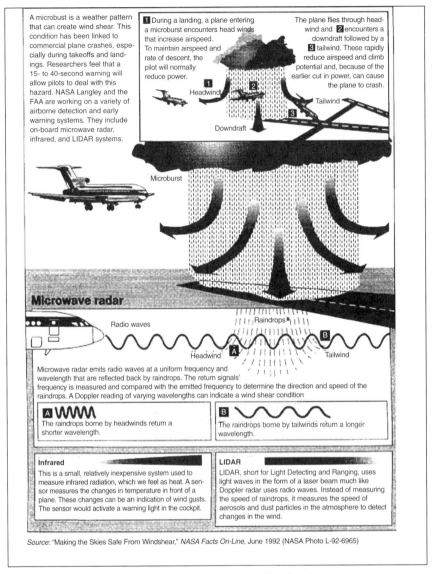

Figure 3–19. On-Board Wind Shear Warning Systems

the stresses placed on the spacecraft to a level beyond its design capability. Although the Rogers Commission officially dismissed wind shear as a contributing cause of the 1986 accident, NASA has increased its level of monitoring of wind shear in the launch pad area.[92]

---

[92]"New Theory on Challenger Disaster," *The Washington Times*, July 8, 1987, reproduced in *NASA Current News*, 87-126; Trudy E. Bell, "Windshear Cited as Likely Factor in Shuttle Disaster," *The Institute of Electrical and Electronics Engineers, Inc.*, May 1987.

Lightning

Lightning is another weather-related phenomenon encountered by aircraft. The effect of lightning strikes on modern aircraft became better understood as a result of a series of flight tests at Langley Research Center during the 1980s. In the Storm Hazards program, which ran from 1980 through 1986, a specially instrumented F-106B jet was repeatedly flown into thunderstorms at various altitudes. The aircraft sustained more than 700 direct lightning strikes during nearly 1,500 storm penetrations.

Newer aircraft made increasingly of composite material did not have the lightning protection provided by older aluminum skins unless they had special conductive fibers embedded during construction. The F-106B examined a variety of protective measures, such as aluminum paint, wire mesh, and diverter strips, while it collected data on lightning and its relationship to other storm hazards.

Icing

Icing is the solidification of moisture that develops on parts of the aircraft, such as the wings, tails, and propellers, in extremely cold weather conditions. Icing usually occurs between ground level and an altitude of 6,100 meters. During World War II, the United States lost more than 100 planes because of icing. Responding to a need expressed by the Army Air Forces and aircraft manufacturers, the National Advisory Committee for Aeronautics (NACA) directed that an Icing Research Tunnel (IRT) be added to the Altitude Wind Tunnel, then under construction at the Aircraft Engine Research Laboratory, the former name of Lewis Research Center (now Glenn Research Center). The first icing test took place there in June 1944.

The IRT is the world's largest refrigerated icing tunnel. It resembles other subsonic wind tunnels in that a wing or other aircraft component placed in the test section can be subjected to various airspeeds, with the airflow being created by a motor-driven fan. However, the IRT has several unique features. To simulate the aircraft icing environment, a heat exchanger and a refrigeration plant to achieve the desired air temperatures and a spray system to generate a cloud of microscopic droplets of unfrozen water were added. The IRT can duplicate the icing conditions (liquid water content, droplet size, and air temperature) that aircraft might encounter, study factors that cause icing, and test proposed anti-icing and de-icing systems.

The advent of jet engines reduced the demand for the facility, and NASA considered closing it. However, new technology and aircraft design and rising fuel costs increased the demand for new ice protection systems. In 1978, NASA re-instituted an icing research program to address the needs for new and future aircraft designs. The facility underwent a $3.6 million renovation in 1986 to cope with its increased workload and to expand its capabilities. In 1987, the American Society of Mechanical

Engineers designated the IRT an "International Historic Mechanical Engineering Landmark" for its leading role in making aviation safer.

The goal of NASA's icing research was to increase the effectiveness of existing ice protection systems and develop advanced concepts for both anti-icing and de-icing systems that were reliable, cost effective, energy efficient, lightweight, and easy to maintain. Researchers paid particular attention to the needs of small planes and helicopters because many of their flights took them into potential icing environments.

One major goal centered on creating computer codes to predict icing and its effects on airfoils and then to validate those predictions experimentally in the IRT and in flight. Lewis Research Center's overall ice accretion code, called LEWICE, approached the question by calculating the flow field around the airfoil and then applying a droplet trajectory code to compute water movement within the flow. An ice accretion code then determined how much of the incoming water would freeze over a specified period. The IRT was used to grow ice accretions on a wide spectrum of fixed-wing and rotorcraft airfoils, as well as engine inlets. Lewis also flew a DHC-6 Twin Otter aircraft equipped with a stereoscopic camera system to photograph ice formations, as well as several standard instruments to measure ice cloud properties.

Crash Survivability

NASA's 1984 Controlled Impact Demonstration program was designed to improve the survivability of crash victims through reducing postcrash fire hazards and improving crash impact protection. The FAA had been evaluating an anti-misting jet-fuel additive that seemed capable of preventing fuel fires in airplane crashes with promising laboratory test results. However, before publishing a "Notice of Proposed Rulemaking" as a first step toward requiring the additive in certain types of jet aircraft, the agency wanted to test the additive in a real airplane crash.

The FAA conducted the test at NASA's Dryden facilities on an old Boeing 720 jetliner with remote controls. The vehicle was fueled with the anti-misting fuel and guided to a remote location on the Rogers Dry Lakebed. The FAA embedded iron posts in the ground to ensure that the fuel tanks would be ripped open. However, upon impact, the plane burst into flames. Plans to use the anti-misting fuel were dropped.[93]

Related research studied how airplanes crash in the hope of finding some basic structural or other design changes that would increase the survivability of crew and passengers in an accident. NASA acquired several small planes that had been condemned as not airworthy because of flood damage but were suitable for research. It deliberately crashed these single- and twin-engine light planes in a carefully controlled, instrumented, and documented series of impacts. Researchers used extensive instru-

---

[93] Wallace, *Flights of Discovery*, pp. 150–51.

mentation inside the aircraft as well as photography outside the planes to acquire data and to document the crashes. They harnessed dummies in crew and passenger positions and assessed their chances of surviving the crashes. These investigations examined energy-absorbing aircraft seat designs and structural design techniques to modify the fuselages to increase their strength and how the progressive destruction of the airframe moved from the point of impact throughout the structure.

Space Research and Technology Programs

Although the space research and technology program also served the needs of non-NASA civil, commercial, and military users of space, the program related more directly to NASA's own priorities than its aeronautics activities. The aeronautics efforts frequently served to further industry's or the military's technology goals as well as NASA's and were often conducted jointly with other agencies or industry. The space research and technology program focused the following:

- Advancing the technology base
- Maintaining technical strength in the scientific and engineering disciplines
- Developing more capable, less costly space transportation systems and large space systems with growth potential
- Promoting scientific and planetary exploration
- Improving understanding of Earth and the solar system
- Supporting the commercial exploitation of space

The program greatly contributed to the Space Shuttle program, developed technologies to be used for the Space Station program, and conducted research into a variety of technological areas with applications in diverse fields. All NASA centers participated in the space research and technology program, and there was significant industry and academic participation.[94]

The program consisted of two parts: the research and technology base program and the focused technology program. The research and technology base program comprised particular disciplines, represented by the divisions, and system technology studies in the areas of propulsion, space energy conversion, aerothermodynamics, materials and structures, controls and guidance, automation and robotics, human factors, computer science, sensors, data systems, and communications.[95]

---

[94]Where no source is specifically cited, information in this section comes from the *Aeronautics and Space Report of the President,* issued annually by NASA.

[95]OAST's division names changed frequently during the 1979–1988 period to reflect their predominant focus. While some of the major headings in this section are also division titles, not all are. Rather, these headings are more descriptive of the types of activities that took place.

In the focused programs, technologies were developed for specific applications, and products were delivered in the form of demonstrated hardware, software, and design techniques and methods. Focused development was most often based on the identified needs and potentials of both current and future programs and missions. Spaceflight experiments carried out aboard the Shuttle were an example of focused development. In addition, the Civil Space Technology Initiative, initiated in 1987, and the Pathfinder program, established in 1988, were focused programs.

The Civil Space Technology Initiative was designed to conduct research in technologies to enable efficient, reliable access to operations in Earth orbit and to support science missions. Its technology focused on space transportation, space science, and space operations. The space transportation thrust centered on providing safer and more efficient access to space. It was involved with the design of a new fleet of space vehicles, including new expendable and partially reusable cargo launch vehicles, fully reusable crewed vehicles, and expendable and reusable space transfer vehicles.

The space science area supported more effective conduct of scientific missions from Earth orbit. Technical programs initiated to address the requirements of future long-term missions included high rate/capacity data systems, sensor technology, precision segmented reflectors, and the control of flexible structures. The technologies to enhance future space operations were designed to lead to increased capability, substantial economies, and improved safety and reliability for ground and space operations. Space operations addressed the technologies of telerobotics, system autonomy, and power.

The Pathfinder program began in 1988. It implemented the new National Space Policy that directed NASA to start planning for potential exploration missions beyond the year 2000.[96] The program aimed at developing technologies that would be required for missions that expanded human presence and activities beyond Earth's orbit into the solar system. Without committing to a specific mission at the current time, the program would focus on developing a broad set of technologies that would enable future robotic or piloted solar system exploration missions. The Pathfinder program called for a significant amount of automation and robotics research on developing a planetary rover that would act semi-autonomously in the place of humans on the Moon and Mars. The rover would effectively be a mobile laboratory with its own instrumentation, tools, and intelligence for self-navigation and rock sample acquisition and analysis.[97]

---

[96]The White House, Office of the Press Secretary, "The President's Space Policy and Commercial Space Initiative to Begin the Next Century," Fact Sheet, February 11, 1988, reproduced in Appendix F-2 of the *Aeronautics and Space Report of the President, 1988 Activities,* pp. 194–96.

[97]"NASA Information Sciences and Human Factors Program, Annual Report, 1988," NASA Technical Memorandum 4126, July 1989, p. 1.

Space Shuttle Development and Support

Early in the Space Shuttle development stage, NASA's lifting-body program, carried out by OAST, provided data that helped select the shape of the orbiter and simulated the landing on the dry lakebed at Edwards Air Force Base. Two of the final landings represented the types of landings that Shuttles would begin making and verified the feasibility of precise, unpowered landings from space.

Data from each lifting-body configuration contributed to the information base NASA used to develop the Shuttles and helped produce energy management and landing techniques used on each Shuttle flight. Lifting-body data led to NASA's decision to build the orbiters without airbreathing jet engines that would have been used during descent and landing operations and that would have added substantially to the weight of each vehicle and to overall program costs.

Because the same airbreathing engines that were eliminated would also have been used to ferry the Shuttle from the landing site back to the launch site, NASA devised the concept of a mothership to carry out the ferry mission. The Boeing 747 Shuttle Carrier Aircraft (SCA) evolved from recommendations by Dryden engineers. The SCA launched the prototype orbiter *Enterprise* during the approach and landing tests in 1977 and has been the standard ferry vehicle since the first Shuttle was launched. (A second 747 was added in 1990.)

The approach and landing tests conducted in 1977 verified orbiter approach and landing characteristics and subsonic airworthiness. The tests revealed a problem with pilot-induced oscillation that was described in the "Aeronautics Research and Technology Programs" section of this chapter. The approach and landing test program also verified that the orbiters could be carried safely on top of the SCA.

OAST also was responsible for the development of the Shuttle's thermal protection system, the solid rocket booster recovery system, flight control system computer software, tests and modifications to the landing gear and braking systems, and, in the 1990s, the drag parachutes that were added to *Endeavour*. In 1977 and 1978, NASA's B-52 was used to test the solid rocket booster parachute recovery system, which allowed empty booster casings to be recovered and reused.[98]

In 1980, using F-15 and F-104 aircraft, NASA pilots flew sixty research flights to test the Space Shuttle's thermal protection tiles under various aerodynamic load conditions. The test tiles represented six locations on the orbiter and were tested up to speeds of Mach 1.4 and dynamic pressures of 1,140 pounds per square foot. The local tests led to several changes to improve bonding and attachment techniques.[99]

---

[98]"B-52 Launch Aircraft," *NASA Facts On-Line,*" FS-1994-11-005-DFRC, Dryden Flight Research Center, November 1994.

[99]"Dryden Contributions to Space Shuttle Development Many," *The X-Press,* NASA Ames Research Center/Dryden Flight Research Facility, April 5, 1991, p. 2.

Before orbital flights took place, NASA conducted an independent analysis of orbiter design structural loads and handling qualities, drawing on experience from the X-15, YF-12, and lifting-body programs. Although the study revealed some minor design deficiencies, it verified the overall adequacy of design to accomplish a successful orbital reentry. NASA also conducted preflight tests, in the Thermostructures Research Facility at Dryden Flight Research Center, of the elevon seals on orbiter wings to assure that hot free-stream air during control surface movement during reentry would not damage the aluminum wing structure.

The B-52 also served as a testbed for drag chute deployment tests that helped verify the drag chute system being installed on the orbiters. The system would allow orbiters to land on shorter runways and help reduce tire and brake wear.

Space research conducted at Langley Research Center, sometimes in conjunction with research at Dryden, also contributed to the Space Shuttle program. These activities included:

- Developing the preliminary Shuttle designs
- Recommending the modified delta wing for the orbiter rather than a conventional straight wing
- Conducting 60,000 hours of wind tunnel tests and analysis
- Conducting structures and materials tests to determine the requirements for various areas of the vehicle
- Investigating and certifying the thermal protection system for the launch environment
- Performing design, analysis, and simulation studies on the control and guidance systems
- Conducting landing tests on the main and nose gear tires and brake systems
- Conducting a runway surface texture test and recommending runway modifications for the Kennedy Space Center runway
- Participating in the redesign of the solid rocket booster components
- Examining launch abort and crew bailout capabilities
- Defining ascent aerodynamic wing loads[100]

OAST was actively involved in the redesign of the Space Shuttle solid rocket motor field joint following the 1986 *Challenger* accident as part of its materials and structures program. A significant part of this effort was directed toward developing a test procedure for qualifying candidate O-ring materials, and a test method was established as the standard for O-ring materials.

The Shuttle's landing gear and tires were another area of investigation. The Shuttle was equipped with four small wheels, two on each main

---

[100] "NASA Langley Research Center Contributions to Space Shuttle Program," *NASA Facts,* Langley Research Center, March 1992.

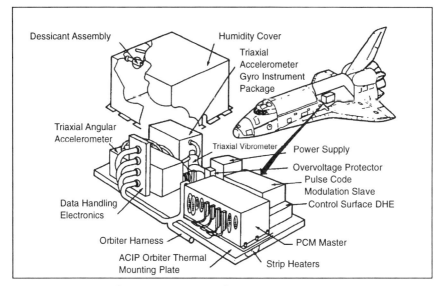

Figure 3–20. Aerodynamic Coefficient Indentification Package (ACIP) Experiment

gear. This contrasted with the eight to sixteen wheels that a commercial airliner of similar weight would have. The difference was because of the extreme temperatures the Shuttle encountered and the difficulty of protecting the landing gear and tires. NASA used the Convair 900 aircraft to test the Shuttle's landing gear components and to learn about tire wear on the Shuttle.

The Space Shuttle as a Research Facility

The Space Shuttle also served as an in-space laboratory to test many of OAST's basic research and technology concepts and to validate technology in the space environment. NASA used the Shuttle as an experimental facility for research in aerodynamics, thermal protection systems, and the payload environment. These included the Orbiter Experiments Program, the OAST-1 payload on STS 41-D, the Assembly Concept for Construction of Erectable Space Structures (ACCESS) and the Experimental Assembly of Structures in Extravehicular Activity (EASE) on STS 61-B, and the Long Duration Exposure Facility (LDEF) on STS 41-C.

Orbiter Experiments Program

The Orbiter Experiments Program consisted of a number of experiments on the early Shuttle missions. These experiments gathered data that assessed Shuttle performance during the launch, boost, orbit, atmospheric entry, and landing phases of the mission. The data verified the accuracy of wind tunnel and other simulations, ground-to-flight extrapolation methods, and theoretical computational methods. Table 3–66 lists the experiments, and Figures 3–20, 3–21, 3–22, 3–23, and 3–24 each depict

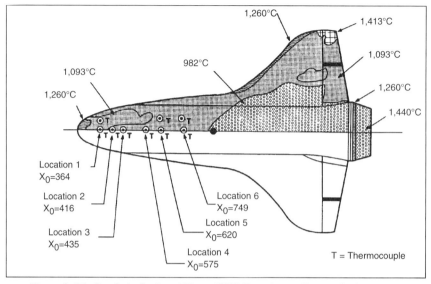

Figure 3–21. Catalytic Surface Effects (CSE) Experiment (Lower Surface View)

one of those experiments. Additional information on the Orbiter Experiments Program can be found in Chapter 3, "Space Transportation/Human Spaceflight," in Volume V of the *NASA Historical Data Book*.

OAST-1

OAST-1 was the primary payload on STS 41-D, which was launched August 30, 1984. Mission objectives were to demonstrate the readiness

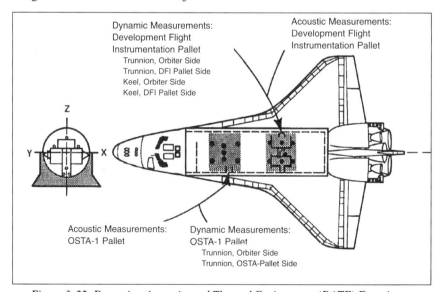

Figure 3–22. Dynamics, Acoustic, and Thermal Environment (DATE) Experiment

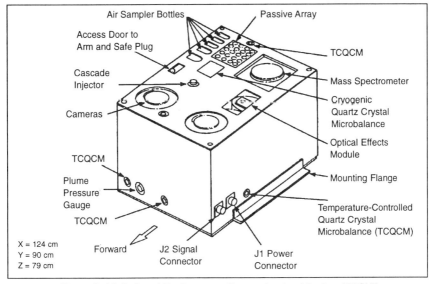

Figure 3–23. Induced Environment Contamination Monitor (IECM)

and determine the performance of large, low-cost, lightweight, deployable/retractable solar array technology, to demonstrate methods to define the structural dynamics of large space structures, and to evaluate solar cell calibration techniques as well as calibrate various types of solar cells. OAST-1 demonstrated the first large, lightweight solar array in space that could be restowed after it had been deployed.

The crew operated OAST-1 from the aft flight deck of the orbiter. The payload carrier was a triangular, truss-like mission support structure that spanned the width of the orbiter cargo bay. The payload consisted of three

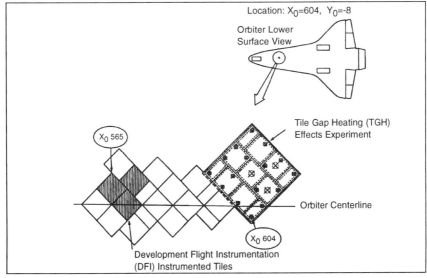

Figure 3–24. Tile Gap Heating (TGH) Effects

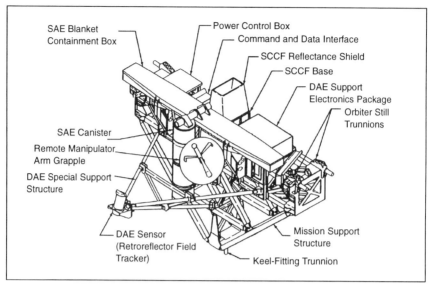

Figure 3–25. OAST-1 Payload Elements
(The OAST-1 payload consisted of three experiment systems that investigated solar energy and large space structures technology, both of which would be vital parts of a space station.)

major experiments and associated equipment: the Solar Array Experiment (SAE), the Dynamic Augmentation Experiment (DAE), and the Solar Cell Calibration Facility (SCCF). Figure 3–25 illustrates the OAST-1 payload elements.

The SAE demonstrated the properties and capabilities of the solar array. The Shuttle crew extended and retracted the solar array several times during the mission and gathered data on system performance. The experiment also measured deflections and bending motions on the fully deployed solar wing and gathered solar cell performance data. The solar array consisted of eighty-four panels and could fold flat. When fully extended, it rose more than ten stories above the cargo bay. When stored for launch and landing, the array folded into a package only seventeen centimeters thick.

The DAE gathered data to validate an on-orbit method to define and evaluate the dynamic characteristics of large space system structures. The SCCF evaluated and validated solar cell calibration techniques then used by the Jet Propulsion Laboratory under contract to NASA. This validation compared the performance of cells on orbit in the facility with the performance of the same cells flown on a high-altitude balloon test flight.

Long Duration Exposure Facility

The Long Duration Exposure Facility (LDEF) was a passive, free-flying reusable structure that accommodated experiments requiring long-term exposure to space. Launched from STS 41-C on April 7, 1984, it was retrieved by STS-32 in January 1990, after nearly six years of service, long after the original plan that had called for retrieval after a useful lifetime of approximately ten months. LDEF was funded by OAST's Systems Technology Program.

The Space Shuttle *Challenger* deployed LDEF on the second day of the mission. Astronaut Terry Hart used the Shuttle's 15.2-meter long remote manipulator arm to engage LDEF and maneuver it out of the payload bay. In the process, a startup signal was sent to electrical systems in the experiments. To move away from LDEF, the Shuttle fired small thrusters.

LDEF carried fifty-seven experiments in eighty-six desktop-sized, open aluminum trays arrayed around the surface of LDEF. Seventy-two experiments were located around the circumference of the facility; six were on the Earth-pointing end and eight on the space-pointing end. Together, all of the trays and their experiments weighed only 6,078 kilograms. The total weight of the structure, trays, and experiments was 9,707 kilograms. The experiments carried more than 10,000 specimens that gathered scientific data and tested the effects of long-term space exposure on spacecraft materials, components, and systems.

All the experiments required free-flying exposure in space but needed no extensive electrical power, data handling, or attitude control systems. The facility was designed to use gravity to be inherently stable in orbit. Thus, an experiment would keep a single orientation with respect to the orbit path. This allowed improved postflight data analysis because impacts and other space environment effects would differ for various orientations. In addition, the constancy of LDEF's drag as it moved through the uppermost traces of Earth's atmosphere also enhanced postflight data analysis.

The experiments fell into four groups: materials and structures, power and propulsion, science, and electronics and optics. They involved 194 principal investigators, representing sixteen U.S. universities, thirteen private companies, eight NASA centers, eight DOD laboratories, and thirty-four similar research organizations in Canada, Denmark, the Federal Republic of Germany (former West Germany), France, Ireland, The Netherlands, Switzerland, and the United Kingdom. NASA designed and built LDEF at Langley Research Center. NASA provided experiment trays to investigators, who built their own experiments, installed them in trays, and tested them.

LDEF had no central power system. Experiments that required power or data recording systems provided their own, although NASA made its Experiment Power and Data System available to investigators. The experiment initiation system, triggered by the orbiter's remote manipulator system, was the only electrical connection between LDEF and the active experiments.

Although LDEF carried a broad range of scientific and technological investigations on this mission, NASA first conceived of it solely as a meteoroid and exposure module (MEM). Langley Research Center proposed MEM in 1970 as the first Shuttle payload. MEM was foreseen as a cylinder sized for the Shuttle's payload bay. The Shuttle would place it in orbit, where its large surface area would collect a comprehensive sample of meteoroid data. MEM was to include thick-skin, thin-skin, and bumper configurations. After several months, the Shuttle would retrieve MEM and return it to Earth for data analysis.

In 1974, MEM was renamed LDEF, and LDEF officially became a NASA project managed by Langley Research Center for OAST. Meteoroid research was still seen as the primary mission. Eventually, however, LDEF also became a vehicle for many other types of studies, tests, and evaluations. Table 3–67 gives a brief chronology of LDEF project development. Table 3–68 describes the facility and mission characteristics.

Access Concept for Construction of Erectable Space Structures

The ACCESS experiment, which flew on STS 61-B along with the Experimental Assembly of Structures in Extravehicular (EASE) activity (launched November 26, 1985), gave astronauts the opportunity to erect the type of structure that would be used for the Space Station. Working in the payload bay at fixed workstations, crew members assembled small components to form larger structures during two spacewalks.

The structure consisted of ninety-three tubular aluminum struts, each 2.54 centimeters in diameter. Thirty-three were 1.37-meter-long struts; sixty were 1.8-meter-long diagonal struts; thirty-three were identical nodal joints; nine struts were used within and between bays; and six struts joined at one node. Once assembled, the structure was 13.7 meters high.

Langley Research Center developed ACCESS and worked with Marshall Space Flight Center in designing both ACCESS and EASE, developing assembly methods in ground-based and neutral buoyancy simulations, as well as assisting in crew training. Following the on-orbit experiment, the ACCESS experiment was repeated in a ground-based laboratory using a teleoperated manipulator. In this demonstration, the teleoperated manipulator system was substituted for one of the astronauts, while a technician assumed the role of the other astronaut. The demonstration proved that current manipulators had sufficient dexterity to assist the flight assembly of Space Station structures.

Experimental Assembly of Structures in Extravehicular Activity

EASE was a geometric structure that resembled an inverted pyramid. It was composed of a small number of large beams and nodes. When completely assembled, the structure was 3.7 meters high. The Massachusetts Institute of Technology developed the structure along with Marshall Space Flight Center. Crew members moved about the payload bay while assembling this structure.

Space Station Development

Many of OAST's space research and technology activities had direct application to Space Station development. Its materials and structures program worked to develop durable materials and design structures that could be erected and serviced in space. OAST provided data to support NASA's selection of the dual-keel configuration for the Space Station. The primary areas being investigated were lightweight structural members, packaging techniques, structurally predictable behavior, and reliable deployment. Its work in developing the technology base for high-performance, long-life power systems also had direct application to the power requirements of the Space Station.

The space human factors program focused on verifying human performance models in long-term weightless conditions such as those that would be encountered in assembling and operating the Space Station. Extravehicular activity (EVA) would be an important part of Space Station operations. OAST gathered quantitative data on the effects of types of spacesuits that could be used for EVA on human capabilities and productivity.

Other Space Research and Technology Activities

In addition to the activities already described, other major space research and technology activities are summarized below.

Space Energy Conversion

This area of research developed the technology base for high-performance, long-life power systems for space applications. It included research in the areas of solar power, space nuclear reactor power systems, batteries, and thermal systems.

In the area of solar power, OAST worked to define the effects of the space environment on space power systems. A space test evaluated the power loss and breakdown phenomena of photovoltaic systems as voltage levels and area varied.[101] This information would be used to correlate interaction phenomena measured in space and ground tests for eventual design guidelines for high-power space systems in low-Earth orbit.

In 1982, researchers identified components that could potentially revolutionize solar cell energy conversion, increasing efficiencies from 16 percent to as much as 50 percent. These concepts included coupling sunlight into the electronic surface charge density, cascading solar cell

---

[101]Photovoltaic describes a technology in which radiant energy from the Sun is converted to direct current electricity. U.S. spacecraft first used photovoltaic cells for power in 1958 (Photovoltaic Systems Assistance Center, U.S. Department of Energy).

junctions for selective spectral utilization, and exploiting the unique properties of the photo-active protein rhodopsin.

Work also continued on high-capacity energy storage for long-range missions. Researchers tested the breadboard model of a solid-polymer-electrolyte fuel-cell-electrolysis system. An alternate energy storage system based on electro-chemistry demonstrated an efficiency of 82 percent over 100 simulated day-night cycles in low-Earth orbit. This technology could reduce the weight of energy storage systems by one-half.

Researchers achieved significant improvements in the power per kilogram, cost, and efficiency of solar array power systems. In the area of low-cost solar arrays, researchers designed, fabricated, and tested a miniature Cassegrain concentrator with a concentration ratio of 100. This had the potential for reducing array costs to about $30 per watt, about one-twentieth the then-current cost.

In 1985, researchers made significant progress in solar cell and solar array technology to improve conversion efficiency, reduce mass and cost, and increase operating life. Over 20 percent conversion efficiency was demonstrated for gallium-arsenide thick-cell technology. Tests of thin-cell gallium arsenide verified cell efficiency exceeding 14 percent.

In 1987, researchers fabricated indium phosphide solar cells—a type that combines good performance and efficiency with improved tolerance to natural radiation. In geosynchronous orbit, conventional silicon solar cells can lose up to 25 percent of their output during a seven-year life. In the radiation belts in low-Earth orbit, the loss can be as high as 80 percent. Measurements showed indium phosphide cell efficiency to be essentially unaffected by natural radiation. This meant that future solar arrays could be smaller and lighter by eliminating the need for oversized systems to accommodate efficiency losses caused by radiation damage.

During the 1980s, NASA participated in the joint Space Nuclear Reactor Power System with the Department of Energy and DOD. Established in 1981, the jointly funded and managed program focused on the barrier technologies for space nuclear reactor power systems. Relating to dynamic energy conversion systems that could be used with the SP-100 nuclear reactor, the program investigated developing space power systems for future lunar and Martian bases. Researchers believed that outer planetary missions could be accomplished using a 120-kilowatt uranium-oxide-fueled reactor and silicon-germanium thermoelectric converters. Related research evaluated thermoelectric, thermionic, and Stirling cycle conversion systems.

In 1985, the Technology Assessment and Advancement Phase (Phase I) of the program was completed. The recommendation that the thermoelectric reactor power system concept be the baseline was approved. The partners executed a memorandum of agreement for Phase II of the program on October 8, 1985. The program also saw significant progress in free-piston Stirling energy conversion technology. In 1987, a 25-kilowatt, free-piston Stirling demonstration engine, the largest of its kind in the world, was built and tested.

More than 95 percent of NASA's Earth-orbiting spacecraft used nickel-cadmium batteries. Increasing the lifetime of these batteries would increase the operating lifetime of many satellites. During 1980, OAST's researchers developed methods to double the operational lifetime of the batteries by developing a technique of deep-discharge reconditioning.

In 1985, researchers were able to change the chemistry and design of nickel-hydrogen batteries, which resulted in a sixfold increase in the cycle life and seemed promising at meeting the 50,000-cycle requirement of systems in low-Earth orbit. As a result of these advances, nickel-hydrogen batteries became a prime candidate for energy storage on the Space Station and on other scientific platforms.

Materials and Structures

This area of research focused on improving the safety of existing vehicles and advancing the technology for future spacecraft, large-area space structures, and advanced space transportation systems. In 1983, a space-environmental-effects facility became operational at Langley Research Center. The facility could simulate the space environment to study effects on materials. It provided ground-based evaluation of the long-term environmental effects of space on materials and helped in developing new materials and protection techniques. The facility allowed for the testing of composite materials and for the observation of changes in structural properties.

This Materials and Structures Division at Langley had a large role in developing and improving thermal protection materials for use on the Space Shuttle. In 1981, the development of advanced ceramic tile made from a new material—fibrous, refractory, composite insulation—promised to offer lower cost, more durable protection. The addition of aluminum borosilicate fibers to the silica fiber already in use formed a new material with unique physical, mechanical, and thermal properties. It had greater strength, greater resistance to damage, and would save approximately 500 kilograms in the weight of each orbiter.

In 1982, laboratory tests demonstrated that another new, low-cost material could increase the life and durability of the orbiter's thermal protection shield. This advanced, flexible, and reusable surface insulation was a quilt-like sandwich with silica on the outside and microquartz felt in the middle. The layers, which were sewn together in the middle to form 2.54-centimeter squares, were for temperatures less than 650 degrees Centigrade. The use of advanced, flexible, and reusable surface insulation on the lee side of the orbiter offered more tolerance to damage, easier maintenance, and lower installation costs than the tiles then in use. In 1984, 2,300 flexible woven ceramic blankets replaced 8,000 existing ceramic tiles on the orbiter *Discovery*.

The Materials and Structures Division also continued investigating advanced thermal protection materials intended for future space transportation systems, such as the Orbital Transfer Vehicle. Silica and silicon carbide (Nicalon) fibers were woven in three-dimensional fabrics for the high-temperature thermal protection system applications required by these vehicles. A chemical vapor deposition approach also showed great promise for producing high-performance ceramic composite thermal protection systems.

Following the *Challenger* accident, the Materials and Structures Division was actively involved in the redesign of the solid rocket motor field joint. It established a test method for qualifying candidate O-ring materials and involved three tests: a resiliency test, a vibrational damage-resistance test, and a test that simulated in-situ conditions, including temperature, gas pressure, and controlled gap closure.

Propulsion

This discipline area focused on developing advanced propulsion systems. The systems were to be used in Earth-to-orbit ascent and planetary transfer vehicles and for orbiting spacecraft auxiliary propulsion systems. Its research emphasized high-performance and extending component life, thus extending maintenance intervals. Researchers developed cooling techniques that were tailored specifically to rocket engines as an alternative to cooling turbine blades with hydrogen fuel, which had proved inadequate. They developed a cryogenic engine-bearing model to determine cooling, lubrication, and bearing design characteristics. Another new model predicted the life of materials subjected to both low-cycle and high-cycle fatigue.

Orbital transfer propulsion research focused on developing high-performance, high-pressure, variable thrust engines that would be stored and fueled in space. Propulsion studies in conjunction with the analysis of orbit transfer vehicle systems indicated that multiple high-performance, low-thrust engines with 13,464 to 33,660 newtons of thrust were appropriate and cost-effective for a space-based, aero-assisted vehicle.

Electrical propulsion research focused on resistojets, arcjets, ion, and magnetoplasmadynamic (MPD) thrusters. In the Auxiliary Propulsion program, researchers determined that electric-powered thrusters known as arcjets offered more than twice the energy level per unit of fuel than conventional chemical systems provided. Arcjet technology objectives included developing high-temperature materials resistant to the electric arc and designing concepts for higher efficiency and longer life. A 1985 memorandum of agreement between NASA and the U.S. Air Force coordinated research activities of the two organizations, and several tests of the Air Force thirty-kilowatt arcjets were conducted in NASA laboratories.

MPD propulsion technology could become an option when megawatt power was available from nuclear power systems. MPD technology was capable of developing the highest specific impulse and relatively high thrust. A main technology goal was long-life cathodes capable of resisting erosion caused by intense heat (more than 1,650 degrees Celsius) and electric plasma. During 1985, tests in an MPD simulator demonstrated that there was less erosion at higher power than at low, indicating electron cooling effects. Compared to chemical systems, the MPD thruster proposed for advanced propulsion systems could potentially provide a two- to fourfold reduction in propellant mass. An MPD thruster was tested that had many advantages as a space propulsion system, being both simple in concept and compact in size.

Automation and Robotics

The Automation and Robotics program was established in FY 1985 in response to congressional interest in the Space Station and to reduce costs and increase the performance of future missions. It developed and demonstrated technology applicable to the Space Station, orbital maneuvering vehicle, orbital transfer vehicle, mobile remote manipulator system, and planetary rovers. The program accomplished major research goals in the areas of operator interface, systems architecture and integration, and planning and reasoning.

The Automation and Robotics program consisted of the Telerobotics program and the Autonomous Systems program. The Telerobotics program achieved its first major technology demonstration through the vision-based de-spin of a spinning satellite (once it was initialized by a human-guided graphic overlay). The Beam Assembly Teleoperator demonstrated three applications: assembling beam elements into a space structure, using a general control structure for coordinated movement of multiple robot arms; and using the Oak Ridge National Laboratory's tele-operated manipulator to recreate the ACCESS experiment.

The Autonomous Systems program moved toward the Space Station Thermal Control Expert System technology demonstration. The program developed an operational readiness prototype expert system for monitoring Space Shuttle communications systems. The program also developed an expert system for aiding the communications officer in the Shuttle Mission Control Room, which was first operational on STS-26 in 1988.

Communications

The objective of the Communications Technology program was to enable data transmission to and from low-Earth orbit, geostationary orbit, and solar and deep space missions. It represented three major research and development discipline areas: microwave and millimeter wave tube components, solid-state monolithic integrated circuits, and free space laser communications components and devices. Its activities ranged from basic research in surface physics to generic research on the dynamics of electron beams and circuits. Researchers investigated advanced semiconductor materials devices for use in monolithic integrated analog circuits, the use of electromagnetic theory in antennas, and the technology needed for the eventual use of lasers for free space communications for future low-Earth, geostationary, and deep space missions that required high data rates with corresponding directivity and reliability.[102]

In 1982, researchers developed the first sixty-gigahertz, low-noise receiver for spacecraft data transfer systems and completed development of a solid-state sixty-gigahertz power amplifier. This would permit high-transfer rates of large quantities of data in millimeter-wave intersatellite communications links—an important characteristic of an advanced, fully integrated ground-to-space communications system. In 1984, researchers demonstrated, for the first time, technology for a fifty-five-meter, offset wrap-rib antenna that would support a nationwide mobile communications system.

Advanced Communications Technology Satellite

OAST developed the fundamental technologies for the experimental Advanced Communications Technology Satellite (ACTS), funded and managed by NASA's Office of Space Science and Applications (see Chapter 2). The satellite demonstrated the critical communications technologies that would be needed for high-capacity operational satellites in the 1990s.

ACTS was a high-capacity domestic communications satellite operating in the Ka-band frequencies (thirty/twenty gigahertz) and was called the "switchboard in the sky." Performing in a largely untapped area of the frequency spectrum, the frequency bandwidth for ACTS in the Ka-band was twice the size of the C-band and Ku-band combined, thereby yielding a greatly increased satellite capacity.[103]

Hardware developments leading up to ACTS began in 1980 with NASA's thirty/twenty gigahertz program. It marked NASA's return to the communications field after an absence that began in 1973. The program formally began in 1984, and launch took place in 1993.

OAST's contributions to ACTS included a multibeam antenna with both fixed and scanned beams that could provide 100 times more power and ten times more bandwidth than other satellite systems. The antenna could produce three stationary and two hopping beams, with each beam encompassing an area approximately 250 kilometers in diameter. The ability to space the beams across expanses of territory allowed the use of the same frequency in many beams, which was called frequency reuse. The use of these high frequencies made wide bandwidth channels available. The satellite also featured a baseband processor, a high-speed programmable switch matrix, a traveling wave tube amplifier, and a low noise receiver.

---

[102]"NASA Information Sciences and Human Factors Program," p. 29.

[103]"Future Satellites to Carry Advanced Technologies," *Space News,* October 22–28, 1990, p. 18.

Computer Science

The Computer Science program was established in 1982 to adapt supercomputer technology, human computer interfaces, and artificial intelligence to aerospace applications, to advance computer technology where NASA requirements push the state of the art, and to provide advanced computational facilities for aerospace research. The program worked at improving knowledge of fundamental aerospace computing principles and advancing computing technology in space applications such as software engineering and information extraction from data collected by scientific instruments in space. Emphasis was placed on producing highly reliable software for critical space applications.

The program included the development of special algorithms and techniques to exploit the computing power provided by high-performance parallel processors and special-purpose architectures. Important areas included computational fluid dynamics, computational chemistry, structural analysis, signal processing, and image processing.

Work in the area of the fundamentals of database logic resulted in the development of a common user interface for accessing data from several databases, even when the databases had very different structures. This work provided the foundation to allow NASA space data users to access multiple databases independently of their physical distribution or structure. It would reduce the cost of investigations and enable database-intensive scientific research that would otherwise be unaffordable.[104]

Researchers in the program were also developing a reconfigurable, fault-tolerant architecture for a space-borne symbolic processor. This effort included addressing the issues of software development environment versus run-time environment, dynamic database maintainability, and an operating system for efficient use of the multiprocessor architecture.

In 1982, researchers developed an experimental computer program for automatically planning and scheduling spacecraft action sequences. The program combined, for the first time, artificial intelligence technology with operations research and discrete-event simulation techniques to perform automatically tasks that usually required many mission operations personnel. Later, in 1984, NASA realized major performance improvements in an automated planning program called DEVISER, which used artificial intelligence techniques to plan and schedule spacecraft operations automatically. The planning system, which was tailored for use by the Voyager spacecraft during its encounter with Uranus, exhibited sufficiently high levels of sophistication and capability for realistic planning of major mission sequences involving as many as 100 distinct tasks.

---

[104]"NASA Information Sciences and Human Factors Program," p. 65.

Controls and Guidance

This program was directed toward enabling future space transportation systems, large future spacecraft, and space systems to have large communications antennas and high-precision segmented reflector astrophysical telescopes. To address the advanced requirements of future systems and spacecraft, the program focused on providing the generic technology base to support the implementation of advanced guidance, navigation, and control. This technology had the capability to reduce the number of people needed to plan and generate mission software and to later provide for mission control.

The area of computational controls was stressed to develop cost-effective, high-speed, and high-fidelity control system simulation and analysis tools. The thrust of the work was to develop methods and software to enable the analysis and real-time simulation of complex spacecraft for control design certification.[105]

Data Systems

The Data Systems program consisted of research and technology focused on controlling, processing, storing, manipulating, and analyzing space-derived data. The objectives of the program were to provide the technology advancements needed to enable the affordable use of space-derived data, to increase substantially the capability for future missions of on-board processing and recording, and to provide high-speed and high-volume computational systems anticipated for missions such as the Space Station and the Earth Observing System.

The program supported fundamental research in areas such as laser diodes, worked to select and provide the appropriate on-board processor technology for future NASA missions, and supported the development of two flight processors with special architectures. The ongoing support for solid-state laser research led directly to the development of a nine-laser diode array used in the Optical Disk Recorder. The laser research also focused on Space Station data handling applications. Also, the Data Systems program focused on providing processors that would work very reliably in the space environment, including missions in polar orbit and some planetary missions that must operate in high-radiation environments.

The NASA End-to-End Data System (NEEDS) was an OAST major data systems program. NEEDS defined system configurations and developed enabling techniques and technology for the NASA-wide information systems of the 1980s. Studies performed as part of this program concluded that space-acquired data in "packet" form should be an integral part of long-term NASA data system architectures.

---

[105]*Ibid.*, p. 127.

Standard packet data would reduce end-to-end data transport costs by allowing a high degree of automation, eliminate the need for unique mission hardware and software for acquisition, staging, and distribution, simplify quality control for all data types, enable deterministic data accountability, allow autonomous instrument formatting, and establish high-level interfaces that were constant throughout the life of an instrument.

One of the major elements of the program was the massively parallel processor (MPP), put into operation at Goddard Space Flight Center in 1983. Built and delivered by Goodyear Aerospace Corporation after four years of development, this "multibillion-operations-per-second" computer consisted of 16,384 processors and was designed for image processing. The MPP permitted modeling of complex space science phenomena not possible with conventional computers. NASA used the MPP for weather and climate modeling and image analysis research.

One of the major applications of the MPP was in the area of image processing from very-high-spatial-resolution image sensors, both active and passive. These sensors generated data at rates up to 10^{13} bits per day, requiring from 10^9 to 10^{10} operations per second for processing. Another application was the assimilation of data for the Global Habitability model that involved the merging of data from various imaging sensors as the Thematic Mapper and the SIR-B synthetic aperture imaging radar and the creation of the images from raw data as required for these sensor systems. Other applications involved signal processor of LIDAR and radar data, infrared and microwave sounder processing, and numerical modeling simulations of climate.[106]

The Advanced Digital Synthetic Aperture Radar (SAR) processor (ADSP) included a special architecture and algorithms to process SAR data. On-board SAR processing was a very challenging technical problem because of the enormous volumes of raw SAR data that the instrument would collect. The Jet Propulsion Laboratory's ADSP, a six-gigaflop processor capable of providing Seasat and SIR-B imagery in real time, was successfully demonstrated in 1986. However, in spite of the successful demonstration, it did not provide the technology required for real-time on-board processing of SAR data because it occupied two meter racks just over two meters and consumed a total of twenty kilowatts. The SAR processor then being developed would use data compression to reduce the data rate and volume problems imposed on the Tracking and Data Relay Satellite System downlink. On-board SAR processing would also allow for the direct downlink of on-board-generated images for users who require images in near real time.[107]

---

[106] "Annual Report, 1983," NASA Computer Science and Electronics Program (no page numbers).

[107] "NASA Information Sciences and Human Factors Program," p. 174.

Human Factors

The Human Factors program focused on developing a technology base for intelligent operator interfaces, especially with autonomous subsystems, and developing a new generation of high-performance spacesuits, gloves, and tools and end effectors to meet the requirements of advanced space systems. Crew station research included the development of methods for the astronaut to supervise, monitor, and evaluate the performance of robotic systems, other space subsystems, and orbital vehicles. A fundamental understanding of the human visual and information integration capabilities provided a technical basis to develop mathematical, anthropometric, and graphical models of human interactions with space systems and equipment. Virtual workstation research demonstrated the initial feasibility to perceive, evaluate, and control robotic assistants as well as computer-generated images of actual systems and space structures. This research also could make it possible to interact with these via computer models.

The development of a new EVA spacesuit and gloves was a second major research area. The completion of the AX-5 hard suit and its initial test for mobility and ease of use were major accomplishments. This suit was a prime candidate for use on the Space Station; it allowed the astronaut to don the spacesuit without extensive prebreathing of oxygen. Also under development was a project to study and develop end-effector mechanisms whereby the EVA-suited astronaut could control and supervise robotic assistants. The research program also included the development of new methods to display information on the spacesuit's visor to allow the astronaut to interact with displayed information by means of voice commands.[108] Researchers also designed an ultrawide field-of-view helmet-mounted display for the visual monitoring of remote operations.

Sensor Systems Technology

The Sensor Systems program provided expertise and technology to advance space remote sensing of terrestrial, planetary, and galactic phenomena through the use of electromagnetic and electro-optic properties of gas, liquid, and solid-state materials technology. The research and development part of this program consisted of research on artificially grown materials such as quantum well and superlattice structures, with the potential for new and efficient means for detecting electromagnetic phenomena. Research was also conducted on unique materials and concepts for detector components and devices for measuring high-energy phenomena such as ultraviolet rays, x-rays, and gamma rays that are required observables in astrophysical and solar physics missions. The focused technology part of the program was balanced among the areas of detector sensors, submillimeter wave sensors, LIDAR/differential absorption sensors, and cooler technology.[109]

---

[108]*Ibid.*, p. 209.
[109]*Ibid.*, p. 231.

Table 3–1. Total Office of Aeronautics and Space Technology Program Funding (in thousands of dollars)

Year	Request	Authorization	Appropriation	Programmed (Actual)
1979	375,400	391,400	a	376,400
1980	427,100 b	433,700	c	426,866
1981	389,500 d	410,000	384,750 e	384,000
1982	344,000 f	414,100	375,800 g	375,800
1983	355,000 h	408,000	403,000	404,500
1984	439,300 i	463,300	440,300	452,300
1985	492,400	502,400	496,000	492,400
1986	522,000 j	520,000	522,000	488,657
1987	592,200 k	599,200	601,200	625,000
1988	691,000 l	687,000	665,000	606,700

a Undistributed. Total 1979 R&D appropriation = $3,477,200,000.
b Amended budget submission. Original budget submission = $419,700,000.
c Undistributed. Total 1980 R&D appropriation = $3,838,500,000.
d Amended budget submission. Original budget submission = $409,500,000.
e Reflected recission.
f Amended budget submission. Original budget submission = $469,000,000.
g Reflected General Supplemental Appropriation of August 13, 1982, which was approved on September 10, 1982.
h Initial budget submission. Revised submission unspecified.
i Revised budget submission. Initial budget submission = $438,300,000.
j Initial budget submission. Revised submission unspecified.
k Amended budget submission. Original budget submission = $601,200,000.
l Amended budget submission unchanged from original submission.

*Table 3–2. Major Budget Category Programmed Funding History
(in thousands of dollars)*

Budget Category/Fiscal Year	1979	1980	1981	1982	1983
Aeronautical Research and Technology	264,100	308,300	271,400	264,800	280,000
Research and Technology Base	109,700	120,767	133,847	172,758	198,475
Systems Technology Programs	154,400	187,533	137,533	92,042	81,525
Space Research and Technology	107,300	115,586	110,700	111,000	124,500
Research and Technology Base	86,277	99,816	100,380	104,646	116,304
Systems Technology Programs	12,023	10,770	8,220	3,354	5,196
Standards and Practices *a*	9,000	5,000	2,100	3,000	3,000
Energy Technology Applications	5,000	3,000	1,000	*b*	—

Budget Category/Fiscal Year	1984	1985	1986	1987	1988
Aeronautical Research and Technology	315,300	342,400	337,257	374,000	332,900
Research and Technology Base	228,450	223,298	228,557	271,111	257,150
Systems Technology Programs	86,850	119,102	108,700	102,889	75,750
Space Research and Technology	137,000	150,000	151,400	206,000	221,300
Research and Technology Base	124,885	136,358	124,200	130,646	107,146
Systems Technology Programs/ Civil Space Technology Initiative	7,515	8,742	27,200	75,354	114,154 *c*
Standards and Practices	4,600	4,900	— *d*	—	—
Transatmospheric Research and Technology	—	—	—	45,000	52,500

a   Formerly called Low Cost Systems Program.
b   Program terminated.
c   Systems Technology Programs terminated, and Civil Space Technology Initiative begun.
d   No programmed amount.

Table 3–3. Total Aeronautical Research and Technology Program Funding (in thousands of dollars)

Year	Request	Authorization	Appropriation	Programmed (Actual)
1979	264,100	275,100	a	264,100
1980	308,000 b	309,300	c	308,300
1981	275,300 d	290,800 e	276,150 f	271,400
1982	264,800 g	284,800 h	264,800	264,800
1983	280,000 i	280,000 j	280,000	280,000
1984	302,300 k	320,300	302,300	315,300
1985	342,400	352,400	342,400	342,400
1986	354,000	354,000	354,000	337,257
1987	376,000 l	376,000	376,000	374,000
1988	375,000	387,000	377,000	332,900

a   Undistributed. Total 1979 R&D appropriation = $3,477,200,000.
b   Amended budget submission. Original budget submission = $300,300,000. The increase resulted from congressional actions that provided additional appropriations of $5,000,000 for advanced rotorcraft technology and $3,000,000 for variable cycle technology.
c   Undistributed. Total 1980 R&D appropriation = $3,838,500,000.
d   Amended budget submission. Original budget submission = $290,300,000.
e   The Senate authorization committee added $20,500,00 to be distributed as follows: Variable Cycle Engine High Temperature Validation—$4,500,000, High Performance Flight Experiment—$5,500,000, High Speed Structures—$4,000,000, Alternative Fuels Utilization—$4,000,000, Alternative Alloys Studies—$1,000,000, and General Aviation Propeller—$1,500,000. Conference committee authorization of $10,500,000 to have $290,800,000 total.
f   Reflects effect of General Provision Section 412. Appended to appropriation on December 15, 1980.
g   Amended budget submission. Original budget submission = $323,600,000.
h   The House authorization committee increased the amount submitted by NASA to be allocated as follows: $3,000,000 for alternative fuels and materials, $8,000,000 for high-speed systems technology, $4,000,000 for large composite structures, $4,000,000 for high-temperature engine core, $12,000,000 for propfan, and a general reduction of $19,800,000 for a total of $272,000,000 (May 8, 1981). Further debate resulted in a House Authorization Bill for $264,000,000 (June 23, 1981), with specific reductions from earlier amounts not specified). The Senate authorization committee added $51,200,000 for various systems technology programs, amounting to a Senate authorization of $316,000,000.
i   Revised budget submission. Initial budget submission = $232,000,000.
j   Increase applied to the Systems Technology Program.
k   Amended budget submission. Original budget submission = $300,300,000.
l   Amended budget submission unchanged from original submission.

Table 3–4. Aeronautics Research and Technology Base Funding History (in thousands of dollars)

Year	Request	Authorization	Appropriation	Programmed (Actual)
1979	109,200	a	b	109,700
1980	119,000 c	117,500	d	120,767
1981	134,100 e	131,100	f	133,847
1982	162,500 g	h	157,800	172,758
1983	182,000	182,000 i	182,000	198,475
1984	215,800 j	205,100 k	217,800 l	228,450
1985	233,300	233,300	223,300	223,298
1986	239,300	239,300	239,000	228,557
1987	272,900	272,900 m	272,900	271,111
1988	285,200	297,200 n	287,200	257,150

a   Undistributed. Total 1979 Aeronautical Research and Technology Program authorization = $275,100,000.
b   Undistributed. Total 1979 R&D appropriation = $3,477,200,000.
c   Amended budget submission. Original budget submission = $117,500,000. Congressional actions resulted in a transfer of $1,500,000 from the Systems Technology Program to the Research and Technology Base Program.
d   Undistributed. Total 1980 R&D appropriation = $3,838,500,000.
e   Amended budget submission. Original budget submission = $131,100,000.
f   Undistributed. Total 1981 Aeronautical Research and Technology Program appropriation = $276,150,000.
g   Amended budget submission. Original budget submission = $160,800,000.
h   Undistributed. However, both the House and Senate authorization committees authorized the identical amount of $157,800,000.
i   The House authorization committee added $6,000,000, but the amount was deleted in the conference committee.
j   Amended budget submission. Original budget submission = $227,800,000.
k   Reduction to offset increases in Systems Technology Program authorization (see Table 3–20).
l   Reduction to offset increases in Systems Technology Program appropriation (see Table 3–20).
m   The Senate authorization committee authorized allocation of funds as follows: Fluid and Thermal Physics Research and Technology—$49,500,000, Applied Aerodynamics Research and Technology—$57,100,000, Propulsion and Power Research and Technology—$35,700,000, Materials and Structures Research and Technology—$39,000,000, Information Sciences Research and Technology—$26,800,000, Controls and Guidance Research and Technology—$24,500,000, Human Factors Research and Technology—$24,000,000, Flight Systems Research and Technology—$21,500,000, and Systems Analysis—$4,800,000.
n   The Senate authorization committee authorized allocation of funds as follows: Fluid and Thermal Physics Research and Technology—$29,000,000, Applied Aerodynamics Research and Technology—$61,000,000, Propulsion and Power Research and Technology—$41,000,000, Materials and Structures Research and Technology—$42,000,000, Information Sciences Research and Technology—$26,000,000, Controls and Guidance Research and Technology—$27,600,000, Human Factors Research and Technology—$26,000,000, Flight Systems Research and Technology—$26,100,000, and Systems Analysis—$6,500,000 for a total of $285,200.000. The conference committee increased authorization to $297,200,000, with the increase unspecified by category.

Table 3–5. Aerodynamics (Fluid and Thermal Physics) Research and Technology Funding History (in thousands of dollars) a

Year (Fiscal)	Submission	Programmed (Actual)
1979	16,500	18,500
1980	22,240	22,587
1981	23,200	23,800
1982	37,100	38,505
1983	43,100	42,665
1984	44,700	43,404
1985	44,000	28,498
1986	30,400	29,210
1987	39,500	39,141
1988	24,600	23,718

a   Redesignated Fluid and Thermal Physics Research and Technology with FY 1981 revised budget estimate.

Table 3–6. Propulsion (and Power) Systems Research and Technology Funding History (in thousands of dollars)

Year (Fiscal)	Submission	Programmed (Actual)
1979	20,900 a	25,500
1980	26,900	26,436
1981	32,400	31,800
1982	37,100	18,616 b
1983	16,600	16,600
1984	20,000	23,500
1985	28,700	33,636
1986	33,800	32,355
1987	38,700	41,365
1988	45,800	46,662

a   Combined Propulsion Environmental Impact Minimization Research and Technology and Propulsion Components Research and Technology budget categories.

b   Proportion of Propulsion Systems Research and Technology transferred to Fluid and Thermal Physics Research and Technology budget category (see Table 3–5).

Table 3–7. Materials and Structures (Aeronautics) Research and Technology Funding History (in thousands of dollars)

Year (Fiscal)	Submission	Programmed (Actual)
1979	17,900 a	15,200
1980	16,100	16,077
1981	19,300	17,800
1982	22,100	21,548
1983	24,700	23,200
1984	23,200	23,900
1985	27,800	27,800
1986	29,500	27,830
1987	39,000	35,536
1988	37,200	28,453

a   Combined Materials Research and Technology and Structures Research and Technology budget categories.

Table 3–8. General Aviation Research and Technology Funding History (in thousands of dollars)

Year (Fiscal)	Submission	Programmed (Actual)
1979	— a	6,800
1980	7,000	7,009
1981	7,500	6,600
1982	7,700	— b

a   No equivalent budget category.
b   Budget category eliminated.

Table 3–9. Applied Aerodynamics Research and Technology Funding History (in thousands of dollars) a

Year (Fiscal)	Submission	Programmed (Actual)
1984	— b	42,300
1985	42,000	50,900
1986	55,300	51,680
1987	56,100	55,885
1988	52,800	56,868

a   Includes programs for high-performance aircraft research and technology, powered-lift research and technology, flight dynamics, supersonic aircraft integration technology, rotorcraft research and technology, laminar flow control research, and subsonic configuration/propulsion/airframe integration.
b   No budget category.

Table 3–10. Low Speed (Subsonic) Aircraft Research and Technology Funding History (in thousands of dollars) a

Year (Fiscal)	Submission	Programmed (Actual)
1979	13,000	11,100
1980	11,200	13,884
1981	11,700	9,600
1982	11,300	13,538
1983	7,500 b	9,000
1984	18,900 c	— d

a Category includes rotorcraft through FY 1982.
b Reduction in amount from previous year reflected reduced activity in materials and structures and aerodynamics systems research.
c Increase over previous year reflected redirected funding from other ongoing research and technology base programs to support research in laminar flow control and advanced transport operations systems. Also supported general aviation aerodynamics and flight dynamics efforts.
d Incorporated into Applied Aerodynamics Research and Technology (see Table 3–9).

Table 3–11. High Speed (High-Performance) Aircraft Research and Technology Funding History (in thousands of dollars)

Year (Fiscal)	Submission	Programmed (Actual)
1979	13,900	14,000
1980	14,800	13,846
1981	16,500	20,600
1982	26,000	29,029
1983	38,000	39,240
1984	37,000	— a

a Incorporated into Applied Aerodynamics Research and Technology (see Table 3–9).

Table 3–12. Rotorcraft Research and Technology Funding History (in thousands of dollars)

Year (Fiscal)	Submission	Programmed (Actual)
1982	—	20,175 a
1983	23,000	23,000
1984	23,000	— b

a First time budget category used in NASA Budget Estimate.
b Incorporated into Applied Aerodynamics Research and Technology (see Table 3–9).

Table 3–13. Avionics and Flight Control (Aircraft Controls and Guidance) Research and Technology Funding History (in thousands of dollars) a

Year (Fiscal)	Submission	Programmed (Actual)
1979	5,200	4,000
1980	4,800	4,804
1981	5,400	5,400
1982	7,000	7,119
1983	11,900	11,900
1984	12,200	19,602
1985	20,500	20,600
1986	22,100	20,653
1987	24,100	22,789
1988	21,200	20,905

a Renamed Aircraft Controls and Guidance with FY 1982 revised budget estimate.

Table 3–14. Human Factors Research and Technology Funding History (in thousands of dollars) a

Year (Fiscal)	Submission	Programmed (Actual)
1979	5,600	5,000
1980	5,700	5,872
1981	6,500	6,147
1982	8,000	8,218
1983	10,200	10,070
1984	10,500	19,934
1985	20,300	20,300
1986	22,000	21,360
1987	24,000	23,954
1988	20,600	20,495

a Formerly called Human-Vehicle Research and Technology.

Table 3–15. Multidisciplinary Research and Technology Funding History (in thousands of dollars) a

Year (Fiscal)	Submission	Programmed (Actual)
1979	— a	3,600
1980	3,760	3,760
1981	4,700	5,000
1982	6,000	7,500
1983	3,500	3,600
1984	3,700	— b

a No equivalent budget category
b Budget category eliminated.

Table 3–16. Transport Aircraft Research and Technology Funding History (in thousands of dollars) a

Year (Fiscal)	Submission	Programmed (Actual)
1979	5,500	6,000
1980	6,500	6,492
1981	7,100	7,100
1982	8,100	— b

a   Formerly called Aircraft Operations and Aviation Safety Research and Technology.
b   No budget category in NASA Budget Estimate.

Table 3–17. Computer Science and Applications (Information Science) Research and Technology Funding History (in thousands of dollars)

Year (Fiscal)	Submission	Programmed (Actual)
1982	—	8,510 a
1983	19,200	19,200
1984	22,300	34,943 b
1985	21,100	21,100
1986	24,900	23,816
1987	23,800	23,800
1988	19,000	19,189

a   First time budget category used in NASA Budget Estimate.
b   Renamed Information Sciences Research and Technology.

Table 3–18. Flight Systems Research and Technology Funding History (in thousands of dollars)

Year (Fiscal)	Submission	Programmed (Actual)
1984	—	17,504 a
1985	16,300	17,864
1986	18,300	17,891
1987	21,900	23,134
1988	24,800	25,400

a   First time budget category used in NASA Budget Estimate.

Table 3–19. System Studies Funding History (in thousands of dollars)

Year (Fiscal)	Submission	Authorization	Appropriation	Programmed (Actual)
1979	3,000	— a	— b	— c
1980	3,200	3,200	— d	— e

a   Undistributed. Total 1979 Aeronautical Research and Technology Program authorization = $275,100,000.
b   Undistributed. Total 1979 R&D appropriation = $3,477,200,000.
c   Incorporated in Aeronautical System Studies Technology Programs budget category (see Table 3–29).
d   Undistributed. Total 1980 R&D appropriation = $3,838,500,000.
e   No budget category in NASA Budget Estimate. See Aeronautical System Studies Technology Programs (Table 3–29).

Table 3–20. Systems Technology Program Funding History (in thousands of dollars)

Year (Fiscal)	Request	Authorization	Appropriation	Programmed (Actual)
1979	85,645	— a	— b	154,400
1980	106,100 c	115,100 d	— e	187,533
1981	141,850 f	159,700	— g	137,553
1982	70,500 h	— i	107,000	92,042
1983	82,300 j	98,000 k	98,000	81,525
1984	86,500 l	115,200 m	84,500 n	86,850
1985	119,100 o	119,100 p	119,100	119,102
1986	113,700 q	114,700 r	114,700	108,700
1987	103,100	103,100 s	103,000	102,889
1988	83,200 t	89,800 u	89,800	75,750

a   Undistributed. Total 1979 Aeronautical Research and Technology Program authorization = $275,100,000.
b   Undistributed. Total 1979 R&D appropriation = $3,477,200,000.
c   NASA Budget Estimate as published by the NASA Comptroller's Office does not break out Experimental Programs and System Studies as appearing in chronological history of congressional action. Total 1980 submission according to Comptroller's Office = $189,900,000.
d   Increase over NASA submission of $9,000,000 for Variable-Cycle Engine Technology ($4,000,000) and Advanced Rotorcraft Technology ($5,000,000).
e   Undistributed. Total 1980 R&D appropriation = $3,838,500,000. Notes indicate that $4,000,000 was to be allocated for Variable-Cycle Engine Technology and $5,000,000 for Advanced Rotorcraft Technology.
f   Amended budget submission. Original budget submission = $159,200,000.
g   Reflected recission.
h   Amended budget submission. Original budget submission = $162,800,000. Deemphasis on Systems Technology reflects the objective of reducing the federal role in areas that directly support industry for product development, while retaining those efforts related to longer range technology and to defense considerations.
i   Undistributed. Total 1982 Aeronautical Research and Technology Program authorization = $284,800,000.
j   Increase was attributable to a congressional increase that provided continuing support for selected ongoing programs in the advanced propulsion, subsonic aircraft, and rotorcraft systems technology areas.
k   The House authorization committee increased amount to $79,100,000 to be applied as follows: Energy Efficient Transport—$1,100,000, Advanced Turboprops—$9,800,000, Energy

AERONAUTICS AND SPACE RESEARCH AND TECHNOLOGY 269

Table 3–20 continued

Efficient Engine—$7,000,000, Terminal Configured Vehicle—$5,000,000, Turbine Engine Hot Section Technology—$4,700,000, Advanced Rotorcraft Technology and Helicopter Transmission Research—$4,500,000, Broad Property Fuels Technology—$4,200,000, Powered Lift and Tilt Rotor Technology—$3,800,000, and Research and Technology Base—$6,000,000. To partially offset the additions, the committee reduced the amounts for activities that were primarily directed at military application: Low Speed Systems Technology and High Speed Systems Technology ($13,000,000). The Senate authorization committee eliminated additional amount for Research and Technology Base but allocated additional funds to Systems Technology as follows: Aeronautical Systems Studies—$2,000,000, Turbine Engine Hot Section—$2,500,000, Broad Property Fuels—$3,000,000, Helicopter Transmission—$1,500,000, Critical Aircraft Resources—$2,200,000, General and Commuter Aviation—$3,000,000, Composite Primary Aircraft Structure—$6,000,000, Energy Efficient Transport—$1,100,000, Terminal Configured Vehicle—$4,600,000, Laminar Flow Control—$3,000,000), Energy Efficient Engine—$7,500,000), and Advanced Turboprop—$27,600,000 for a total of $114,000,000. The conference committee reallocated the authorized funds so that the amounts added to the NASA request went to acceleration of advanced turboprop ($15,000,000), composite primary aircraft structures ($6,000,000), general and commuter aviation including small engine component technology ($3,000,000), broad property fuels ($3,000,000), energy efficient engine ($7,000,000), energy efficient transport ($3,000,000), and terminal configured vehicle ($5,000,000). The remaining $6,000,000 was to be applied to those projects that NASA considered most feasible.

l   Revised budget submission. Original budget submission = $72,000,000.

m   The House appropriations committee added $20,000,000 for Advanced Turboprop and $14,000,000 for Advanced Transport Operating System (ATOPS). The Senate authorization committee increased the amount for ATOPS to fund a total of $22,700,000 for Systems Technology. Advanced Turboprop remained at $20,000,000.

n   The House authorization committee added $20,000,000 for Advanced Turboprop (taking $10,000,000 from Research and Technology Base) and reduced the amount for Numerical Aerodynamic Simulation by $5,000,000 and the amount for ATOPS by $5,000,000. The Senate appropriations committee reduced the amount for Numerical Aerodynamic Simulation by $3,000,000 and added $10,000,000 to Advanced Propulsion and Composite Materials. The appropriations conference committee reduced the amount for Numerical Aerodynamic Simulation by $3,000,000 and added $15,000,000 to Advanced Turboprop, Composite Materials, and Laminar Flow.

o   Revised budget submission. Original budget submission = $109,100,000.

p   The Senate authorization committee authorized allocation of Systems Technology Program funds as follows: Rotorcraft Systems Technology—$26,600,000, High-Performance Aircraft Systems Technology—$21,000,000, Subsonic Aircraft Systems Technology—$19,000,000, Advanced Propulsion Systems Technology—$31,100,000, and Numerical Aerodynamic Simulation—$26,500,000 for a total of $124,100,000. The conference committee returned to the $19,100,000 amount. Of the total authorization, $24,000,000 was authorized only for activities that were designed to lead to a flight test of a single rotation or counter-rotation turboprop concept no later than 1987 (and for supporting research and technology).

q   Revised budget submission. Original budget submission = $114,700,000.

r   The Senate authorization committee authorized allocation of Systems Technology Program funds as follows: Rotorcraft Systems Technology—$20,500,000, High-Performance Aircraft Systems Technology—$21,800,000, Subsonic Aircraft Systems Technology—$0 (to be terminated at the end of 1985), Advanced Propulsion Systems Technology—$44,200,000, and Numerical Aerodynamic Simulation—$28,200,000.

s   The authorization committee allocated funds as follows: Rotorcraft Systems Technology—$18,700,000, High-Performance Aircraft Systems Technology—$26,000,000, Advanced Propulsion Systems Technology—$28,400,000, and Numerical Aerodynamic Simulation—$30,000,000.

t   Revised budget submission. Original budget submission = $3,200,000.

u   The Senate authorization committee allocated funds as follows: Rotorcraft Systems Technology—$5,000,000, High-Performance Aircraft Systems Technology—$14,000,000, Advanced Propulsion Systems Technology—$30,500,000, and Numerical Aerodynamic Simulation—$39,700,000.

Table 3–21. Aircraft Energy Efficiency Funding History (in thousands of dollars) a

	1979	1980	1981	1982	1983	1984	1985	Total b
Engine Component Improvement	12.9	6.1	—	—	—	—	—	39.5
Energy Efficient Engine	43.2	55.6	48.7	22.9	—	—	—	198.0
Energy Efficient Transport	13.9	25.4	15.3	9.1	—	—	—	85.0
Composite Primary Aircraft Structures	20.5	15.9	6.4	22.9	—	—	—	94.3
Laminar Flow Control	3.7	10.3	11.6	4.5	—	—	—	37.2
Advanced Turboprops Phase I	3.2	2.8	—	—	—	—	—	8.0
Advanced Turboprops Phase II	—	—	3.0	6.0	8.0	9.0	6.0	32.0
Total (NASA Share)	**97.4**	**116.1**	**85.0**	**46.7**	**9.6**	**9.0**	**6.0**	**494.0**
Industry Share	(7.9)	(10.6)	(7.2)	(2.6)	(—)	(—)	(—)	(37.2)

a   Aircraft Energy Efficiency appeared as a supplement in the 1980 Budget Estimate only (although some of the subcategories appeared at other times). It combined particular subcategories to form a new initiative. The Aircraft Energy Efficiency budget categories were subcategories to the Transport Aircraft Systems Technology and Advanced Propulsion Systems budget categories. Presumably, the FY 1979 figures are actuals, and the FY 1980 and future figures are estimates.

b   Includes funding prior to FY 1979.

Table 3–22. Materials and Structures Systems Technology Funding History (in thousands of dollars)

Year (Fiscal)	Submission	Programmed (Actual)
1979	4,500	3,300
1980	5,555	5,553
1981	9,600	8,715
1982	6,600	1,600
1983	— a	—
1987 b	—	(7,200) c
1988	8,800	8,818

a   The integrated program for aerospace vehicle design continued under Computer Science and Applications Research and Technology budget category. Other activities concluded in FY 1982.

b   No Materials and Structures Systems Technology budget category from FY 1983 to FY 1987.

c   Budget category reinstated. The Advanced High-Temperature Engine Materials Technology Program was transferred from the Advanced Propulsion Systems Technology Program to Materials and Structures Systems Technology.

AERONAUTICS AND SPACE RESEARCH AND TECHNOLOGY

Table 3–23. Low Speed (Subsonic) Aircraft Systems Technology Funding History (in thousands of dollars) a

Year (Fiscal)	Submission	Programmed (Actual)
1979	14,545 b	14,970
1980	23,250	23,175
1981	24,300	23,511
1982	25,600	27,022
1983	17,000	16,975
1984	5,000 c	5,000
1985	19,000	19,000

a Category includes rotorcraft through FY 1982.
b Combined Aircraft Operating Systems Technology and Rotorcraft Systems Technology budget categories.
c Reduced request because of the completion of several activities.

Table 3–24. High Speed (High-Performance) Aircraft Systems Technology Funding History (in thousands of dollars)

Year (Fiscal)	Submission	Programmed (Actual)
1979	11,060	9,800
1980	14,800	14,695
1981	16,700	16,615
1982	7,700	13,800
1983	15,000	14,950
1984	19,900	19,900
1985	21,500	21,530
1986	20,800	17,800
1987	26,000	25,985
1988	12,800	5,430

Table 3–25. Propulsion Systems Technology Funding History (in thousands of dollars)

Year (Fiscal)	Submission	Programmed (Actual)
1979	28,400 a	3,600
1980	6,700	6,700
1981	4,900	4,400
1982	500	500
1983	— b	—

a The large difference between submission and programmed amounts reflected new Advanced Propulsion Systems budget category and the relocation of some other propulsion systems functions in other budget categories.
b Technology efforts continued under the Propulsion Systems Research and Technology Program.

Table 3–26. Avionics and Flight Control Systems Technology Funding History (in thousands of dollars)

Year (Fiscal)	Submission	Programmed (Actual)
1979	2,400	3,000
1980	2,850	1,206
1981	1,200	1,200
1982	1,300	1,300
1983	— a	—

a Activities were transferred to Aircraft Controls and Guidance Research and Technology budget category.

Table 3–27. Transport Aircraft Systems Technology Funding History (in thousands of dollars)

Year (Fiscal)	Submission	Programmed (Actual)
1979	19,140 a	44,750
1980	58,545	57,891
1981	33,100	32,746
1982	13,400	— b

a Incorporated Advanced Civil Aircraft Systems Technology and Aerodynamic Vehicle Systems Technology activities. No equivalent budget category.
b Activities transferred to Subsonic Aircraft budget category.

Table 3–28. Advanced Propulsion Systems Technology Funding History (in thousands of dollars)

Year (Fiscal)	Submission	Programmed (Actual)
1979	— a	66,255
1980	72,500	72,278
1981	47,800 b	46,196
1982	15,400 c	26,155
1983	28,000	27,300
1984	17,000	17,000
1985	26,100 d	26,100
1986	44,200	42,200
1987	28,400	28,220
1988	18,000 e	17,955

a No submission in this category.
b Significant drop from prior year because of the completion of several activities.
c Reduction in submission reflects the descoping of the energy efficient engine program and the advanced turboprop effort, as well as the completion of other activities.
d Increase reflected realignment of $10,000,000 to the advanced turboprop program from the research and technology base program for efforts leading to an initial flight test in 1987, as directed by Congress.
e Reflected the transfer of the advanced high-temperature engine materials technology program to Materials and Structures Systems Technology (see Table 3–22).

AERONAUTICS AND SPACE RESEARCH AND TECHNOLOGY 273

Table 3–29. Aeronautical System Studies Technology Funding History (in thousands of dollars) a

Year (Fiscal)	Submission	Programmed (Actual)
1979	— b	4,825
1980	4,100	4,134
1981	3,200	3,125
1982	— c	—

a   Formerly System Studies (see Table 3–19).
b   See Table 3–19.
c   Program terminated.

Table 3–30. Numerical Aerodynamic Simulation Funding History (in thousands of dollars)

Year (Fiscal)	Submission	Programmed (Actual)
1984	17,000	17,000
1985	26,500	26,472
1986	28,200	28,200
1987	30,000	29,984
1988	39,000	39,018

Table 3–31. Advanced Rotorcraft Technology Funding History (in thousands of dollars)

Year (Fiscal)	Submission	Programmed (Actual)
1982	— a	21,665
1983	22,300	22,300
1984	27,600	27,950
1985	26,000	26,000
1986	20,500	20,500
1987	18,700	18,700
1988	4,600 b	4,529

a   No budget category.
b   Reduction reflected the elimination of funding for the Technology-for-Next-Generation Rotorcraft Program.

Table 3–32. Experimental Programs Funding History
(in thousands of dollars)

Year (Fiscal)	Submission	Authorization	Appropriation	Programmed (Actual)
1979	66,255	— a	— b	— c
1980	73,500	73,500	— d	— e

a   Undistributed. Total 1979 Aeronautical Research and Technology Program authorization = $275,100,000.
b   Undistributed. Total 1979 R&D appropriation = $3,477,200,000.
c   No budget category in NASA Budget Estimate.
d   Undistributed. Total 1980 R&D appropriation = $3,838,500,000.
e   No budget category in NASA Budget Estimate.

Table 3–33. Space Research and Technology Funding History
(in thousands of dollars)

Year (Fiscal)	Submission	Authorization	Appropriation	Programmed (Actual)
1979	107,300 a	111,300	— b	107,300
1980	115,800 c	119,400	— d	115,586
1981	110,200 e	115,200	110,700 f	110,700
1982	125,300 g	129,300	111,000	111,000
1983	123,000	128,000	123,000	124,500
1984	137,000 h	143,000	138,000	137,000
1985	150,000	150,000	154,000	150,000
1986	168,000	166,000	168,000	151,400
1987	171,000 i	183,200	185,200 j	206,000
1988	250,000 k	234,000	235,000	221,300

a   Amended budget submission. Original budget submission = $108,300,000.
b   Undistributed. Total 1979 R&D appropriation = $3,477,200,000.
c   Amended budget submission. Original budget submission = $116,400,000.
d   Undistributed. Total 1980 R&D appropriation = $3,838,500,000.
e   Amended budget submission. Original budget submission = $115,200,000.
f   Reflected recission. Unchanged from earlier appropriated amount that reflected the effect of General Provision Sec. 412 appended to appropriation on December 15, 1980.
g   Amended budget submission. Original budget submission = $141,000,000.
h   Amended budget submission. Original budget submission = $138,000,000.
i   Amended budget submission. Original budget submission = $180,200,000.
j   Additional amount was the result of congressional action.
k   Amended budget submission unchanged from original submission.

AERONAUTICS AND SPACE RESEARCH AND TECHNOLOGY 275

Table 3-34. Space Research and Technology Base Funding History (in thousands of dollars)

Year (Fiscal)	Submission	Authorization	Appropriation	Programmed (Actual)
1979	71,795 a	71,700	— b	86,277
1980	99,785 c	77,100	— d	99,816
1981	100,300 e	105,300 f	101,100 g	100,380
1982	115,300 h	117,300 i	— j	104,646
1983	115,100 k	120,600 l	115,600	116,304
1984	125,400 m	131,300 n	126,200	124,885
1985	150,000 o	136,000	140,000	136,358
1986	132,800 p	140,000	140,000	124,200
1987	133,600	133,600 q	133,600	130,646
1988	115,900	115,900 r	115,900	107,146

a   Amended budget submission. Original budget submission = $71,700,000.
b   Undistributed. Total 1979 R&D appropriation = $3,477,200,000.
c   Amount in *1980 Chronological History* = $77,100,000. The amount in the table (as published in the NASA Budget Estimate) reflects different split between the Research and Technology Base and Systems Technology Programs, with $99,785,000 allocated for Research and Technology Base and $11,015,000 allocated for Systems Technology. The amount agrees with the sum of amounts for individual programs/budget categories. (The *Chronological Budget History* does not provide amounts for individual programs.)
d   Undistributed. Total 1980 R&D appropriation = $3,838,500,000.
e   Amended budget submission. Original budget submission = $103,400,000.
f   Amended budget increased $3,000,000 to enhance advanced chemical propulsion technology activities and to accelerate expander cycle dual-thrust engine technology, as well as $2,000,000 for enhancements of space platform and large space structures advanced technology activities.
g   Reflected recission. Unchanged from earlier appropriated amount that reflected the effect of General Provision Sec. 412 appended to appropriation on December 15, 1980.
h   Amended budget submission. Original budget submission = $124,800,000.
i   House action increased authorization $2,000,000 for chemical propulsion technology. The Senate restored $5,000,000, which included an additional amount of $200,000 for space power and electric propulsion.
j   Undistributed. Total 1982 Space Research and Technology appropriation = $111,000,000.
k   Amended budget submission. Original budget submission = $115,600,000.
l   A total of $5,000,000 was added for propulsion research and technology activities.
m   Revised budget submission. Original budget submission = $126,200,000.
n   The House authorization committee added $5,000,000 for university research instrumentation and lab equipment ($2,500,000) and to augment advanced chemical propulsion technology ($2,500,000).
o   Amended budget submission. Original budget submission = $136,000,000.
p   Revised budget submission. Original budget submission = $140,000,000.
q   The Senate authorization committee authorized the allocation of funds as follows: Aerothermodynamics Research and Technology—$11,200,000, Space Energy Conversion Research and Technology—$20,400,000, Propulsion Research and Technology—$21,000,000, Materials and Structures Research and Technology—$18,900,000, Space Data and Communications Research and Technology—$13,600,000, Information Sciences Research and Technology—$10,200,000, Controls and Guidance Research and Technology—$7,500,000, Human Factors Research and Technology—$2,300,000, Space Flight Research and Technology—$22,400,000, and Systems Analysis—$6,100,000.

Table 3-34 continued

r The Senate authorization committee authorized the allocation of funds as follows: Aerothermodynamics Research and Technology—$11,100,000, Space Energy Conversion Research and Technology—$14,600,000, Propulsion Research and Technology—$14,500,000, Materials and Structures Research and Technology—$17,900,000, Space Data and Communications Research and Technology—$8,900,000, Information Sciences Research and Technology—$8,000,000, Controls and Guidance Research and Technology—$6,300,000, Human Factors Research and Technology—$4,900,000, Space Flight Research and Technology—$23,200,000, and Systems Analysis—$6,500,000.

Table 3-35. Materials and Structures (Space) Research and Technology Funding History (in thousands of dollars)

Year (Fiscal)	Submission	Programmed (Actual)
1979	14,700 a	16,400
1980	16,400	25,376
1981	14,000	14,000
1982	14,100	14,565
1983	14,700	13,245
1984	13,900	16,694
1985	18,800	18,800
1986	18,600	18,126
1987	18,900	20,877
1988	15,900	17,215

a Combined Materials Research and Technology and Structures Research and Technology budget categories.

Table 3-36. Space Power and Electric Propulsion (Space Energy Conversion) Research and Technology Funding History (in thousands of dollars) a

Year (Fiscal)	Submission	Programmed (Actual)
1979	9,200 b	17,000
1980	19,750	19,364
1981	19,200	18,900
1982	18,500	18,080 c
1983	17,400	17,900
1984	22,100	22,006
1985	22,500	22,312
1986	21,200	19,955
1987	20,400	20,922
1988	12,500	12,154

a Renamed Space Energy Conversion Research and Technology with FY 1983 revised estimate.
b Included only Electric Propulsion activities.
c Renamed Space Energy Conversion Research and Technology with Revised FY 1983 Budget Estimate.

Table 3–37. Platform Systems (Systems Analysis) Research and Technology Funding History (in thousands of dollars)

Year (Fiscal)	Submission	Programmed (Actual)
1982	—	2,649 a
1983	5,100	6,020
1984	8,800	7,200 b
1985	6,610	6,788
1986	6,800	6,438
1987	6,100	6,576
1988	5,700	5,376

a   Funded primarily from Spacecraft Systems budget category. Included systems analysis, operations technology, and crew and life support technology.
b   Descoped to include only Systems Analysis.

Table 3–38. Information Systems (Space Data and Communications) Research and Technology Funding History (in thousands of dollars)

Year (Fiscal)	Submission	Programmed (Actual)
1979	— a	16,308
1980	20,600	21,847
1981	21,300	21,100
1982	22,900	16,902 b
1983	18,100	16,609
1984	17,800	17,802
1985	16,500	16,500
1986	16,000	15,384
1987	13,600	13,252
1988	7,900	7,765

a   No equivalent category.
b   Most funding was used for the new Space Data and Communications budget category.

Table 3–39. Computer Sciences and Electronics (Information Sciences) Research and Technology Funding History (in thousands of dollars)

Year (Fiscal)	Submission	Programmed (Actual)
1982	— a	14,130
1983	15,700	16,165
1984	16,100	16,001 b
1985	17,600	17,590
1986	9,900 c	12,462
1987	10,200	8,827
1988	7,700	7,428

a   No budget category.
b   Renamed Information Sciences Research and Technology.
c   Reduction from prior year because of the transfer of Automation Robotics funding to Systems Technology and a $300,000 reduction in Information Sciences to support Transatmospheric Technology efforts.

Table 3–40. Electronics and Automation Research and Technology Funding History (in thousands of dollars)

Year (Fiscal)	Submission	Programmed (Actual)
1979	— a	8,200
1980	8,550	8,123
1981	7,900	7,700
1982	8,100	— b

a   No budget category.
b   Activities moved to Computer Sciences and Electronics budget category.

Table 3–41. Transportation Systems (Space Flight) Research and Technology Funding History (in thousands of dollars)

Year (Fiscal)	Submission	Programmed (Actual)
1979	—	7,074
1980	10,235	10,725
1981	12,400	8,900
1982	8,200	7,073
1983	7,800	7,300
1984	7,400	6,800 a
1985	11,450	11,468
1986	17,000 b	14,054
1987	22,200	20,096
1988	21,400	21,052

a   Renamed Space Flight Systems.
b   Increase included the consolidation of funds from other Research and Technology Base programs for Control of Flexible Structures, Transatmospheric Technology, Aerospace Industry/University Space Flight Experiments, and Cryogenic Fluid Management Technology Activities. In addition, the aero-assist portion of the Orbital Transfer Vehicle systems technology program was transferred to this Research and Technology Base program from Systems Technology.

Table 3–42. (Chemical) Propulsion Research and Technology Funding History (in thousands of dollars)

Year (Fiscal)	Submission	Programmed (Actual)
1979	9,200	8,600
1980	8,900	8,900
1981	12,400	12,400
1982	13,700	12,956
1983	15,400	16,600
1984	16,400	19,497
1985	20,500	20,500
1986	22,300	18,156
1987	21,000	18,844
1988	13,300	12,679

Table 3–43. Spacecraft Systems Research and Technology Funding History (in thousands of dollars)

Year (Fiscal)	Submission	Programmed (Actual)
1979	— a	5,495
1980	7,250	7,437
1981	9,000	8,900
1982	9,100	5,071
1983	3,500	4,520
1984	5,200	— b

a   No budget category.
b   No budget category.

Table 3–44. Fluid Physics (Aerothermodynamics) Research and Technology Funding History (in thousands of dollars)

Year (Fiscal)	Submission	Programmed (Actual)
1979	5,800 a	5,200
1980	5,400	5,400
1981	7,800	7,800
1982	7,900	7,894
1983	8,500	8,385
1984	8,400	8,480
1985	10,100	10,100
1986	10,800	10,490
1987	11,400	11,678
1988	10,300	10,170

a   Formerly called Entry Research and Technology.

Table 3–45. Control and Human Factors (Controls and Guidance) Research and Technology Funding History (in thousands of dollars)

Year (Fiscal)	Submission	Programmed (Actual)
1982	—	2,964 a
1983	6,800	7,460
1984	8,300	7,402 b
1985	8,600	8,600
1986	7,500	7,035
1987	7,500	7,300
1988	5,500	5,260

a   New budget category funded primarily from Spacecraft Systems.
b   Renamed Controls and Guidance Research and Technology. Human Factors became a separate budget category.

Table 3–46. Human Factors Research and Technology Funding History (in thousands of dollars)

Year (Fiscal)	Submission	Programmed (Actual)
1984	—	3,003
1985	3,700	3,700
1986	2,300	2,100
1987	2,300	2,274
1988	4,200	4,047

Table 3–47. System Studies (Space) Funding History (in thousands of dollars)

Year (Fiscal)	Submission	Authorization	Appropriation	Programmed (Actual)
1979	2,000	— a	— b	— c
1980	2,200	2,200	— d	— e

a   Undistributed. Total 1979 Space Research and Technology authorization = $111,300,000.
b   Undistributed. Total 1979 R&D appropriation = $3,477,200,000.
c   No budget category in NASA Budget Estimate.
d   Undistributed. Total 1980 R&D appropriation = $3,383,500,000.
e   No budget category in NASA Budget Estimate.

Table 3–48. Systems Technology Program (Civil Space Technology Initiative) Funding History (in thousands of dollars) a

Year (Fiscal)	Submission	Authorization	Appropriation	Programmed (Actual)
1979	7,900 b	10,900	— c	12,023
1980	11,015	19,000 d	— e	10,770
1981	7,800 f	7,800	7,500 g	8,220
1982	2,800 h	9,000 i	— j	3,354
1983	4,900 k	4,400	3,000	5,196
1984	7,000 l	7,200	7,200	7,515
1985	8,750 m	9,100	9,100	8,742
1986	27,200 n	20,000	20,000	27,200
1987	37,400	37,400 o	37,400	75,354
1988	115,200 p	118,100 q	119,100	114,154

a   Designated as Civil Space Technology Initiative (CSTI) program starting with the FY 1988 budget submission.
b   The Systems Technology Program budget categories provided at the time of the FY 1979 budget requests as listed in the NASA Budget Estimate were not equivalent to the categories provided for the programmed amounts as listed in the FY 1981 NASA Budget Estimate.
c   Undistributed. Total 1979 R&D appropriation = $3,477,200,000.
d   The Senate authorization committee increased the amount $3,000,000 for Large Space Structures.
e   Undistributed. Total 1980 R&D appropriation = $9,700,000.
f   Amended budget submission. Original budget submission = $103,400,000.
g   Reflected recission. Unchanged from earlier appropriated amount that reflected the effect of General Provision Sec. 412 appended to appropriation on December 15, 1980.
h   Amended budget submission. Original budget submission = $13,200,000. Reflects intention to eliminate the Systems Technology Program.
i   The House authorization committee increased Information Systems Technology by $2,000,000.
j   Undistributed. Total 1982 Space Research and Technology appropriation = $111,000,000.
k   Revised budget submission. Initial budget submission = $4,400,000. No budget categories from prior years were included.
l   Revised budget submission. Initial budget submission = $7,200,000.
m   Revised budget submission. Initial budget submission = $9,100,000.
n   Revised budget submission. Original budget submission = $27,200,000.
o   The Senate authorization committee authorized the allocation of funds as follows: Chemical Propulsion Systems Technology—$8,100,000, Control of Flexible Structures Flight Experiment—$11,300,000, and Automation and Robotics Technology—$18,000,000.
p   Space Systems Technology programs were incorporated into the CSTI program in FY 1988.
q   The Senate authorization committee authorized the allocation of funds as follows: Propulsion—$1,200,000, Vehicle—$15,000,000, Propulsion Research and Technology—$21,000,000, Materials and Structures Research and Technology—$18,900,000, Information—$17,400,000, Large Structures and Control—$22,800,000, Power—$14,000,000, and Automation and Robotics—$96,100,000. The conference committee increased the total amount.

Table 3–49. Space Systems Studies Funding History
(in thousands of dollars)

Year (Fiscal)	Submission	Programmed (Actual)
1979	— a	2,000
1980	2,200	2,323
1981	2,000	2,083
1982	— b	—

a   No equivalent budget category.
b   Amended budget submission. The deletion of the original $500,000 budget request reflected a decision to eliminate Space Systems Studies as an independent line item and to conduct necessary studies within the Research and Technology Base or specific Systems Technology programs as appropriate.

Table 3–50. Information (Systems) Technology Funding History
(in thousands of dollars)

Year (Fiscal)	Submission	Programmed (Actual)
1979	— a	—
1980	2,600	1,500
1981	4,100	4,026
1982	— b	—
1983 c	16,500	16,310

a   No equivalent budget category.
b   Amended budget submission. The deletion of the original $9,400,000 estimate reflected a decision to eliminate this program as an independent line item and to consolidate the remaining elements into the Information Systems program in the Research and Technology Base.
c   The budget category was reinstated as part of CSTI.

Table 3–51. Space Flight Systems Technology Funding History
(in thousands of dollars)

Year (Fiscal)	Submission	Programmed (Actual)
1982	—	3,354 a
1983	4,900	5,196
1984	7,000	7,515
1985	6,650	6,642
1986	6,200	11,200 b
1987	11,300	11,254

a   Included Space Flight Experiments, the Long Duration Exposure Facility, and the Ion Auxiliary Propulsion System.
b   Included Control of Flexible Structures funding.

Table 3–52. Spacecraft Systems Technology Funding History (in thousands of dollars)

Year (Fiscal)	Submission	Programmed (Actual)
1979	a	10,023
1980	6,215	6,947
1981	1,400 b	2,075
1982	2,800	— c

a   No equivalent budget category.
b   Reduction in FY 1981 submission from prior year reflected the completion and delivery of most Spacelab experiments and the development and completion of most Long Duration Exposure Facility experiments during FY 1981, as well as the delivery of flight hardware for the experimental test of the eight-centimeter ion engine auxiliary propulsion system to the U.S. Air Force in FY 1980.
c   No budget category.

Table 3–53. Automation and Robotics Funding History (in thousands of dollars)

Year (Fiscal)	Submission	Programmed (Actual)
1986	10,200	10,200
1987	18,000	18,000
1988	25,100	25,332

Table 3–54. (Chemical) Propulsion Systems Technology Funding History (in thousands of dollars)

Year (Fiscal)	Submission	Programmed (Actual)
1985	2,100	2,100
1986	5,800	5,800
1987	8,100	46,100 a
1988	38,800	23,600

a   Increase reflected the expansion of research on Earth-to-orbit technology aimed at assuring a mid-1990 capability to enable the development of reusable, high-performance, liquid oxygen/hydrogen, and high-density fuel propulsion systems for next-generation space transportation vehicles beyond the Shuttle. Also reflected a new Booster Technology program.

Table 3–55. Vehicle Funding History (in thousands of dollars)

Year (Fiscal)	Submission	Programmed (Actual)
1988	15,000	15,000

Table 3-56. Large Structures and Control Funding History
(in thousands of dollars)

Year (Fiscal)	Submission	Programmed (Actual)
1988	22,000	22,158

Table 3-57. High-Capacity Power Funding History
(in thousands of dollars)

Year (Fiscal)	Submission	Programmed (Actual)
1988	12,800	12,754

Table 3-58. Experimental Programs Funding History
(in thousands of dollars)

Year (Fiscal)	Submission	Authorization	Appropriation	Programmed (Actual)
1979	17,700	17,700	— a	— b
1980	18,100	18,100	— c	— d

a   Undistributed. Total 1979 R&D appropriation = $3,477,200,000.
b   No budget category in NASA Budget Estimate.
c   Undistributed. Total 1980 R&D appropriation = $3,383,500,000.
d   No budget category in NASA Budget Estimate.

Table 3-59. Standards and Practices Funding History
(in thousands of dollars) a

Year (Fiscal)	Submission	Authorization	Appropriation	Programmed (Actual)
1979	9,000	9,000	— b	9,000
1980	5,000 c	3,000	— d	5,000
1981	2,100 e	2,100	2,100 f	2,100
1982	3,000 g	3,000	— h	3,000
1983	3,000	3,000	3,000	3,000
1984	4,600	4,600	4,600	4,600
1985	4,900	4,900	4,900	4,900
1986	8,000	8,000	8,000	— i
1987	9,200	9,200	9,200	— j

a   Formerly named Low Cost Systems.
b   Undistributed. Total 1979 R&D appropriation = $3,477,200,000.
c   Amended budget submission. Original budget submission = $3,000,000.
d   Undistributed. Total 1980 R&D appropriation = $3,383,500,000.
e   Amended budget submission. Original budget submission = $4,000,000.
f   Reflected recission. Unchanged from earlier appropriated amount that reflected the effect of General Provision Sec. 412 appended to appropriation on December 15, 1980.
g   Amended submission unchanged from original submission.
h   Undistributed. Total Space Research and Technology = $111,000,000.
i   No programmed amount.
j   No programmed amount.

Table 3–60. Energy Technology Applications Funding History (in thousands of dollars)

Year (Fiscal)	Submission	Authorization	Appropriation	Programmed (Actual)
1979	3,000	5,000	— a	5,000
1980	3,000	5,000 b	— c	3,000
1981	4,000 d	4,000	1,900 e	1,900
1982	4,400	0 f	0	—

a   Undistributed. Total 1979 R&D appropriation = $3,477,200,000.
b   The Senate authorization committee increased the amount for Energy Technology Verification and Identification by $2,000,000.
c   Undistributed. Total 1980 R&D appropriation = $3,383,500,000.
d   Amended budget submission unchanged from original submission.
e   Reflected recission.
f   No authorization or appropriation passed for FY 1982.

Table 3–61. Transatmospheric Research and Technology Funding History (in thousands of dollars) a

Year (Fiscal)	Submission	Authorization	Appropriation	Programmed (Actual)
1987	35,000 b	40,000	40,000	45,000
1988	52,500 c	66,000	53,000 d	52,500

a   Budget category to fund the development of the technology base for a potential national aerospace plane. The program was initiated in FY 1986, by FY 1986 funding was included in ongoing Research and Technology Base funding ($16,000,000).
b   Revised budget submission. Original budget submission = $45,000,000.
c   Revised budget submission. Original budget submission = $45,000,000.
d   General reduction reduced the appropriation from $66,000,000 to $53,000,000.

Table 3-62. Helicopter and Compound RSRA Configurations

Feature	Helicopter Configuration	Compound Configuration
Gross Weight	9,200 kilograms	13,100 kilograms
Power Plant	Sikorsky S-61 rotor and drive system powered by twin General Electric T58-GE-5/100 gas turbine engines generating 1,044 shaft kilowatts each	Additional General Electric TF34-GE-44A wing and auxiliary thrust jet engines rated at 41,255 newtons thrust each
Horizontal Stabilizer	"T" tail with a 4.1-meter span and 2.4-square-meter area	Additional 6.5-meter span stabilizer and a rudder and associated controls
Wing Span	None	14 meters

Table 3-63. HiMAT Characteristics

First Flight	July 27, 1979	
First Supersonic Flight	May 11, 1982	
Length	7 meters	
Wing Span	4.6 meters	
Height	1.3 meters	
Weight at Launch	1,543 kilograms	
Thrust	22,240 newtons	
Maximum Speed	Mach 1.4	
Engine	General Electric J-85 turbojet	
Composition (% of total structural weight)	Graphite	26
	Fiberglass	3
	Aluminum	26
	Titanium	18
	Steel	9
	Sintered Tungsten	4
	Miscellaneous	14
Prime Contractor	Rockwell International	
Program Responsibility	NASA Dryden Flight Research Center	

Table 3–64. X-29 Characteristics

Length of Aircraft	14.7 meters
Width of Wing	8.3 meters
Height	4.3 meters
Power Plant	One General Electric F404-GE-400 engine producing 71,168 newtons of thrust
Empty Weight	6,170 kilograms
Takeoff Weight	7,983 kilograms
Maximum Operating Altitude	15,240 meters
Maximum Speed	Mach 1.6
Flight Endurance Time	1 hour
External Wing Structure	Composites
Wing Substructure	Aluminum and titanium
Basic Airframe Structure	Aluminum and titanium

Table 3–65. Boeing 737 Transport Systems Research Vehicle Specifications

Model	Boeing 737-130 (aircraft was a 737-100, but given customer designation of 737-130 when modified to NASA specifications) Serial no. 19437 Boeing designation PA-099 (Prototype Boeing 737)
Date of Manufacture	1967
First Flight	April 9, 1967
Description	Twin-jet, short-range transport
Total Flight Hours:	
Upon Arrival at Langley	978
At End of FY 1993	2,936
Engines	Two Pratt & Whitney JT8D-7s
Thrust	62,272 newtons each
Wing Span	28.3 meters
Length	28.65 meters
Wing Area	91 square meters
Tail Height	11.3 meters
Gross Takeoff Weight	44,362 kilograms
Maximum Payload	13,154 kilograms
Cruising Speed	925 kilometers per hour
Range	3,443 kilometers
Service Ceiling	10,668 meters

Source: Lane E. Wallace, *Airborne Trailblazer: Two Decades With NASA Langley's 737 Flying Laboratory* (Washington, DC: NASA SP-4216, 1994), p. 147.

Table 3-66. Experiments of the Orbiter Experiments Program

Experiment Name	Principal Technologist	STS Flights	Description
Aerodynamic Coefficient Identification Package (ACIP) (Figure 3–20)	D.B. Howes, Johnson Space Center	STS-1, STS-6, STS-8	ACIP provided a way to collect aerodynamic data during the launch, entry, and landing phases of the Shuttle flight. It established an extensive aerodynamic database for the verification of and correlation with ground-based data, including assessments of the uncertainties of such data. It also provided flight dynamics data in support of other technology areas, such as aerothermal and structural dynamics. ACIP incorporated three groups of instruments: dual-range linear accelerometers, angular accelerometers, and rate gyros.
Catalytic Surface Effects (CSE) (Figure 3–21)	D. Stewart, Ames Research Center	STS-2, STS-5	CSE determined the effects of thermal protection system coating catalytic efficiency on orbiter flight convective heating and maximum temperature reduction.
Dynamic, Acoustic, and Thermal Environment (DATA) (Figure 3–22)	W. Bangs, Goddard Space Flight Center	STS-2, STS-5	DATA acquired environmental response and input data for predicting environments for future payloads. The environments were neither constant nor consistent throughout the payload bay and were influenced by interactions between cargo elements. The experiment consisted of accelerometers and force gauges, microphones, and thermal sensors that were installed on the payload components and on the carrying structures.

AERONAUTICS AND SPACE RESEARCH AND TECHNOLOGY 289

Table 3–66 continued

Experiment Name	Principal Technologist	STS Flights	Description
Induced Environment Contamination Monitor (IECM) (Figure 3–23)	Marshall Space Flight Center	STS-2, STS-4	The IECM measured and recorded concentration levels of gases and particulate contamination emitted by the Shuttle during all phases of the mission to verify that contamination associated with the orbiter would not preclude or seriously interfere with the gathering of data preparing for or during the orbital flight. The IECM was a self-contained aluminum unit and contained ten and support systems mounted on the Development Flight Instrument Unit.
Tile Gap Heating (TGH) Effects (Figure 3–24)	F. Centolanzi, Ames Research Center	STS-2, STS-5	TGH Effects evaluated the thermal response of different tile gaps and provided optimum tile gap designs for the orbiter thermal protection system. The experiment consisted of a removable carrier panel with eleven thermal protection system tiles of baseline material located on the underside of the orbiter fuselage. The gap spacing and depth between tiles were controlled to assure heating rates no higher than baseline, with the primary objective of identifying optimum heating rates. Thermocouples fitted to the tile surfaces and in the gaps measured the temperature during entry.

Table 3-67. Long Duration Exposure Facility (LDEF) Mission Chronology

Date	Event
1970	Langley Research Center proposes the conceptual forerunner of LDEF, called Meteoroid and Exposure Module (MEM), to be the first Shuttle payload.
June 1974	The LDEF project is formally under way, managed by Langley for NASA's OAST.
1976–August 1978	The LDEF structure is designed and fabricated at Langley.
Summer of 1981	LDEF preparations are under way for the December 1983 target launch date.
September 1981	The first international meeting of LDEF experimenters is held at Langley.
1982	The LDEF structure is tested for its ability to withstand Shuttle-induced loads.
June 1983	LDEF is shipped from Langley to Kennedy Space Center and placed in the Spacecraft Assembly and Encapsulation Facility.
April 7, 1984	During the STS 41-C mission, at 12:26 p.m., EST, the Space Shuttle *Challenger* placed LDEF in nearly circular orbit.
March 1985	The planned LDEF retrieval (via STS 51-D) is deferred to a later Shuttle flight.
January 1986–September 1988	LDEF's stay in space is extended indefinitely when all Shuttle operations were suspended because of the loss of *Challenger*.
1987–1988	Solar activity intensity threatens to accelerate the decay of LDEF's orbit and thus influences retrieval planning. The retrieval target is set for July 1989.
June 1989	The LDEF retrieval flight date, after slipping from July and then November, is set for the December 18 launch of the Space Shuttle *Columbia*.
December 18, 1989	The STS-32 launch is postponed until the second week of January.
January 1990	STS-32 is launched January 9. LDEF is retrieved 9:16 a.m., CST, January 12. *Columbia* lands at Edwards Air Force Base, California, January 20.
January 26, 1990	*Columbia,* with LDEF still in the payload bay, is returned to Kennedy via a ferry flight from Edwards Air Force Base.
January 30–31, 1990	LDEF is removed from *Columbia* in Kennedy's Orbiter Processing Facility, placed in a special payload canister, and transported to the Operations and Checkout Building.
February 1–2, 1990	LDEF is placed in its special transporter, the LDEF Assembly and Transportation System, and moved to the Spacecraft Assembly and Encapsulation Facility for experiment deintegration.
February 5–22, 1990	Deintegration preparation activities take place, including extensive inspection and photo-documentation.
February 23–March 29, 1990	Trays are removed, closely inspected, individually photo-documented, packed, and shipped to home institutions for comprehensive data analysis.
April–May 1990	Deintegration wrap-up occurs, including the comprehensive investigation and photo-documentation of the LDEF structure itself.

Table 3-68. Long Duration Exposure Facility (LDEF) Characteristics

Launch Date/Range	April 6, 1984/Kennedy Space Center
Date of Reentry	Retrieved January 12, 1990, on STS-32 (*Columbia*)
Launch Vehicle	STS 41-C (*Challenger*)
Customer/Sponsor	NASA/OAST
Responsible (Lead) NASA Center	Langley Research Center
Mission Objectives	Provide a low-cost means of space access to a large experiment group
Instruments and Experiments	

Materials and Structures

1. Growth of Crystals From Solution in Low Gravity attempted to grow single crystals of lead sulfide, calcium carbonate, and synthetic metals in low gravity.
2. Atomic Oxygen-Stimulated Outgassing investigated the effect of atomic oxygen impingement on thermal control surfaces in orbit.
3. Interaction of Atomic Oxygen With Solid Surfaces determined the measurable effects of impingement of high fluxes of atomic oxygen on various solid surfaces, investigated the mechanisms of interaction in several materials (some not chemically affected by oxygen), and altered the exposure, angle of incidence, and temperature of the substrates by their position on the spacecraft and experimental design.
4. Mechanical Properties of High-Toughness Graphite-Epoxy Composite Material tested the effect of space exposure on the mechanical properties of a specially toughened graphite-epoxy composite material.
5. Space-Based Radar Phased-Array Antenna evaluated the space effects on candidate polymeric materials for space-based radar phased-array antennas, degradation mechanisms caused by thermal cycling, ultraviolet and charged particle irradiation, applied load and high-voltage plasma interaction.
6. Composite Materials for Large Space Structures evaluated the space effects on physical and chemical properties of laminated continuous-filament composites and composite resin films for large structures and advanced spacecraft.
7. Epoxy Matrix Composites Thermal Expansion and Mechanical Properties detected possible variation in coefficient of thermal expansion of composite samples in space, detected possible change in the mechanical integrity of composite products, and compared the behavior of two epoxy resins commonly used in space structure production.
8. Composite Materials tested different materials to determine actual useful life and integration of histories of thermal and mechanical characteristics into models of composite structures.

Table 3–68 continued

9. Microwelding of Various Metallic Materials Under Ultravacuum examined metal surfaces representative of mechanism-constituent metals for microwelds after space exposure.
10. Graphite-Polymide and Graphite-Epoxy Mechanical Properties accumulated operational data on space exposure of graphite-polymide and graphite-epoxy material.
11. Polymer Matrix Composite Materials investigated the effect of space exposure on the mechanical properties of polymer matrix composite materials.
12. Spacecraft Materials analyzed the materials specimens to understand changes in properties and structures in space, including structural materials, solar power components, thermal control materials, laser communications components, laser mirror coatings, laser-hardened materials, antenna materials, and advanced composites.
13. Balloon Materials Degradation assessed space exposure effects on balloon films, tapes, and lines.
14. Thermal Control Coatings examined the validity of ground simulations of the space environment to study degradation of satellite thermal control coatings.
15. Spacecraft Coatings determined the space effects on new coatings being developed for spacecraft thermal control. Paint, other coatings, and second-surface mirror samples were exposed—some to all mission environments and some to specific ones. Sample spectral reflectance was measured before and after the mission.
16. Thermal Control Surfaces determined the effects of space on new coatings being developed for spacecraft thermal control. Samples were mounted on an indexing wheel, where a reflectometer periodically recorded reflectance values.
17. Ion-Beam-Textured and Coated Surfaces measured launch and space effects on optical properties of ion-beam-textured high-absorptance solar thermal control surfaces, optical and electrical properties of ion-beam-sputtered conductive solar thermal control surfaces, and weight loss of ion-beam-deposited oxide-polymer films.
18. Cascade Variable-Conductance Heat Pipe verified the ability of a variable-conductance heat pipe system to provide precise temperature control of long-life spacecraft without needing a feedback heater or other power source for temperature adjustment, under conditions of widely varying power input and ambient environment.

Table 3–68 continued

19. Low-Temperature Heat Pipe Experiment Package evaluated the performance in space of a fixed-conductance transporter heat pipe, a thermal diode heat pipe, and a low-temperature phase-change material.
20. Transverse Flat-Plate Heat Pipe evaluated the zero-gravity performances of a transverse flat-plate heat pipe, including heat transport capability, temperature drop, and ability to maintain temperature over varying duty cycles and environments.
21. Thermal Measurements System measured the average LDEF flight temperature and temperature time history of selected components and representative experiment boundary conditions.

Power and Propulsion
22. Space Plasma High-Voltage Drainage determined the long-term current drainage properties of dielectric films subjected to high-level electric stress in the presence of space plasma and solar radiation.
23. Solar Array Materials evaluated the synergistic effects of space on mechanical, electrical, and optical properties of solar array materials, such as solar cells, cover slips with various anti-reflectance coatings, adhesives, encapsulants, reflector materials, substrate strength materials, mast and harness materials, structural composites, and thermal control treatments.
24. Advanced Photovoltaic Experiment investigated the space effects on new solar cell and array materials and evaluated their performance and measured long-term variations in spectral content of sunlight and calibration of solar cells for space use.
25. Critical Surface Degradation Effects on Coatings and Solar Cells Developed in Germany investigated the radiation and contamination effects on thermal coatings and solar cells, with and without conductive layers, and provided design criteria, techniques and test methods for the control of space and spacecraft effects.
26. Space Aging of Solid Rocket Materials determined the space effects on various mechanical and ballistic properties of solid rocket propellants, liners, insulation materials, and case and nozzle materials.

Table 3-68 continued

Science

27. Interstellar Gas analyzed the interstellar noble gas atoms (helium and neon) that penetrate the heliosphere near Earth.
28. High-Resolution Study of Ultra-Heavy Cosmic-Ray Nuclei studied charge and energy spectra of cosmic-ray nuclei, superheavy nuclei, and heavy anti-nuclei to help understand the physical processes of cosmic-ray nuclei production and acceleration in interstellar space. It also obtained data on nucleosynthesis.
29. Heavy Ions in Space investigated three components of heavy nuclei in space: low-energy nuclei of nitrogen, oxygen, and neon; heavy nuclei in the Van Allen radiation belts; and ultraheavy nuclei of galactic radiation.
30. Trapped-Proton Energy Spectrum Determination measured the flux and energy spectrum of protons trapped on Earth's magnetic field lines as part of the inner radiation belt and examined neutron and proton radioactivity, microsphere dosimetry, flux measurement by ion trapping, and elemental and isotopic abundances of heavy cosmic ray nuclei.
31. Measurement of Heavy Cosmic-Ray Nuclei measured the elemental and isotopic abundances of certain heavy cosmic-ray nuclei and of chemical and energy spectra for particles.
32. Linear Energy Transfer Spectrum Measurement measured the linear energy transfer spectrum behind different shieldings, which were increased in small increments to provide data for future spacecraft designs and other LDEF experiments.
33. Multiple Foil Microabrasion Package provided a passive evaluation of the near-Earth micrometeoroid environment.
34. Meteoroid Impact Craters on Various Materials studied the impact microcraters made by micrometeoroids on metals, glasses, and minerals made into thick targets.
35. Attempt at Dust Debris Collection With Stacked Detectors investigated the feasibility of using multilayer, thin-film detectors as energy sorters to collect micrometeoroids—if not in original shape, at least as fragments suitable for chemical analysis.
36. Chemistry of Micrometeoroids conducted a chemical analysis of a significant number of micrometeoroids for data on density, shape, and mass flux.
37. Secondary Ion Mass Spectrometry of Micrometeoroids measured the chemical and isotopic composition of certain interplanetary dust particles for most expected major elements.
38. Interplanetary Dust measured the impact rate and direction of micrometeoroids in near-Earth space.

Table 3–68 continued

39. Space Debris Impact exposed passive targets to impacts by meteoroid and artificial space debris to determine the type and degree of damage expected on future spacecraft.
40. Meteoroid Damage to Spacecraft gathered examples of meteoroid impact damage to typical spacecraft components to help establish designs that would reduce the effects of meteoroid damage to future spacecraft.
41. Free-Flyer Biostack investigated the biological effectiveness of cosmic radiation, especially individual very heavy ion effects, including a quantitative assessment of the human hazards of heavy ion particles in space to establish radiation protection guidelines for human and biological experiments in spaceflights.
42. Seeds in Space Experiment evaluated the survivability of seeds stored in space and determined possible mutants and changes in mutation rates.
43. Space-Exposed Experiment Developed for Students used seeds returned from the Seeds in Space Experiment in a national education program for several million students in science and related subjects.

Electronics and Optics

44. Fiber Optics Space Effects Experiment investigated approaches and selected components of spacecraft fiber-optic transmission links to evaluate space radiation in terms of permanent degradation and transient (noise) effects.
45. Passive Exposure to Earth Radiation Budget Experiment Components measured solar and Earth-flux radiation to provide information on the amounts and sources of radiation and how it is influenced by such environmental phenomena as the "greenhouse effect" that may be unduly warming Earth's atmosphere.
46. Holographic Data Storage Crystals tested the effect of space on electro-optic crystals for use in ultrahigh-capacity space data storage and retrieval systems.
47. High-Performance Infrared Multilayer Filters and Materials exposed to space radiation infrared multilayer interference filters of novel design, construction, and manufacture and used to sense atmospheric temperature and composition.
48. Pyroelectric Infrared Detectors determined the effect of launch and space exposure on pyroelectric detectors.
49. Thin Metal Film and Multilayers tested the space behavior of optical components (extreme ultraviolet thin films, ultraviolet gas filters, and ultraviolet crystal filters).
50. Vacuum-Deposited Optical Coatings investigated the stability of several vacuum-deposited optical coatings used in spacecraft optical and electro-optical instruments.

Table 3–68 continued

51. Ruled and Holographic Gratings investigated the stability of various ruled and holographic gratings used in spacecraft optical and electro-optical instruments.
52. Optical Fiber and Components examined the radiation effects of fiber-optic waveguides that have become important components in new communications systems, opto-electronic circuits, and data links. Comparisons of radiation-induced damages in flight with samples irradiated in laboratory tests would determine the validity of irradiation tests with radioactive sources.
53. Solar Radiation Effects on Glasses determined solar radiation and space effects on optical, mechanical, and chemical properties of various glasses.
54. Radiation Sensitivity of Quartz Crystal Oscillators gathered data on the prediction and improvement of quartz crystal oscillator radiation sensitivity and compared space radiation effects with results from a transmission electron microscope.
55. Fiber Optics Systems assessed fiber-optic data link design performance for application in future spacecraft systems and documented and analyzed space effects on link and component performance.
56. Space Environment Effects examined the effects of space exposure on advanced electro-optical sensor and radiation sensor components.
57. Active Optical System Components measured space effects on the performance of lasers, radiation detectors, and other optical components to identify any degradation and to establish guidelines for selecting space electro-optical system components.

Orbit Characteristics:	
Apogee (km)	483
Perigee (km)	473
Inclination (deg.)	28.5
Period (min.)	94.3
Weight (kg)	9,707
Dimensions	Diameter of 4.3 meters; length of 9.1 meters
Shape	Twelve-sided structure
Power Source	LDEF had no power system. Any experiment that required a power or data system provided its own.
Prime Contractor	Langley Research Center
Results	Because LDEF was left in orbit much longer than anticipated, NASA officials estimated that 70 percent of the experiments had been degraded significantly, 15 percent were enhanced by the extended stay, and another 15 percent were unaffected.

CHAPTER FOUR
TRACKING AND DATA ACQUISITION/ SPACE OPERATIONS

CHAPTER FOUR
TRACKING AND DATA ACQUISITION/ SPACE OPERATIONS

Introduction

NASA's tracking and data acquisition program provided vital support for all NASA flight projects. NASA also supported, on a reimbursable basis, projects of the Department of Defense, other government agencies, commercial firms, and other countries and international organizations engaged in space research activities.

The tracking and data acquisition program supported sounding rockets and balloons, research aircraft, Earth orbital and suborbital missions, planetary spacecraft, and deep space probes. The support included:

- Tracking to determine the position and trajectory of vehicles in space
- Acquisition of scientific and Earth applications data from on-board experiments and sensors
- Acquisition of engineering data on the performance of spacecraft and launch vehicle systems
- Transmission of commands from ground stations to spacecraft
- Communication with astronauts
- Communication of information among the various ground facilities and central control centers
- Processing of data acquired from launch vehicles and spacecraft
- Reception of television transmission from space vehicles

NASA established three types of support capabilities:

- The Spaceflight Tracking and Data Network (STDN) supported low-Earth orbital missions.
- The Deep Space Network (DSN) supported planetary and interplanetary flight missions. It also supported geosynchronous and highly elliptical missions and those in low-Earth orbit not compatible with the Tracking and Data Relay Satellite System (TDRSS).
- The TDRSS provided low-Earth orbital mission support and reduced NASA's need for an extensive network of ground stations.

By the late 1980s, a worldwide network of NASA ground stations and two tracking and data relay satellites in geosynchronous orbit tracked and acquired data from spaceflight projects. The two tracking and data relay satellites worked with a single highly specialized ground station at White Sands, New Mexico. Ground communications lines, undersea cables, and communications satellite circuits, which were leased from domestic and foreign communications carriers, interconnected the ground stations.

Together, NASA referred to the STDN and the DSN as the Ground Network. TDRSS was called the Space Network. NASA was able to phase out a number of the STDN ground stations when the TDRSS had three spacecraft in place—two operational and one spare.

NASA also maintained computation facilities to provide real-time information for mission control and to process into meaningful form the large amounts of scientific, applications, and engineering data collected from flight projects. In addition, instrumentation facilities provided support for sounding rocket launchings and flight testing of aeronautical and research aircraft.

The Last Decade Reviewed

Three types of networks operated from 1969 to 1978: the Manned Spaceflight Network (MSFN), the STDN, and the DSN. The MSFN supported the Apollo program. It was consolidated with the Space Tracking and Data Acquisition Network (STADAN) in 1972 to form the STDN. NASA supplemented the ground stations with a fleet of eight Apollo Range Instrument Aircraft for extra support during orbital injection and reentry.

When the MSFN and STADAN consolidated into the STDN, the network acquired use of the tracking stations and equipment that had been used by the MSFN as well as added some new facilities. In 1972, the total network consisted of seventeen stations. By the end of 1978, fourteen stations remained in operation. During that time, NASA added new hardware to several of the stations. Goddard Space Flight Center in Greenbelt, Maryland, which managed and operated the STDN, improved the facilities at the center, adding a new telemetry processing system, modifying the control center to allow participating project scientists to manipulate their experiments directly, and improving the Image Processing Facility with new master data units.

The DSN continued to be operated by the Jet Propulsion Laboratory in Pasadena, California. The network depended primarily on three stations—at Canberra, Australia, in the Mojave Desert in California (Goldstone), and near Madrid, Spain. The network was equipped with a variety of antennas; the largest could communicate with spacecraft near the most distant planets.

NASA also began developing the TDRSS in the 1970s. Planned to support the Space Shuttle and other Earth-orbiting satellites, the TDRSS would rely on two synchronous orbit satellites and an on-orbit spare rather than a network of ground stations. Planners anticipated that this system would reduce the dependence on ground stations. Feasibility studies were completed during the 1970s, and contracts for the user antenna system and the three multiplexer-demultiplexers were awarded in 1976. Western Union Space Communications, Inc., was selected as the prime contractor for the system.

Management of the Tracking and Data Acquisition Program

The management organization of NASA's tracking and data acquisition activities could be considered in two phases: the first before the establishment of the Space Network and the second phase following the establishment of the Space Network. During both phases, NASA's tracking and data acquisition activities were centered in the Office of Space Tracking and Data Systems (OSTDS), designated as Code T. In 1987, NASA reorganized the OSTDS into the Office of Space Operations (OSO).

Phase I—Pre-Space Network

William Schneider led OSTDS from 1978 until April 1980, when Robert E. Smylie replaced him. Smylie led the office until Robert O. Aller took over as associate administrator in November 1983.

Three program divisions were in place in 1979: Network Operations and Communications, Network Systems Development, and Tracking and Data Relay Satellite System. The Network Operations and Communications Division was led by Charles A. Taylor. Frederick B. Bryant led the Network Systems Development Division. Robert Aller headed the TDRSS Division.

A 1980 reconfiguration eliminated Network Operations and Communications and Network Systems Development Divisions and established the Network Systems Division, led by Charles Taylor, and the Communications and Data Systems Division, headed by Harold G. Kimball. TDRSS continued as a division, and Robert Aller remained with the program until November 1983 when he became OSTDS associate administrator.

In April 1981, NASA established the Advanced Systems Office under the direction of Hugh S. Fosque. In January 1982, H. William Wood replaced Taylor as head of the Network Systems Division. He remained at that post until May 1984, when Charles T. Force was appointed to the position.

When Robert Aller became associate administrator in 1983, Wood was also given responsibility as acting TDRSS director until May 1984, when Jack W. Wild became director of that division. He remained with TDRSS until 1987. Figure 4–1 shows the organizational configuration during most of Phase I.

Phase II—The Space Network Becomes Operational

In 1984, OSTDS was reorganized to reflect the increasing importance of TDRSS. The new Space Networks Division replaced the TDRSS Division and had responsibility for implementing and operating TDRSS, for acquiring, operating, and maintaining the TDRSS ground terminals,

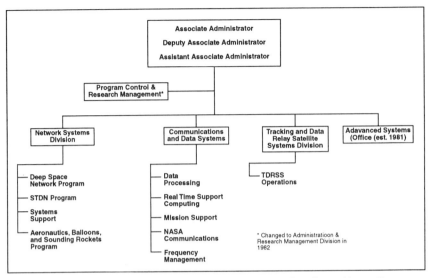

Figure 4–1. OSTDS Organizational Configuration (Pre–Space Network)

and for handling other activities and functions in support of the Space Network. The Network Systems Division was replaced by the Ground Networks Division, which had responsibility for all NASA ground networks, including the Goddard Space Flight Center Ground Network, the Jet Propulsion Laboratory Deep Space Network, the Wallops Orbital Tracking Station, and the Dryden Flight Research Facility and other tracking and data acquisition facilities. The Communications and Data Systems Division continued with its responsibility for all communications and data systems services for mission operations. Figure 4–2 shows the organizational configuration during this period.

In August 1984, Charles Force moved to the Ground Networks Division. He stayed there until 1987. Harold Kimball left the Communications and Data Systems Division in 1984, and the position remained vacant until S. Richard Costa became division director in November 1986.

Another reorganization took place in January 1987 when OSTDS became the Office of Space Operations (OSO). The office was responsible for developing a plan to manage NASA's increasingly complex space operations, with initial priority given to human-related space operations. The functions of OSTDS were integrated into OSO.

Also in 1987, Eugene Ferrick took over as director of TDRSS, and Robert M. Hornstein became acting division director for the Ground Networks program. In late 1988, S. Richard Costa left the Communications and Data Systems Division; John H. Roeder became acting division director.

The Mission Operations and Data Systems Directorate (MO&DSD) at Goddard Space Flight Center managed and operated the Ground Network, the Space Network (TDRSS), NASA worldwide communications, and

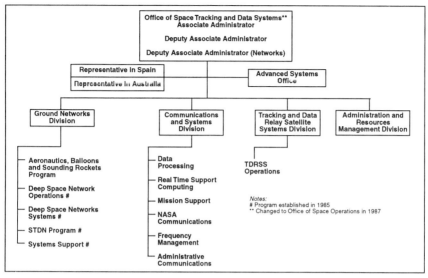

Figure 4–2. OSTDS Configuration During the Space Network Period

other functions necessary to the communications, data capture, processing and distribution, and orbit and attitude computations in support of space exploration and related activities. The Flight Dynamics Division performed orbital computations, spacecraft attitude determination, and flight maneuvering. The Operations Control Center commanded the spacecraft and monitored their health and safety. The Space and Ground Networks Division provided tracking services and relayed commands to and data from the user spacecraft through the Network Control Center. The NASA Communications (NASCOM) network provided data transport services. Data Processing captured and processed raw data to create usable information products for end users. The Technology Applications Division advanced the quality and effectiveness of the Data System by applying state-of-the-art technologies to system enhancements.

The Jet Propulsion Laboratory managed and operated the DSN. Its Office of Telecommunications and Data Acquisition had responsibility for twelve deep space stations located at three deep space communications complexes, the Network Operations Control Center at Pasadena, California, and the network's Ground Communications Facility.

Money for Tracking and Data Acquisition/Space Operations

The budget for NASA's tracking and data acquisition activities increased more than two and a half times from 1979 to 1988. This growth exceeded the rate for the entire NASA budget, which less than doubled, during the decade, the combined Research and Development (R&D) and Space Flight, Control, and Data Communications (SFC&DC) budget.

In 1988, tracking and data acquisition activities totaled approximately 10 percent of the NASA budget and 12.5 percent of the combined R&D

and SFC&DC budget. This increased from 1979, when tracking and data acquisition activities totaled approximately 6.5 percent of the NASA budget and 8.7 percent of the R&D budget. The growth can be primarily attributed to the costs associated with the TDRSS.

A comparison of most other budget elements cannot be done at a meaningful level because of the extensive reordering of budget categories that occurred during the decade. Budget items that had been in the Operations or Systems Implementation Program, when all tracking and data acquisition activities were in the R&D appropriation, were combined and put into one of three major categories (Space Networks, Ground Networks, or Communications and Data Systems) when the SFC&DC appropriation was established in 1984. However, it is possible to look at the budget activity in two groups: one before the SFC&DC appropriation began and one after that time.

The reader should note that all budget amounts reflect the value of the money at the time the budget was submitted and approved and funds were allocated. See Tables 4–1 through 4–37.

Tracking and Data Acquisition System Description

From 1979 to 1988, NASA's space tracking and data systems transitioned from a totally ground-based network mode of operation to a system with space-based capabilities for monitoring and commanding low-Earth orbital spacecraft and ground-based capabilities for deep space missions and particular types of low-Earth orbital missions. The following sections describe the Ground Network and the Space Network as they existed from 1979 to 1988.

Ground Network

The NASA Ground Network consisted of the STDN, the DSN, and the Aeronautics, Balloons, and Sounding Rocket Network. From 1979 to 1988, the Ground Network reduced the number of tracking stations while adding to the facilities and increased the capabilities at the remaining stations. Table 4–38 summarizes the locations and capabilities of the tracking stations.

Spaceflight Tracking and Data Network

The STDN was operated, maintained, and controlled by the Networks Division of the Goddard Space Flight Center in Greenbelt, Maryland. It provided tracking, data acquisition, and associated support for low-Earth orbital missions. The network was operated through NASA contracts and interagency and international agreements that provided staffing and logistical support. The Networks Division also operated the Network Control Center and NASA Ground Terminal. The division was responsible for testing, calibrating, and configuring network resources to ensure network support capability before each mission. It coordinated, scheduled, and directed all network activity and provided the necessary interface among Goddard elements and other agencies, centers, and networks.

The STDN was composed of the White Sands Ground Terminal and the NASA Ground Terminal in White Sands, New Mexico; NASCOM, the Flight Dynamics Facility, and the Simulation Operations Center at Goddard; and the ground network. The ground elements were linked by voice and data communications services provided by NASCOM. The prime operational communications data were formatted into 4,800-bit blocks and transmitted on the NASCOM wideband data and message switching system. Other communications traveled by teletype and facsimile facilities. Each ground station in the network provided coverage for approximately 20 percent of a satellite's or spacecraft's orbit and was limited to brief periods when the satellite or spacecraft was within the line of sight of a given tracking station. The various antennas at each STDN site accomplished a specific task, usually in a specific frequency band.

To provide reliable, continuous, and instantaneous communications support to the Space Shuttle, NASA added new sites and upgraded some of its existing facilities and capabilities for the Shuttle test phase and early Shuttle flights. In 1981, new sites for UHF air-to-ground voice were added in Dakar in Senegal, Botswana, and Yarragardee in Australia. Also added were three Shuttle-unique stations in Florida, California, and New Mexico. Department of Defense tracking and telemetry elements also supported the Shuttle flights. The Dakar UHF air-to-ground voice station was upgraded in 1982, before the STS-4 mission, to an S-band telemetry, voice, and command station. The change allowed for continuous telemetry data coverage between Bermuda and Hawaii for all due-east launches beginning with STS-4. This mid-point station allowed for the analysis of initial Orbital Maneuvering System burn data and provided for crew updates in case of an abort.

The network for the Shuttle orbital flight test program (STS-1 through STS-4) consisted of seventeen ground stations equipped with 4.26-, 9.14-, 12.19-, and 25.9-meter S-band antenna systems and C-band radar systems, NASCOM augmented by fifteen Department of Defense geographical locations providing C-band support, and one Department of Defense 18.3-meter S-band antenna system. In addition, six major computing interfaces—the Network Operations Control Center at Goddard, the Western Space and Missile Center at Vandenberg Air Force Base in California, the Air Force Satellite Control Facility in Sunnyvale, California, the White Sands Missile Range in New Mexico, and the Eastern Space and Missile Center in Florida—provided real-time network computational support. The stations that closed during this period were at Winkfield, England, at Rosman, North Carolina, which was turned over to the Department of Defense at the start of 1981, and at Quito, Ecuador, which closed in 1982 and transferred its equipment to the Dakar station.

The STDN stations were at Ascension Island, Bermuda, Botswana (beginning with STS-3), Buckhorn (Dryden Flight Research Facility) in

California, Dakar (beginning with STS-3), Fairbanks in Alaska, Goddard Space Flight Center, Goldstone (Ft. Irwin, California), Guam, Kokee in Hawaii, Madrid in Spain, Merritt Island in Florida, Orroral Valley (Canberra, Australia), Ponce de Leon in Florida (added for Shuttle program), Quito (closed in 1982), Santiago in Chile, Seychelles in the Indian Ocean (added for Shuttle program), Tula Peak in New Mexico, Wallops Orbital Tracking Station in Virginia, and Yarragardee (added for Shuttle program). Tula Peak, which initiated operations in 1979, was designated as a tracking support site for Shuttle orbital flight test landing activities. It initially suspended operations following STS-2, because of budget restrictions, but it was forced to reactivate its facilities on very short notice when STS-3 had to land at White Sands, New Mexico, rather than at Edwards Air Force Base in California because of bad weather in California.

Several instrumented U.S. Air Force aircraft, referred to as advanced range instrumentation aircraft, also supported the STDN. They were situated on request at various locations around the world where ground stations could not support Space Shuttle missions.

The Merritt Island, Florida, S-band station provided data to the Launch Control Center at Kennedy Space Center and the Mission Control Center at Johnson Space Center during prelaunch testing and terminal countdown. During the first minutes of launch, the Merritt Island and Ponce de Leon, Florida, S-band and Bermuda S-band stations, respectively, provided tracking data, both high speed and low speed, to the control centers at Kennedy and Johnson. The C-band stations located at Bermuda, Wallops Island in Virginia, the Grand Bahamas, Grand Turk, Antigua, and Cape Canaveral and Patrick Air Force Base in Florida also provided tracking data.

The Madrid, Indian Ocean Seychelles, Australian Orroral and Yarragardee, and Guam stations provided critical support to the Orbital Maneuvering System burns. During the orbital phase, all the S- and C-band stations that saw the Space Shuttle orbiter at 30 degrees above the horizon provided appropriate tracking, telemetry, air-ground, and command support to the Johnson Mission Control Center through Goddard.

During the nominal reentry and landing phase planned for Edwards Air Force Base, California, the Goldstone and Buckhorn, California, S-band stations and the C-band stations at the Pacific Missile Test Center, Vandenberg Air Force Base, Edwards Air Force Base, and Dryden Flight Research Center provided tracking, telemetry, command, and air-ground support to the orbiter. These locations also sent appropriate data to the control centers at Johnson and Kennedy. The tracking station at Ponce de Leon Inlet, Florida, provided support for the Space Shuttle during powered flight because of attenuation problems from the solid rocket booster motor plume.

In 1983, after supporting the STS-8 night landing, the Buckhorn special-purpose tracking station at Dryden Flight Research Center in California was phased out and operations terminated. This station had been established to support the Space Shuttle approach and landing tests and the operational flight test landings. Equipment from the Buckhorn site was moved a short distance to the Aeronautical Training Facility at Dryden, which already had been used to support NASA's aeronautics activities. This site was then also used to support STS missions.

When the first Tracking and Data Relay Satellite (TDRS-1) began tracking Shuttle missions in 1984, the White Sands Ground Terminal acquired the ground terminal communications relay equipment for the command, telemetry, tracking, and control equipment of the TDRSS (see the "Space Network" section below). The NASA Ground Terminal was co-located with the White Sands Ground Terminal. The NASA Ground Terminal, in combination with NASCOM, was NASA's physical and electrical interface with the TDRSS. The NASA Ground Terminal provided the interfaces with the common carrier, monitored the quality of the service from the TDRSS, and provided remote data quality to the Network Control Center.

The STDN consolidated its operations as the TDRSS took over the function of tracking most Earth-orbiting satellites. The facilities at Fairbanks, Alaska, were transferred to the National Oceanic and Atmospheric Administration in 1984. The STDN relinquished its Goldstone, Madrid, and Canberra stations and transferred them to the DSN sites. It gave the DSN support responsibility for spacecraft above the view of the TDRSS and for older spacecraft that were incompatible with the TDRSS. If the second TDRS had been successfully placed in orbit in 1986 as planned, the closure of additional STDN tracking stations would have occurred. However, the loss of the spacecraft in the *Challenger* explosion delayed the TDRS deployment by two years, and the reduction in STDN facilities was put on hold until the launch of TDRS-3 in September 1988.

At the time of the TDRS-3 launch, STDN tracking stations remained at Ascension Island, Bermuda, Canberra, Dakar, Guam, Kauai, Merritt Island, Ponce de Leon, Santiago, and Wallops Flight Facility. After the TDRSS was declared operational in 1989, the STDN decreased to stations at Wallops Island, Bermuda, Merritt Island, Ponce de Leon, and Dakar.

Deep Space Network

In 1988, the DSN consisted of twelve stations positioned at three complexes: Goldstone in southern California's Mojave Desert, near Madrid, and near Canberra. The Network Operations Control Center, at the Jet Propulsion Laboratory in Pasadena, California, controlled and monitored operations at the three complexes, validated the performance of the DSN for flight project users, provided information for configuring and controlling the DSN, and participated in DSN and mission testing. The DSN's Ground Communications Facility provided and managed the communications circuits that linked the complexes, the control center in Pasadena, and the remote flight project operations centers. The NASCOM network at Goddard Space Flight Center leased the communications circuits from

308   NASA HISTORICAL DATA BOOK

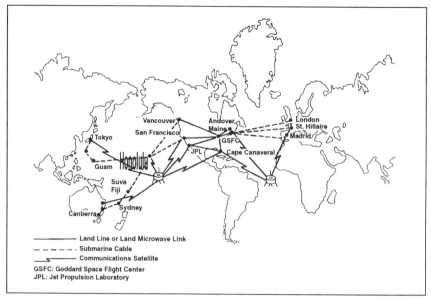

*Figure 4–3. The Deep Space Network Ground Communications Facility
Used Land Lines, Microwave Links, Satellite Cables, and
Communications Satellites to Link the Network's Elements*

common carriers and provided them as needed to all NASA projects, centers, and facilities. Figure 4–3 shows the elements that the Ground Communications Facility used to link elements of the network.

The DSN supported the unmanned spaceflight projects that NASA's Office of Space Science and Applications or other space agencies managed and controlled. The DSN received the telemetry signals from the spacecraft, transmitted commands that controlled the spacecraft operating modes, and generated the radio navigation data that were used to locate and guide each spacecraft to its destination. The DSN was also used for flight radio science, radio and radar astronomy, very long baseline interferometry, and geodynamics measurements.

The locations of the DSN complexes were approximately 120 degrees apart in longitude. This ensured continuous observation and suitable overlap for transferring the spacecraft radio link from one complex to the next. Each complex was situated in semi-mountainous, bowl-shaped terrain to shield against radio-frequency interference.

Each complex consisted of four deep space stations equipped with ultrasensitive receiving systems and large parabolic dish antennas. Equipment included two thirty-four-meter diameter S- and X-band antennas that had been converted from twenty-six-meter S-band antennas in 1980, one twenty-six-meter antenna, and one seventy-meter antenna. Figure 4–4 shows a twenty-six-meter antenna at Goldstone. In Canberra and Madrid, the seventy-meter antennas were extended in 1987 from their original sixty-four-meter-diameter configurations in preparation for the 1989 Voyager 2 encounter with Neptune. The extension of the sixty-four-

Figure 4–4. Twenty-Six-Meter Antenna at Goldstone

meter antenna at Goldstone took place in 1988. One of the 34-meter antennas at each complex was a new-design, high-efficiency antenna that provided improved telemetry performance needed for outer-planet missions.

The thirty-four- and seventy-meter stations were remotely operated from a centralized Signal Processing Center, which housed the electronic subsystems that pointed and controlled the antennas, received and processed the telemetry, generated and transmitted the commands, and produced the spacecraft navigation data. The twenty-six-meter stations required on-location operation.

Each antenna size formed separate subnets with different communications capabilities. The seventy-meter antenna subnet was the most sensitive and supported deep space missions. The twenty-six-meter subnet supported spacecraft in near-Earth orbit that were incompatible with the TDRSS, Shuttle flights, and geostationary launch service for space agencies worldwide. The two thirty-four-meter subnets supported both deep space and near-Earth orbital missions. The twenty-six-meter antenna stations were originally part of the STDN and were consolidated into the DSN in 1985, when it assumed that the added tracking responsibility for spacecraft in high elliptical Earth orbits that could not be supported by the TDRSS.

DSN support for inner-planet exploration began in 1962 with NASA's Mariner series of missions to Venus, Mars, and Mercury. Support for the first outer-planet missions, the Pioneer 10 and 11 flybys of Jupiter and Saturn, began in 1972–1973.

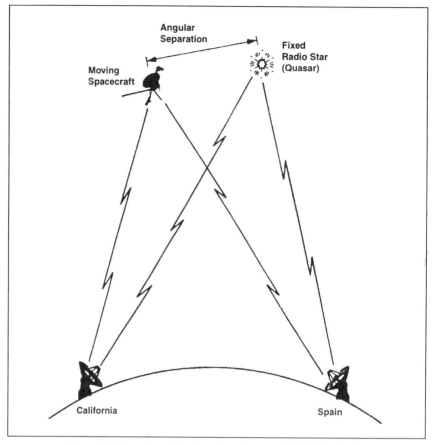

Figure 4–5. Very Long Baseline Interferometry Radio Navigation

The DSN's most complex support was for the Voyager mission. In the Voyager's 1981 encounter with Saturn, the DSN used differential very long baseline interferometry navigation to supplement the conventional Doppler and ranging navigation technique (Figure 4–5). The interferometry technique used two widely separated DSN stations on different continents to simultaneously receive signals from the spacecraft and from an angular nearby natural radio source (quasar) whose celestial coordinates were very well known. The data taken by the two stations were then correlated to provide a precise measurement of the angular separation between the spacecraft and the quasar. These measurements proved to have a repeatable precision of approximately fifty nanoradians, or five to ten times the angle measurement accuracy of the Doppler and range technique. This technology was especially important when missions required a close flyby of a planet to get an assist from that planet's gravity to alter trajectory and reduce travel time.

Another technique that improved the capability of tracking equipment over greater distances and the return data rates of planetary spacecraft was "arraying." This technique used two separate antennas to collect data from the spacecraft and then electronically added the signals together, producing the effect of a single antenna of larger diameter. NASA used this technique experimentally during the Voyager encounter with Jupiter and the Pioneer 11 encounter with Saturn, both in 1979. In 1980, NASA installed low-noise X-band maser antennas on existing sixty-four-meter antennas and integrated the enhancements by arraying the two antennas through a real-time combined assembly at each station complex. Electronically combining the spacecraft signals received by the two antennas provided about 35 percent more images from Saturn than could be obtained with a single sixty-four-meter antenna.[1]

In the 1986 Voyager 2 encounter with Uranus, the DSN carried the arraying technology farther by combining DSN antennas at each complex and augmenting the sixty-four-meter antenna at Canberra with the large radio telescope at Parkes, Australia. The weaker signals received at each antenna were combined into a single stronger signal, resulting in an increase of approximately fifty percent in the reception of Uranus data.[2] The construction of new-design, high-efficiency thirty-four-meter antennas at each complex permitted a three-element array, consisting of two thirty-four-meter antennas and one sixty-four-meter antenna. (The sixty-four-meter antennas were enlarged and redesigned as high-efficiency seventy-meter antennas before the Voyager encounter with Neptune.)

NASA would also use the arraying technique for the Voyager encounter with Neptune in 1989, combining two antennas at Goldstone with the Very Large Array (twenty-seven antennas) at Socorro, New Mexico.[3] Arraying Goldstone with the Very Large Array would result in more than doubling Goldstone's capability on its own. The arraying with the Very Large Array would also result in the most antennas ever arrayed anywhere at once, the largest fully steerable equivalent aperture ever used for a communications link (151 meters), the longest array (19,300 kilometers) ever used for communications, and the first arraying for telemetry via satellite.[4]

Radio interferometry was more commonly used to detect details of celestial objects. In 1987, NASA used the technique to observe Supernova 1987-A. A DSN antenna located in Tidbinbilla, Australia, was connected by microwave to Parkes Radio Telescope (of Australia's Commonwealth Scientific and Industrial Research Organization)

---

[1] *Aeronautics and Space Report of the President, 1980* (Washington, DC: U.S. Government Printing Office, 1981), p. 32.

[2] "DSN Fact Sheet," Jet Propulsion Laboratory, NASA (on-line).

[3] The Very Large Array was operated by the National Radio Astronomy Observatory and sponsored by the National Science Foundation. It consisted of twenty-seven antennas, each twenty-five meters in diameter, configured in a "Y" arrangement on railroad tracks over a twenty-kilometer area.

[4] Edward C. Posner, Lawrence L. Rauch, and Boyd D. Madsen, "Voyager Mission Telecommunication Firsts," *IEEE Communications Magazine* 28(9) (September 1990): 23.

200 miles away. The two linked antennas formed a theoretical receiver the size of the distance between the two antennas. (The link had been put in place to observe the Voyager's encounter with Uranus.) NASA formed an even wider network using very long baseline interferometry that connected four antennas: Tidbinbilla, Parkes, a Landsat ground station at Alice Springs in central Australia, and a twenty-six-meter antenna at the University of Tasmania on the island of Tasmania.

NASA's DSN also supported international missions. In 1985, as part of a French-led international tracking network, the DSN tracked two Soviet-French balloon experiments that studied Venus' atmosphere and provided information on its weather dynamics. Two Soviet Vega spacecraft on their way to Halley's comet inserted the meteorological balloons into the Venusian atmosphere as they passed the planet. The tracking stations used very long baseline interferometry to measure the balloons' velocity and, therefore, the wind velocity with a precision of about 3 kilometers per hour at Venus' distance (about 108 million kilometers from Earth).

The DSN also provided navigation support to five international spacecraft that encountered Halley's comet in March 1986. The DSN supported Japanese efforts to track their two spacecraft, provided backup tracking of the European Space Agency's Giotto spacecraft, and tracked the Soviet Vega spacecraft as they approached the comet.

The Soviet Phobos project also received DSN support. Phobos 1 and 2 were launched in July 1988. The DSN tracked the spacecraft as well as the Martian moon Phobos to permit the landers to land on the moon. Scientists used very long baseline interferometry as well as Doppler and range tracking to pinpoint the position and motions of Phobos.

Aeronautics, Balloons, and Sounding Rockets

The Aeronautics, Balloons, and Sounding Rockets (AB&SR) Program provided fixed and mobile instrumentation systems to meet the tracking, data acquisition, and range safety requirements of NASA research vehicles using primarily suborbital vehicles. The principal facilities supporting this program were Wallops Island, the Dryden Flight Research Center and Moffett Field, the Poker Flats Research Facility, White Sands Missile Range, and the National Scientific Balloon Facility. Mobile facilities were used worldwide to meet varied scientific requirements.

In February 1987, the AB&SR program responded to the supernova discovery by establishing a sounding rocket capability in Australia, launching both balloons and sounding rockets. In 1988, the Office of Space Operations program continued support to the supernova program with balloon and sounding rocket launches from Australia. Major AB&SR activities during 1988 included thirty-three large rockets and forty-six large balloons with scientific payloads launched worldwide.

Space Network

The NASA Space Network was a space-based communications system composed of the TDRSS and supporting ground elements. These elements included a space-to-ground single ground terminal at White Sands, New Mexico, comprised of the Network Control Center, NASA Ground Terminal, Flight Dynamics Facility, and Simulation Operations Center. The system had a constellation of three data relay satellites—two operational and one spare—in geosynchronous orbit.[5]

The Space Network provided tracking, telemetry, and command services to the Space Shuttle, to other low-Earth-orbiting spacecraft, and to some suborbital platforms that had been supported by a number of ground stations throughout the world. From 1979 to 1988, two Tracking and Data Relay Satellites (TDRS) were deployed and became operational. The first, TDRS-1, was launched from STS-6 in 1983. TDRS-3 was launched from STS-26 when the Shuttle program returned to operational status in 1988. TDRS-B was lost in the *Challenger* explosion in 1986.

A third TDRS (TDRS-4) was launched in 1989 and was positioned as TDRS-East. At that time, TDRS-1 was moved to the spare position. NASA launched later tracking and data relay satellites in 1991, 1993, and 1995.

The system did not perform processing of user traffic. Rather, it operated as a "bent-pipe" repeater—that is, it relayed signals and data between the user spacecraft and ground terminal in real time. The system was characterized by its unique ability to provide bi-directional high data rates as well as position information to moving objects in real time nearly everywhere around the globe. The satellites were the first designed to handle telecommunications services through three frequency bands—S-, Ku-, and C-bands. They could carry voice, television, and analog and digital data signals. The tracking and data relay satellites could transmit and receive data and track a user spacecraft in a low-Earth orbit for a minimum of 85 percent of each spacecraft's orbit.

Background

NASA's satellite communications system was initiated following studies in the 1970s. These studies showed that a system of telecommunications satellites operated from a single ground station could support the Space Shuttle and scientific application mission requirements planned for the space program better than ground-based tracking stations. Ground-based tracking stations could track a satellite during only about

---

[5] Deep space probes and Earth-orbiting satellites above approximately 5,700 kilometers used the three ground stations of the DSN, operated for NASA by the Jet Propulsion Laboratory in Pasadena, California. The DSN stations were at Goldstone in California, Madrid in Spain, and Canberra in Australia.

20 percent of its orbit and only when that satellite was in direct line of sight with a tracking station. Consequently, it was necessary to have ground stations around the globe and to continually "hand off" a satellite from one station to another.

In addition, the system was viewed as a way to halt the growing costs of upgrading and operating a network of tracking and communications ground stations around the world. It was planned that when the TDRSS became fully operational, ground stations of the worldwide STDN would be closed or consolidated, resulting in savings in personnel and operating and maintenance costs. The Merritt Island, Ponce de Leon, and Bermuda ground stations would remain open to support the launch and landing of the Space Shuttle at the Kennedy Space Center in Florida.

It was also decided that leasing a system was more desirable than purchasing it. In December 1976, NASA awarded a contract to Western Union Space Communications (Spacecom), which would own and operate the system. The principal subcontractors were TRW for the satellite development and system integration and The Harris Corporation for ground terminal development. The contract provided for ten years of service to NASA and included both space and ground segments of the system.[6] It also established a joint government-commercial program with one satellite intended to provide domestic communications services commercially. The development was to be financed with loans provided to the contractor by the Federal Financing Bank, an arm of the U.S. Treasury. NASA would make loan repayments to the bank once service began. Public Law 95-76, dated July 30, 1977, provided permanent legislation for the TDRSS.

In 1980, the contract was transferred to a partnership of Western Union, Fairchild, and Continental Telephone. In 1983, Western Union sold its share of the business to the other two partners, and in 1985, Fairchild sold its share, leaving Continental Telecom (Contel) as the sole owner of Spacecom. In 1990, a new contract transferred ownership of the system to NASA but retained Contel as the operator.[7]

The Tracking and Data Relay Satellite System

The full TDRSS network consisted of three satellites in geosynchronous orbits. TDRS-East was positioned at forty-one degrees west longitude. TDRS-West was positioned at 171 degrees west longitude. A third TDRS was positioned as a backup above a central station just west of South America at sixty-two degrees west longitude. The positioning of

---

[6] See Linda Neuman Ezell, *NASA Historical Data Book, Volume III: Programs and Projects, 1969–1978* (Washington, DC: NASA SP-4012, 1988), for further information on the development of the TDRSS program.

[7] Donald H. Martin, *Communications Satellites, 1958–1992* (El Segundo, CA: The Aerospace Corporation, December 31, 1991), pp. 186–89.

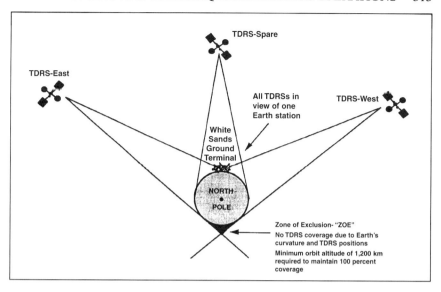

Figure 4–6. TDRSS Coverage Area

two tracking and data relay satellites 130-degree spacing reduced the ground station requirements to one station instead of the two stations required for 180-degree spacing. Figure 4–6 shows the coverage area of the TDRSS. The satellites were positioned in geosynchronous orbits above the equator at an altitude of approximately 35,880 kilometers. At that altitude, because the speed of the satellites were the same as the rotational speed of Earth, they remained fixed in orbit over the same location.

The TDRSS network had three primary capabilities: tracking, telemetry and data, and command. Network tracking determined the precise location of orbiting user spacecraft by measuring range (distance) and range rate (velocity) with respect to the known position of the TDRS. Ground-based stations determined the TDRS position.

The user spacecraft transmitted telemetry signals indicating certain operational parameters, such as power level and temperature. It also transmitted data signals that corresponded to the scientific or applications information collected by the spacecraft instruments. The tracking and data relay satellites relayed the telemetry and data signals from the user spacecraft to the White Sands Ground Terminal for use by the Goddard Space Flight Center and the user community. The White Sands Ground Terminal sent the raw data directly by domestic communications satellite to NASA control centers at Johnson Space Center (for Space Shuttle operations) and Goddard, which scheduled TDRSS operations and controlled a large number of satellites. Figure 4–7 shows the user data flow.

The White Sands Ground Terminal sent command signals via the tracking and data relay satellites to user spacecraft, ordering the spacecraft to perform certain functions. The commands originated from Goddard for unmanned spacecraft or from Johnson for manned spacecraft. Figure 4–8 shows the entire system.

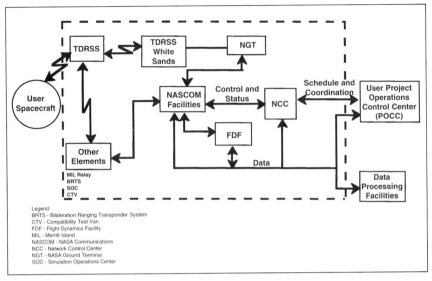

Figure 4–7. User Data Flow

The White Sands Ground Terminal was located at a longitude with a clear line of sight to the tracking and data relay satellites and very little rain, because rain could interfere with the Ku-band uplink and downlink channels. It was one of the largest and most complex communications terminals ever built. Many command and control functions ordinarily found in the space segment of a system were performed by the ground station, such as the formation and control of the receiver beam of the TDRS multiple-access phased-array antenna and the control and tracking functions of the TDRS single-access antennas.

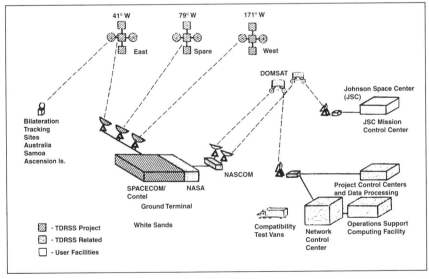

Figure 4–8. TDRSS Elements

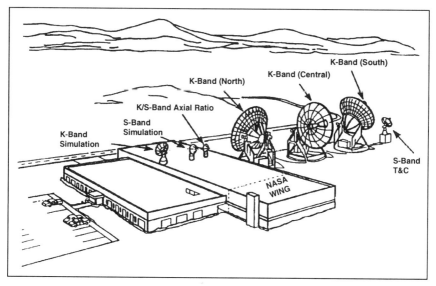

Figure 4–9. White Sands Ground Terminal

The most prominent features of the ground station were three eighteen-meter Ku-band dish antennas used to transmit and receive user traffic. Several other antennas were used for S- and Ku-band communications. NASA developed sophisticated operational control facilities at Goddard and next to the White Sands Ground Terminal to schedule TDRSS support of each user and to distribute user data from White Sands to the user (Figure 4–9).

In the mid-1980s, NASA identified the need for a second TDRSS ground terminal at White Sands. This ground terminal would provide a backup to the White Sands Ground Terminal in the event of a catastrophic failure or planned outages for system upgrades or repair. It would also provide expanded capability for the increased user demand that was expected for the 1990s. The TDRSS program office initiated competitive Definition Phase studies for the development of a second ground terminal in 1987. In 1988, General Electric Company received the contract to develop, fabricate, install, and test the second TDRSS ground terminal communications hardware and software. The complex was dedicated in 1990.

In addition to the Space Shuttle, the TDRSS could support up to twenty-six user satellites simultaneously. It provided two types of service: a multiple-access service, which could relay data from as many as twenty low-data-rate user satellites simultaneously, and a single-access service, which provided two high-data-rate communications relays. Tables 4–39 and 4–40, respectively, provide single-access and multiple-access link summaries.

The tracking and data relay satellites were deployed from the Space Shuttle at an altitude of approximately 296 kilometers, and inertial upper stage (IUS) boosters propelled them to geosynchronous orbit. The antennas and solar panels unfolded, and the satellite then separated from the

IUS and drifted in orbit to a position free from radio-frequency interference, where it was checked out. The TDRS was allowed to drift to the final orbital location, where it was maintained, monitored, and commanded by the ground segment of the Space Network.

Three-axis stabilization aboard the TDRS maintained attitude control. Body-fixed momentum wheels combined with body-fixed antennas pointing constantly at Earth, while the satellite's solar arrays tracked the Sun. Monopropellant hydrazine thrusters were used for TDRS positioning and north-south, east-west stationkeeping.

The tracking and data relay satellites were the largest privately owned telecommunications satellites ever built. They were composed of three modules: an equipment module, a communications payload module, and an antenna module. The equipment module consisted of the attitude control, electrical power, and thermal control subsystems. The attitude control subsystem stabilized the satellite so that the antennas had the proper orientation toward Earth and the solar panels were properly aimed toward the Sun. The electrical power subsystem consisted of two solar panels that provided a ten-year life span of approximately 1,700 watts of power. The thermal control subsystem consisted of surface coatings and controlled electric heaters. The communications payload module was composed of the electronic equipment and associated antennas required for linking the user spacecraft with the ground terminal.

The antenna module housed four antennas. For single-access services, each TDRS had two dual-feed S-band/Ku-band deployable parabolic antennas. They were just under five meters in diameter, unfurled like a giant umbrella when deployed, and attached on two axes that could move horizontally or vertically (gimbal) to focus the beam on orbiting satellites below. They were fabricated of woven molybdenum mesh and plated with 14K gold. When deployed, the antenna's 18.8 square meters of mesh were stretched tautly on sixteen supporting tubular ribs by fine thread-like quartz cords. The entire antenna structure, including the ribs, reflector surface, a dual-frequency antenna feed and the deployment mechanisms needed to fold and unfold the structure, weighed approximately twenty-three kilograms. Their primary function was to relay communications to and from user spacecraft. The high-bit-rate service made possible by these antennas was available to users on a time-shared basis. Each antenna simultaneously supported two user satellites or spacecraft (one on S-band and one on Ku-band) if both users were within the antenna's bandwidth.

For multiple-access service, the multi-element S-band phased array of helical radiators was mounted on the satellite body. The multiple-access forward link (between the TDRS and the user spacecraft) transmitted command data to the user spacecraft. In the return link, the signal outputs from the array elements were sent separately to the White Sands Ground Terminal parallel processors.

A fourth antenna, a two-meter parabolic reflector, provided the prime link for relaying transmissions to and from the ground terminal at Ku-band. The antenna was used to control the TDRS while it was in transfer orbit to geosynchronous altitude. Table 4–41 provides the TDRS characteristics.

TDRS-1. TDRS-A and its IUS were carried aboard the Space Shuttle *Challenger* on the April 4, 1983, STS-6 mission.[8] After it was deployed and the first-stage boost of the IUS solid rocket motor was completed, the second-stage IUS motor malfunctioned and left TDRS-1 in an elliptical orbit far short of the planned geosynchronous altitude. Also, the satellite was spinning out of control at a rate of thirty revolutions per minute until the contractor flight control team recovered control and stabilized it.

Later, the contractors and the NASA TDRSS program officials devised a procedure for using the small hydrazine-fueled Reaction Control System thrusters on TDRS-1 to raise its orbit. The thrusting, which began on June 6, 1983, required thirty-nine maneuvers to raise TDRS-1 to geosynchronous orbit. The maneuvers consumed approximately 408 kilograms of the satellite's propellant, leaving approximately 226 kilograms of hydrazine for the ten-year on-orbit operations.

During the maneuvers, overheating caused the loss of one of the redundant banks of twelve thrusters and one thruster in the other bank. The flight control team developed procedures to control TDRS-1 properly in spite of the thruster failures.

TDRS-1 was turned on for testing on July 6, 1983. The tests proceeded without incident until October 1983, when one of the Ku-band single-access-link diplexers failed. Shortly afterward, one of the Ku-band traveling-wave-tube amplifiers on the same single-access antenna failed, and the forward link service was lost. On November 19, 1983, one of the Ku-band traveling-wave-tube amplifiers serving the other single-access antenna failed. TDRS-1 testing was completed in December 1984. Although the satellite could provide only one Ku-band single-access forward link, it could still function.

Later tracking and data relay satellites were identical to TDRS-1 except for modifications to correct the malfunctions that occurred on TDRS-1 and a modification of the C-band antenna feeds. The C-band minor modification improved coverage for providing government point-to-point communications.

TDRS-2. Originally scheduled for launch in March 1985, a problem in the timing circuitry associated with the command system resulted in a launch delay. The spacecraft was subsequently lost on the STS 51-L (*Challenger*) mission.

TDRS-3. The launch of TDRS-3 went smoothly, and the IUS successfully boosted the spacecraft to the required orbit. When it was positioned at 171 degrees west longitude, it provided coverage to the eastern United States and westward into central China. The successful deployment of

---

[8] Further information on the STS-6 mission can be found in Chapter 3, "Space Transportation/Human Spaceflight," in Volume V of the *NASA Historical Data Book*.

TDRS-3 allowed NASA to continue with the shutdown of additional tracking ground stations.

Communications and Data Systems

The elements of NASA's Communications and Data Systems Program linked the data acquisition stations and users. These elements included communications, mission control, data capture and processing, and frequency management and were organized into two major programmatic areas. The Communications program provided for communications required to link remote tracking stations with mission control and data processing facilities and for administrative services for NASA centers and Headquarters. The Data Systems program provided for real-time operational and postflight data processing support and mission operations crucial to determining the condition of spacecraft and payloads and to the generation of commands for spacecraft and payload control.

Communications Program

Two networks comprised NASA's communications facilities: the NASA Communications Network (NASCOM) and the Program Support Communications Network (PSCN). Other systems also provided communications support.

NASA Communications Network. NASCOM linked the elements of the Ground and Space Networks. The NASCOM network was a worldwide complex of communications services, including data, voice, teletype and video systems that were a mixture of government-owned and -leased equipment as well as leased services. The major NASCOM switching centers were at the Goddard Space Flight Center. From Goddard, personnel directed overall network operations, including those at supporting NASCOM switching centers in Madrid, Canberra, and the Jet Propulsion Laboratory in Pasadena, California. In addition, support activities were provided by Air Force communications centers at Cape Canaveral, Florida, and Vandenberg Air Force Base, California.

The communications network consisted of more than 2 million circuit miles of diversely routed communications channels. It used domestic and international communications satellites, submarine cable and terrestrial landlines and microwave radio systems to interconnect the tracking stations, launch and orbital control centers, and other supporting locations.

Numerous computers at the different ground tracking stations controlled the tracking antennas, handled commands, and processed data for transmission to the control centers at Johnson Space Center and Goddard. Mission data from all the tracking stations were funneled into the main switching computers at Goddard and rerouted to the users without delay by domestic communications satellites. Commands were transmitted to the main switching computers at Goddard and switched to the proper tracking station for transmission to the Space Shuttle or other spacecraft.

The Shuttle flights implemented a key change in the communications network. For the first time, two simultaneous air-ground S-band voice circuits in addition to UHF radio capability were provided. In previous Apollo missions, only one S-band circuit was provided. Telemetry data circuitry from tracking stations was increased in size to handle 128,000 bits per second (128 kilobits per second) in real time versus the 14–21 kilobits per second in previous programs. Correspondingly, the command data circuit to a station was increased from 7.2 kilobits per second to a 56-kilobyte-per-second capability.

A station conferencing and monitoring arrangement allowed various traffic managers to hold conferences with as many as 220 different voice terminals throughout the United States and abroad with talking and listening capability. The system was redundant, with a mission support reliability record of 99.6 percent. All Space Shuttle voice traffic was routed through this arrangement at Goddard.

Program Support Communications Network. The PSCN, which became operational in 1986, connected NASA centers, Headquarters, and major contractors to provide programmatic and administrative information. Its services included voice, voice conferencing, data, and facsimile. It also linked the NASA supercomputers at the Ames Research Center with those at other centers. It was a fully digital backbone network supporting both circuit switching and pocket switching over digital transmission facilities.

Time Division Multiple Access. This system, which also became operational in 1986, used advanced technologies developed by the communications industry. It provided operational circuits by way of satellites that could be used by NASA as workloads required.

Data Systems Program

This program planned, designed, developed, and operated systems that processed spacecraft telemetry for the worldwide science community. One of its major systems was the Spacelab Data Processing Facility. This facility processed and delivered extensive data received from the Spacelab missions. During the 1980s, the Data Systems program was also preparing to support the Hubble Space Telescope mission and, in 1985, completed the development of its Data Capture Facility.

Table 4–1. Total Office of Space Tracking and Data Systems Funding (in thousands of dollars) a

Year	Request	Authorization	Appropriation	Programmed (Actual)
1979	305,400	305,400	b	299,900
1980	332,400 c	332,800	d	332,100
1981	341,100 e	349,750	341,100 f	339,900
1982	402,100 g	408,180	415,200	402,100
1983	498,900 h	503,900	508,900	485,600
1984	688,200 i	700,200	690,200 j	674,000
1985	811,000 k	811,000 l	811,000 m	795,700
1986	675,900 n	717,500 o	717,500 p	660,400
1987	880,000 q	878,000	878,000	845,900
1988	902,500 r	943,000 s	912,000 t	879,400

a   Beginning in FY 1984, the Office of Space Tracking and Data Systems (OSTDS) became part of the Space Flight, Control, and Data Communications (SFC&DC) appropriation. All major programs moved to SFC&DC except for Advanced Systems, which remained in R&D.

b   Undistributed. Total 1979 R&D appropriation = $3,477,200,000.

c   Amended budget submission. Original budget submission = $332,800,000. The reduction results from the congressional general reduction in the FY 1980 NASA R&D appropriation request. The revised submission reflects adjustments between the Operations and the Systems Implementation categories to consolidate funding for the more significant capabilities being implemented in the Space Tracking and Data Systems program.

d   Undistributed. Total 1980 R&D appropriation = $4,091,086,000.

e   Amended budget submission. Initial submission = $359,000,000.

f   Appended appropriation (6-3-81) reflects the effect of General Provision, Sec. 412. Basic appropriation was $349,000,000.

g   Amended budget submission. Initial submission = $435,200,000. The decrease reflects the general congressional reduction of the FY 1982 appropriation and FY 1983 decreases, including the closing of the Quito and Tula Peak tracking stations, the closure of the deep space twenty-six-meter antenna subnetwork, a reduction in staffing at a number of STDN tracking stations, an adjustment in the deep space systems implementation program based on requirements for the Deep Space Network configuration, a reprogramming to the Construction of Facilities appropriation for two thirty-four-meter antenna facilities, the decision to lease major computer replacement systems, and the rephasing of the space telescope data capture system.

h   Amended budget submission. Initial submission = $508,900,000. The decrease reflects the application of a portion of the general congressional reduction in the FY 1983 appropriation request. The major portion of the reduction occurred in the Space Network budget line because of a revision in the date for the initiation of TDRSS loan repayments and a decrease in the projected amount to be borrowed under the Federal Financing Bank loan. A second portion of the decrease occurred in the Ground Network budget line item from decreased staffing and related support requirements in the STDN.

i   This includes both the SFC&DC and R&D budget categories controlled by OSTDS. Revised SFC&DC amount = $674,000,000 (initial − $700,200,000); R&D amount − $14,200,000 (no revision). The decrease reflects a reduction in the payment to the Federal Financing Bank consistent with the FY 1984 HUD-Independent Agencies Appropriations Conference Agreement and the application of a portion of the general appropriation reduction to this program. Within the initial operating plan, adjustments were made primarily to accommodate the impact on the program resulting from the failure of the inertial upper stage to properly deploy the first TDRS to geosynchronous orbit in April 1983.

j   This includes amounts for both the new SFC&DC appropriations category and R&D appropriations category. All Space Tracking and Data Acquisition activities moved to SFC&DC except Advanced Systems, which remained in R&D.

Table 4–1 continued

k   This includes both the SFC&DC and R&D budget categories controlled by OSTDS. SFC&DC amount = $795,700,000; R&D amount = $15,300,000.

l   This includes both the SFC&DC and R&D budget categories controlled by OSTDS. Revised SFC&DC amount = $660,400,000 (initial SFC&DC amount = $808,300,000); Revised R&D amount = $15,500,000 (initial R&D amount = $16,200,000).

m   This includes both the SFC&DC and R&D budget categories controlled by OSTDS. SFC&DC amount = $795,700,000; R&D amount = $15,300,000.

n   This includes both the SFC&DC and R&D budget categories controlled by OSTDS. SFC&DC amount = $808,300,000; R&D amount = $16,200,000.

o   This includes both the SFC&DC and R&D budget categories controlled by OSTDS. SFC&DC amount = $701,300,000; R&D amount = $16,200,000.

p   This includes both the SFC&DC and R&D budget categories controlled by OSTDS. SFC&DC amount = $701,300,000; R&D amount = $16,200,000.

q   This includes both the SFC&DC and R&D budget categories controlled by OSTDS. Revised SFC&DC amount = $862,900,000 (initial SFC&DC budget submission = $798,900,000); R&D budget submission = $17,100,000 (no change between revised and initial submission).

r   This includes both the SFC&DC and R&D budget categories controlled by OSTDS. Revised SFC&DC amount = $884,400,000 (initial SFC&DC amount = $948,900,000); R&D amount = $18,100,000. The reduction reflects a reduction of $40 million for the TDRSS Replacement and a general reduction consistent with congressional direction.

s   Reductions from budget submission in SFC&DC budget categories. No change to R&D budget submission.

t   Reductions from budget submission in SFC&DC budget categories. No change to R&D budget submission.

Table 4–2. Major Budget Category Programmed Funding History (in thousands of dollars)

Budget Category/Fiscal Year	1979	1980	1981	1982	1983
Tracking and Data Acquisition (R&D)	299,900	332,100	339,900	402,100	485,500
Operations	249,903	264,400	266,496		
System Implementation	40,497	57,100	62,105		
Advanced Systems	9,500	10,600	11,300	12,500	13,400
Space Network				21,800	104,300
Ground Network				237,457	242,920
Communications and Data Systems				130,343	138,280

Budget Category/Fiscal Year	1984	1985	1986	1987	1988
Space and Ground Networks, Communication, and Data Systems (SFC&DC)	674,000	795,700	660,400	845,400	879,400
Space Network	259,300	378,300	273,700	404,300	433,400
Ground Network	249,300	233,200	210,400	237,200	231,000
Communications and Data Systems	165,600	184,200	176,300	203,900	215,000
Tracking and Data Advanced Systems (R&D)	14,200	14,800	15,500	17,100	17,900
Advanced Systems	14,200	14,800	15,500	17,100	17,900

Table 4–3. Operations Funding (in thousands of dollars)

Year	Request	Authorization	Appropriation	Programmed (Actual)
1979	254,200	254,200	a	249,903
1980	264,500 b	275,800	c	264,400
1981	267,100 d	270,750	267,100 e	266,495
1982	300,500 f	g	305,500	h
1983	338,200	i	338,200	j

a   Undistributed. Total 1979 R&D appropriation = $3,477,200,000.
b   Amended budget submission. Original budget submission = $275,800,000.
c   Undistributed. Total 1980 R&D appropriation = $$4,091,086,000.
d   Amended budget submission. Initial submission = $271,500,000.
e   Appended appropriation (6-3-81) reflected the effect of General Provision, Sec. 412. Basic appropriation was $270,000,000.
f   Amended budget submission. Initial submission = $309,800,000.
g   Undistributed. Total 1982 Tracking and Data Acquisition authorization = $408,180,000.
h   Budget categories changed in FY 1984 Budget Estimate, which included FY 1982 programmed amounts.
i   Undistributed. Total 1983 Tracking and Data Acquisition authorization = $503,900,000.
j   Reordering of budget categories split Operations among Space Network (Table 4–17), Ground Network (Table 4–24), and Communications and Data Systems (Table 4–31).

Table 4–4. Spaceflight Tracking and Data Network (STDN) Funding History (in thousands of dollars)

Year (Fiscal)	Submission	Programmed (Actual)
1979	129,100 a	127,068
1980	136,400 b	130,530
1981	130,400 c	130,652
1982	143,600 d	e

a   Revised budget submission. Original submission = $129,900,000.
b   Revised budget submission. Original submission = $141,200,000. The decrease results primarily from the delay in the Shuttle orbital flight test schedule and termination of ATS-6 support.
c   Revised budget submission. Original submission = $133,300,000. The decrease reflects the general reduction to FY 1981 appropriations that would result in the closure of telemetry links at Alaska and the Network Test and Training Facility, a one-shift reduction at the Hawaii station, and the consolidation of a mission control activity for HEAO-2 and HEAO-3.
d   Revised budget submission. Original submission = $149,100,000.
e   This became part of new Ground Network budget category. Most STDN costs moved to STDN Operations (Table 4–20). TDRSS-related costs moved to new TDRSS budget categories (Tables 4–18, 4–19, and 4–20).

Table 4–5. Deep Space Network Funding History
(in thousands of dollars)

Year (Fiscal)	Submission	Programmed (Actual)
1979	49,800	51,032
1980	49,500 a	58,200
1981	56,000 b	54,427
1982	63,400 c	d

a Revised budget submission. Original submission = $55,600,000. The decrease reflects the transfer of funding for Jet Propulsion Laboratory engineering efforts associated with specific projects from the Operations budget to the Systems Implementation budget so that budgeting for specific projects resides in one program area.
b Revised budget submission. Original submission = $54,100,000.
c Revised budget submission. Original submission = $64,400,000.
d See Table 4–28.

Table 4–6. Aeronautics and Sounding Rocket Support Operations
Funding History (in thousands of dollars)

Year (Fiscal)	Submission	Programmed (Actual)
1979	4,900	4,516
1980	5,300 a	4,830
1981	5,500	6,025
1982	6,600 b	c

a Revised budget submission. Original submission = $4,800,000. The increase provides for operational communications in support of the tilt rotor program and Shuttle at Dryden Flight Research Center and the rehabilitation of heavy mechanical subsystems for the radar and antenna pedestals at Wallops Flight Facility.
b Revised budget submission. Original submission = $6,800,000.
c See Table 4–30.

Table 4–7. Communications Operations Funding History
(in thousands of dollars)

Year (Fiscal)	Submission	Programmed (Actual)
1979	34,800 a	34,027
1980	37,700 b	35,130
1981	37,400 c	37,531
1982	39,800 d	e

a Revised budget submission. Original submission = $37,900,000.
b Revised budget submission. Original submission = $39,800,000. The decrease reflects the delay in the Shuttle orbital flight test schedule, which allowed for a delay in ordering up the wideband and video communications circuits required for Shuttle support and lower than originally estimated prices for some overseas wideband circuits.
c Revised budget submission. Original submission = $39,300,000.
d Revised budget submission. Original submission = $40,500,000.
e See Table 4–33.

TRACKING AND DATA ACQUISITION/SPACE OPERATIONS

Table 4–8. Data Processing Operations Funding History, 1979–1982 (in thousands of dollars)

Year (Fiscal)	Submission	Programmed (Actual)
1979	33,000 a	33,260
1980	35,600 b	35,890
1981	37,800	37,860
1982	47,100 c	d

a Revised budget submission. Original submission = $31,700,000.
b Revised budget submission. Original submission = $34,400,000. The increase reflects greater-tan-expected costs associated with bringing the Image Data Processing Facility into full operation and the implementation of a domsat terminal for rapid handling of Landsat data.
c Revised budget submission. Original submission = $44,700,000.
d See Table 4–37.

Table 4–9. Systems Implementation Funding History (in thousands of dollars)

Year (Fiscal)	Submission	Authorization	Appropriation	Programmed (Actual)
1979	41,300	41,300	a	40,497
1980	57,300 b	46,400	c	57,100
1981	62,700 d	67,700	62,700 e	62,105
1982	89,100 f	g	97,200	h
1983	96,000	i	96,000	j

a Undistributed. Total 1979 R&D appropriation = $3,477,200,000.
b Amended budget submission. Original budget submission = $46,400,000.
c Undistributed. Total 1980 R&D appropriation = $$4,091,086,000.
d Amended budget submission. Initial submission = $76,200,000.
e Appended appropriation (June 3, 1981) reflected the effect of General Provision, Sec. 412. Basic appropriation was $67,700,000.
f Amended budget submission. Initial submission = $112,900,000.
g Undistributed. Total 1982 Tracking and Data Acquisition authorization = $408,180,000.
h Budget categories changed in FY 1984 Budget Estimate, which included FY 1982 programmed amounts.
i Undistributed. Total 1983 Tracking and Data Acquisition authorization = $503,900,000.
j Budget category split among Space Network, Ground Network, and Communications and Data Systems budget categories.

*Table 4–10. Spaceflight Tracking and Data Network (STDN)
Implementation Funding History (in thousands of dollars)*

Year (Fiscal)	Submission	Programmed (Actual)
1979	13,000	14,085
1980	22,400 a	19,320
1981	20,700 b	22,775
1982	26,000 c	d

a   Revised budget submission. Original submission = $15,100,000.
b   Revised budget submission. Original submission = $22,600,000.
c   Revised budget submission. Original submission = $22,900,000.
d   Funding moved to STDN Operations and STDN Implementation budget categories. See Tables 4–25 and 4–26.

*Table 4–11. Deep Space Network Implementation Funding History
(in thousands of dollars)*

Year (Fiscal)	Submission	Programmed (Actual)
1979	12,500 a	10,115
1980	22,400 b	15,000
1981	23,100	20,165
1982	36,900 c	d

a   Revised budget submission. Original submission = $14,800,000.
b   Revised budget submission. Original submission = $15,100,000. The increase reflects the transfer of funding for the Jet Propulsion Laboratory engineering efforts for specific projects from the Operations budget to the Systems Implementation budget.
c   Revised budget submission. Original submission = $41,800,000.
d   See Table 4–27.

*Table 4–12. Aeronautics and Sounding Rocket Support Systems
Implementation Funding History (in thousands of dollars)*

Year (Fiscal)	Submission	Programmed (Actual)
1979	3,500	4,052
1980	4,000 a	3,850
1981	3,500 b	3,345
1982	6,200 c	d

a   Revised budget submission. Original submission = $3,700,000. The increase provides for increased costs associated with the upgrading of radar systems.
b   Revised budget submission. Original submission = $4,100,000.
c   Revised budget submission. Original submission = $6,400,000.
d   See Table 4–29.

Table 4–13. Communications Implementation Funding History (in thousands of dollars)

Year (Fiscal)	Submission	Programmed (Actual)
1979	5,100	4,815
1980	4,400 a	5,030
1981	3,100 b	3,100
1982	4,400	c

a   Revised budget submission. Original submission = $3,700,000. The increase reflects greater-than-expected costs for the status and control system for the TDRSS multiplexer and fifty-megabyte-per-second transmission capability for the support of Spacelab and Landsat-D.
b   Revised budget submission. Original submission = $5,600,000.
c   See Table 4–30.

Table 4–14. Data Processing Systems Implementation Funding History, 1979–1982 (in thousands of dollars)

Year (Fiscal)	Submission	Programmed (Actual)
1979	6,400 a	7,430
1980	9,500 b	13,900
1981	12,300	12,720
1982	15,600 c	d

a   Revised budget submission. Original submission = $4,900,000.
b   Revised budget submission. Original submission = $6,900,000. The increase results from the provision of redundant capability for critical parts of the Spacelab data processing system at Goddard Space Flight Center to ensure a reliable data processing capability.
c   Revised budget submission. Original submission = $21,700,000.
d   See Table 4–36.

Table 4–15. Advanced Systems Funding History
(in thousands of dollars)

Year (Fiscal)	Submission	Authorization	Appropriation	Programmed (Actual)
1979	9,900	9,900	a	9,500
1980	10,600	10,600	b	10,600
1981	11,300	11,300	11,300	11,300
1982	12,500	c	12,500	12,500
1983	13,400	d	13,400	13,400
1984	14,200	14,200	14,200	14,200
1985	15,300	15,300	15,300	14,800
1986	15,000 e	16,200	16,200	15,500
1987	17,100	17,100	17,100	17,100
1988	18,100	18,100	18,100	17,900

a   Undistributed. Total 1979 R&D appropriation = $3,477,200,000.
b   Undistributed. Total 1980 R&D appropriation = $$4,091,086,000.
c   Undistributed. Total 1982 Tracking and Data Acquisition authorization = $408,180,000.
d   Undistributed. Total 1983 Tracking and Data Acquisition authorization = $503,900,000.
e   Revised budget submission. Original submission = $16,200,000.

Table 4–16. Initial TDRSS Funding History (in thousands of dollars) a

Year (Fiscal)	Submission	Authorization	Appropriation	Programmed (Actual)
1983	61,300	b	61,300	c

a   The TDRSS was included as a major budget category for only one fiscal year. It became part of the Space Network budget category beginning with FY 1984. See Table 4–17.
b   Undistributed. Total 1983 Tracking and Data Acquisition authorization = $503,900,000.
b   See Table 4–18.

Table 4–17. Space Network Funding History (in thousands of dollars)

Year (Fiscal)	Submission	Authorization	Appropriation	Programmed (Actual)
1982		a		21,800
1983	498,900 b	—	—	104,300 c
1984	259,100 d	294,700	284,700 e	259,100
1985	378,300 f	386,500	386,500	378,300
1986	273,700 g	293,800 h	293,800	273,700
1987	407,300 i	374,300	407,300	404,300
1988	435,700 j	457,500	426,500 k	433,400

a No submission, authorization, or appropriation in this program category.
b Budget submission reflects reordering of budget categories and inclusion of Space Networks in SFC&DC appropriation that began at the time of the FY 1984 budget estimate (and revised FY 1983 budget estimate). This budget category included items from both former Operations and Systems Implementation categories. Authorization and appropriations did not yet reflect the new budget category.
c This reflects only the original R&D budget categories.
d Revised budget submission. Initial submission = $294,700,000.
e Moved to SFC&DC appropriations category. The reduction of $10,000,000 reflects a payment to the Federal Financing Bank.
f Revised budget submission. Initial submission = $386,500,000. The reduction results from the impact of the addition slip in the launch schedule of the TDRS-B and -C because of the inertial upper stage anomaly on TDRS-1 and adjustments in the operation of the NASA ground elements of the Space Network.
g Revised budget submission. Original submission = $400,800,000. The reduction reflects the net effect of the congressional direction to defer the FY 1986 principal payment of $107.0 million for the TDRSS to the Federal Financing Bank and the reallocation of $7.5 million to Ground Network and Communications and Data Systems. The adjustment was need to continue operation of the ground station network and necessary communications into late FY 1986.
h The $59,000,000 reduction from NASA's budget submission agreed to by both the House and Senate authorization committees in separate deliberations reflects the deferral of the scheduled $107 million principal payment to the Federal Financing Bank, an additional authorization of $48,000,000 to the TDRSS program, and the implementation of a general reduction of $4,000,000. The Conference Committee further reduced the authorization to $293,800,000 (a reduction of $107,000,000), eliminating the additional authorization to the TDRSS program.
i Revised budget submission. Initial submission = $374,300,000.
j Revised budget submission. Initial submission = $481,500,000.
k This reduction reflects a reduction of $40,000,000 from the amount requested for the replacement of a tracking and data relay satellite lost on *Challenger* and a reduction of $15,000,000 for general tracking and data acquisition activities.

Table 4–18. Tracking and Data Relay Satellite System Funding History (in thousands of dollars)

Year (Fiscal)	Submission	Programmed (Actual)
1983	51,300 a	41,000
1984	204,300 b	204,300
1985	316,600 c	316,600
1986	210,500 d	205,600
1987	301,500	285,098
1988	318,900 e	318,900

a   This reflects amounts from the TDRSS (Table 4–16) and Spaceflight Tracking and Data Network budget category.

b   Revised budget submission. Original submission = $242,900,000. The decrease resulted from the restructuring of the TDRSS loans with the Federal Financing Bank and the schedule impact of an inertial upper stage anomaly. Included in the deferred activities because of the schedule impact are testing and some launch-related items.

c   Revised budget submission. Initial submission = $319,900,000. The decrease resulted from a delay in the launch of TDRS-B and -C because of the anomaly experienced during the first launch.

d   Revised budget submission. Initial submission = $335,600,000. The decrease includes a $107 million reduction in the payment of principal on the TDRSS loans to the Federal Financing Bank, consistent with congressional direction. The balance of the reduction resulted from adjustments in launch and production schedules because of the delay in the launch of the second and third spacecraft.

e   Revised budget submission. Initial submission = $320,900,000. The decrease reflects a detailed reassessment of support requirements, leading to greater-than-anticipated savings during the STS standdown period.

Table 4–19. Space Network Operations Funding History
(in thousands of dollars)

Year (Fiscal)	Submission	Programmed (Actual)
1982	a	8,400
1983	34,400 b	42,500
1984	31,300 c	31,300
1985	35,900 d	36,151
1986	40,500 e	40,500
1987	43,700	35,700
1988	42,700 f	40,400

a   See Table 4–4.
b   Revised budget submission. Initial submission = $33,500,000.
c   Revised budget submission. Initial submission = $31,800,000. The decrease resulted from revised operational requirements for the Network Control Center.
d   Revised budget submission. Initial submission = $40,800,000. The decrease was caused by the launch delay and changes in operational support requirements, primarily for the Operations Support Computing Facility and the Network Control Center. The delay in the TDRSS program, along with schedule slips in user programs, resulted in a reassessment of support requirements and a "stretchout" in the projected support workload.
e   Revised budget submission. Initial submission = $37,100,000. The increase reflects the operational support requirements in the Space Network caused by the delay of the TDRSS program.
f   Revised budget submission. Initial submission = $43,900,000. The decrease reflects revised budget estimate results from reduced contractor support required during the period, principally as a result of the STS standdown.

Table 4–20. Systems Engineering and Support Funding History
(in thousands of dollars)

Year (Fiscal)	Submission	Programmed (Actual)
1982	a	13,400
1983	18,100 b	20,800
1984	23,500 c	23,500
1985	25,800	25,549
1986	22,700 d	21,300
1987	28,100	26,404
1988	26,700 e	26,700

a   See Table 4–4.
b   Revised budget submission. Initial submission = $18,000,000.
c   Revised budget submission. Initial submission = $20,000,000. The increase was because of additional engineering and software support for the Network Control Center, a higher rate switching capability for the NASA Ground Terminal, and an additional transponder required for the Bilateration Ranging Transponder System.
d   Revised budget submission. Initial submission = $28,100,000.
e   Revised budget submission. Initial submission = $25,600,000. The increase reflects the necessary advanced planning to support the development of space station operational concepts, interface definition for data handling and distribution, and support requirements definition.

Table 4–21. TDRS Replacement Spacecraft Funding History (in thousands of dollars)

Year (Fiscal)	Submission	Programmed (Actual)
1986	a	4,900
1987	33,000	50,398
1988	35,800 b	35,800

a No submission.
b Revised budget submission. Previous submission = $75,800,000. The decrease reflects congressional action on the NASA FY 1988 budget request.

Table 4–22. Second TDRS Ground Terminal Funding History (in thousands of dollars)

Year (Fiscal)	Submission	Programmed (Actual)
1986		1,400
1987	1,000	2,700
1988	7,600 b	7,600

a No submission.
b Revised budget submission. Previous submission = $9,100,000. The decrease reflects a rephasing of procurement activities planned for FY 1988 into FY 1989.

Table 4–23. Advanced TDRSS Funding History (in thousands of dollars)

Year (Fiscal)	Submission	Programmed (Actual)
1987	a	4,000
1988	4,000	7,600

a No submission.

Table 4–24. Ground Network Funding History (in thousands of dollars)

Year (Fiscal)	Submission	Authorization	Appropriation	Programmed (Actual)
1982		a		237,457
1983	242,400 b	—	—	242,920
1984	249,300 c	231,500	231,500 d	249,300
1985	233,200 e	223,600	223,600	233,200
1986	210,400 f	219,300	219,300	210,400
1987	250,100 g	222,000	250,100	237,200
1988	232,200 h	257,100	257,100	231,000

a   No submission, authorization, or appropriation in this program category.
b   Revised budget submission. Initial submission = $243,500,00. This reflects reordered budget categories at the time of the revised submission. Congressional committees acted on former R&D budget categories.
c   Revised budget submission. Initial submission = $231,500,000.
d   Moved to SFC&DC appropriations category.
e   Revised budget submission. Initial submission = $233,600,000. The increase resulted from program adjustments made to accommodate an additional six months of STDN station operations from April 1 through September 30, 1985. This extension resulted from the additional delay in the launch of TDRS-B and –C and in the availability of two fully operational TDRS spacecraft.
f   Revised budget submission. Initial submission = $219,300,000.
g   Revised budget submission. Initial submission = $222,000,000.
h   Revised budget submission. Original submission = $257,100,000.

Table 4–25. Spaceflight Tracking and Data Network (STDN) Systems Implementation Funding History (in thousands of dollars)

Year (Fiscal)	Submission	Programmed (Actual)
1982	a	3,900
1983	6,000	6,000
1984	8,600 b	8,500
1985	6,600 c	6,400
1986	3,000 d	3,000
1987	3,900	3,800
1988	3,200 e	3,200

a   See Table 4–10.
b   Revised budget submission. Original submission = $8,100,000.
c   Revised budget submission. Original submission = $6,300,000. The increase was for sustaining STDN systems and facilities for an extended time period because of the delay in the station closure dates.
d   Revised budget submission. Original submission = $2,700,000.

*Table 4–26. Spaceflight Tracking and Data Network (STDN)
Operations Funding History (in thousands of dollars)*

Year (Fiscal)	Submission	Programmed (Actual)
1982	a	120,536
1983	118,200 b	118,500
1984	120,800 c	119,800
1985	93,000 d	91,447
1986	53,200 e	53,960
1987	81,400	78,000
1988	70,100 f	68,000

a   See Table 4–4.
b   Revised budget submission. Original submission = $118,200,000.
c   Revised budget submission. Original submission = $102,500,000. The increase results from the additional six months of tracking operations for Shuttle support in FY 1984 caused by the delay in the TDRSS reaching operational status (because of inertial upper stage problems), thus requiring the ground stations to provide Shuttle and other support until the TDRSS becomes operational.
d   Revised budget submission. Initial submission = $83,300,000. The increase reflects the additional six months of tracking operations in FY 1985 for Shuttle and other support brought about by the delay in the TDRSS reaching operational status.
e   Revised budget submission. Initial submission = $58,700,000.
f   Revised budget submission. Initial submission = $84,600,000. The decrease reflects program adjustments made to accommodate a portion of the general reduction specified by Congress and a reallocation of funds for increased communications support requirements.

*Table 4–27. Deep Space Network Systems Implementation Funding
History (in thousands of dollars)*

Year (Fiscal)	Submission	Programmed (Actual)
1982	a	36,900
1983	45,300 b	44,300
1984	38,100	38,800
1985	37,100	37,100
1986	43,000 c	42,765
1987	44,000	40,000
1988	46,200 d	46,200

a   See Table 4–11.
b   Revised budget submission. Original submission = $44,800,000.
c   Revised budget submission. Original submission = $44,400,000.
d   Revised budget submission. Original submission = $49,000,000. The decrease reflects the decision to defer various system upgrades.

Table 4–28. Deep Space Network Operations Funding History (in thousands of dollars)

Year (Fiscal)	Submission	Programmed (Actual)
1982	a	63,296
1983	61,300	61,300
1984	65,500	65,500
1985	76,800	77,661
1986	85,700 b	85,301
1987	93,300	87,700
1988	88,000 c	88,900

a   See Table 4–5.
b   Revised budget submission. Original submission = $88,900,000.
c   Revised budget submission. Original submission = $94,100,000. The decrease reflects a reallocation of funds for increased communications support requirements and the rephasing of activities to accommodate a portion of the general reduction.

Table 4–29. Aeronautics, Balloon, and Sounding Rocket Support Systems Implementation Funding History (in thousands of dollars)

Year (Fiscal)	Submission	Programmed (Actual)
1982	a	6,255
1983	3,800	4,120
1984	8,100	8,600
1985	8,200	8,965
1986	10,500 b	10,434
1987	11,200	11,200
1988	8,200 c	8,200

a   See Table 4–12.
b   Revised budget submission. Original submission = $11,400,000.
c   Revised budget submission. Original submission = $8,400,000.

Table 4–30. Aeronautics, Balloon, and Sounding Rocket Support Operations Funding History (in thousands of dollars)

Year (Fiscal)	Submission	Programmed (Actual)
1982	a	6,570
1983	7,800	8,700
1984	8,200	8,152
1985	11,500 b	11,627
1986	15,000 c	14,940
1987	16,300	16,500
1988	16,500 d	16,500

a   See Table 4–6.
b   Revised budget submission. Original submission = $11,900,000.
c   Revised budget submission. Original submission = $13,200,000.
d   Revised budget submission. Original submission = $16,300,000. The increase reflects support of the supernova sounding rocket campaigns in Australia.

*Table 4–31. Communications and Data Systems Funding History
(in thousands of dollars)*

Year (Fiscal)	Submission	Authorization	Appropriation	Programmed (Actual)
1982		a		130,343
1983	139,300 b	—	—	138,280
1984	165,600 c	159,800	159,800 d	165,600
1985	184,200 e	185,600	185,600	184,200
1986	176,300 f	188,200	188,200	176,300
1987	205,500 g	202,600	205,500	203,900
1988	216,500 h	210,300	210,300	215,000

a   No submission, authorization, or appropriation in this budget category.
b   New SFC&DC budget category replaced former Communications Systems Operations (Table 4–7), Communications Systems Implementation (Table 4–13), and Data Processing (Table 4–8) budget categories.
c   Revised budget submission. Original submission = $159,800,000.
d   Moved to SFC&DC appropriations category.
e   Revised budget submission. Initial submission = $185,600,000. The reduction reflects the net result of adjustments in program support and equipment deferrals required to fund an additional six months of communications costs for operating the STDN stations and the increases in the Ground Network.
f   Revised budget submission. Original submission = $188,200,000.
g   Revised budget submission. Initial submission = $202,600,000.
h   Revised budget submission. Initial submission = $210,300,000.

*Table 4–32. Communications Systems Implementation Funding History
(in thousands of dollars)*

Year (Fiscal)	Submission	Programmed (Actual)
1982	a	4,250
1983	5,600	5,600
1984	5,900 b	5,912
1985	6,500	6,500
1986	5,500 c	5,500
1987	7,400	6,800
1988	6,300	7,000

a   See Table 4–13.
b   Revised budget submission. Initial submission = $5,300,000.
c   Revised budget submission. Original submission = $6,500,000.

Table 4–33. Communications Operations Funding History
(in thousands of dollars)

Year (Fiscal)	Submission	Programmed (Actual)
1982	a	39,731
1983	45,600	45,700
1984	64,600 b	64,600
1985	73,000 c	73,000
1986	82,100 d	82,049
1987	91,700	16,500
1988	109,500 e	109,400

a   See Table 4–7.
b   Revised budget submission. Initial submission = $59,700,000. The increase reflects the need to provide communications with the overseas tracking sites for Shuttle support longer than planned because of the delay in the TDRSS becoming fully operational.
c   Revised budget submission. Original submission = $68,200,000. The increase reflects the need to provide communications with the overseas tracking sites for Shuttle support longer than planned because of the delay in the TDRSS becoming fully operational.
d   Revised budget submission. Original submission = $75,700,000. The increase reflects the need to provide communications with the STDN tracking sites for Shuttle support longer than planned because of the delay in TDRS launches.
e   Revised budget submission. Original submission = $95,700,000. The increase reflects increased requirements in the Program Support Communications Network (PSCN) to meet user demands. The PSCN increase was partially offset by further NASCOM savings associated with the STS standdown.

Table 4–34. Mission Facilities Funding History
(in thousands of dollars)

Year (Fiscal)	Submission	Programmed (Actual)
1982	a	8,900
1983	10,900	10,900
1984	12,900 b	13,545
1985	12,400	12,675
1986	13,800 c	13,820
1987	12,200	12,200
1988	11,500 d	9,900

a   No budget category.
b   Revised budget submission. Original submission = $18,600,000.
c   Revised budget submission. Original submission = $27,100,000.
d   Revised budget submission. Original submission = $27,000,000. The decrease reflects the rephasing of mission support requirements.

Table 4–35. Mission Operations Funding History (in thousands of dollars)

Year (Fiscal)	Submission	Programmed (Actual)
1982	a	14,838
1983	17,800	16,200
1984	19,100 b	18,260
1985	21,900	21,200
1986	18,900 c	18,900
1987	23,700	23,300
1988	25,000 d	25,400

a No budget category.
b Revised budget submission. Original submission = $18,600,000.
c Revised budget submission. Original submission = $27,100,000.
d Revised budget submission. Original submission = $27,000,000. The decrease reflects the rephasing of mission support requirements.

Table 4–36. Data Processing Systems Implementation Funding History, 1982–1988 (in thousands of dollars)

Year (Fiscal)	Submission	Programmed (Actual)
1982	a	15,492
1983	19,500	20,580
1984	22,400	23,683
1985	24,400 b	25,016
1986	21,100 c	21,100
1987	28,400	24,385
1988	21,500	22,200

a See Table 4–14.
b Revised budget submission. Original submission = $26,600,000. The decrease reflects a reduction in the number of Nimbus and Landsat data products to be processed.
c Revised budget submission. Original submission = $24,100,000.

*Table 4–37. Data Processing Operations Funding History, 1982–1988
(in thousands of dollars)*

Year (Fiscal)	Submission	Programmed (Actual)
1982	a	47,082
1983	39,900	39,300
1984	40,700 b	39,600
1985	46,000 c	45,809
1986	34,900 d	34,931
1987	42,100	41,700
1988	42,700 e	41,100

a See Table 4–8.
b Revised budget submission. Original submission = $42,300,000.
c Revised budget submission. Original submission = $50,000,000.
d Revised budget submission. Original submission = $41,500,000.
e Revised budget submission. Original submission = $46,700,000. The decrease reflects the termination of Nimbus mission support and the rephasing of procurements planned for future Spacelab mission support.

Table 4–38. Tracking and Data Acquisition Stations (1979–1988)

Station (Location)	Code Name or Number	Latitude Longitude	Type of Station STDN	Type of Station DSN	Est.	Out	Phased Equipment/Capabilities	Remarks
Alaska (near Fairbanks)	ALASKA	64°59'N 147°31'W	X		1962	1984	GRARR and MOTS; SATAN receivers and command; dish antennas (12, 14 and 26 m)	It was transferred to NOAA in 1984.
Ascension Island (South Atlantic)	ACN	7°57'S 14°35'W	X		1966	1989	9-m USB command and receive; SATAN VHF telemetry links; FM remoting telemetry; decommutators; telemetry recording; data processing; communications (voice, UHF air-to-ground, teletype, video, and high-speed data)	
Bermuda (Atlantic)	BDA	32°15'N 64°50'W	X		1961		C-band radar; two 9-m USB command and receive antennas (single link with dual antennas); telemetry links; FM remoting telemetry; decommutators; telemetry recording; data processing; communications (voice, UHF air-to-ground, teletype, video, and high-speed data)	Data received at Bermuda were crucial in making the go/no go decision for orbital insertion. The station also provided reentry tracking for Atlantic recovery situations.
Buckhorn Lake, CA (Dryden)	BUC	34°56'N 117°54'W	X		1975	1983	Two 4.3-m UHF air-to-ground, C-band radars for Shuttle support	The site was established to support Shuttle approach and landing test and flight test landings. It was closed after the STS-8 night landing. One 4.3-m antenna was moved to Dryden in 1985.

TRACKING AND DATA ACQUISITION/SPACE OPERATIONS   343

Table 4–38 continued

Station (Location)	Code Name or Number	Latitude Longitude	Type of Station STDN	Type of Station DSN	Est.	Out	Phased Equipment/Capabilities	Remarks
Canberra (southeastern Australia)	CAN	35°24'S 148°59'E	X		1965	1984	FM remoting telemetry; decommutators; telemetry recording; data processing; communications (voice, video, teletype, and high-speed data)	This was transferred to DSN in 1984. It was also called Tidbinbilla. The 26-m antenna moved from closed DSS 44 to the main DSN complex in Tidbinbilla.
	42 Weemala	35°24'S 148°58'E		X			34-m antenna; transmit and receive	This supported the Voyager spacecraft. It expanded from the 26-m antenna.
	43 Ballima	35°24'S 148°58'E		X	1973		70-m antenna; transmit and receive	The 70-m antenna was extended from the original 64-m antenna in 1987 to support the Voyager 2 encounter with Neptune.
	44			X	1965	1982	USB 26-m antenna	The antenna moved to Tidbinbilla in 1984 and became DSS 46. It originally was the Honeysuckle Creek antenna.
	45	35°23'S 148°58'E		X	1985		34-m antenna; receive only	This supported the Voyager spacecraft.
	46	35°24'S 148°58'E		X	1984		USB 26-m antenna	The 26-m antenna moved from Honeysuckle Creek in 1984. This Transferred from STDN in 1985.

Table 4–38 continued

Station (Location)	Code Name or Number	Latitude Longitude	Type of Station STDN	Type of Station DSN	Est.	Out	Phased Equipment/Capabilities	Remarks
Dakar, Senegal	DKR	14°43'N 17°08'W	X		1982		4.3-m command and receive antenna; UHF voice air-to-ground, command and receive antenna	
Dryden Flight Research Facility ATF-2	DFRF	34°56'N 117°54'W	X		1978		Two C-band radars on Western range to provide Shuttle support; 4.3-m S-band antenna; 7-m antenna; communications (voice, UHF air-to-ground, teletype, video, and high-speed data)	This began operations in 1978 for the Shuttle program. It was providing aeronautics tracking earlier. It received one 4.3-m S-band antenna from the Buckhorn Lake site in 1985, making a total of two 4.3-m antennas at Dryden.
Gaberone, Botswana (southern Africa)	BOT	24°42'S 25°52'E	X		1981	1986	UHF air-to-ground voice system; Yagi command and receive	It was also called Kgale. Located near South Africa, Botswana could pick up many of the functions formerly handled by Johannesburg, which closed in 1975. Surplus equipment was donated to the Botswana National Museum.
Goldstone (California)								Located in the Mojave Desert, Goldstone has the largest concentration of NASA tracking and data acquisition equipment. All 26-m antennas were principally dedicated to deep space programs. Coverage was also available for transfer orbit operations and tracking and data acquisition for near-Earth orbital spacecraft.

Table 4-38 continued

Station (Location)	Code Name or Number	Latitude Longitude	Type of Station STDN	Type of Station DSN	Est.	Out	Phased Equipment/Capabilities	Remarks
	GDS	35°20'N	X		1967	1985	26-m antenna; 9-m antenna; FM remoting telemetry; decommutators; telemetry recording; data processing; communications (voice, teletype, video, and high-speed data)	This was turned over to DSN in 1985 and redesignated as DSS 16 and DSS 17.
	ECHO 12	35°17'N 116°44'W		X	1960		34-m antenna; transmit and receive	The original 26-m antenna was extended to 34 m in 1979. This supported the Voyager spacecraft.
	MARS 14	35°25'N 116°44'W		X	1966		70-m antenna; transmit and receive	The 70-m antenna replaced the 64-m antenna in 1988 to support the Voyager 2 encounter with Uranus.
	PIONEER 11	35°23'N 116°44'W		X	1958	Early 1980s	26-m antenna	
	VENUS 13	35°23'N 116°53'W		X	1962		9-m antenna closed; 26-m antenna	This was a research and development facility.
	URANUS 15	35°25'N 116°42'W		X	1984		34-m antenna; receive only	This was used for the first time in 1986 to support the Voyager spacecraft.
	16	35°20'N 116°53'W		X	1965		USB 26-m antenna	The equipment was moved from STDN in 1985.
	17	35°20'N 116°53'W		X	1967		USB 9-m antenna; backup for DSS 16	This was transferred from STDN in 1985.

Table 4–38 continued

Station (Location)	Code Name or Number	Latitude Longitude	Type of Station		Est.	Out	Phased Equipment/Capabilities	Remarks
			STDN	DSN				
Greenbelt, MD (NTTF)	BLT	38°59'N 76°51'W	X	X	1966	1986	*18-m antenna used for IUE support given to the Naval Academy; 9-m antenna moved Wallops in 1986	Located at Goddard Space Flight Center, this became part of the operational network in 1974. In 1986, all orbital satellite tracking/telemetry operations were transferred to Wallops.
Guam (Pacific)	GWM	13°18'N 144°44'E	X		1966	1990	9-m USB command and receive antenna; VHF telemetry links; FM remoting telemetry; decommutators; telemetry recording; data processing; communications (voice, UHF air-to-ground, teletype, video, and high-speed data)	
Kohee Park (Hawaii)	KAUAIH or HAW	22°07'N 157°40'W	X		1961	1989	Two Yagi command; 4.3-m antenna; C-band radar; 9-m USB command and receive antenna; VHF telemetry links; FM remoting telemetry recording; data processing; communications (voice, VHF air-to-ground, teletype, video, and high-speed data)	The Manned Space Flight Network station began operations in 1961; it became the Space Tracking and Data Acquisition Network in 1965.
Madrid (Spain)								There were three individual NASA complexes near Madrid.
	MAD (became RID in 1984)	40°27'N 4°10'W	X		1967		26-m antenna; FM remoting telemetry; decommutators; telemetry recording; data processing; communications (voice, video, teletype, and high-speed data)	This was transferred to DSN in 1985 and redesignated as DSS 66.

TRACKING AND DATA ACQUISITION/SPACE OPERATIONS   347

Table 4–38 continued

Station (Location)	Code Name or Number	Latitude Longitude	Type of Station STDN	Type of Station DSN	Est.	Out	Phased Equipment/Capabilities	Remarks
	61 Robledo-I	40°26'N 4°15'W		X	1965		34-m antenna; transmit and receive	This antenna was expanded from 26 m.
	62	40°27'N 4°22'W		X	1967	1969	26-m antenna	This was known as the Cebreros DSN station.
	63 Robledo-II	40°26'N 4°15'W		X	1973		70-m antenna; transmit and receive	This supported the Voyager spacecraft. The antenna expanded to 70 m from 64 m in 1987.
	65 Robledo-III	40°25'N 5°16'W		X	1987		34-m antenna; transmit and receive	This supported the Voyager spacecraft.
	66 Robledo-IV	40°25'N 5°16'W		X	1985		26-m antenna; transmit and receive	This was transferred from STDN in 1985.
Merritt Island (Florida)	MIL	28°25'N 80°40'W	X		1973		9-m command and receive antenna; C-band radar; VHF telemetry links; FM remoting telemetry; decommutators; telemetry recording; data processing; communications (voice, UHF air-to-ground, teletype, video, and high-speed data)	This was located near the Kennedy Space Center launch complex.
Orroral Valley (southeastern Australia)	ORR	35°38'S 148°57'E	X		1965	1984	26-m antenna; two SATAN receivers and command; YAGI command; MOTS	The 26-m antenna was relocated to the University of Tasmania when Orroral closed in 1984.
Ponce de Leon (Florida)	PDL	29°04'N 279°05'W	X		1978		4.3-m command and receive antenna	This was used for Shuttle support only.

Table 4–38 continued

Station (Location)	Code Name or Number	Latitude Longitude	Type of Station STDN	Type of Station DSN	Est.	Out	Equipment/Capabilities	Phased Remarks
Quito (Ecuador)	QUITOE	37°S 78°35'W	X		1957	1981	12-m antenna; SATAN receivers and command; three Yagi command; MOTS	
Rosman, NC	ROSMAN	35°12'N 82°52'W	X		1963	1981	Two 26-m antennas; GRARR; three SATAN receivers and command; MOTS; ATS telemetry and command	This received high-data-rate telemetry from observatory-class satellites. It was turned over to the Department of Defense in 1981.
Santiago (Chile)	AGO	33°09'S 70°40'W	X		1957	1988	9-m antenna; 12-m antenna; GRARR; USB; two SATAN receivers; one SATAN command; Yagi command; MOTS	
Seychelles (Indian Ocean)	IOS	04°40'S 55°28'E	X		1981		18-m antenna	This military site supported the Shuttle.
Tula Peak (New Mexico)	TULA	33°01'N 117°09'W	X		1979	1982	4.3-m antenna; telemetry and command; UHF voice	This was used for STS-1 and –2. It was shut down after STS-2 but reopened to provide emergency support to STS-3.
Wallops Island (Virginia)	WOTS	37°55'N 75°31'W	X		1959		Two 7.3-m receive-only antennas; 6-m command-only antenna; 9-m command and receive antenna; 18-m command and receive antenna; SCAMP VHF command only (broadside array antenna—150 megacycles); SATAN VHF command only (broadside array antenna—150 megacycles); two SATAN VHF receive only	This began providing support for scientific satellites in 1986.

TRACKING AND DATA ACQUISITION/SPACE OPERATIONS 349

Table 4–38 continued

Station (Location)	Code Name or Number	Latitude Longitude	Type of Station STDN	Type of Station DSN	Est.	Out	Phased Equipment/Capabilities	Remarks
White Sands (New Mexico)	WHS	32°21'N 106°22'W	X		1961		Three 18-m Ku-band antennas; two S-band telemetry and command antennas; C-band radar; communications (voice and teletype)	Located on the U.S. Army's White Sands Missile Range, the station was equipped with Defense Department radar and NASA-owned acquisition aids. This was the site of the TDRSS ground station.
Winkfield (United Kingdom)	WNKFLD	52°27'N 00°42'W	X		1961	1981	4.3-m antenna; SATAN receivers and command; Yagi command; MOTS	This was operated by British personnel.
Yarragardee (Australia)	YAR	29°08'S 115°21'E	X		1980	1991	UHF voice air-to-ground command and receive antenna	This provided STS deorbit burn coverage.

Legend: ATF—Aeronautical Training Facility; ATS—Applications Technology Satellite; DSN—Deep Space Network; FM—frequency modulation; GRARR—Goddard range and range rate; MOTS—minitrack optical tracking system; SATAN—satellite automatic tracking antenna; SCAMP—satellite command antenna on medium pedestal; STDN—Spaceflight Tracking and Data Network; TDRSS—Tracking and Data Relay Satellite System; UHF—ultrahigh frequency; USB—unified S-band; and VHF—very high frequency.

Table 4–39. Single-Access Link Summary

Band	Forward Link (White Sands Ground Terminal to User Spacecraft)	Data Rate	Return Link (User Spacecraft to White Sands Ground Terminal)	Data Rate
S	2	0.1 to 300 kilobits per second	2	6 megabits per second maximum
K	2	1 kilobit per second to 25 megabits per second	2	300 megabits per second maximum
	Used for: 1. Commanding user spacecraft 2. Transmitting PN code for range and range rate		Used for: 1. Receiving spacecraft telemetry 2. Receiving PN code turnaround for range and range rate measurements	

PN—Pseudorandom noise.

Table 4–40. Multiple-Access Link Summary

Number of links per TDRS	Forward Link (White Sands Ground Terminal to User Spacecraft)	Data Rate	Return Link (User Spacecraft to White Sands Ground Terminal)	Data Rate
1		0.1 to 10 kilobits per second		
20		Command bit rate capability		0.1 to 50 kilobits per second
	Used for: 1. Commanding user spacecraft 2. Transmitting PN code for range and range rate		Used for: 1. Receiving spacecraft telemetry 2. Receiving PN code turnaround for range and range rate measurements	

PN—Pseudorandom noise.

Table 4–41. TDRS Characteristics

	TDRS-1	TDRS-3
Launch Date	April 4, 1983	September 29, 1988
Launch Vehicle	STS-6 (*Challenger*)/IUS	STS-26 (*Discovery*)/IUS
Range	Kennedy Space Center	Kennedy Space Center
Program Objectives	Establish a three-satellite geosynchronous orbit telecommunications satellite system to provide improve tracking and data acquisition services to spacecraft in low-Earth orbit, procured by NASA through a lease service contract with Spacecom	Establish and provide improved tracking and data acquisition services to spacecraft in low-Earth orbit through a system of two telecommunications satellites in geosynchronous orbit with one additional orbiting satellite to serve as a system space, procured by NASA through a lease service contract with a wholly owned subsidiary of Contel
Mission Objectives	Deliver the first of three TDRS satellites to a stationary geosynchronous orbit location with sufficient stationkeeping propulsion fuel on board to meet the NASA support requirements and initiate TDRS-A user support services	Deliver the second of three TDRS satellites to a stationary geosynchronous orbit location with sufficient stationkeeping propulsion fuel on board to meet the NASA support requirements and initiate dual TDRS satellite user support
Owner	Spacecom, leased by NASA	Contel, leased by NASA
Orbit Characteristics		
Apogee (km)	35,779	35,804
Perigee (km)	35,777	35,764
Inclination (deg.)	2.3	0.1
Period (min.)	1,436	1,434.8
Location	41 degrees W longitude over the equator	171 degrees W longitude over the equator
Weight (kg)	2,268 at launch (with IUS); 2,146 when built	2,224 at launch (with IUS); 2,103 when built
Dimensions	17.4 m across the solar arrays; 14.2 m across the antennas; two 4.9-m-diameter high-gain parabolic antennas	17.4 m across the solar arrays; 14.2 m across the antennas; two 4.9-m-diameter high-gain parabolic antennas
Communications	Two single-access S-/Ku-band antennas, C-band dish, S-band omni antenna, K-band space-to-ground dish antenna, 30-element multiple-access array, and additional K-band antenna mounted on the platform	Two single-access S-/Ku-band antennas, C-band dish, S-band omni antenna, K-band space-to-ground dish antenna, 30-element multiple-access array, and additional K-band antenna mounted on the platform
Power Source	Solar panels and nickel cadmium batteries that provide 1,700-watt peak power in sunlight and support an eclipse period average load of 1,400 watts	Solar panels and nickel cadmium batteries that provide 1,700-watt peak power in sunlight and support an eclipse period average load of 1,400 watts

Table 4–41 continued

	TDRS-1	TDRS-3
Primary Contractors	Spacecom; spacecraft—TRW Space and Technology Group; ground terminal equipment and antennas—Harris Government Communications Systems Division	Contel; spacecraft—TRW Space and Technology Group; ground terminal equipment and antennas—Harris Government Communications Systems Division
Remarks	A malfunction of the IUS left TDRS-1 in an elliptical orbit. A sequence of thruster firings raised the satellite to its proper altitude. It was placed into a geosynchronous orbit located at 67 degrees west longitude over the equator above northwest Brazil and later moved to its operating location at 41 degrees west.	The launch and positioning of TDRS-3 went without problems. The satellite was positioned over the Pacific to give NASA one satellite over each ocean.

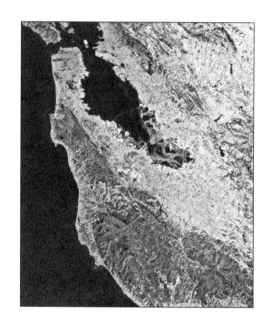

CHAPTER FIVE
COMMERCIAL PROGRAMS

CHAPTER FIVE
COMMERCIAL PROGRAMS

Introduction

Although NASA historically collaborated with the private sector through its aeronautics programs, academic grants, commercial satellite launch support, and dissemination of remote-sensing and other data, cooperation between NASA and the private sector grew in the late 1970s and through the 1980s. More than once, President Ronald Reagan stated his belief that NASA should encourage private-sector involvement in space and that the agency should remove obstacles to that involvement.

In 1984, in response to the Reagan administration's 1984 National Policy on the Commercial Use of Space, NASA established the Office of Commercial Programs (OCP). This office encouraged the private sector to become more involved in using space for commercial purposes and increased NASA's efforts to find private-sector uses for NASA-developed technology. This chapter describes the establishment and activities of OCP through 1988.

Management of the Office of Commercial Programs

NASA Administrator James Beggs established OCP in September 1984. He appointed Isaac (Ike) Gillam IV, who had been director of the Dryden Flight Research Facility, as its first associate administrator. Gillam served until his retirement in September 1987. James T. Rose was appointed to the position in October 1987.

OCP consisted of the Program Support Office, Technology Utilization Division, Commercial Development Division, and Plans, Policy and Evaluation Division. The Technology Utilization Division was charged with enhancing the transfer of NASA-developed technologies to U.S. industry through cooperative agreements, joint ventures, and information dissemination. Henry J. Clarks was the division director until 1987, when Raymond Whitten followed as acting division director. Clarks returned as division director in August 1988.

The Commercial Development Division negotiated and coordinated the bilateral/multilateral agreements with aerospace and nonaerospace companies that sought access to NASA-developed technologies and facilities. Gary E. Krier was named division director in 1985 until he left in 1987. Henry Clarks then served as acting division director and was appointed division director in 1988. He remained there until August 1988 when he returned to the Technology Utilization Division. Richard H. Ott became director of the Commercial Development Division in 1988.

The Plans, Policy and Evaluation Division conducted long-term strategic planning, supporting the development of agency policies to expand private-sector investment in civil space and space-related activities. Peter T. Eaton was acting division director, followed by Barbara A. Luxenberg in 1986. Barbara A. Stone succeeded Luxenberg in 1987.

In 1985, the Small Business Innovation Research Division was moved to OCP from the Office of Aeronautics and Space Technology. Harry W. Johnson was named division director.

Money for Commercial Programs

Before the establishment of OCP in 1984, funding for commercialization activities was located in the Office of Space and Terrestrial Applications or the Office of External Relations. Until the end of FY 1985, the only major program category was Technology Utilization. In FY 1985, OCP established the Commercial Use of Space program, and total OCP funding became considerably greater. Initially, programmed funding for the Commercial Use of Space program was drawn from the Space Station, Physics and Astronomy, and Space Research and Technology programs within the Research and Development (R&D) appropriation, while Space Transportation Operations, Space and Ground Network, and Communications and Data Systems programs fell within the Space Flight Control and Data Communications (SFC&DC) appropriation. Appropriated funding for the Commercial Use of Space program began with FY 1986. See Table 5–1 through 5–13.

NASA's Commercial Programs (1979–1988)

In 1979, NASA Administrator Robert A. Frosch released guidelines aimed at increasing private-sector use of NASA's resources. These guidelines provided early direction to NASA's efforts to support private-sector development of space technologies and were based on provisions of the Space Act of 1958, which called for the preservation of U.S. leadership in space science and technology. Administrator Frosch stated, "Since substantial portions of the U.S. technological base and motivation reside in the U.S. private sector, NASA will enter into transactions and take necessary and proper actions to achieve the objective of national technological superiority through joint action with United States domestic concerns."[1]

---

[1] Robert A. Frosch, "NASA Guidelines Regarding Early Usage of Space for Industrial Purposes," June 25, 1979, NASA Historical Reference Collection, NASA History Office, NASA Headquarters, Washington, DC.

The Frosch guidelines named three incentives to interest the private sector in joining NASA in research activities:

- Providing flight time on the Space Shuttle
- Providing technical advice, consultation, data, equipment, and facilities
- Entering into joint research and demonstration programs with NASA and the private sector partner funding their own efforts

The Reagan administration set the overall direction for U.S. space policy, including the encouragement of commercial space activities, in its 1982 National Space Policy. The administration declared that a goal of the U.S. space policy was to "expand United States private sector investment and involvement in civil space and space related activities." It went on to say that the United States encouraged "domestic commercial exploitation of space capabilities, technology, and systems for national economic benefit" and that the government would "provide a climate conducive to expanded private sector investment and involvement in space activities. . . ."

Congress stated its support of commercial space activities during its deliberations in the spring of 1983. A House Committee on Science and Technology report stated that "we should establish a policy which would encourage commercialization of space technology to the maximum extent feasible." The Senate Committee on Commerce, Science and Transportation also stated its support of "efforts by the private sector to invest and seek commercial opportunities in space."[2]

In mid-1983, NASA formed the NASA Space Commercialization Task Force, chaired by L.J. Evans, Jr., who reported to NASA Associate Administrator Philip E. Culbertson. NASA charged the task force with examining the opportunities or impediments to expanded commercial activities in space and developing a policy for NASA's commercialization efforts and an implementation plan for putting the policy to work. The task force consisted of representatives from NASA Headquarters and field centers, advisory groups from industry and academia, private contractors, and a NASA senior management steering committee.

In early 1984, the task force completed its efforts to develop a NASA-wide policy and plan to enhance the agency's ability to encourage and stimulate free enterprise in space. The task force concluded that private enterprise should help the nation retain its lead in science and technology, as well as modify or eliminate natural and bureaucratic barriers to the commercial use of space. A partnership among government, industry, and academia could result in great benefits to the United States. The task force recommended the implementation of a NASA Commercial Space

---

[2] Committee on Science and Technology, U.S. House of Representatives, report, April 15, 1983. Committee on Commerce, Science and Transportation, U.S. Senate, report, May 15, 1983.

Policy "to expedite the expansion of self-sustaining, profit-earning, tax-paying, jobs-providing commercial space activities."

Congress expressed its endorsement of this policy by amending the NASA Space Act of 1958, on July 16, 1984, to include the following provision: "The general welfare of the United States of America requires that the National Aeronautics and Space Administration seek and encourage, to the maximum extent possible, the fullest commercial use of space."[3] Within days of passage and signing of the amendment, the Reagan administration announced a National Policy on the Commercial Use of Space to encourage private enterprise in space. The national policy was designed to actively support commercial space ventures in the areas of new commercial high-technology ventures, new commercial application of existing space technology, and unsubsidized initiatives aimed at transferring existing space programs to the private sector. The policy stated that "private sector investment and involvement is essential if the enormous commercial potential of space is to be developed." It defined steps in four areas to be taken to benefit commercial involvement in space: (1) economic and tax-related issues, (2) legal and regulatory issues, (3) research and development initiatives, and (4) initiatives to implement the policy. It also spelled out specific ways in which NASA and other agencies could "facilitate the commercial use of space" and called for the establishment of a Cabinet Council on Commerce and Trade Working Group on the Commercial Use of Space, with a NASA representative serving as vice chairman.[4]

NASA established the Office of Commercial Programs (OCP) to support the National Space Policy and to translate the conclusions of the 1983 NASA task force into working policies and programs. NASA Administrator James Beggs stated that OCP "will provide a focus for and facilitate efforts within NASA to expand U.S. private sector investment and involvement in civil space related activities."[5] Beggs was also looking toward the proposed space station and the opportunity for private-sector investment and involvement.

NASA released its Commercial Space Policy in October 1984. The plan, drawn up by representatives from NASA Headquarters and field centers, was a detailed policy and implementation plan aimed to foster commercial involvement in space. It stated that NASA encouraged "free enterprise to participate in space by inviting industries and other private entities to finance and conduct business in space." It stated NASA's support for commercial space activities by reducing the technical, financial, and institutional risks to levels competitive with conventional investments and by establishing new links with the private sector to stimulate

---

[3] Subsection added by the National Aeronautics and Space Administration Authorization Act, 1985, Public Law 98–361, July 16, 1984.

[4] The White House, Office of the Press Secretary, "National Policy on the Commercial Use of Space," Fact Sheet, July 20, 1984.

[5] "Special Announcement," NASA, September 11, 1984.

the development of private businesses in space. The policy specified initiatives for involving the private sector in research and development, the use of NASA facilities, patent rights and procedural issues, organizations designed for commercial involvement in space, and NASA's outreach program.[6]

This new OCP absorbed two existing programs: the Small Business Innovation Research (SBIR) program from the Office of Aeronautics and Space Technology and the Technology Utilization program from the Office of External Relations. OCP focused its activities on two major areas: technology utilization and commercial use of space.

Technology Utilization

The Technology Utilization program concentrated on technology transfer activities. The program involved cooperation and collaboration with industry, primarily through its nationwide network of Industrial Applications Centers (IACs), the dissemination of publications and computer software, conferences and seminars on the subject of technology transfer, and technology applications projects. OCP funds were used for the NASA field centers to conduct research and develop technology in response to needs that the private sector identified.

The university-based IACs disseminated NASA-developed technology to a broad range of industrial clients by providing them access to nearly 100 million scientific and technical documents in the NASA data bank. They also provided access to more than 600 other computerized data banks. In 1988, NASA had ten IACs.

A 1987 agreement between NASA and the Federal Laboratory Consortium, established by the Federal Technology Transfer Act of 1986, connected IACs and their affiliates to the consortium's network of 500 research and development laboratories and its clearinghouse. This agreement enabled U.S. industries and entrepreneurs, using access points within their home states, to learn about relevant federal technology available in federal laboratories throughout the country.

The Federal Technology Transfer Act of 1986 amended the Stevenson-Wydler Technology Act of 1980. It permitted each federal agency to allow its laboratories to "enter into cooperative research and development agreements with other federal agencies," state and local governments, industry, and nonprofit organizations and to negotiate licensing agreements.[7] It also established the Federal Laboratory

---

[6] "Preamble" to NASA Commercial Space Policy, October 1984, as printed as part of Document III-27 in John M. Logsdon, gen. ed., with Dwayne A. Day and Roger D. Launius, *Exploring the Unknown: Selected Documents in the History of the U.S. Civil Space Program, Volume II: External Relations* (Washington, DC: NASA SP-4407, 1996), pp. 573–74.

[7] Federal Technology Transfer Act of 1986, Public Law 99–502.

Consortium for Technology Transfer, which was charged with developing and administering "techniques, training courses, and materials concerning technology transfer" for federal employees and using "the expertise and services of . . . the National Aeronautics and Space Administration" and other agencies.

Commercial Use of Space

The Commercial Use of Space program, managed by the Commercial Development Division, focused on increasing private-sector awareness of space opportunities and encouraging increased industry investment and participation in high-technology space-based research and development. The program also promoted the development of new markets for the Space Transportation System and other NASA space services. It worked to facilitate private-sector space activities through improved access to available NASA resources and encouraged increased private-sector investment in the commercial use of space independent of NASA funding.

Centers for the Commercial Development of Space

In 1985, NASA initiated the Centers for the Commercial Development of Space. This program was designed to increase private-sector interest and investment in using space for commercial activities. Through 1988, OCP selected and funded sixteen Centers for the Commercial Development of Space at a level of up to $1 million per center. These centers received additional financial and in-kind contributions from industrial affiliates. NASA also provided the centers with scientific and technical expertise, opportunities for cooperative activities, and other forms of continuing assistance. The centers performed basic space research activities and received funding for up to five years. Table 5–14 lists the centers and their research areas.

Joint Agreements

When NASA Administrator Frosch issued his 1979 guidelines calling for more involvement with the private sector, NASA also defined and developed three types of cooperative agreements into which the agency could enter with the private sector: the Joint Endeavor Agreements (JEA), the Technical Exchange Agreement (TEA), and Industrial Guest Investigator (IGI) Agreement. These agreements all involved research efforts in the Materials Processing in Space (MPS) program area and called for using the microgravity environment of space for a variety of experimental purposes in the MPS area that hopefully would lead to commercial activities in the space environment.

In each of these joint legal arrangements, the government would not fund any of the work done by the private-sector partner but would provide other incentives, such as space on the Space Shuttle at no cost. The JEA had the objective of encouraging early space ventures and demonstrating the usefulness of space technology to meet marketplace needs. The TEA was an agreement to exchange technical information and cooperate in ground-based research programs. In an IGI Agreement, a company arranged for an industry scientist to collaborate with a NASA investigator on a spaceflight MPS experiment and to become a member of the investigative team. NASA also used the already established memorandum of understanding and memorandum of agreement arrangements to structure additional agreements. In 1985, NASA added the Space Systems Development Agreement (SSDA) to the cooperative agreements available to the private sector. The SSDA enabled a company to acquire Shuttle launch and launch-related services under special terms, such as a deferred payment schedule, tailored to the particular venture.

NASA signed the first JEA in December 1979 with McDonnell Douglas and began implementing the agreement in January 1980. The agency was fairly successful in inducing industry to become partners in space research on the Space Shuttle until the *Challenger* accident in 1986. That event forced NASA to delay implementing the agreements, and the agency found it difficult to regain momentum when the Shuttle flew again. Table 5–15 lists the various cooperative agreements entered into by NASA and the private sector from 1979 to 1988.

Small Business Innovation Research

NASA established the Small Business Innovation Research (SBIR) program in 1983 in response to the Small Business Innovation Development Act of 1982. Congress passed reauthorization legislation in 1986 that extended the program until October 1, 1993.[8] The objectives were to stimulate technological innovation, use small business to meet federal research and development needs, foster and encourage participation by minority and disadvantaged persons in technological innovation, and increase private-sector commercialization innovations derived from federal research and development. The program enabled the government to use the innovation and efficiency of small, high-technology firms and research institutions to accomplish agency mission objectives.

The legislation initially placed SBIR funding for agencies of NASA's size at 1.25 percent of its extramural research and development budget. Funding came from an assessment of each NASA organization's research and development budget. In July 1985, NASA transferred its SBIR office administratively to OCP from the Office of Aeronautics and Space Technology.

---

[8] Small Business Innovation Development Act of 1982, Public Law 97–219, July 22, 1982, and Reauthorization Amendment extending program until October 1, 1993, Public Law 99–443, enacted October 6, 1986.

Small businesses could participate in the program in two phases. Phase I SBIR activities were to determine the scientific and technical merit and feasibility of ideas submitted in response to SBIR program solicitations. The commercial potential of the ideas were considered when NASA selected the winning proposals. Phase II projects were selected from the Phase I participants. To be selected, the commercial potential of a project must be apparent in addition to meeting the requirements of Phase I projects.

Table 5–16 shows funding that NASA made available to the SBIR program by fiscal year. Table 5–17 provides the number of proposals and contract awards resulting from each annual solicitation. Table 5–18 shows the SBIR projects chosen in each topic area through 1988, and Table 5–19 shows the cumulative SBIR awards in each topic area and the percentage selected in each area.[9]

---

[9] These four tables are derived from a Small Business Innovation Research (SBIR) program/Small Business Technology Transfer (STTR) program presentation, NASA, SBIR data base, 1996.

*Table 5–1. Total Commercial Programs Funding
(in thousands of dollars)* a

Year	Request	Authorization	Appropriation	Programmed (Actual)
1979	9,100	12,000	b	9,100
1980	12,100	12,100	c	11,980
1981	11,800 d	12,600	8,800	8,800
1982	8,000 e	12,680	f	8,000
1983	4,000 g	9,000	4,000	9,000
1984	4,000	10,000 h	9,000	9,000 i
1985	9,500	9,500	9,500 j	17,100
1986	41,100 k	28,100	28,100	26,800
1987	41,300 l	40,300	41,300	40,900
1988	54,000	49,000	50,000	48,700

a   The Office of Commercial Programs (OCP) was established in September 1984. Prior to that, the Technology Utilization program was located in the Office of Space and Terrestrial Applications (1979–1983) and the Office of External Relations (1984).
b   Undistributed. Total 1979 R&D appropriation = $3,477,200,000.
c   Undistributed. Total 1980 R&D appropriation = $4,091,086,000.
d   Amended budget estimate. Original estimate = $13,100,000.
e   Amended budget estimate. Original estimate = $14,600,000.
f   Undistributed. Total 1982 appropriation = $4,740,900,000.
g   The Technology Utilization program was part of the Office of Space and Terrestrial Applications.
h   The Technology Utilization program was part of the Office of External Relations.
i   This was the last time the programmed amount was for Technology Utilization only. See Table 5–6.
j   This was the final time the appropriation for commercial activities was for Technology Utilization only. See Table 5–6 for the Technology Utilization line item after OCP was established.
k   Commercial activities added the Commercial Use of Space program.
l   Revised budget request. Original request = $45,300,000.

*Table 5–2. Major Budget Category Programmed Funding History
(in thousands of dollars)*

Budget Category/Fiscal Year	1979	1980	1981	1982	1983
Technology Utilization a	9,100	11,980	8,800	8,000	9,000
Technology Dissemination	3,200	3,700	2,594	5,700	5,800
Technology Applications	4,500	4,400	3,858	2,300	3,200
Program Control and Evaluation b	1,400	1,480	1,148	c	

Budget Category/Fiscal Year	1984	1985	1986	1987	1988
Technology Utilization d	9,000	9,500	10,580	15,700	19,000
Technology Dissemination	e				
Technology Applications	f				
Commercial Applications R&D		(6,350)			
Commercial Development Support		(1,250)			
Commercial Use of Space			16,220	25,200	29,700

a   The Technology Utilization program was located in the Office of Space and Terrestrial Applications in fiscal years 1979 to 1983.
b   The budget category was changed to Program Evaluation and Support.
c   The budget category was eliminated.
d   The Technology Utilization program was located in the Office of External Relations prior to the establishment of the Office of Commercial Programs in September 1984.
e   The budget category was eliminated.
f   The budget category was eliminated.

*Table 5–3. Technology Dissemination Funding
(in thousands of dollars)*

Year	Request	Authorization	Appropriation	Programmed (Actual)
1979	3,600 a	— b	—	3,200
1980	3,800	3,800	c	3,700
1981	4,000 d	4,100	2,400	2,594
1982	5,700 e	f	g	5,700
1983	5,700 h	3,200	3,200	5,800
1984	5,500 i	2,200	2,200	5,500
1985	5,800	5,800	5,800	5,800
1986	6,300	6,300	6,300	j
1987	7,600	7,600	7,600	k

a   Revised budget estimate. Original estimate = $3,715,000.
b   No authorization or appropriation for budget category.
c   Undistributed. Total 1980 R&D appropriation = $4,091,086,000.
d   Revised budget estimate. Original estimate = $4,100,000.
e   Amended budget request. Original request = $4,600,000.
f   Undistributed. Total Technology Utilization authorization = $12,600,000.
g   Undistributed. Total 1982 R&D appropriation = $4,740,900,000.
h   Revised budget estimate. Original estimate = $3,200,000.
i   Amended budget request. Original request = $2,200,000. The increase reflects congressional action to provide for dissemination activities at approximately fiscal year 1983 level and to allow the continuation of ongoing as well as some new efforts in technology applications. The funding increase also reflects increased dissemination activities in support of small business activities.
j   No programmed amount for budget category.
k   No programmed amount for budget category.

*Table 5–4. Program Control and Evaluation Funding
(in thousands of dollars) a*

Year	Request	Authorization	Appropriation	Programmed (Actual)
1979	1,400 b	1,27	c	1,400
1980	1,500 d	1,500	e	1,480
1981	1,600	1,600	400	1,148 f

a   This was called Program Evaluation and Support beginning with the fiscal year 1980 request.
b   Amended budget estimate. Original estimate = $1,275,000,000.
c   Undistributed. Total 1979 R&D appropriation = $3,477,200,000.
d   Amended budget estimate. Original estimate = $1,600,000.
e   Undistributed. Total 1980 R&D appropriation = $4,091,086,000.
f   The program concluded. The activity formerly funded under this budget category was incorporated into the Technology Dissemination budget category.

Table 5–5. Technology Applications Funding
(in thousands of dollars)

Year	Request	Authorization	Appropriation	Programmed (Actual)
1979	4,100 a	b	c	4,500
1980	4,400	4,400	d	4,400
1981	3,800 e	4,300	2,800	3,858
1982	2,300 f	g	h	2,300
1983	3,300 i	5,800	800	3,200
1984	3,500 j	1,800	1,800	3,500
1985	3,700	3,700	3,700	3,700
1986	4,800	4,800	4,800	4,580
1987	5,700	5,700	5,700	6,000
1988	7,000 k	6,620	6,620	7,000

a   Revised budget request. Original request = $4,110,000.
b   Undistributed. Total 1979 R&D Technology Utilization Authorization = $12,100,000.
c   Undistributed. Total 1979 R&D appropriation = $3,477,200,000.
d   Undistributed. Total 1980 R&D appropriation = $4,091,086,000.
e   Revised budget request. Original request = $4,800,000.
f   Amended budget request. Original request = $2,100,000.
g   Undistributed. Total Technology Utilization authorization = $12,600,000.
h   Undistributed. Total 1982 R&D appropriation = $4,740,900,000.
i   Amended budget estimate. Original estimate = $800,000.
j   Amended budget estimate. Original estimate = $1,800,000.
k   Amended budget estimate. Original estimate = $6,600,000.

Table 5–6. Technology Utilization Funding
(in thousands of dollars)

Year	Request	Authorization	Appropriation	Programmed (Actual)
1985	a	—	—	9,500 b
1986	11,100	11,100	11,100	10,580
1987	15,700 c	13,300	15,700	15,700
1988	18,300	18,300	18,300	19,000

a   Prior to fiscal year 1986, all commercialization activity was funded from the Technology Utilization budget. See Table 5–1.
b   This budget category became a line item under the Office of Commercial Programs.
c   Amended budget request. Original request = $13,300,000.

Table 5–7. Product Development Funding History (in thousands of dollars)

Year	Request	Authorization	Appropriation	Programmed (Actual)
1986				1,140
1987	1,500 a	— b	—	1,500
1988	1,400 c	1,920	1,920	1,400

a   Amended budget estimate. Original estimate = $1,470,000.
b   No authorization or appropriation.
c   Amended budget estimate. Original estimate = $2,000,000.

Table 5–8. Acquisition, Dissemination, and Network Operations Funding History (in thousands of dollars)

Year	Request	Authorization	Appropriation	Programmed (Actual)
1986				3,270
1987	4,100	— a	—	4,100
1988	4,665 b	4,730	4,730	4,700

a   No authorization or appropriation.
b   Amended budget estimate. Original estimate = $4,700,000.

Table 5–9. Program Development, Evaluation, and Coordination Funding History (in thousands of dollars)

Year	Request	Authorization	Appropriation	Programmed (Actual)
1986				1,590
1987	1,780	— a	—	1,700
1988	2,600 b	2,380	2,380	2,600

a   No authorization or appropriation.
b   Amended budget estimate. Original estimate = $1,800,000.

Table 5–10. Industrial Outreach Funding History (in thousands of dollars)

Year	Request	Authorization	Appropriation	Programmed (Actual)
1987	2,370	— a	—	2,400
1988	2,035 b	2,650	2,650	2,000

a   No authorization or appropriation.
b   Amended budget estimate. Original estimate = $2,600,000.

Table 5–11. Commercial Use of Space Funding
(in thousands of dollars)

Year	Request	Authorization	Appropriation	Programmed (Actual)
1985				7,600 a
1986	30,000	17,000	17,000	16,220
1987	25,600 b	27,000	26,500 c	25,200
1988	35,700	30,700	31,700	29,700

a   New budget category. The amount was drawn from the Space Station, Physics and Astronomy, and Space Research and Technology programs in the Research and Development (R&D) appropriation and the Space and Ground Network and Communications and Data Systems programs in the Space Flight Control and Data Communications (SFC&DC) appropriation.
b   Amended budget estimate. Original estimate = $32,000,000.
c   This reflects a general reduction of $6,400,000.

Table 5–12. Commercial Applications R&D Funding
(in thousands of dollars)

Year	Request	Authorization	Appropriation	Programmed (Actual)
1985				(6,350)
1986	15,500 a	— b	—	12,940
1987	22,600 c	30,100	30,100	22,200
1988	22,200 d	26,000	31,000	25,200

a   Amended budget estimate. Original estimate = $28,500,000.
b   No authorization or appropriation.
c   Amended budget request. Original request = $31,100,000.
d   Amended budget estimate. Original estimate = $31,000,000.

Table 5–13. Commercial Development Support Funding
(in thousands of dollars)

Year	Request	Authorization	Appropriation	Programmed (Actual)
1985				(1,250)
1986	1,500	— a	—	3,280
1987	3,000 b	1,900	1,900	3,000
1988	2,400 c	4,700	4,700	4,599

a   No authorization or appropriation.
b   Amended budget request. Original request = $1,900,000.
c   Amended budget estimate. Original estimate = $1,400,000.

Table 5–14. Centers for the Commercial Development of Space

Center	Technical Discipline
Center for Advanced Materials, Battelle Columbus Laboratories	Materials Processing
Center for Macromolecular Crystallography, University of Alabama at Birmingham	Life Sciences
Consortium for Materials Development in Space, University of Alabama at Huntsville	Materials Processing
ITD Space Remote Sensing Center, Institute for Technology Development	Remote Sensing
Center for Space Processing of Engineering Materials, Vanderbilt University	Materials Processing
Center for Mapping, Ohio State University	Remote Sensing
Wisconsin Center for Automation and Robotics, University of Wisconsin	Automation and Robotics
Center for Development of Commercial Crystal Growth in Space, Clarkson University	Materials Processing
Space Vacuum Epitaxy Center, University of Houston	Materials Processing
Center for Advanced Space Propulsion, University of Tennessee Space Institute	Space Propulsion
Center for the Commercial Development of Space Power and Advanced Electronics, Auburn University	Space Power
Center for Autonomous and Man-Controlled Robotic and Sensing Systems, Environmental Research Institute of Michigan	Automation and Robotics
Center for Cell Research, Pennsylvania State University	Life Sciences
Center for Bioserve Research, University of Colorado at Boulder	Life Sciences
Center for Materials for Space Structures, Case Western Reserve University	Space Structures and Materials
Center for Space Power, Texas A&M University	Space Power

Table 5–15. Cooperative Agreements Between NASA and the Private Sector

Date	Company	Description	Type of Agreement	Remarks
January 1980	McDonnell Douglas	Production of pharmaceuticals in space	JEA	NASA provided Shuttle flights to McDonnell Douglas Astronautics Company (MDAC). Although greater purity was achieved by processing in space, MDAC abandoned the project because alternative ground-based processes were more cost-effective.
June 1981/ June 1984	John Deere & Co.	Ground-based research on graphite formation in cast iron and improved metallic alloys	TEA	No space-based research was undertaken.
June 1981	DuPont	Ground-based research on catalytic materials	TEA	Research was completed but never led to research in space.
August 1981	INCO	Ground-based research on electroplating devices	TEA	Research was completed but never led to research in space.
April 1982	Defense Systems, Inc.	Shuttle deployment of experiments leading to a network of small communications satellites being deployed in low-Earth orbit	JEA	The company abandoned the project prior to any activity on the Shuttle.
July 1982	GTI Corporation	Development of Shuttle-based furnace for processing materials	JEA	The company abandoned the project prior to building any hardware

COMMERCIAL PROGRAMS

Table 5–15 continued

Date	Company	Description	Type of Agreement	Remarks
February 1983	A.D. Little	Ground-based research on long-term blood storage	MOU	The project never materialized because the company was unable to form the necessary consortium of firms.
February 1983	Honeywell	Research on mercury-cadmium-telluride materials (semiconductor materials)	TEA	There was no real activity by the company under this agreement.
April 1983	Orbital Sciences Corp.	Shuttle launch of a Transfer Orbit Stage (TOS) for boosting payloads from the Shuttle cargo bay to geosynchronous orbit	JEA	The company never built the TOS because of an insufficient market. However, based on other successful projects, Orbital Sciences went from a start-up company to become a viable NASA and Defense Department contractor.
April 1983	Micro-Gravity Research Associates	Production of gallium arsenide crystals (semiconductor material) in space	JEA	The company did not provide any experiments for flight on the Shuttle.
August 1983	Fairchild Industries, Inc.	Shuttle deployment of an unpressurized space laboratory platform to house experiments and manufacturing equipment in space	JEA	The company built no hardware because it could not justify the project based on its subsequent market analysis.
December 1983	Spaceco, Ltd.	Shuttle flight of equipment for monitoring the space environment	JEA	The company provided no hardware for flight.

Table 5–15 continued

Date	Company	Description	Type of Agreement	Remarks
January 1984	C² Spacelines	Shuttle cargo booking service	MOU	The company never initiated business operations.
January 1984/ July 1986/ December 1986	3M Corporation	MOU: Research on thin film materials and organic crystal materials TEA: Ground-based experiments on polymers JEA: Shuttle flights of organic and polymer chemistry experiments	MOU, TEA, JEA	The MOU led to two JEAs. NASA provided Shuttle flights. 3M abandoned the project. As for the TEA, 3M subsequently abandoned interest in commercial utilization of the space environment.
February 1984	Space Industries, Inc.	Shuttle deployment of an Industrial Space Facility for housing experiments and manufacturing facilities	MOU	The MOU led to the first Space Systems Development Agreement (SSDA).
October 1984	Martin Marietta Corporation	Shuttle flights of propellant fluids management equipment	JEA	NASA provided Shuttle flights of the equipment, but no commercial business resulted based on the device.
November 1984/April 1986/March 1987	Rockwell International	TEA: Ground-based materials science MOU/JEA: Shuttle flights of fluids experiment equipment regarding float zone crystal growth technology	MOU, TEA, JEA	The MOU and TEA led to the follow-on JEA. NASA provided Shuttle flights, but the results did not lead to a commercial business venture for Rockwell.

COMMERCIAL PROGRAMS

Table 5–15 continued

Date	Company	Description	Type of Agreement	Remarks
November 1984/May 1986	Boeing Aerospace Corporation	Shuttle flights of materials processing experiments leading to commercial materials manufacturing in space	MOU, JEA	The MOU led to a follow-on JEA. Boeing completed its Shuttle-based research under the JEA. The work did not lead to a commercial venture.
February 1985/October 1986	Grumman Space Systems	Gallium arsenide crystal growth in space	MOU, JEA	The MOU lead to a follow-on JEA. The company abandoned the project before any Shuttle flights were provided.
June 1985	Instrumentation Technology Associates	Standardized carrier for housing materials processing experiments	JEA	The apparatus was built and flown; the equipment proved to be only marginally successful, and no commercial venture resulted from the work.
August 1985	Sperry Corporation	Magnetic Isolation System to buffer materials processing equipment from vibrations	MOU	The company abandoned the project.
August 1985	Space Services, Inc.	See Space Industries, Inc., MOU above	SSDA	The company was to pay for the Shuttle flights and associated NASA services on a deferred payment basis. The company never developed the Industrial Space Facility.
September 1985	GTE	Organic and polymer crystal growth in space	MOU	The company never developed hardware.

Table 5–15 continued

Date	Company	Description	Type of Agreement	Remarks
October 1985	Scott Science and Technology, Inc.	NASA technical assistance in the design and development of a transfer vehicle to deliver satellites to a geosynchronous transfer orbit from the Shuttle cargo bay	MOA	The company never built any hardware. The *Challenger* accident diminished investor interest.
November 1985	Geostar Corporation	Shuttle deployment of satellites regarding a commercial position determination/nonvoice communications system	SSDA	Geostar was to reimburse NASA for Shuttle flights and associated services on a deferred payment basis. The company went out of business because the technology became obsolete, and no Shuttle flights were provided.
December 1985	Rantek	Shuttle-based research equipment to study rheumatoid arthritis	MOU	The company never built any hardware.
January 1986/ August 1988	SPACEHAB, Inc.	Shuttle middeck augmentation module for housing and servicing materials processing research and manufacturing facilities	MOU, SSDA	The MOU led to the SSDA with NASA. The commercial module was built and flown on the Shuttle. The economic viability of the venture was tied directly to a subsequent NASA lease of the module for government-sponsored research. SPACEHAB was unsuccessful in developing a commercial customer base; accordingly, it was not a real commercial venture.

Table 5–15 continued

Date	Company	Description	Type of Agreement	Remarks
February 1986	David A. Mouat	Satellite and aircraft remote sensing equipment for mineral and hydrocarbon exploration	TEA	No activity under this agreement was undertaken by the firm.
March 1986	Earth Data Corporation	Development of a commercial remote sensing system	TEA	No commercial activity resulted.
March 1986	Institute for Technology Development	Commercial remote sensing technology	MOA	No commercial activity resulted.
May 1986	Abex Corporation	Microgravity experiments aboard the KC 135 aircraft on the subsequent development of a materials processing furnace	TEA	There was no follow-on commercial venture.
May 1986	Union Oil Company	Research to develop remote sensing equipment	MOU	There were no follow-on commercial activities.
October 1986	General Sciences Corporation	Development of a commercial orbiting power source to service the Shuttle, free-flying platforms, and the space station	MOU	No commercial activity resulted.
November 1986	Hercules Corporation	Materials processing research	TEA	There was no follow-on commercial activity by the company.

Table 5–15 continued

Date	Company	Description	Type of Agreement	Remarks
October 1988	United Technologies Corporation	Ground-based materials experiments	TEA	There was no follow-on joint activity with NASA by the company.

Legend: JEA—Joint Endeavor Agreement, MOA—Memorandum of Agreement, MOU—Memorandum of Understanding, SSDA—Space Systems Development Agreement, and TEA—Technical Exchange Agreement.

Table 5–16. SBIR Funding by Fiscal Year
(in millions of dollars)

Fiscal Year	1983	1984	1985	1986	1987	1988
Funding	4.9	13.2	24.2	32.6	38.9	40.7

Table 5–17. Proposals and Contract Awards
From Each Annual Solicitation

Program Year	1983	1984	1985	1986	1987	1988
Phase I						
Proposals	1,000	920	1,164	1,631	1,827	2,379
Awards	102	127	150	172	204	228
Phase II						
Proposals	92	113	129	154	179	203
Awards	58	71	84	86	98	112

Table 5–18. NASA SBIR Awards by Phase and Topic Area (1983–1988)

Topic	1983 I	1983 II	1984 I	1984 II	1985 I	1985 II	1986 I	1986 II	1987 I	1987 II	1988 I	1988 II
Aero Propulsion and Power	7	4	9	5	9	3	10	6	9	3	9	6
Aerodynamics and Acoustics	12	8	12	3	10	4	11	8	21	10	19	10
Aircraft Systems	8	5	6	3	10	5	11	5	13	7	16	8
Materials and Structures	12	6	11	5	16	13	15	6	21	8	23	7
Teleoperators and Robotics	4	1	5	1	8	5	10	3	20	11	21	9
Computer Sciences	6	2	7	3	11	4	11	7	11	6	11	5
Information Systems	3	2	7	3	8	4	14	8	11	7	12	5
Instrumentation and Sensors	18	11	16	15	19	15	31	20	35	17	37	21
Spacecraft Systems	6	5	15	10	17	8	16	5	12	8	18	9
Space Power	6	2	7	5	5	3	4	2	6	4	13	5
Space Propulsion	4	3	3	2	2	2	5	3	5	2	6	3
Space Habitability and Biology	4	3	10	6	13	6	8	3	16	6	16	8
QA, Safety and Checkout	4	3	7	2	8	6	10	5	10	4	9	4
Space Communications	6	3	7	4	9	3	8	3	7	2	11	7
Commercial Space Applications	2	0	5	4	5	3	8	2	7	3	7	5
Total	102	58	127	71	150	84	172	86	204	98	228	112

Table 5–19. Cumulative NASA SBIR Funding Awards by Topic Area (1983–1988)

Topic	Total Award Number	Total Award Percent
Aero Propulsion and Power	80	5.4
Aerodynamics and Acoustics	128	8.6
Aircraft Systems	97	6.5
Materials and Structures	143	9.6
Teleoperators and Robotics	98	6.6
Computer Sciences	84	5.6
Information Systems	84	5.6
Instrumentation and Sensors	255	17.1
Spacecraft Systems	129	8.6
Space Power	62	4.2
Space Propulsion	40	2.7
Space Habitability and Biology	99	6.6
QA, Safety and Checkout	72	4.8
Space Communications	70	4.7
Commercial Space Applications	51	3.4
Total	**1,492**	**100.0**

CHAPTER SIX
FACILITIES AND INSTALLATIONS

CHAPTER SIX
FACILITIES AND INSTALLATIONS

Introduction

By 1979, NASA's facilities and installations were no longer increasing in number or size. Although the agency added to and upgraded its equipment and increased the number of buildings and other structures, the number of NASA installations remained the same throughout the decade. The only change was the consolidation of two centers with two others. In an October 1981 agency reorganization, Dryden Flight Research Center was consolidated with Ames Research Center and renamed Dryden Flight Research Facility, and Wallops Flight Center became part of Goddard Space Flight Center and was renamed Wallops Flight Facility.

In addition to NASA Headquarters, in 1979, NASA consisted of ten field installations and the contractor-operated Jet Propulsion Laboratory. Five of the installations—Ames Research Center, Dryden Flight Research Center, Langley Research Center, Lewis Research Center, and Wallops Flight Center—had been facilities of the National Advisory Committee for Aeronautics (NACA). These installations were transferred to NASA in 1958 when the agency was established. Within the next few years, Goddard Space Flight Center, Kennedy Space Center, Marshall Space Flight Center, and the Jet Propulsion Laboratory were transferred to NASA from the U.S. military space program. NASA established the National Space Technology Laboratories as a NASA center in 1974 and renamed it the Stennis Space Center in 1988. Figure 6–1 diagrams the locations of the NASA installations, and Figure 6–2 chronicles the establishment of these installations.

Each NASA installation focused its resources on particular major programs and mission areas. Table 6–1 lists these areas of concentration.

The first part of this chapter reviews NASA's aggregate facilities: the value and size of its total holdings and the installations' holdings grouped together for easy comparison. The second part of the chapter describes NASA Headquarters and the individual NASA installations that existed during all or part of the years from 1979 through 1988. It briefly describes their history and mission and also provides tables that characterize the property, personnel, funding, and procurement activity of each installation during this period.

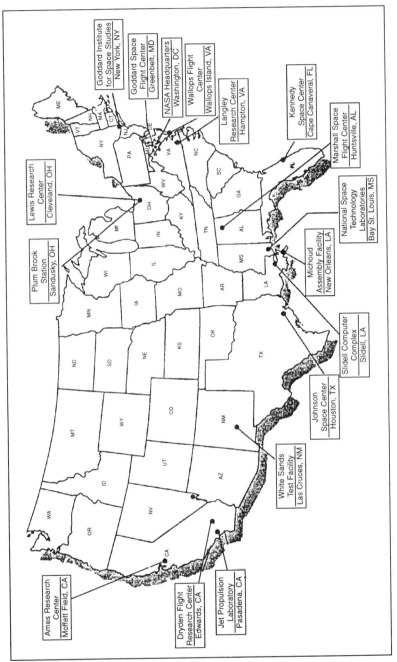

Figure 6–1. NASA Facilities (1980)

FACILITIES AND INSTALLATIONS

Installation	1979	1980	1981	1982	1983	1984	1985	1986	1987	1988
Ames Research Center	Authorized 1939; dedicated June 1940									
Dryden Flight Research Center	Authorized 1952; named NACA High Speed Flight Station and made autonomous July 1954; designated Flight Research Center September 1959; renamed Dryden Flight Research Facility March 1976; consolidated with Ames Research Center October 1981 as Dryden Flight Research Facility									
Goddard Space Flight Center	Authorized 1958; dedicated March 1961									
Jet Propulsion Laboratory	Organized 1944									
Johnson Space Center	Established January 1961 as Space Task Group; major occupancy of Clear Lake site February 1964; renamed Johnson Space Center February 1973									
Kennedy Space Center	Established March 1962; effective July 1962 as Launch operations Center; redesignated Kennedy Space Center December 1963									
Langley Research Center	Authorized 1917; dedicated June 1920									
Lewis Research Center	Authorized 1940; groundbreaking 1942									
Marshall Space Flight Center	Established March 1960; transfer of personnel from U.S. Army effective July 1960; dedicated September 1960									
Stennis Space Center	Established as an independent NASA field installation June 1974; redesignated Stennis Space Center May 1988									
Wallops Flight Center	Established under Langley Research Center 1945; became autonomous May 1959; renamed Wallops Flight Center 1974; consolidated with Goddard Space Flight Center October 1981									

Figure 6–2. NASA Installations (1979–1988)

Facilities

Definition of Terms

The discussion of facilities and installations uses several terms with which the reader may not be familiar. The following definitions come from NASA Management Instructions and NASA Management Handbooks:

Buildings: Facilities with the basic function of enclosing usable space.

Capital Equipment: An item of equipment with an acquisition cost of $5,000 or more that has an estimated service life of two years or more, that will not be consumed in an experiment, and that most generally will be identified as an independently operable item (identified as "capitalized equipment" in this chapter).

Collateral Equipment: Building-type equipment, built-in equipment, and large, substantially affixed equipment that are normally acquired and installed as a part of a facility project.

Component Installation: An installation, office, or other NASA organizational element that is located geographically apart from a NASA installation and that, pursuant to delegations from the Administrator, is assigned for management purposes to the official-in-charge of a Headquarters office, the director of a field installation, or an immediate subordinate of these officials.

Equipment: An item of real or personal property generally in the configuration of a mechanical, electrical, or electronic apparatus or tool, normally costing in excess of $100, that may perform a function independently or in conjunction with other equipment or components.

Facility: A generic term used to encompass real property and related integral and collateral equipment of a capital nature. The term does not encompass operating materials, supplies, and noncapitalized equipment. The term is used in connection with land, buildings, structures, and other real property improvements.

Field Installation: A NASA organizational element located geographically apart from NASA Headquarters and headed by a director.

Fixed Assets: Assets of a permanent character having a continuing value, such as land, buildings, and other structures and facilities, including collateral and noncollateral equipment meeting the criteria for capitalization.

Installation: A NASA organizational element, including both Headquarters and field installations.

Integral Equipment: Equipment that is normally required to make a facility useful and operable as a facility and that is built in or permanently affixed to it in such a manner that removal would impair the usefulness, safety, or comfort of the facility.

Investment Value: A figure representing the total of real property value (including land, buildings, and other structures and facilities), leasehold improvements value, capitalized equipment value, and assets-in-progress value.

Land: A category of real property that includes all acquired interests in land (for example, owned, leased, or acquired by permit) but excludes NASA-controlled easements and rights-of-way that are under leasehold improvements.

Leased Property: Property under the control of NASA through lease, administrative agreement, temporary permit, license, or other arrangements.

Leasehold Improvements: NASA-funded long-term capital improvements to leases, rights, interests, and privileges relating to land not owned by NASA, such as easements, rights-of-way, permits, use agreements, water rights, air rights, and mineral rights.

Noncollateral Equipment: Equipment that imparts to the facility or test apparatus its particular character at the time—for example, furniture in an office building, laboratory equipment in a laboratory, and so forth. Such equipment, when acquired and used in a facility or a test apparatus, can be severed and removed after erection or installation without substantial loss of value or damage thereto or to the premises where installed.

Other Structures and Facilities: A category of real property that includes facilities having the basic function of research or operational tools or activities as distinct from buildings (includes items such as airfield pavements, power production facilities and distribution systems, flood control and navigation aids, storage, industrial service, and research and development facilities other than buildings, and communications systems).

Personal Property: Property of any kind, including equipment, materials, and supplies, but excluding real property.

Real Property: Land, buildings, structures, utilities systems, and improvements and appurtenances thereto, permanently annexed to land.

NASA Facility Property Statistics

Tables 6–2 through 6–18 include statistics for 1979 through 1988 on NASA property, real property, investment values, land, buildings, other structures, capitalized equipment, and tracking and data acquisition stations.

Installations

The following pages list the directors and deputy directors of each NASA installation and provide a brief center history and description of each installation's mission. Table 6–19 details the 1988 budget plan by installation and program office. The tables for each installation list the property holdings and their value, personnel levels and characteristics, funding levels, and procurement activity of each installation (see Tables 6–21 through 6–78).

NASA Headquarters

Location

From 1979 to 1988, NASA Headquarters was housed in two federal buildings in Washington, D.C. One location was on Independence Avenue, SW; the second was on Maryland Avenue, SW.

Administrator

James C. Fletcher (May 1986–April 1989)
William R. Graham, Acting (December 1985–May 1986)
James Beggs (July 1981–December 1985)[1]
Alan M. Lovelace, Acting (January 1981–July 1981)
Robert A. Frosch (June 1977–January 1981)
James C. Fletcher (April 1971–May 1977)
George M. Low, Acting (September 1970–April 1971)
Thomas O. Paine (March 1969–September 1970)
James E. Webb (February 1961–October 1968)
T. Keith Glennan (August 1958–January 1961)

Deputy Administrator

Dale D. Myers (October 1986–May 1989)
William R. Graham (November 1985–October 1986)
Hans Mark (July 1981–September 1984)
Alan Lovelace (July 1976–December 1980)
George M. Low (December 1969–June 1976)
Thomas O. Paine (March 1968–March 1969)
Robert C. Seamans, Jr. (December 1965–January 1968)
Hugh L. Dryden (September 1959–December 1965)

History

The development of NASA Headquarters before 1979 is detailed in Volumes I and IV of the *NASA Historical Data Book*. During the decade from 1979 through 1988, several changes occurred in Headquarters-level organizations that reflected the changing priorities of the agency. A detailed description is given in the appropriate chapters of this book and of Volume V. The following paragraphs provide a brief overview.

In 1979, NASA established the Office of Space Transportation Systems Operations, and the former Office of Space Transportation

---

[1] James Beggs went on an indefinite leave of absence beginning December 4, 1985, while he was answering charges of fraud alleged to have occurred while he was executive vice president of General Dynamics Corporation. He was later cleared of all charges.

Systems was renamed Office of Space Transportation Systems Acquisition. This office focused on completing the Space Shuttle's production. The Office of Space Transportation Systems Operations focused on preparing for the Shuttle system once it was fully tested, scheduling flights, developing pricing policies and launch service agreements, and managing the Spacelab and expendable launch vehicle programs.

The only major Headquarters-level reorganization of the decade occurred in November 1981. Since 1978, the field installations had been under the direct control of the NASA Administrator. The 1981 reorganization placed each field installation under the administrative control of the associate administrator of a NASA program office that corresponded to the installation's major mission areas. The Office of Space Science and Applications took over management of Goddard Space Flight Center and the Jet Propulsion Laboratory. The Office of Aeronautics and Space Technology administered Ames Research Center, Langley Research Center, and Lewis Research Center. The Office of Space Flight managed Johnson Space Center, Kennedy Space Center, Marshall Space Flight Center, and the National Space Technology Laboratories/Stennis Space Center. Each Headquarters associate administrator was responsible for program content and execution and program and institutional resources for their respective centers.

This reorganization also merged the Office of Space Science and the Office of Space and Terrestrial Applications into the Office of Space Science and Applications. In addition, NASA established the Office of Management as a new organization to handle part of the functions performed by the Office of Management Operations as well as some of the duties performed by the comptroller.

The associate deputy administrator moved outside the Office of the Administrator to the staff office level. The offices of the chief engineer and chief scientist remained as staff positions in the Office of the Administrator.

In 1982, at the end of the Shuttle's developmental flights and the beginning of initial operations, the Office of Space Transportation Systems Operations and the Office of Space Transportation Systems Acquisitions merged into the Office of Space Flight. In 1984, the agency established the Office of Space Station, reflecting NASA's and the nation's commitment to the program. Also that year, NASA created the Office of Commercial Programs in response to federal legislation and the recommendation of an agency task force on the commercialization of space.

Following the *Challenger* accident in 1986 and at the recommendation of the Rogers Commission, NASA established an independent Office of Safety, Reliability, Maintainability, and Quality Assurance. The agency also established an Office of Exploration in 1987 as a focus for long-term goals.

Mission

NASA Headquarters was responsible for the overall planning, coordination, and control of NASA programs. Headquarters was composed of program offices, which planned and directed agency-wide research and development programs and management and administrative processes; staff offices, which provided agency-wide leadership in certain administrative and specialized areas; and the Office of the Associate Deputy Administrator. In addition, other offices with specific functions, such as the Office of the Chief Engineer, the Office of the Chief Scientist, and the Office of Policy, were established and disestablished over time. Each of these offices reported directly to the Administrator. Table 6–20 lists the Headquarters major organizations in 1979 and 1988.

Ames Research Center

Location

Ames Research Center is located at the south end of San Francisco Bay, approximately fifty-six kilometers southeast of San Francisco, California. It is adjacent to the U.S Naval Air Station at Moffett Field, California.

Director

Dale L. Compton, Acting (July 1989–December 1990)
William F. Ballhaus, Jr. (January 1984–February 1988;
    February 1989–July 1989)
Clarence A. Syvertson (April 1978–January 1984)
Clarence A. Syvertson, Acting (August 1977–April 1978)
Hans Mark (February 1969–August 1977)
H. Julian Allen (October 1965–February 1969)
Smith J. De France (October 1958–October 1965)

Deputy Director

Dale Compton (January 1985–July 1989)
Angelo Guastaferro (October 1980–January 1985)
A. Thomas Young (February 1979–February 1980)
Clarence A. Syvertson (February 1969–April 1978)

History

The National Advisory Committee for Aeronautics (NACA) founded what would become Ames Research Center as an aircraft research laboratory in 1940. The original name was the Moffett Field Laboratory. In

1944, the center was renamed the Ames Aeronautical Laboratory in honor of Joseph S. Ames, chair of NACA from 1927 to 1939, former president of the Johns Hopkins University, and a leading authority on aerodynamics. In 1958, Ames became part of NASA and was renamed Ames Research Center. In 1981, NASA merged Ames with the Dryden Flight Research Center. The two installations were then referred to as Ames-Moffett and Ames-Dryden.

Mission

Ames-Moffett's major program areas were computer science and applications, computational and experimental aerodynamics, flight simulation, flight research, hypersonic aircraft, rotorcraft and powered-lift technology, aeronautical and space human factors, life sciences, space sciences, solar system exploration, airborne science and applications, and infrared astronomy. The center also supported military programs, the Space Shuttle, and various civil aviation projects. In addition, it emphasized meeting the needs of the U.S. aerospace industry.

Dryden Flight Research Center

Location

Dryden Flight Research Center is located at Edwards, California, in the Mojave Desert, approximately 130 kilometers north of the Los Angeles metropolitan area. It was adjacent to Edwards Air Force Base and Rogers Dry Lake, a 168-square-kilometer natural surface for landing.

Director

Isaac T. Gillam IV (June 1978–September 1981)
Isaac T. Gillam IV, Acting (October 1977–June 1978)
David R. Scott (August 1977–October 1977)
David R. Scott, Acting (April 1975–August 1977)
Lee R. Scherer (October 1971–January 1975)
Paul F. Bikle (September 1959–May 1971)
Walter C. Williams (October 1958–August 1959)

Deputy Director

Angelo Guastaferro (October 1980–October 1981)
Robert P. Johannes (December 1979–October 1980)
John Boyd (January 1979–December 1979)
Isaac T. Gillam IV (August 1977–June 1978)
David R. Scott (August 1973–August 1977)
D.E. Beeler (April 1961–August 1973)

History

The U.S. Army originally used the Dryden area as a bombing and gunnery range before World War II. In July 1942, the Army established a formal air base near the town of Muroc, California. The first NACA contingent of engineers, technicians, and support staff arrived at Muroc from Langley Research Center in 1946 on temporary assignment. In early 1947, the contingent was made a permanent facility, known as the NACA Muroc Flight Test Unit, under Langley management. The group used Muroc as a test site when it designed and built a research aircraft to break the sound barrier.

In 1949, Muroc was renamed Edwards Air Force Base. Also that year, the name of the NACA facility was changed to the NACA High Speed Flight Research Station. It became an autonomous facility in 1954, reporting directly to NACA headquarters. In March 1976, the center became the Hugh L. Dryden Flight Research Center, in honor of Hugh L. Dryden, the internationally renowned aerodynamicist who had been the NACA's director from 1947 to 1957. In October 1981, the center's independent status was removed, and it was redesignated as Dryden Flight Research Facility under the administration of the Ames Research Center.

Mission

Ames-Dryden's mission was to research, develop, verify, and transfer advanced aeronautics, space, and related technologies. It was NASA's prime installation for aeronautical flight research. The facility was also actively involved in supporting the Space Shuttle program as a backup landing site and as a facility to test and validate design concepts and systems used in the development and operation of the orbiters. It participated in the approach and landing tests of the Space Shuttle orbiter *Enterprise* and supported Shuttle orbiter landings from space as well as processing for ferry flights to the launch site.

Goddard Space Flight Center

Location

Goddard Space Flight Center is located in Greenbelt, Maryland, sixteen kilometers northeast of Washington, D.C. In addition to its main site, until FY 1981, Goddard leased 620 acres of nearby land from the Department of Agriculture, where the Goddard Antenna Test Range, the Magnetic Test Facility, the Optical Tracking and Ground Plane Test Facility, the Bi-Propellant Test Facility, and the Network Test and Training Facility were located. In 1981, 544 acres were transferred to Goddard. The remaining land stayed with the Department of Agriculture.

Goddard also managed the Goddard Institute for Space Studies in New York City. Established in 1961, the institute conducted basic research in space and Earth sciences in support of Goddard programs by working cooperatively with New York area universities and research organizations. The program focused particularly on the study of global change, in particular long-range climate, biogeochemical cycles, and planetary atmospheres. Since October 1981, Goddard Space Flight Center's facilities included the Wallops Flight Facility on Wallops Island on the Eastern Shore of Virginia.

Director

John W. Townsend, Jr. (June 1987–June 30, 1990)
Noel W. Hinners (June 1982–June 1987)
Leslie H. Meredith, Acting (March 1982–June 1982)
A. Thomas Young (February 1980–March 1982)
Robert E. Smylie, Acting (June 1979–February 1980)
Robert Cooper (August 1976–June 1979)
John F. Clark (May 1966–August 1976)
Harry J. Goett (September 1959–July 1965)

Deputy Director

John J. Quann (September 1982–January 1988)
John McElroy (September 1980–September 1982)
Robert E. Smylie (December 1976–February 1980)
Donald P. Hearth (April 1970–September 1975)
Vacant (July 1968–April 1970)
John W. Townsend (July 1965–July 1968)

History

Goddard Space Flight Center was created on January 15, 1959, and was the first facility built for NASA. It was named in commemoration of Robert H. Goddard, the American pioneer in rocket research. The first 157 employees were from the Vanguard project, transferred from the Naval Research Laboratory in Washington, D.C.

From early in NASA's history, Goddard also managed the facilities of the Space Tracking and Data Network (STDN). Located around the world, the number of tracking stations first increased, decreased during the early days of the Shuttle program, and then decreased as the Tracking and Data Relay Satellite System (TDRSS) became operational. However, the STDN continued to support missions that could not be tracked by the TDRSS and provided additional Shuttle tracking support. See Chapter 4 for a description of the STDN.

Mission

Goddard Space Flight Center's mission was to expand knowledge of Earth and its environment, the solar system, and the universe through the development and use of near-Earth-orbiting spacecraft. It was responsible for supporting NASA's role in space and Earth sciences, conducting research and applying technology for sensors, instruments, and information systems, planning and executing spaceflight projects for scientific research, and tracking Earth satellites through a worldwide communications system.

Jet Propulsion Laboratory

Location

The Jet Propulsion Laboratory is located in Pasadena, California, approximately thirty-two kilometers northeast of Los Angeles.

Director

Lew Allen, Jr. (July 1982–December 31, 1990)
Bruce C. Murray (April 1976–June 1982)
William H. Pickering (October 1958–March 1976)

Deputy Director

Peter T. Lyman (September 1987–July 1992)
Robert J. Parks (January 1984–September 1987)
Charles H. Terhune, Jr. (1969–December 1983)

History

The Jet Propulsion Laboratory (JPL) was transferred from Army jurisdiction to NASA's control in December 1958. NASA's contractual arrangement with the California Institute of Technology for the performance of research and development at JPL dates from 1962. The NASA Management Office in Pasadena administered the contract. JPL is a government-owned facility.

Mission

JPL's primary mission was to explore the solar system with automated spacecraft. Its major programs involved exploring Earth and the solar system, managing the Deep Space Network for communications, data acquisition, mission control, and radio-science space study, and performing basic and applied scientific and engineering research.

Johnson Space Center

Location

Johnson Space Center is located at Clear Lake, near Houston, Texas. Additional facilities are located at Ellington Air Force Base, approximately eleven kilometers north of the main facility.

Director

Aaron Cohen (October 1986–August 20, 1993)
Jesse Moore (January 1986–October 1986)
Robert Goetz, Acting (January 1986—two weeks' duration)
Gerald Griffin (August 1982–January 1986)
Christopher C. Kraft, Jr. (January 1972–August 1982)
Robert R. Gilruth (November 1961–January 1972)

Deputy Director

Paul J. Weitz (December 1986–April 1994)
Robert C. Goetz (July 1983–October 1986)
Charles E. Charlesworth (August 1979–May 1983)
Sigurd A. Sjoberg (January 1972–May 1979)
Christopher C. Kraft, Jr. (November 1969–January 1972)
George S. Trimble (October 1967–September 1969)
George M. Low (February 1964–April 1967)
James C. Elms (November 1963–February 1964)

History

Johnson Space Center was established in September 1961 as NASA's Manned Spacecraft Center. It was NASA's primary center for the design, development, and testing of spacecraft and associated systems for human spaceflight, the selection and training of astronauts, the planning for conducting human spaceflight missions, and extensive participation in the medical, engineering, and scientific experiments carried aboard spaceflights. It was renamed Lyndon B. Johnson Space Center in February 1973.

The White Sands Test Facility, a component installation of Johnson Space Center, was established in 1962 at Las Cruces, New Mexico, for testing Apollo propulsion and power systems. It became the primary ground terminal for the Tracking and Data Relay Satellite System in 1983.

Mission

Johnson Space Center had program management responsibility for the Space Shuttle program. It also had a major responsibility for the development of the space station.

Kennedy Space Center

Location

John F. Kennedy Space Center is located on the east coast of Florida, immediately north and west of Cape Canaveral. It is approximately 241 kilometers south of Jacksonville and eighty kilometers east of Orlando.

Director

Forrest McCartney (October 1986–December 1991)
Thomas E. Utsman, Acting (July 1986–October 1986)
Richard G. Smith (September 1979–July 1986)
Lee R. Scherer (January 1975–September 1979)
Kurt H. Debus (March 1962–October 1964)

Deputy Director

Thomas E. Utsman (August 1985–January 1990)
Horace Lamberth (Acting) November 1984–August 1985)
George F. Page (July 1982–October 1984)
Gerald D. Griffin (July 1977–August 1981; on Headquarters assignment July 1980–May 1981)
Miles Ross (June 1970–May 1977)

History

The site of Kennedy Space Center, halfway between Miami and Jacksonville, Florida, had been used as a missile launching ground since the late 1940s. In 1951, it was used for test flights of the U.S. Army's Redstone intermediate-range ballistic missile. In January 1953, its name was changed from the Long-Range Proving Ground to the Missile Firing Laboratory. In July 1960, it became part of NASA's Marshall Space Flight Center's Launch Operations Directorate. The directorate was disbanded in March 1962.

The U.S. Congress approved the development of the strip of land on Florida's east coast called Cape Canaveral in 1961, shortly after President John F. Kennedy announced plans to fly American astronauts to the Moon. In July 1962, the site was established as a separate NASA installation and renamed the Launch Operations Center. NASA built the Atlantic Missile Range at Cape Canaveral, adjacent to the northern part of Merritt Island, where Kennedy Space Center was eventually located. Later, the Cape Canaveral peninsula became the Eastern Test Range, site of the Mercury and Gemini launches. NASA began acquiring land across the Banana River from Cape Canaveral in 1962.

President Lyndon Johnson renamed the facility the John F. Kennedy Space Center in November 1963, less than a week after the death of President Kennedy.

By 1967, Kennedy Space Center's Complex 39 was operational. The complex was strategically located next to a barge site and consisted of a variety of structures including a vehicle assembly building, processing facilities, press site, crawlerways to Complex 39 launch pads, and the Launch Control Center.

Twelve Saturn V/Apollo missions were launched from Kennedy between 1967 and 1972, and in 1973, the Skylab space station was placed into a high circular orbit, followed by three-member crews aboard Saturns later that year. The Saturn/Apollo era ended in 1975 with the launch of a Saturn IV/Apollo crew on a joint mission with the Soviet Union.

In 1979, a three-mile-long Shuttle Landing Facility and an Orbiter Processing Facility were built, and the Orbital Flight Test Program began at Kennedy Space Center. NASA launched the first Shuttle mission from Kennedy on April 12, 1981.

Mission

Kennedy Space Center had primary responsibility for ground turn-around and support operations, prelaunch checkout, and launch of the Space Shuttle and its payloads, including NASA's eventual space station. The center's responsibility also extended to the facilities and ground operations at Vandenberg Air Force Base in California and designated landing sites.

Langley Research Center

Location

Langley Research Center is located at Langley Field in Hampton, Virginia, approximately 241 kilometers southeast of Washington, D.C.

Director

Richard H. Petersen (November 1984–December 2, 1991)
Donald P. Hearth (September 1975–November 1984)
Edgar M. Cortright (May 1968–September 1975)
Floyd L. Thompson (May 1960–May 1968)
Henry J.E. Reid (October 1958–May 1960)

Deputy Director

Paul F. Holloway (February 1985–October 14, 1991)
Richard H. Petersen (July 1980–November 1984)
Oran W. Nicks (November 1970–July 1980)
Charles J. Donlan (November 1967–May 1968)

History

In 1916, the NACA selected a site near Hampton, Virginia, for Langley Field, its experimental air station. It was named after Samuel Pierpont Langley, the third secretary of the Smithsonian Institution and an aeronautical pioneer. Construction of the Langley Memorial Aeronautical Laboratory, the first national civil aeronautics laboratory began in 1917.

Until 1940, Langley was the only NACA laboratory. In 1948, the NACA changed the laboratory's name to the Langley Aeronautical Laboratory. When NASA was formed in 1958, it was renamed Langley Research Center.

In 1958, NASA selected Langley to manage Project Mercury, the first U.S. human spaceflight project. Heading the project was Langley's Space Task Group, a group of NASA employees that led the original seven astronauts through the initial phases of their spaceflight training. The group later expanded and moved on to become the Manned Spacecraft Center (later Johnson Space Center). Since 1959, Langley managed the Scout launch vehicle program, a four-stage solid fuel satellite system capable of launching a 175-kilogram satellite into an 800-kilometer orbit. The first Scout launch took place in 1960.

The center was also responsible for NASA's Lunar Orbiter project in the 1960s and the Viking project that orbited and landed spacecraft on Mars in 1976. In the late 1960s, environmental space science became a major research thrust at Langley. Its goal was to preserve Earth's ecological balance and prevent undesirable environmental conditions.

In the early stages of the Space Shuttle program, Langley conducted thousands of hours of wind tunnel testing on the orbiter. The center also was responsible for optimizing the design of the Shuttle's thermal protection system.

Langley also investigated technologies necessary for the design and operation of the space station. The Long Duration Exposure Facility, launched from the Shuttle in 1984, was conceived, designed, and developed at Langley (see Chapter 3, "Aeronautics and Space Research and Technology," for a description of this project).

In 1985, the U.S. Department of the Interior designated five Langley facilities as National Historic Landmarks: Variable-Density Tunnel (built in 1921), Full-Scale Tunnel (1930), Eight-Foot High-Speed Tunnel (1935), Rendezvous Docking Simulator (1963), and Lunar Landing Research Facility (1965). Langley has also received five Robert J. Collier Trophies: in 1929 for the low-drag engine cowling, in 1946 for de-icing research, in 1947 for supersonic flight research, in 1951 for the slotted throat transonic wind tunnel, and in 1954 for the transonic area rule.

Mission

Langley Research Center's primary mission was the research and development of advanced concepts and technology for future aircraft and spacecraft systems, with particular emphasis on environmental effects, performance, range, safety, and economy. Langley also had responsibility for systems analysis and independent evaluation and assessment of NASA programs prior to the commitment of major development funding. The center was the NASA expert for airborne systems, aerodynamics, mission and systems analysis, and hypersonic technologies.

Lewis Research Center

Location

Lewis Research Center is located approximately thirty-two kilometers southwest of Cleveland, Ohio, adjacent to the Cleveland Hopkins International Airport. Additional facilities were located at Plum Brook Station, about five kilometers south of Sandusky, Ohio.

Director

John M. Klineberg (May 1987–July 1, 1990)
John M. Klineberg, Acting (June 1986–May 1987)
Andrew J. Stofan (July 1982–June 1986)
John F. McCarthy (October 1978–July 1982)
Bernard Lubarsky, Acting (August 1977–October 1978)
Bruce T. Lundin (November 1969–August 1977)
Abe Silverstein (November 1961–October 1969)
Eugene J. Manganiello, Acting (January 1961–October 1961)
Edward R. Sharp (October 1958–December 1960)

Deputy Director

Lawrence J. Ross (December 1987–July 1, 1990)
John M. Klineberg (July 1979–May 1987)
Bernard Lubarsky (1974–July 1979)
Eugene J. Manganiello (December 1961–1972)

History

In 1940, the NACA selected Cleveland as the site of the new NACA aircraft engine research laboratory. Groundbreaking took place in 1941, and the NACA Aircraft Research Laboratory was officially dedicated in 1943. During World War II, the laboratory concentrated on investigating the problems of aircraft reciprocating, or piston, engines. Lewis engineers also contributed to solving engine cooling problems on the Super Fortress (B-29) bomber. Before the end of the war, the turbojet engine began to revolutionize the field of aircraft propulsion. The Altitude Wind Tunnel, completed in 1944, contributed to the early testing of American-built jet engines and started the center on what would become its major focus: jet propulsion studies.

In 1948, the name of the laboratory was changed to NACA Lewis Flight Propulsion Laboratory, in memory of George W. Lewis, the NACA's director of research from 1924 to 1947. The center broadened its scope of research to include turbojet engines, ramjets, and rockets and constructed new facilities, including two supersonic wind tunnels and the Propulsion Systems Laboratory.

The center became one of the original NASA centers when the agency was established in 1958. The Centaur rocket was one of the most important contributions Lewis made to the space program.

During the energy crisis of the 1970s, Lewis worked with the U.S. Department of Energy to investigate wind and solar power and to improve the fuel efficiency of automobile engines. Engineers also began work on the advanced turboprop engine. In 1987, a government-contractor team won the Collier Trophy for its work on the advanced turboprop project.

Mission

Lewis Research Center defined and developed new propulsion, power, and communications technologies for aeronautical and space applications. It managed a launch vehicle program and cooperated with other NASA activities and other research organizations in managing and supporting research and programs of national interest. The center managed research and technology development programs relevant to advanced aeronautical engines and complete propulsion systems for both civilian and military applications. Lewis also had responsibility for developing the electrical space power system for the life support systems and research experiments on the space station.

Marshall Space Flight Center

Location

George C. Marshall Space Flight Center is located at the U.S. Army's Redstone Arsenal in Huntsville, Alabama. Marshall also manages the Michoud Assembly Facility in New Orleans, Louisiana, and the Slidell Computer Complex in Slidell, Louisiana.

Director

James R. Thompson, Jr. (September 1986–July 6, 1989)
William R. Lucas (June 1974–July 1986)
Rocco A. Petrone (January 1973–March 1974)
Eberhard F.M. Rees (March 1970–January 1973)
Wernher von Braun (July 1960–January 1970)

Deputy Director

Thomas J. (Jack) Lee (December 1980–July 6, 1989)
Richard G. Smith (November 1974–August 1978)

History

Marshall Space Flight Center was established in 1960 and named in honor of General George C. Marshall. The center became active on July

1, 1960, with the transfer of buildings, land, space projects, property, and personnel from the U.S. Army Ballistic Missile Agency. Dr. Wernher von Braun was the center's first director.

In 1961, Marshall's Mercury-Redstone launch vehicle boosted America's first astronaut, Alan B. Shepard, on a suborbital flight. The center's first major program was the development of the Saturn rockets, the largest of which boosted a NASA crew to the Moon in 1969. Saturns also lifted Skylab and the Apollo spacecraft into Earth orbit for the linkup with the Russian Soyuz spacecraft in 1975.

Other Marshall projects included Pegasus in 1965, the Lunar Roving Vehicle in 1971 for transporting astronauts on the lunar surface, Skylab in 1973 (the first U.S. crewed orbiting space station, the three High Energy Astronomy Observatories to study stars and star-like objects, and the Hubble Space Telescope. Marshall was the lead center for NASA's Spacelab missions. It also had responsibility for the definition and preliminary design of pressurized common modules, environmental control, life support, and propulsive systems, and other elements of the space station.

A Marshall-developed propulsion system launched the Space Shuttle. Marshall provided the Shuttle orbiter's engines, the external tank that carried liquid hydrogen and liquid oxygen for those engines, and the solid rocket boosters that assisted in lifting the orbiter from the launch pad.

Mission

Marshall Space Flight Center was a management, scientific, and engineering center and emphasized multiple projects involving scientific investigation and application of space technology to the solution of problems on Earth.

National Space Technology Laboratories/Stennis Space Center

Location

The National Space Technology Laboratories/Stennis Space Center is located approximately eighty-eight kilometers northeast of New Orleans in Bay St. Louis, Mississippi.

Director

Jerry I. Hlass (September 1, 1976–January 23, 1989)[2]
Jackson M. Balch (June 1974–August 1976)

---

[2]Prior to the 1988 renaming of the National Space Technology Laboratories to Stennis Space Center, the director and deputy director held the titles of "manager" and "deputy manager," respectively. Therefore, Hlass was "manager" until May 20, 1988, when he became "director."

Deputy Director

Roy Estess (August 1980–January 23, 1989)
Harry Auter (1963–February 1979)

History

Stennis Space Center began as Mississippi Test Operations in October 1961 when the federal government selected the area in Hancock County, Mississippi, for the site of a static test facility for launch vehicles to be used in the Apollo lunar landing program. The center's name was changed to the Mississippi Test Facility in 1965. It received independent NASA installation status in 1974 and became the National Space Technology Laboratories. In May 1988, it was renamed the John C. Stennis Space Center in honor of U.S. Senator John C. Stennis, a staunch supporter of the space program.

The center has evolved into a multidisciplinary facility made up of NASA and twenty-two other resident agencies engaged in space and environmental programs and the national defense. The U.S. Navy was the center's largest resident agency. During the early years, the center flight-certified all first and second stages of the Saturn V rocket for the Apollo program. When the Space Shuttle program got under way, the center flight-certified all the engines used to boost the Shuttle into low-Earth orbit.

Mission

The primary mission of Stennis Space Center was to provide the capacity to test rocket propulsion engines, systems, and vehicles. Its major test program was the development and flight certification of the Space Shuttle Main Engine, which powered the Shuttle during its first minutes of flight.

Wallops Flight Center

Location

Wallops Flight Center is located on Wallops Island, off the Delmarva Peninsula in Virginia, and on additional nearby property on the Virginia mainland. It is approximately eighty kilometers southeast of Salisbury, Maryland.

Director

Abraham D. Spinak, Acting (August 1981–November 1981)
Robert Kreiger (June 1948–August 1981) (actually retired in February 1980 but stayed on at the request of the NASA Administrator to help with the consolidation with Goddard)

Associate Director

Abraham D. Spinak (August 1966–November 1981)

History

The NACA established Wallops Flight Center in 1945 when it authorized Langley Research Center to proceed with the development of Wallops Island as a site for research with rocket-propelled models and as a center for aerodynamic research. It is one of the oldest launch sites in the world. Before NASA's establishment, Wallops helped provide the foundation for aerodynamic and heat transfer research through the establishment of a high-speed aeronautics launch site that used rockets to propel aircraft models. The facility allowed researchers to overcome the limited capabilities offered by the wind tunnels of the day.

With the establishment of NASA in 1958, research conducted at Wallops included developing components for the human space program, including capsule escape techniques, maximum pressure tests, and recovery systems. Wallops provided range support for research in reentry and life support systems, Scout launch vehicles, and mobile research projects. It also expanded its scope to include Earth studies of ocean processes and used the Wallops Research Airport for runway surface and aircraft noise reduction studies.

Wallops met its requirements for propulsion with relatively small solid rockets staged in various ways to meet the needs of the research task. The largest and most sophisticated of the launch vehicles was the Scout four-stage solid-fuel vehicle that could launch small scientific satellites, space probes, and reentry missions.

The center was consolidated with Goddard Space Flight Center on October 19, 1981. At that time, it became Wallops Flight Facility, with the designation Suborbital Projects and Operations Directorate.

Mission

The Wallops mission included managing and implementing NASA's sounding rocket and balloon programs, conducting observational Earth sciences studies, providing flight services for scientific investigations, and operating the Wallops Test Range and Orbital Tracking Station.

Table 6–1. NASA Centers' Major Programs and Mission Areas

Installation	Major Programs	Mission Areas
Ames Research Center Moffett Field, California	Aeronautics/Space Science and Applications	Computational fluid dynamics Aircraft flight simulation testing Astrophysical and biological sciences Rotorcraft technology
Dryden Flight Research Center Edwards, California	Aeronautics/Space Transportation Systems	Aeronautics flight testing Shuttle landing site
Goddard Space Flight Center Greenbelt, Maryland	Space Science and Applications Space Tracking and Data Systems	Atmospheric and Earth science Physics astronomy Near-Earth space tracking
Jet Propulsion Laboratory (JPL) Pasadena, California	Space Science and Exploration Space Tracking and Data Systems	Planetary exploration Space sciences Deep Space Network
Johnson Space Center Houston, Texas	Space Transportation Systems	Shuttle program management Space station program management Integration and thermal systems Flight and mission operations Space medicine and crew systems
Kennedy Space Center Cape Canaveral, Florida	Space Transportation Systems	Cargo processing and checkout Launch operations Primary STS landing site
Langley Research Center Hampton, Virginia	Aeronautics/Space Technology	Aerodynamics Materials and structures Guidance and control Environmental quality
Lewis Research Center Cleveland, Ohio	Aeronautics/Space Technology	Power and propulsion technology Space station power systems Space data and communications
Marshall Space Flight Center Huntsville, Alabama	Space Transportation Systems	Propulsion systems development Space station propulsion/ common modules
Stennis Space Center Bay St. Louis, Mississippi	Space Transportation Systems Space Applications	Liquid rocket engine and system testing Remote sensing
Wallops Flight Facility Wallops Island, Virginia	Space Science	Scout expendable launch vehicle launches Space science research and development Space tracking

FACILITIES AND INSTALLATIONS   405

Table 6–2. Property: In-House and Contractor-Held (FY 1979–1988)
(at end of fiscal year; money amount in thousands)

Category	1979	1980	1981	1982	1983	1984	1985	1986	1987	1988
1. Total real property value	2,938,901	3,084,767	3,155,917	3,242,295	3,343,279	3,414,309	3,603,821	3,757,100	3,888,427	4,050,072
Percentage change	3.7	5.0	2.3	2.7	3.1	2.1	5.6	4.3	3.5	4.2
Land value	115,189	115,257	115,304	116,476	117,931	118,248	117,199	117,222	117,217	117,210
Percentage change	-1.2	0.06	a	1.0	1.2	0.3	-0.9	a	a	a
Building value	1,617,211	1,660,624	1,680,692	1,723,724	1,791,689	1,819,173	1,967,307	2,049,784	2,161,979	2,281,595
Percentage change	2.0	2.7	1.2	2.6	3.9	1.5	8.1	4.2	5.5	5.5
Other structures and facilities' value	1,205,831	1,307,295	1,358,036	1,400,152	1,431,697	1,474,873	1,517,286	1,567,960	1,607,096	1,649,097
Percentage change	6.4	8.4	3.9	3.1	2.3	3.0	2.9	3.3	2.5	2.6
Leasehold improvements value	670	1,591	1,885	1,943	1,962	2,015	2,029	2,134	2,135	2,170
Percentage change	-7.2	137.4	18.4	3.1	1.0	2.7	0.7	5.2	0.05	1.6
2. Capitalized equipment value	3,210,678	3,431,887	3,496,070	4,111,986	4,291,371	2,969,391	3,201,147	3,537,506	3,167,448	3,187,004
Percentage change	8.6	6.9	1.9	17.6	4.4	-31.0	7.8	10.5	-10.5	0.6
3. Fixed assets-in-progress value	291,727	304,528	380,400	381,455	435,965	495,795	379,769	426,517	439,513	467,673
Percentage change	29.1	4.4	24.9	0.3	14.3	13.7	-23.4	12.3	3.1	6.4
4. Total investment value (1+2+3)	6,441,306	6,821,182	7,032,387	7,735,736	8,070,615	6,879,495	7,184,737	7,721,123	7,495,488	7,704,749
Percentage change	7.1	5.9	3.1	10.0	4.3	-14.8	4.4	7.5	-2.9	2.8

Table 6–2 continued

Category	1979	1980	1981	1982	1983	1984	1985	1986	1987	1988
5. Number of acres of land	133,892.4	133,754.6	134,300.7	136,680.4	136,738.0	136,491.5	134,939.7	134,939.9	134,820.5	134,827.8
Percentage change	-1.6	0.1	0.4	1.8	*	-0.2	-1.1	*	0.09	*
Number of buildings	2,429	2,464	2,499	2,550	2,600	2,760	2,644	2,660	2,679	2,653
Percentage change	0.2	1.4	1.4	2.0	2.0	6.2	-4.2	0.6	0.7	-0.1
Number of square feet of buildings	32,571,723	32,679,347	32,876,733	33,096,988	33,524,976	33,886,388	34,065,483	34,582,023	35,601,147	35,736,863
Percentage change	-0.2	0.3	0.6	0.7	1.3	1.1	0.5	1.5	2.9	0.4

a Less than 0.05 percent.

Source: NASA Budget Estimates, 1979–1988, Summary Tables, "Recorded Value of Capital Type Property In-House and Contractor-Held," and "NASA Locations by Accounting Installations," Annual Reports, 1979–1988, Office of Management Systems and Facilities, Facilities Engineering Division.

Table 6–3. *Value of Real Property Components as a Percentage of Total Real Property: In-House and Contractor-Held (FY 1979–1988) (at end of fiscal year; total real property value in thousands)*

Component	1979	1980	1981	1982	1983	1984	1985	1986	1987	1988
Land	3.9	3.7	3.7	3.6	3.5	3.5	3.3	3.0	3.0	2.9
Buildings	55.0	53.8	53.2	53.1	53.6	53.3	54.6	55.0	55.6	56.3
Other structures and facilities	41.0	42.4	43.0	43.2	42.9	43.2	42.1	42.3	41.3	40.7
Total real property value	2,938,901	3,084,767	3,155,917	3,242,295	3,343,279	3,414,309	3,603,821	3,757,100	3,888,427	4,050,072

Note: Percentages may not add up to 100 percent due to rounding.
Source: Table 6–1.

Table 6–4. *NASA Facilities Total Investment Value (FY 1979): In-House and Contractor-Held (at end of fiscal year; in thousands of dollars)*

Facility	Total Real Property Value [a]	Capitalized Equipment	Fixed Assets-in-Progress	Total Investment	Percentage of NASA Total Investment
NASA Headquarters	0	17,877	6,644	25,451	3.8
Office of Space Flight					
Kennedy Space Center	751,095	967,416	117,914	1,836,425	28.5
Johnson Space Center	266,907	442,347	24,623	733,877	11.4
Marshall Space Flight Center	338,283	396,707	3,522	738,512	11.5
National Space Technology Laboratories	280,984	29,296	0	310,280	4.8
Total	1,637,269	1,835,766	146,059	3,619,094	56.1
Office of Aeronautics and Space Technology					
Ames Research Center	225,781	165,781	56,247	447,809	6.9
Dryden Flight Research Center	21,251	63,564	3,095	87,910	1.7
Langley Research Center	365,178	183,710	17,665	566,553	8.8
Lewis Research Center	287,781	143,194	17,971	449,036	6.9
Total	899,991	556,249	94,978	1,551,308	24.1
Office of Space Science and Applications					
Goddard Space Flight Center	161,591	511,020	19,555	692,166	10.7
Jet Propulsion Laboratory	159,171	233,821	19,408	412,400	6.4
Wallops Flight Center (Facility)	80,789	56,015	5,083	141,887	2.2
Total	401,551	800,856	44,046	1,246,453	19.4
NASA Total	2,938,901	3,210	291,727	6,441,306	100.0

[a] Total Real Property Value includes land, buildings, real property and other structures and facilities, and leasehold improvements.

Source: NASA Budget Estimates, 1979–1988, Summary Tables, "Recorded Value of Capital Type Property In-House and Contractor-Held."

Table 6-4A. NASA Facilities Total Investment Value (FY 1980–1982): In-House and Contractor-Held
(at end of fiscal year; in thousands of dollars)

Facility	Total Real Property Value			Capitalized Equipment		
	1980	1981	1982	1980	1981	1982
NASA Headquarters	0	4	0	16,710	22,358	17,496
Office of Space Flight						
Kennedy Space Center	827,579	841,837	881,176	934,382	935,986	1,549,138
Johnson Space Center	270,671	271,917	275,858	608,577	614,839	602,774
Marshall Space Flight Center	354,983	355,526	363,635	421,361	430,850	418,113
National Space Technology Laboratories	282,618	281,554	280,956	24,134	25,475	28,680
Total	1,735,851	1,750,834	1,801,625	3,619,094	2,007,150	2,598,705
Office of Aeronautics and Space Technology						
Ames Research Center	227,961	251,698	235,144	198,403	300,338	253,841
Dryden Flight Research Center	22,348	a	22,723	57,914	b	69,779
Langley Research Center	394,545	425,093	437,162	184,587	189,088	193,620
Lewis Research Center	294,474	295,281	310,466	153,298	153,463	160,603
Total	939,328	972,072	1,005,495	594,202	642,889	677,843
Office of Space Science and Applications						
Goddard Space Flight Center c	165,944	267,030	265,852	528,745	565,964	556,690
Jet Propulsion Laboratory	161,265	165,977	169,323	246,751	257,709	261,252
Wallops Flight Center (Facility) d	82,379	—	—	57,025	—	—
Total	409,588	433,007	435,175	832,521	823,673	582,815
NASA Total	3,084,767	3,155,917	3,242,295	3,415,177	3,496,070	4,111,986

a Included in amounts for Ames Research Center.
b Included in amounts for Ames Research Center.
c Beginning in FY 1981, Goddard Space Flight Center includes Wallops Flight Center amounts.
d Beginning in FY 1981, amounts for Wallops Flight Center were included with Goddard Space Flight Center.

Table 6–4A continued

Facility	Fixed Assets-in-Progress			Total Investment			Percentage of NASA Total Investment [a]		
	1980	1981	1982	1980	1981	1982	1980	1981	1982
NASA Headquarters	6,644	6,640	0	23,354	29,002	17,496	0.3	0.4	0.2
Office of Space Flight									
Kennedy Space Center	72,612	88,983	72,092	1,834,573	1,866,806	2,502,406	26.9	26.5	32.3
Johnson Space Center	25,731	31,592	31,208	904,979	918,348	909,840	13.3	13.1	11.8
Marshall Space Flight Center	2,195	1,504	718	778,539	787,880	782,466	11.4	11.2	10.1
National Space Technology Laboratories	0	0	0	306,752	307,029	309,636	4.5	4.4	4.0
Total	100,538	122,079	104,018	3,824,843	3,880,063	4,504,348	56.1	55.2	58.2
Office of Aeronautics and Space Technology									
Ames Research Center	91,918	145,218	164,695	518,282	697,254	653,680	7.6	9.9	8.4
Dryden Flight Research Center	3,040	—	4,955	83,302	—	97,457	1.2	—	1.3
Langley Research Center	29,481	27,697	44,309	608,613	641,878	675,091	8.9	9.1	8.7
Lewis Research Center	15,487	16,080	14,545	463,259	465,605	485,614	6.8	6.7	6.3
Total	139,926	188,995	228,504	1,673,456	1,804,737	1,911,842	24.5	25.7	24.7
Office of Space Science and Applications									
Goddard Space Flight Center	29,529	31,977	26,624	724,218	864,971	849,166	10.6	12.3	11.0
Jet Propulsion Laboratory	21,933	29,929	22,309	429,949	453,615	452,884	6.3	6.4	5.9
Wallops Flight Center (Facility)	5,958	—	—	145,362	—	—	2.1	—	—
Total	57,420	61,906	48,933	1,299,529	1,318,586	1,302,050	19.1	18.8	16.8
NASA Total	304,528	380,400	381,455	6,821,182	7,032,387	7,737,736	99.7	100.0	99.7

[a] Percentage totals may not add up to 100 percent due to rounding.

Table 6-4B. *NASA Facilities Total Investment Value (FY 1983–1985): In-House and Contractor-Held (at end of fiscal year; in thousands of dollars)*

Facility	Total Real Property Value			Capitalized Equipment		
	1983	1984	1985	1983	1984	1985
NASA Headquarters	0	0	0	27,749	28,896	31,144
Office of Space Flight						
Kennedy Space Center	905,163	947,111	976,524	1,576,544	423,636	605,005
Johnson Space Center	289,235	292,163	293,822	649,879	551,269	525,849
Marshall Space Flight Center	390,082	395,769	406,201	416,751	397,902	430,488
National Space Technology Laboratories	282,239	294,104	291,698	30,306	31,252	33,356
Total	1,866,719	1,929,147	1,968,245	2,673,480	1,404,059	1,594,698
Office of Aeronautics and Space Technology						
Ames Research Center a	263,644	265,618	367,280	348,604	343,836	355,441
Langley Research Center	447,534	455,664	489,258	217,078	216,123	222,046
Lewis Research Center	312,487	315,445	218,181	171,350	159,139	182,380
Total	1,023,665	1,036,727	1,074,719	737,032	719,098	759,867
Office of Space Science and Applications						
Goddard Space Flight Center b	272,286	275,454	268,599	583,843	533,488	248,239
Jet Propulsion Laboratory	180,609	182,981	194,484	269,267	283,850	300,364
Total	452,895	458,435	463,083	853,110	817,338	548,603
NASA Total	3,343,279	3,414,309	3,603,821	4,291,371	2,969,391	3,201,147

a Includes Dryden.
b Includes Wallops.

Table 6–4B continued

Facility	Fixed Assets-in-Progress			Total Investment			Percentage of NASA Total Investment [a]		
	1983	1984	1985	1983	1984	1985	1983	1984	1985
NASA Headquarters	0	0	0	27,749	28,896	31,144	0.3	0.4	0.4
Office of Space Flight									
Kennedy Space Center	81,369	72,585	95,994	2,563,076	1,443,332	1,677,523	31.8	20.1	23.3
Johnson Space Center	42,687	47,142	69,208	981,801	890,574	828,157	12.2	12.9	11.5
Marshall Space Flight Center	837	0	0	807,670	793,671	836,689	10.0	11.5	11.6
National Space Technology Laboratories	0	0	0	312,545	315,356	325,054	3.9	4.6	4.5
Total	124,893	119,727	165,202	4,665,092	3,442,933	3,667,423	57.8	50.0	51.0
Office of Aeronautics and Space Technology									
Ames Research Center	184,611	206,711	132,730	796,859	816,165	855,451	9.9	11.9	11.9
Langley Research Center	43,879	61,073	53,946	708,491	732,060	765,250	8.8	10.6	10.7
Lewis Research Center	21,574	23,298	25,479	505,411	497,882	523,814	6.3	7.2	7.3
Total	250,064	291,082	212,155	2,010,761	2,046,107	2,144,515	24.9	29.7	29.8
Office of Space Science and Applications									
Goddard Space Flight Center	24,942	31,756	37,892	881,071	840,698	821,565	10.9	12.2	11.4
Jet Propulsion Laboratory	36,066	53,230	25,242	485,942	520,061	520,090	6.0	7.6	7.2
Total	61,008	84,986	63,134	1,367,013	1,360,759	1,341,655	16.9	19.8	18.7
NASA Total	435,965	495,795	379,769	8,070,615	6,879,495	7,184,737	99.6	99.5	99.5

[a] Percentage totals may not add up to 100 percent due to rounding.

Table 6–4C. *NASA Facilities Total Investment Value (FY 1986–1988): In-House and Contractor-Held*
(at end of fiscal year; in thousands of dollars)

Facility	Total Real Property Value			Capitalized Equipment		
	1986	1987	1988	1986	1987	1988
NASA Headquarters	0	0	0	39,525	13,541	39,378
Office of Space Flight						
Kennedy Space Center	1,049,146	1,108,018	1,116,251	765,008	708,750	623,870
Johnson Space Center	298,145	311,809	316,545	470,726	449,727	316,545
Marshall Space Flight Center	427,014	447,152	459,030	484,959	465,347	459,030
National Space Technology Laboratories [a]	300,882	302,790	321,360	349,923	27,719	29,876
Total	2,075,187	2,169,769	2,213,186	1,755,616	1,651,543	1,429,321
Office of Aeronautics and Space Technology						
Ames Research Center	375,829	387,298	456,244	371,797	340,456	348,975
Langley Research Center	514,025	509,576	552,379	306,845	200,700	205,899
Lewis Research Center	320,337	325,196	326,670	189,082	150,367	163,617
Total	1,210,191	1,222,070	1,335,293	867,724	691,523	718,491
Office of Space Science and Applications						
Goddard Space Flight Center	270,473	282,047	286,517	533,178	427,030	458,185
Jet Propulsion Laboratory	201,249	214,541	215,076	339,463	383,811	376,271
Total	471,722	496,588	501,593	872,641	810,841	834,456
NASA Total	3,757,100	3,888,427	4,050,072	3,537,506	3,167,448	3,187,004

[a] Name changed to John C. Stennis Space Center in 1988.

Table 6-4C continued

Facility	Fixed Assets-in-Progress			Total Investment			Percentage of NASA Total Investment [a]		
	1986	1987	1988	1986	1987	1988	1986	1987	1988
NASA Headquarters	0	0	0	39,525	13,541	39,378	0.5	0.2	0.5
Office of Space Flight									
Kennedy Space Center	79,980	38,007	23,797	1,894,134	1,854,775	1,763,918	24.5	24.7	22.9
Johnson Space Center	7,530	4,824	16,706	776,401	766,360	780,400	10.1	10.2	10.1
Marshall Space Flight Center	0	973	2,690	911,973	913,472	955,504	11.9	12.2	12.4
National Space Technology Laboratories	0	0	0	337,805	330,509	351,234	4.4	4.1	4.6
Total	87,510	43,804	43,193	3,920,313	3,865,116	3,851,056	50.8	51.6	50.0
Office of Aeronautics and Space Technology									
Ames Research Center	179,610	187,008	120,039	927,244	914,762	975,258	12.0	12.2	12.7
Langley Research Center	40,230	53,600	27,780	861,100	763,876	786,058	11.2	10.2	10.2
Lewis Research Center	27,196	37,947	54,510	536,615	513,510	544,797	6.9	6.9	7.0
Total	247,036	278,555	202,329	2,324,959	2,192,148	2,306,113	30.1	29.2	29.9
Office of Space Science and Applications									
Goddard Space Flight Center	41,774	44,449	75,318	845,425	753,526	820,020	10.9	10.1	10.6
Jet Propulsion Laboratory	50,189	72,805	96,833	590,901	671,157	688,180	7.7	9.0	8.9
Total	91,963	117,254	172,151	1,436,326	1,424,683	1,508,200	18.5	19.1	19.6
NASA Total	426,517	439,613	467,673	7,721,123	7,495,488	7,704,749	99.4	99.9	99.5

[a] Percentage totals may not add up to 100 percent due to rounding.

Table 6-5. Land Owned by Installation and Fiscal Year in Acres: In-House and Contractor-Held
(at end of fiscal year)

Installation	1979	1980	1981	1982	1983	1984	1985	1986	1987	1988
Ames Research Center	429.9	429.9	432.0	432.0	432.0	432.0	432.0	432.0	432.0	432.0
Dryden Flight Research Center	—	—	—	—	—	—	—	—	—	—
Goddard Space Flight Center	12,002.7	11,864.9	12,408.9 a	12,636.5	12,660.6	12,410.6	12,410.6	12,410.6	12,410.6	12,410.6
Jet Propulsion Laboratory	155.8	155.8	155.8	155.8	155.8	155.8	155.8	155.8	155.8	155.8
Johnson Space Center	1,785.9	1,785.9	1,785.9	1,785.9	1,819.4	1,822.9	1,822.9	1,823.1	1,823.1	1,821.0
Kennedy Space Center	82,943.0	82,943.0	82,943.0	82,943.0	82,943.0	82,943.0	82,943.0	82,943.0	82,943.0	82,943.0
Langley Research Center	897.6	897.6	897.6	897.6	897.6	897.6	897.6	897.6	787.6 b	787.6
Lewis Research Center	6,204.6	6,204.6	6,204.6 c	8,356.7 d	8,356.7	8,356.7	6,804.8 e	6,804.8	6,804.8	6,804.8
Marshall Space Flight Center	409.5	409.5	409.5	409.5	409.5	409.5	409.5	409.5	409.5	409.5
National Space Technology Laboratories	20,642.2	20,642.2	20,642.2	20,642.2	20,642.2	20,642.2	20,642.2	20,642.2	20,642.2	20,642.2
Wallops Flight Center (Facility)	6,165.8	6,165.8	6,165.8	6,165.8	6,165.8	6,165.8	6,165.8	6,165.9	6,165.9	6,165.9
Total f	133,892.4	133,754.6	134,300.7	136,680.4	136,738.0	136,491.5	134,939.7	134,939.9	134,829.9	134,827.8

a Increase in acreage reflects the transfer of land held under permit from the Beltsville Agricultural Research Center to Goddard.
b Decrease in acreage reflects condemnation of property.
c This does not include 2,152.15 acres of land declared excess to the General Services Administration (GSA).
d Increase in acreage reflects the reinstatement of excess land to Lewis.
e Decrease in acreage reflects the transfer of excess land to GSA.
f Total includes Michoud Assembly Facility (New Orleans, Louisiana), Slidell Computer Complex (Slidell Louisiana), and White Sands Test Facility (Las Cruces, New Mexico).

Source: "NASA Locations by Accounting Installations," Annual Reports, 1979–1988, Office of Management Systems and Facilities, Facilities Engineering Division.

Table 6–6. *Number of Buildings Owned by Installation and Fiscal Year: In-House and Contractor-Held (at end of fiscal year)* [a]

Installation	1979	1980	1981	1982	1983	1984	1985	1986	1987	1988
Ames Research Center	144	147	145	145	145	146	179	181	182	156
Dryden Flight Research Center	63	63	65	65	71	81	76	66	66	66
Goddard Space Flight Center [b]	260	261	264	242	255	263	247	228	214	210
Jet Propulsion Laboratory [c]	341	341	341	344	337	338	353	335	333	327
Johnson Space Center	182	182	190	190	202	204	204	204	217	212
Kennedy Space Center	350	353	442	500	532	678	514	556	572	566
Langley Research Center	151	151	153	159	159	159	165	166	171	170
Lewis Research Center	250	250	250	259	258	256	256	256	256	258
Marshall Space Flight Center	176	168	166	166	166	165	167	174	175	194
National Space Technology Laboratories	120	120	115	111	104	106	111	117	118	120
Wallops Flight Center (Facility)	282	284	258	257	258	261	259	259	256	253
Total [d]	2,429	2,426	2,499	2,550	2,600	2,760	2,644	2,660	2,679	2,653

[a] Changes in the number of buildings frequently reflect the erection or dismantling of trailers or other temporary buildings.
[b] This includes tracking stations.
[c] This includes tracking stations.
[d] Total includes Michoud Assembly Facility, Slidell Computer Complex, and White Sands Test Facility.

Source: "NASA Locations by Accounting Installations," Annual Reports, 1979–1988, Office of Management Systems and Facilities, Facilities Engineering Division.

Table 6–7. Number of Square Feet of Buildings Owned by Installation and Fiscal Year: In-House and Contractor-Held
(at end of fiscal year)

Installation	1979	1980	1981	1982	1983	1984	1985	1986	1987	1988
Ames Research Center	2,224,372	2,143,077	2,275,686	2,291,262	2,290,812	2,305,866	2,334,684	2,371,954	2,391,559	2,683,184
Dryden Flight Research Center	459,447	504,856	501,578	501,778	508,198	521,548	450,463	549,620	549,620	549,620
Goddard Space Flight Center	2,709,230	2,694,672	2,787,364	2,881,871	2,923,577	2,907,556	2,873,898	2,846,842	3,219,489	3,219,495
Jet Propulsion Laboratory	2,014,712	2,014,112	2,017,972	2,027,990	2,037,004	2,041,774	2,078,225	2,056,581	2,087,472	2,075,528
Johnson Space Center	4,534,939	4,551,432	4,555,826	4,561,147	4,771,805	4,793,677	4,794,341	4,817,208	4,886,903	4,892,190
Kennedy Space Center	5,337,276	5,312,570	5,375,291	5,431,336	5,507,621	5,766,191	5,875,980	6,041,808	6,640,212	6,720,200
Langley Research Center	2,140,135	2,085,380	2,098,203	2,068,679	2,066,812	2,098,215	2,110,851	2,141,362	2,160,326	2,180,360
Lewis Research Center	3,105,064	3,176,921	3,178,851	3,203,085	3,205,185	3,212,567	3,213,383	3,213,383	3,213,383	3,215,691
Marshall Space Flight Center	3,823,813	3,814,283	3,814,817	3,820,069	3,820,069	3,819,288	3,820,375	3,839,696	3,756,193	3,783,472
National Space Technology Laboratories	1,178,177	1,212,226	1,220,982	1,247,031	1,250,970	1,272,435	1,361,392	1,456,829	1,458,841	1,546,685
Wallops Flight Center (Facility)	1,057,344	1,064,064	1,068,312	1,058,045	1,083,545	1,087,893	1,091,268	1,094,093	1,083,382	1,083,362
Total a	32,571,723	32,679,347	32,876,733	33,096,988	33,524,976	33,886,388	34,065,483	34,582,023	35,601,147	35,736,863

a Total includes Michoud Assembly Facility, Slidell Computer Complex, and White Sands Test Facility.
Source: "NASA Locations by Accounting Installations," Annual Reports, 1979–1988, Office of Management Systems and Facilities, Facilities Engineering Division.

FACILITIES AND INSTALLATIONS 417

Table 6–8. Total Real Property Value by Installation and Fiscal Year: In-House and Contractor-Held
(at end of fiscal year; in thousands of dollars)

Installation	1979	1980	1981	1982	1983	1984	1985	1986	1987	1988
Headquarters	0	0	4	0	0	0	0	0	0	0
Ames Research Center	225,781	227,961	251,698	235,144	263,644	265,618	367,280	375,029	387,298	456,244
Dryden Flight Research Center	21,251	22,348	a	22,723	b					
Goddard Space Flight Center	161,591	165,944	267,030	265,852	272,286	275,454	268,599	270,473	282,047	286,517
Jet Propulsion Laboratory	159,171	161,265	165,977	169,323	180,609	182,981	194,484	201,249	214,541	215,076
Johnson Space Center	266,907	270,671	271,917	275,858	289,235	292,163	293,822	298,145	311,809	316,545
Kennedy Space Center	751,095	827,579	841,837	881,176	905,163	947,111	976,524	1,049,146	1,103,018	1,116,251
Langley Research Center	365,178	394,545	425,093	437,162	447,534	455,664	489,258	514,025	509,576	552,379
Lewis Research Center	287,871	294,474	295,281	310,466	312,487	315,445	315,955	320,337	325,196	326,670
Marshall Space Flight Center	338,283	354,983	355,526	363,635	390,082	395,769	395,153	427,014	447,152	459,030
National Space Technology Laboratories	280,984	282,618	281,554	280,956	282,239	284,104	291,698	300,882	302,790	321,360
Wallops Flight Center (Facility)	80,789	82,379	c							
Total	2,938,901	3,084,767	3,155,917	3,242,295	3,343,279	3,414,309	3,603,821	3,757,100	3,888,427	4,050,072

a Included with Ames Research Center.
b Included with Ames through 1988.
c Included with Goddard Space Flight Center through 1988.

Source: Table 6–4.

Table 6–9. Land Value by Installation and Fiscal Year: In-House and Contractor-Held
(at end of fiscal year; in thousands of dollars)

Installation	1979	1980	1981	1982	1983	1984	1985	1986	1987	1988
Headquarters	0	0	0	0	0	0	0	0	0	0
Ames Research Center	2,928	2,928	2,928	2,928	2,929	2,929	2,929	2,928	2,929	2,929
Dryden Flight Research Center	0	0	a	0	b					
Goddard Space Flight Center	1,706	1,685	3,111	2,862	2,860	2,860	2,840	2,857	2,857	2,856
Jet Propulsion Laboratory	1,188	1,188	1,188	1,188	1,188	1,188	1,188	1,188	1,188	1,188
Johnson Space Center	9,107	9,107	9,115	9,115	10,571	10,888	10,889	10,889	10,889	10,883
Kennedy Space Center	71,345	71,345	71,345	71,345	71,345	71,345	71,345	71,345	71,345	71,345
Langley Research Center	162	162	162	162	162	162	162	162	156	156
Lewis Research Center	2,230	2,230	2,230	3,651	3,651	3,651	2,621	2,621	2,621	2,621
Marshall Space Flight Center	7,157	7,160	7,164	7,164	7,164	7,164	7,164	7,171	7,171	7,171
National Space Technology Laboratories	18,061	18,061	18,061	18,061	18,061	18,061	18,061	18,061	18,061	18,061
Wallops Flight Center (Facility)	1,305	1,391	c							
Total	115,189	115,257	115,304	116,476	117,931	118,248	117,199	117,222	117,217	117,210

a Included with Ames Research Center.
b Included with Ames through 1988.
c Included with Goddard Space Flight Center through 1988.

Source: NASA Budget Estimates, 1979–1988, Summary Tables.

Table 6–10. *Building Value by Installation and Fiscal Year: In-House and Contractor-Held*
(at end of fiscal year; in thousands of dollars)

Installation	1979	1980	1981	1982	1983	1984	1985	1986	1987	1988
Headquarters	0	0	0	0	0	0	0	0	0	0
Ames Research Center	213,479	214,687	231,128	220,584	237,965	239,847	338,495	345,008	357,016	424,012
Dryden Flight Research Center	14,447	15,387	—	15,534	—	—	—	—	—	—
Goddard Space Flight Center	101,711	105,220	143,105	145,696	149,811	119,745	153,875	156,069	167,649	116,424
Jet Propulsion Laboratory	89,871	90,916	94,968	97,311	102,760	104,266	106,548	113,595	121,924	122,092
Johnson Space Center	195,875	199,120	198,397	201,533	210,687	212,379	213,725	215,869	221,976	224,437
Kennedy Space Center	349,073	373,472	377,650	390,154	407,436	415,255	432,633	455,310	508,917	526,623
Langley Research Center	147,434	135,586	135,762	137,318	142,321	147,046	156,937	168,061	171,213	182,064
Lewis Research Center	209,128	215,619	215,746	226,576	227,958	230,067	231,269	233,927	235,022	235,953
Marshall Space Flight Center	200,979	213,544	215,161	219,216	241,842	244,620	253,259	272,412	288,168	297,646
National Space Technology Laboratories	68,629	69,893	68,775	69,802	70,909	72,844	80,566	88,733	90,094	102,344
Wallops Flight Center (Facility)	26,585	27,180	—	—	—	—	—	—	—	—
Total	1,617,211	1,660,624	1,680,692	1,723,724	1,791,697	1,819,173	1,967,307	2,049,784	2,161,979	2,281,595

Table 6–11. Other Structures and Facilities Value by Installation and Fiscal Year: In-House and Contractor-Held (at end of fiscal year; in thousands of dollars)

Installation	1979	1980	1981	1982	1983	1984	1985	1986	1987	1988
Headquarters	0	0	0	0	0	0	0	0	0	0
Ames Research Center	9,374	10,346	17,642	11,362	27,750	22,842	25,856	27,093	27,353	29,303
Dryden Flight Research Center	6,804	6,961	—	7,189	—	—	—	—	—	—
Goddard Space Flight Center	58,174	59,039	120,814	117,294	119,615	119,745	111,884	111,547	111,541	117,237
Jet Propulsion Laboratory	67,578	67,706	68,076	69,017	74,835	75,648	84,855	84,573	89,535	89,867
Johnson Space Center	61,925	62,444	64,405	65,210	67,977	68,896	69,208	71,282	78,839	81,120
Kennedy Space Center	330,677	382,762	392,842	419,677	426,382	415,255	472,546	522,491	527,756	518,283
Langley Research Center	217,582	258,797	289,169	299,682	305,051	308,456	332,159	345,802	338,207	370,159
Lewis Research Center	76,377	76,489	77,169	80,103	80,742	81,591	81,929	83,653	87,417	87,960
Marshall Space Flight Center	130,147	134,279	133,201	137,255	141,076	143,985	145,778	147,431	151,813	154,213
National Space Technology Laboratories	194,294	194,664	194,718	193,093	193,269	193,199	193,071	194,088	194,635	200,955
Wallops Flight Center (Facility)	52,899	53,808	—	—	—	—	—	—	—	—
Total	1,205,831	1,307,295	1,358,036	1,400,152	1,431,697	1,474,873	1,517,286	1,567,960	1,607,096	1,649,097

Source: Table 6–4.

Table 6–12. Capitalized Equipment Value by Installation and Fiscal Year: In-House and Contractor-Held
(at end of fiscal year; in thousands of dollars)

Installation	1979	1980	1981	1982	1983	1984	1985	1986	1987	1988
Headquarters	17,807	16,710	22,358	17,496	27,749	28,896	31,144	39,525	13,541	39,378
Ames Research Center	165,781	198,403	300,338	253,841	348,604	343,836	355,441	371,797	340,456	348,975
Dryden Flight Research Center	63,564	57,914	—	69,779	—	—	—	—	—	—
Goddard Space Flight Center	511,020	528,725	565,964	556,690	583,843	533,488	515,074	533,178	240,265	458,185
Jet Propulsion Laboratory	233,821	246,751	257,709	261,252	269,267	283,850	300,364	339,463	383,811	376,271
Johnson Space Center	442,347	608,577	614,839	602,774	649,879	551,269	525,849	470,726	449,727	447,149
Kennedy Space Center	967,416	934,382	935,986	1,549,138	1,576,544	423,636	605,005	765,008	708,750	623,870
Langley Research Center	183,710	184,587	189,088	193,620	217,078	216,123	222,046	306,845	200,700	205,899
Lewis Research Center	143,194	153,298	153,463	160,603	171,350	159,139	182,380	189,082	150,367	163,617
Marshall Space Flight Center	396,707	421,361	430,850	418,113	416,751	397,902	430,488	484,959	465,347	493,784
National Space Technology Laboratories	29,296	24,134	25,475	28,680	30,306	31,252	33,356	36,923	27,719	29,876
Wallops Flight Center (Facility)	56,015	57,025	—	—	—	—	—	—	—	—
Total	3,210,678	3,431,887	3,496,070	4,111,986	4,291,371	2,969,391	3,201,147	3,537,506	3,167,448	3,187,004

Source: Table 6–4.

Table 6–13. Land Value as a Percentage of Total Real Property Value by Installation and Fiscal Year: In-House and Contractor-Held (at end of fiscal year)

Installation	1979	1980	1981	1982	1983	1984	1985	1986	1987	1988
Ames Research Center	1.3	1.3	1.2	1.1	1.1	1.1	0.8	0.8	0.8	0.6
Dryden Flight Research Center	0.0	0.0	—	0.0	—	—	—	—	—	—
Goddard Space Flight Center	1.1	1.0	1.2	1.1	1.1	1.0	0.1	1.0	1.0	1.0
Jet Propulsion Laboratory	0.7	0.7	0.7	0.7	0.7	0.7	0.6	0.6	0.6	0.6
Johnson Space Center	3.4	3.4	3.4	3.4	3.7	3.7	3.7	3.7	3.5	3.4
Kennedy Space Center	9.5	8.6	8.5	8.1	7.8	7.5	7.3	6.8	6.5	6.4
Langley Research Center	a	a	a	a	a	a	a	a	a	a
Lewis Research Center	0.8	0.8	0.8	1.2	1.2	1.2	0.8	0.8	0.8	0.8
Marshall Space Flight Center	2.1	2.0	2.0	2.0	1.9	1.8	2.1	1.7	1.6	1.6
National Space Technology Laboratories	6.4	6.4	6.4	6.4	6.4	6.4	6.2	6.0	6.0	5.6
Wallops Flight Center (Facility)	1.6	1.7	—	—	—	—	—	—	—	—

a Less than 0.05 percent.
Source: Tables 6–4 and 6–9.

Table 6–14. *Building Value as a Percentage of Total Real Property Value by Installation and Fiscal Year: In-House and Contractor-Held (at end of fiscal year)*

Installation	1979	1980	1981	1982	1983	1984	1985	1986	1987	1988
Ames Research Center	94.6	94.2	91.8	93.8	90.3	90.3	92.2	93.9	92.2	92.9
Dryden Flight Research Center	68.0	68.9	—	68.4	—	—	—	—	—	—
Goddard Space Flight Center	62.9	63.4	63.4	54.8	55.0	43.5	57.3	57.7	59.4	40.6
Jet Propulsion Laboratory	56.5	56.4	56.4	57.5	57.0	57.0	54.8	56.4	56.8	56.8
Johnson Space Center	73.4	73.6	73.0	73.1	72.8	73.0	72.7	72.4	71.2	49.0
Kennedy Space Center	46.5	45.1	45.1	44.3	45.0	44.0	44.3	43.4	45.9	47.2
Langley Research Center	40.4	34.4	34.4	31.4	30.7	32.3	32.1	32.7	33.6	33.0
Lewis Research Center	72.6	73.2	73.2	73.0	72.9	78.7	73.2	73.0	72.3	72.2
Marshall Space Flight Center	59.4	60.2	60.5	60.3	62.0	61.8	74.9	63.8	64.4	64.8
National Space Technology Laboratories	24.4	24.7	24.4	24.8	25.1	25.6	27.6	29.5	29.8	31.8
Wallops Flight Center (Facility)	32.9	33.0	—	—	—	—	—	—	—	—

Source: Tables 6–4 and 6–10.

Table 6–15. *Other Structures and Facilities Value as a Percentage of Total Real Property Value by Installation and Fiscal Year: In-House and Contractor-Held (at end of fiscal year)*

Installation	1979	1980	1981	1982	1983	1984	1985	1986	1987	1988
Ames Research Center	4.2	4.5	7.0	4.9	10.5	9.0	7.0	7.2	7.1	6.4
Dryden Flight Research Center	32.0	31.1	—	31.6	—	—	—	—	—	—
Goddard Space Flight Center	36.0	35.6	45.2	44.1	43.9	43.4	41.5	41.2	39.5	58.1
Jet Propulsion Laboratory	42.5	35.6	41.0	40.8	41.4	41.3	43.6	42.0	41.7	41.8
Johnson Space Center	23.2	23.1	24.0	23.6	23.5	23.6	23.6	23.9	25.3	25.6
Kennedy Space Center	44.0	42.0	46.7	47.6	50.6	43.8	48.4	49.8	47.6	46.4
Langley Research Center	60.0	65.6	68.0	68.6	68.2	67.7	67.8	67.3	66.4	70.6
Lewis Research Center	27.0	26.0	26.1	25.8	25.8	25.9	25.9	26.1	27.5	26.9
Marshall Space Flight Center	38.5	37.8	37.5	37.7	36.2	36.4	43.1	34.5	34.0	33.6
National Space Technology Laboratories	69.1	68.9	69.2	68.7	68.5	68.0	66.2	64.5	64.3	62.5
Wallops Flight Center (Facility)	65.5	65.3	—	—	—	—	—	—	—	—

Source: Tables 6–4 and 6–11.

Table 6–16. *Real Property Value of Installations Ranked as a Percentage of Total Real Property Value: In-House and Contractor-Held (at end of fiscal year; selected years)*

Ranking	1979		1982		1986		1988	
1	Kennedy	25.6	Kennedy	27.2	Kennedy	27.9	Kennedy	27.5
2	Langley	12.4	Langley	13.5	Langley	13.7	Langley	13.5
3	Marshall	11.5	Marshall	11.2	Marshall	11.4	Marshall	11.3
4	Lewis	9.8	Lewis	9.6	Ames [a]	10.0	Ames	11.3
5	Nat'l Labs	9.6	Nat'l Labs	8.7	Lewis	8.5	Lewis	8.1
6	Johnson	9.1	Johnson	8.5	Nat'l Labs	8.0	Nat'l Labs [b]	8.0
7	Ames	7.7	Goddard [c]	8.2	Johnson	7.9	Johnson	7.3
8	Goddard	5.5	Ames	7.3	Goddard	7.2	Goddard	7.1
9	Jet Prop. Lab	5.4	Jet Prop. Lab	5.2	Jet Prop. Lab	5.4	Jet Prop. Lab	5.3
10	Wallops	2.7	Dryden	0.7				
11	Dryden	0.7						
Total [d]		100.0		100.1		100.0		100.1

[a] This included Dryden.
[b] This was renamed John C. Stennis Space Center.
[c] This included Wallops.
[d] Totals may not add up to 100 percent due to rounding.

Source: Table 6–4.

Table 6–17. *Capitalized Equipment Value of Installations Ranked as a Percentage of Total Capitalized Equipment Value: In-House and Contractor-Held (at end of fiscal year; selected years)*

Ranking	1979		1982		1986		1988	
1	Kennedy	30.1	Kennedy	17.7	Kennedy	21.7	Kennedy	19.6
2	Goddard	15.9	Johnson	14.7	Goddard	15.1	Marshall	15.5
3	Johnson	13.8	Goddard a	13.5	Marshall	13.7	Goddard	14.4
4	Marshall	12.4	Marshall	10.2	Johnson	13.3	Johnson	14.0
5	Jet Prop. Lab	7.3	Jet Prop. Lab	6.4	Ames b	10.5	Jet Prop. Lab	11.8
6	Langley	5.7	Ames	6.2	Jet Prop. Lab	9.6	Ames	10.9
7	Ames	5.2	Langley	4.7	Langley	8.7	Langley	6.5
8	Lewis	4.5	Lewis	3.9	Lewis	5.3	Lewis	5.1
9	Dryden	2.0	Dryden	1.7	Headquarters	1.1	Headquarters	1.2
10	Wallops	1.7	Nat'l Labs	0.7	Nat'l Labs	1.0	Nat'l Labs c	0.9
11	Nat'l Labs	0.9	Headquarters	0.4				
12	Headquarters	0.6						
Total d		100.1		100.1		100.0		99.9

a   This included Wallops.
b   This included Wallops.
c   This was renamed John C. Stennis Space Center.
d   Totals may not add up to 100 percent due to rounding.

Source: Table 6–4.

*Table 6–18. NASA Tracking and Data Acquisition Stations
(at end of fiscal year; value in thousands of dollars)*

Fiscal Year	Buildings	Acres of Land	Value of Facilities
1979	317	11,448.4	129,616
1980	310	11,312.2	129,136
1981	313	11,312.2	137,644
1982	293	11,539.8	129,099
1983	303	11,539.8	133,712
1984	311	11,289.8	134,095
1985	309	11,289.8	134,605
1986	273	11,289.8	128,053
1987	263	11,289.8	133,163
1988	255	11,289.8	133,631

Source: "NASA Locations by Accounting Installations," Annual Reports, 1979–1988, Office of Management Systems and Facilities, Facilities Engineering Division.

Table 6–19. Distribution of Research and Development and Space Flight Control and Data Communications Budget Plan by Installation and Program Office: FY 1988 (in thousands of dollars; percentage of total budget plan in parentheses)

Installation	Space Station	Space Flight a	Space Science and Applications	Aeronautics and Space Technology	Space Tracking and Data Systems b	Commercial Programs	Total Budget Plan c
Ames Research Center	1,130	6,100	89,500	163,453	10,500	1,155	271,838
	(0.4)	(2.2)	(33.0)	(60.1)	(3.9)	(0.4)	(100.0)
Goddard Space Flight Center	44,304	31,800	443,078	7,790	442,135	1,450	970,557
	(4.6)	(3.3)	(45.7)	(0.8)	(45.6)	(0.1)	(100.1)
Jet Propulsion Laboratory	11,393	3,800	425,449	31,774	142,031	1,070	615,517
	(1.9)	(0.6)	(69.1)	(5.2)	(23.1)	(0.2)	(100.1)
Johnson Space Center	135,189	1,029,500	55,929	9,695	50	2,602	1,232,965
	(11.0)	(83.5)	(4.5)	(0.8)	d	(0.2)	(100.0)
Kennedy Space Center	10,898	782,200	10,880	402	—	1,036	805,416
	(1.4)	(97.1)	(1.4)	d	—	(0.1)	(100.0)
Langley Research Center	2,553	1,200	19,968	160,423	—	1,215	185,359
	(1.4)	(0.6)	(10.8)	(87.2)	—	(0.7)	(100.0)
Lewis Research Center	31,821	5,400	93,280	115,914	—	2,575	248,991
	(12.8)	(2.2)	(37.5)	(46.6)	—	(1.0)	(100.1)
Marshall Space Flight Center	54,701	1,574,100	281,675	37,887	51,200	4,853	2,004,416
	(2.7)	(78.5)	(14.1)	(1.9)	(2.6)	(0.2)	(100.0)
Stennis Space Center	331	28,900	723	—	—	5,325	35,279
	(0.9)	(81.9)	(2.0)	—	—	(15.1)	(99.9)
NASA Headquarters	99,980	56,700	161,318	26,862	251,384	27,419	623,663
	(16.0)	(9.1)	(25.9)	(4.3)	(40.3)	(4.4)	(100.0)
Total	392,300	3,519,700	1,581,800	554,200	897,300	48,700	6,994,000
	(5.7)	(51.0)	(22.9)	(8.8)	(12.8)	(0.7)	(100.0)

a Space Tracking and Data Systems includes both Research and Development and Space Flight Control and Data Communications appropriation categories.
b Space Flight includes both Research and Development and Space Flight Control and Data Communications appropriation categories.
c Total percentages may no: add up to 100 percent due to rounding.
d Less than 0.05 percent.

Source: NASA Budget Estimates, 1990.

Table 6–20. NASA Headquarters Major Organizations

Code	Title
1979	
A	Office of the Administrator
B	Office of the Comptroller
C	Office of Legislative Affairs
D	Office of the Chief Engineer
E	Office of Space and Terrestrial Applications
G	Office of General Counsel
H	Office of Procurement
L	Office of External Affairs
M	Office of Space Transportation Systems
N	Office of Management Operations
P	Office of the Chief Scientist
R	Office of Aeronautics and Space Technology
S	Office of Space Science
T	Office of Space Tracking and Data Systems
U	Office of Equal Opportunity Programs
W	Office of Inspector General
1988	
A	Office of the Administrator
B	Office of the Comptroller
C	Office of Commercial Programs
D	Office of Headquarters Operations
E	Office of Space Science and Applications
G	Office of General Counsel
H	Office of Procurement
K	Office of Small and Disadvantaged Business Utilization
L	Office of Communications
M	Office of Space Flight
N	Office of Management
P	Office of the Chief Scientist
Q	Office of Safety, Reliability, Maintainability and Quality Assurance
R	Office of Aeronautics and Space Technology
S	Office of Space Station
T	Office of Space Operations
U	Office of Equal Opportunity Programs
W	Office of Inspector General
X	Office of External Relations
Z	Office of Exploration

Table 6–21. Headquarters Capitalized Equipment Value (at end of fiscal year; in thousands of dollars)

1979	1980	1981	1982	1983	1984	1985	1986	1987	1988
17,807	16,710	22,358	17,496	27,749	29,896	31,144	39,525	27,719	39,378

Source: Table 6–4.

Table 6–22. Headquarters Personnel (at end of fiscal year)

Category	1979	1980	1981	1982	1983	1984	1985	1986	1987	1988
Paid Employees										
Permanent	1,414	1,516	1,504	1,431	1,492	1,396	1,383	1,362	1,532	1,653
Temporary	120	142	207	183	144	130	170	106	116	176
Total Paid Employees	1,534	1,658	1,638	1,614	1,636	1,526	1,553	1,468	1,648	1,829
Occupational Code Groups (permanent only)										
200, 700, and 900	368	382	382	365	363	340	314	309	387	462
600 and 500	1,033	1,123	1,110	1,054	1,117	1,043	1,057	1,042	1,137	1,182
300	5	4	4	4	4	5	4	4	3	4
100	8	7	8	8	8	8	8	7	5	5
Excepted and Supergrade	173	194	187	180	194	188	173	189	198	225
Minority Permanent Employees	299	312	317	301	354	338	349	351	378	395
Female Permanent Employees	559	608	609	568	612	576	585	587	674	718

Source: Tables 7–14 through 7–17.

FACILITIES AND INSTALLATIONS

Table 6–23. Headquarters Funding by Fiscal Year (in millions of dollars)

Appropriation Title	1979	1980	1981	1982	1983	1984	1985	1986	1987	1988
Research and Development	115.7	135.9	144.5	137.2	239.7	160.7	179.6	204.5	328.1	374.6
Space Flight Control and Data Communications	—	—	—	—	—	240.4	263.7	211.0	860.9	202.9
Research and Program Management	84.5	89.6	96.4	109.8	111.0	108.2	116.9	74.0	139.4	212.5
Total	200.2	225.5	240.9	247.0	350.7	509.3	560.2	489.5	1,328.4	870.0

Source: NASA Budget Estimates.

Table 6–24. Headquarters Total Procurement Activity by Fiscal Year (in millions of dollars)

	1979	1980	1981	1982	1983	1984	1985	1986	1987	1988
Net Value of Contract Awards	157.1	181.1	201.3	221.9	277.5	419.1	476.0	435.1	507.5	600.7
Percentage of NASA Total	3.7	3.7	3.7	3.8	4.1	5.7	5.7	5.3	5.9	6.9

Source: Annual Procurement Reports.

Table 6–25. Ames in-House and Contractor-Held Property (at end of fiscal year; money amounts in thousands of dollars)

Category	1979	1980	1981	1982	1983	1984	1985	1986	1987	1988
In-House and Contractor-Held Property										
Land (acres)	429.9	429.9	432.0	432.0	432.0	432.0	432.0	432.0	432.0	432.0
Number of Buildings	144	147	145	145	216	227	255	247	248	222
Area of Buildings (square feet)	2,224,372	2,143,077	2,275,686	2,291,262	2,799,010	2,855,486	2,785,147	2,921,574	2,941,179	3,232,804
Value of in-House and Contractor-Held Property										
Land	2,928	2,928	2,928	2,928	2,929	2,929	2,929	2,928	2,929	2,929
Buildings	213,479	214,687	231,128	220,584	237,965	239,847	338,495	345,808	357,016	424,012
Other Structures and Facilities	9,374	10,346	17,642	11,632	22,750	22,842	25,856	27,093	27,353	29,303
Total Real Property Value	225,781	227,961	251,698	235,144	263,644	265,618	367,280	375,029	387,298	456,244
Capitalized Equipment Value	165,781	198,403	300,338	253,841	348,604	343,836	355,441	371,797	340,456	348,975

Note: Beginning with FY 1983, figures include amounts for Dryden.
Source: Table 6–5 through 6–12.

Table 6–26. Ames Value of Real Property Components as a Percentage of Total (total real property value in thousands of dollars)

Component	1979	1980	1981	1982	1983	1984	1985	1986	1987	1988
Land	1.3	1.3	1.2	1.1	1.1	1.1	0.8	0.8	0.8	0.6
Buildings	94.6	94.2	91.8	93.8	90.3	90.3	92.2	93.9	92.2	92.9
Other Structures and Facilities	4.2	4.5	7.0	4.9	10.5	9.0	7.0	7.2	7.1	6.4
Total Real Property Value	225,781	227,961	251,698	235,144	263,644	265,618	367,280	375,029	387,298	456,244

Source: Tables 6–8 and 6–13 through 6–15.

Table 6–27. Ames Personnel (at end of fiscal year)

Category	1979	1980	1981	1982	1983	1984	1985	1986	1987	1988
Paid Employees										
Permanent	1,664	1,651	1,606	2,041	2,033	2,043	2,052	2,072	2,079	2,101
Temporary	49	62	46	123	105	102	107	81	82	68
Total Paid Employees	1,713	1,713	1,652	2,164	2,138	2,145	2,159	2,153	2,161	2,169
Occupational Code Groups (permanent only)										
200, 700, and 900	847	856	823	1,011	1,037	1,052	1,047	1,061	1,085	1,102
600 and 500	301	300	391	462	458	469	486	502	506	369
300	149	105	106	216	175	177	166	153	146	142
100	267	290	286	352	363	345	353	356	342	342
Excepted and Supergrade	29	22	24	36	35	34	33	35	38	37
Minority Permanent Employees	263	285	284	372	377	391	409	419	425	425
Female Permanent Employees	329	244	404	402	412	436	450	478	487	494

Source: Tables 7–14 through 7–17, 7–22, and 7–27.

Table 6–28. Ames Funding by Fiscal Year (in millions of dollars)

Appropriation Title	1979	1980	1981	1982	1983	1984	1985	1986	1987	1988
Research and Development	141.5	147.9	160.4	175.3	180.1	189.7	217.3	245.8	282.1	261.2
Space Flight Control and Data Communications	—	—	—	—	—	9.8	12.3	15.4	16.3	15.4
Research and Program Management	62.7	67.4	94.8	101.1	107.2	113.9	120.3	123.3	133.6	165.2
Construction of Facilities	9.7	2.9	13.9	18.5	3.5	4.7	13.6	7.8	22.2	23.4
Total	213.9	218.2	269.1	294.9	290.9	318.1	363.5	392.3	454.2	465.2

Note: Beginning with FY 1981, figures include funding for Dryden.
Source: NASA Budget Estimates.

Table 6–29. Ames Total Procurement Activity by Fiscal Year (in millions of dollars)

	1979	1980	1981	1982	1983	1984	1985	1986	1987	1988
Net Value of Contract Awards	219.8	236.2	207.0	222.1	247.8	281.2	340.4	361.3	467.0	431.5
Percentage of NASA Total	4.6	4.3	3.4	3.8	3.6	3.8	4.1	4.4	5.4	4.5

Source: Annual Procurement Reports.

Table 6–30. Dryden in-House and Contractor-Held Property (at end of fiscal year; money amounts in thousands of dollars)

Category	1979	1980	1981	1982
In-House and Contractor-Held Property				
Land (acres)	—	—	—	—
Number of Buildings	63	63	65	65
Area of Buildings (square feet)	459,447	504,856	501,578	501,778
Value of in-House and Contractor-Held Property				
Land	0	0	— a	0
Buildings	14,447	15,387	—	15,534
Other Structures and Facilities	6,804	6,961	—	7,189
Total Real Property Value	21,251	22,348	—	22,723
Capitalized Equipment Value	63,564	57,914	—	69,779

a Amounts are included with Ames.
Source: Table 6–5 through 6–12.

Table 6–31. Dryden Value of Real Property Components as a Percentage of Total (total real property value in thousands of dollars)

Component	1979	1980	1981 a	1982
Land	0.0	0.0	—	0.0
Buildings	68.0	68.9	—	68.4
Other Structures and Facilities	32.0	31.1	—	31.6
Total Real Property Value	21,251	22,348	—	22,723

a   Amounts are included with Ames.
Source: Tables 6–8 and 6–13 through 6–15.

Table 6–32. Dryden Personnel (at end of fiscal year)

Category	1979	1980	1981
Paid Employees			
Permanent	468	465	446
Temporary	30	34	45
Total Paid Employees	498	499	491
Occupational Code Groups (permanent only)			
200, 700, and 900	183	186	181
600 and 500	97	87	78
300	185	192	183
100	3	—	4
Excepted and Supergrade	9	11	12
Minority Permanent Employees	65	70	75
Female Permanent Employees	66	67	67

Note: Personnel for Dryden were included with Ames beginning in 1982.
Source: Tables 7–14 through 7–17, 7–22, and 7–27.

Table 6–33. Dryden Funding by Fiscal Year (in millions of dollars)

Appropriation Title	1979	1980
Research and Development	13.1	16.6
Research and Program Management	19.1	20.4
Construction of Facilities	—	—
Total	32.2	37.0

Source: NASA Budget Estimates.

Table 6–34. Dryden Total Procurement Activity by Fiscal Year (in millions of dollars)

	1979	1980	1981
Net Value of Contract Awards	25.0	26.0	25.0
Percentage of NASA Total	0.6	0.5	0.5

Note: Beginning in FY 1982, Dryden procurements were included with Ames.
Source: Annual Procurement Reports.

Table 6–35. *Goddard in-House and Contractor-Held Property*
(at end of fiscal year; money amounts in thousands of dollars)

Category	1979	1980	1981	1982	1983	1984	1985	1986	1987	1988
In-House and Contractor-Held Property										
Land (acres)	12,002.7	11,408.9	12,408.9	12,636.5	18,826.4	18,576.4	18,576.5	18,576.5	18,576.5	18,576.5
Number of Buildings	260	261	264	242	513	524	506	487	470	463
Area of Buildings (square feet)	2,709,230	2,694,672	2,787,364	2,881,871	4,007,122	3,995,449	3,965,166	3,940,935	4,302,871	4,302,857
Value of in-House and Contractor-Held Property										
Land	1,706	1,685	3,111	2,862	2,860	2,860	2,840	2,857	2,857	2,856
Buildings	101,711	105,220	143,105	145,696	149,811	152,869	153,875	156,069	167,549	166,424
Other Structures and Facilities	58,174	59,039	120,814	117,294	119,615	119,745	111,884	111,547	111,541	117,237
Total Real Property Value	161,591	165,944	267,030	265,852	272,286	275,454	268,599	270,473	282,347	286,517
Capitalized Equipment Value	511,020	528,745	565,964	556,690	583,843	533,488	515,074	533,178	427,030	458,185

Note: Beginning with FY 1983, figures include amounts for Wallops.
Source: Table 6–5 through 6–12.

Table 6–36. Goddard Value of Real Property Components as a Percentage of Total
(total real property value in thousands of dollars)

Component	1979	1980	1981	1982	1983	1984	1985	1986	1987	1988
Land	1.1	1.0	1.2	1.1	1.1	1.0	1.0	1.1	1.0	1.0
Buildings	62.9	63.4	53.6	54.8	55.0	55.5	57.3	57.7	59.4	58.1
Other Structures and Facilities	36.0	35.6	45.2	44.1	43.9	43.5	41.7	41.2	39.6	40.9
Total Real Property Value	161,591	165,944	267,030	265,853	272,286	275,454	268,599	270,473	282,047	286,517

Source: Tables 6–8 and 6–13 through 6–15.

Table 6–37. Goddard Personnel (at end of fiscal year)

Category	1979	1980	1981	1982	1983	1984	1985	1986	1987	1988
Paid Employees										
Permanent	3,482	3,436	3,319	3,621	3,668	3,541	2,629	3,679	3,648	3,626
Temporary	80	99	112	125	126	106	109	106	98	101
Total Paid Employees	3,562	3,535	3,431	3,746	3,794	3,647	3,738	3,785	3,746	3,727
Occupational Code Groups (permanent only)										
200, 700, and 900	1,726	1,707	1,675	1,725	1,774	1,738	1,830	1,895	1,893	1,901
600 and 500	1,112	1,126	1,063	1,191	1,204	1,123	1,158	1,152	1,150	1,134
300	497	463	443	563	557	536	513	500	478	468
100	147	140	138	142	133	125	128	132	127	123
Excepted and Supergrade	45	47	45	49	51	49	46	41	45	50
Minority Permanent Employees	385	431	410	455	476	453	469	506	527	541
Female Permanent Employees	750	803	821	834	875	865	914	962	973	990

Source: Tables 7–14 through 7–17, 7–22, and 7–27.

FACILITIES AND INSTALLATIONS

Table 6–38. Goddard Funding by Fiscal Year (in millions of dollars)

Appropriation Title	1979	1980	1981	1982	1983	1984	1985	1986	1987	1988
Research and Development	519.3	552.0	571.2	744.0	730.1	361.6	399.0	461.6	483.9	506.8
Space Flight Control and Data Communications	—	—	—	—	—	431.0	431.1	331.2	416.9	464.3
Research and Program Management	127.9	133.5	162.4	169.1	180.6	186.8	196.9	199.5	213.8	242.8
Construction of Facilities	5.6	—	—	—	4.7	—	2.2	3.6	15.2	19.8
Total	650.3	685.5	717.8	913.1	915.4	979.4	1,029.4	995.9	1,129.8	1,233.7

Source: NASA Budget Estimates.

Table 6–39. Goddard Total Procurement Activity by Fiscal Year (in millions of dollars)

	1979	1980	1981	1982	1983	1984	1985	1986	1987	1988
Net Value of Contract Awards	634.2	694.0	790.4	864.0	952.4	953.8	1,076.8	1,265.4	1,313.3	1,356.3
Percentage of NASA Total	14.4	13.7	14.0	14.7	14.0	13.0	13.0	15.5	15.3	14.2

Source: Annual Procurement Reports.

Table 6–40. JPL in-House and Contractor-Held Property
(at end of fiscal year; money amounts in thousands of dollars)

Category	1979	1980	1981	1982	1983	1984	1985	1986	1987	1988
In-House and Contractor-Held Property										
Land (acres)	155.8	155.8	155.8	155.8	155.8	155.8	155.8	155.8	155.8	155.8
Number of Buildings	341	341	341	344	337	338	353	335	333	327
Area of Buildings (square feet)	2,014,712	2,014,112	2,017,972	2,027,990	2,037,004	2,041,774	2,078,225	2,056,581	2,087,472	2,075,528
Value of in-House and Contractor-Held Property										
Land	1,188	1,188	1,188	1,188	1,188	1,188	1,188	1,188	1,188	1,188
Buildings	89,871	90,916	94,698	97,311	102,760	104,266	106,548	113,595	121,924	122,092
Other Structures and Facilities	67,578	67,706	68,076	69,017	74,835	75,648	84,855	84,573	89,535	89,867
Total Real Property Value	159,171	161,265	165,977	169,323	180,609	182,981	194,484	201,249	214,541	215,076
Capitalized Equipment Value	233,821	246,751	257,709	261,252	269,267	283,850	300,364	339,463	383,811	376,271

Source: Table 6–5 through 6–12.

FACILITIES AND INSTALLATIONS 441

Table 6–41. JPL Value of Real Property Components as a Percentage of Total
(total real property value in thousands of dollars)

Component	1979	1980	1981	1982	1983	1984	1985	1986	1987	1988
Land	0.7	0.7	0.7	0.7	0.7	0.7	0.6	0.6	0.6	0.6
Buildings	56.5	56.4	57.2	57.5	56.9	57.0	54.8	56.4	56.8	56.8
Other Structures and Facilities	42.5	42.0	41.0	40.8	41.4	41.3	43.6	42.0	41.7	41.8
Total Real Property Value	159,171	161,265	165,977	169,323	180,609	182,981	194,484	201,249	214,541	215,076

Source: Tables 6–8 and 6–13 through 6–15.

Table 6–42. JPL Funding by Fiscal Year (in millions of dollars)

Appropriation Title	1979	1980	1981	1982	1983	1984	1985	1986	1987	1988
Research and Development	235.1	315.4	258.7	316.2	304.9	248.3	338.7	412.7	452.4	486.3
Space Flight Control and Data Communications	—	—	—	—	—	97.2	111.0	116.2	124.3	131.5
Research and Program Management [a]	—	—	—	—	—	—	—	—	—	—
Construction of Facilities	4.6	—	3.5	1.0	—	4.3	12.2	9.4	15.3	7.2
Total	239.7	315.4	262.2	317.2	304.9	349.8	461.9	538.3	593.0	625.0

[a] JPL was staffed entirely by contractor personnel so no funds from the Research and Program Management appropriation were allocated to the facility.

Source: NASA Budget Estimates.

Table 6–43. JPL Total Procurement Activity by Fiscal Year (in millions of dollars)

	1979	1980	1981	1982	1983	1984	1985	1986	1987	1988
Net Value of Contract Awards	338.6	397.2	410.8	426.3	454.9	523.1	724.6	895.4	1,008.8	986.0
Percentage of NASA Total	8.1	8.2	7.6	7.2	6.7	7.3	8.7	11.0	11.7	10.3

Source: Annual Procurement Reports.

Table 6–44. Johnson in-House and Contractor-Held Property (at end of fiscal year; money amounts in thousands of dollars)

Category	1979	1980	1981	1982	1983	1984	1985	1986	1987	1988
In-House and Contractor-Held Property										
Land (acres)	1,785.9	1,785.9	1,785.9	1,785.9	1,819.4	1,822.9	1,822.9	1,823.1	1,823.1	1,821.0
Number of Buildings	182	182	190	190	202	204	204	204	217	212
Area of Buildings (square feet)	4,534,939	4,551,432	4,555,826	4,561,147	4,771,805	4,793,677	4,794,341	4,817,208	4,886,903	4,892,190
Value of in-House and Contractor-Held Property										
Land	9,107	9,107	9,115	9,115	10,571	10,888	10,889	10,889	10,889	10,883
Buildings	195,875	199,120	198,397	201,533	210,687	212,379	213,725	215,869	221,976	224,437
Other Structures and Facilities	61,925	62,444	64,405	65,210	67,977	68,896	69,208	71,282	78,839	81,120
Total Real Property Value	266,907	270,671	271,917	275,858	289,235	292,163	293,822	298,145	311,089	316,545
Capitalized Equipment Value	442,347	608,577	614,839	602,774	649,879	551,269	525,849	470,726	449,727	447,149

Source: Table 6–5 through 6–12.

Table 6–45. Johnson Value of Real Property Components as a Percentage of Total (total real property value in thousands of dollars)

Component	1979	1980	1981	1982	1983	1984	1985	1986	1987	1988
Land	3.4	3.4	3.4	3.3	3.7	3.7	3.7	3.7	3.5	3.4
Buildings	73.4	73.6	73.0	73.1	72.8	72.7	72.7	72.4	71.2	70.9
Other Structures and Facilities	23.2	23.0	23.6	23.6	23.5	23.6	23.6	23.9	25.3	25.6
Total Real Property Value	266,907	270,671	271,917	275,858	289,235	292,163	293,822	298,145	311,089	316,545

Source: Tables 6–8 and 6–13 through 6–15.

FACILITIES AND INSTALLATIONS

Table 6–46. Johnson Personnel (at end of fiscal year)

Category	1979	1980	1981	1982	1983	1984	1985	1986	1987	1988
Paid Employees										
Permanent	3,481	3,508	3,380	3,268	3,235	3,227	3,330	3,269	3,349	3,399
Temporary	82	108	118	177	176	125	119	93	114	99
Total Paid Employees	3,563	3,616	3,498	3,445	3,411	3,352	3,449	3,362	3,463	3,498
Occupational Code Groups (permanent only)										
200, 700, and 900	2,185	2,212	2,157	2,087	2,078	2,013	2,086	2,069	2,135	2,204
600 and 500	976	989	926	900	893	958	1,001	573	1,002	986
300	297	287	273	259	245	239	228	220	205	202
100	23	24	24	22	19	17	15	7	7	7
Excepted and Supergrade	59	57	54	52	56	55	50	48	50	50
Minority Permanent Employees	395	463	431	320	425	458	475	476	510	533
Female Permanent Employees	687	755	717	705	724	824	910	930	996	1,034

Source: Tables 7–14 through 7–17, 7–22, and 7–27.

Table 6–47. Johnson Funding by Fiscal Year (in millions of dollars)

Appropriation Title	1979	1980	1981	1982	1983	1984	1985	1986	1987	1988
Research and Development	1,151.2	1,388.0	1,523.3	1,598.5	1,550.0	175.8	239.7	244.7	325.1	327.5
Space Flight Control and Data Communications	—	—	—	—	—	1,303.2	1,159.7	988.2	2,337.9	907.8
Research and Program Management	152.9	164.7	176.1	186.5	195.2	201.1	214.8	206.0	227.9	281.9
Construction of Facilities	—	—	—	0.7	—	2.3	3.2	—	18.4	11.1
Total	1,304.1	1,552.7	1,717.8	1,699.4	1,785.2	1,682.4	1,671.4	1,439.0	2,909.3	1,528.3

Source: NASA Budget Estimates.

444 NASA HISTORICAL DATA BOOK

Table 6–48. Johnson Total Procurement Activity by Fiscal Year (in millions of dollars)

	1979	1980	1981	1982	1983	1984	1985	1986	1987	1988
Net Value of Contract Awards	1,197.6	1,429.1	1,625.8	1,690.0	1,737.8	1,636.2	1,719.1	1,359.5	1,627.4	1,806.4
Percentage of NASA Total	28.4	29.5	30.0	38.7	25.5	22.1	20.7	16.6	18.9	18.9

Source: Annual Procurement Reports.

Table 6–49. Kennedy in-House and Contractor-Held Property
(at end of fiscal year; money amounts in thousands of dollars)

Category	1979	1980	1981	1982	1983	1984	1985	1986	1987	1988
In-House and Contractor-Held Property										
Land (acres)	82,943.0	82,943.0	82,943.0	82,943.0	82,943.0	82,943.0	82,943.0	82,943.0	82,943.0	82,943.0
Number of Buildings	350	353	442	500	532	678	514	556	572	566
Area of Buildings (square feet)	5,337,276	5,312,570	5,375,291	5,431,336	5,507,621	5,766,191	5,875,980	6,041,808	6,640,212	6,720,200
Value of in-House and Contractor-Held Property										
Land	71,345	71,345	71,345	71,345	71,345	71,345	71,345	71,345	71,345	71,345
Buildings	349,073	373,472	377,650	390,154	407,436	415,255	432,633	455,310	508,917	526,623
Other Structures and Facilities	330,677	382,762	392,842	419,677	426,382	415,255	472,546	522,491	527,756	518,283
Total Real Property Value	751,095	827,579	841,837	881,176	905,163	947,111	976,524	1,049,146	1,108,018	1,116,251
Capitalized Equipment Value	967,416	934,382	935,986	1,549,138	1,576,544	423,636	605,005	765,008	708,750	623,870

Source: Table 6–5 through 6–12.

Table 6–50. *Kennedy Value of Real Property Components as a Percentage of Total*
(total real property value in thousands of dollars)

Component	1979	1980	1981	1982	1983	1984	1985	1986	1987	1988
Land	9.5	8.6	8.4	8.0	7.9	7.6	7.3	6.8	6.4	6.4
Buildings	46.5	45.1	45.0	44.3	45.0	43.8	44.3	43.4	50.0	47.2
Other Structures and Facilities	44.0	46.3	46.7	47.7	47.1	48.6	48.4	49.8	47.6	46.4
Total Real Property Value	751,095	827,579	841,837	881,176	905,163	947,111	976,524	1,049,146	1,108,018	1,116,251

Source: Tables 6–8 and 6–13 through 6–15.

Table 6–51. *Kennedy Personnel (at end of fiscal year)*

Category	1979	1980	1981	1982	1983	1984	1985	1986	1987	1988
Paid Employees										
Permanent	2,192	2,201	2,138	2,104	2,084	2,067	2,081	2,051	2,188	2,236
Temporary	72	90	86	95	96	64	84	69	90	94
Total Paid Employees	2,264	2,291	2,224	2,199	2,180	2,131	2,165	2,120	2,278	2,330
Occupational Code Groups (permanent only)										
200, 700, and 900	1,205	1,191	1,165	1,725	1,169	1,738	1,830	1,895	1,893	1,901
600 and 500	760	778	752	740	718	705	718	695	701	694
300	224	228	217	202	193	188	176	185	194	223
100	3	4	4	3	4	5	5	5	4	5
Excepted and Supergrade	29	29	25	26	26	64	66	71	32	31
Minority Permanent Employees	186	169	171	177	192	195	211	222	257	268
Female Permanent Employees	442	474	474	486	495	511	555	563	504	618

Source: Tables 7–14 through 7–17, 7–22, and 7–27.

Table 6–52. Kennedy Funding by Fiscal Year (in millions of dollars)

Appropriation Title	1979	1980	1981	1982	1983	1984	1985	1986	1987	1988
Research and Development	233.5	300.6	365.1	486.1	529.3	53.0	45.7	69.0	53.8	86.0
Space Flight Control and Data Communications	—	—	—	—	—	439.9	383.1	448.5	642.1	719.7
Research and Program Management	123.3	133.2	150.2	156.0	161.3	172.6	184.5	192.2	200.2	241.9
Construction of Facilities	—	5.8	0.8	1.7	10.7	58.6	36.9	—	11.9	26.9
Total	356.8	439.6	512.1	643.8	701.3	724.1	650.2	709.7	908.0	1,704.5

Source: NASA Budget Estimates.

Table 6–53. Kennedy Total Procurement Activity by Fiscal Year (in millions of dollars)

	1979	1980	1981	1982	1983	1984	1985	1986	1987	1988
Net Value of Contract Awards	357.3	395.9	527.9	585.4	724.7	814.2	977.9	1,026.6	883.4	1,069.2
Percentage of NASA Total	8.5	8.2	9.8	9.9	10.7	11.1	11.8	12.6	10.3	11.2

Source: Annual Procurement Reports.

Table 6–54. Langley In-House and Contractor-Held Property (at end of fiscal year; money amounts in thousands of dollars)

Category	1979	1980	1981	1982	1983	1984	1985	1986	1987	1988
In-House and Contractor-Held Property										
Land (acres)	897.6	897.6	897.6	897.6	897.6	897.6	897.6	897.6	787.6	787.6
Number of Buildings	151	151	153	159	159	159	165	166	171	170
Area of Buildings (square feet)	2,140,135	2,085,380	2,098,203	2,068,679	2,066,812	2,098,215	2,110,851	2,141,362	2,160,326	2,180,360
Value of in-House and Contractor-Held Property										
Land	162	162	162	162	162	162	162	162	156	156
Buildings	147,434	135,586	135,762	137,318	142,321	147,046	156,937	168,061	171,213	182,064
Other Structures and Facilities	217,582	258,797	289,169	299,682	305,051	308,456	332,159	345,802	338,207	370,159
Total Real Property Value	365,178	394,545	425,093	437,162	447,534	455,664	489,258	514,025	509,576	552,379
Capitalized Equipment Value	183,710	184,587	189,088	193,620	217,078	216,123	222,046	306,845	200,700	205,899

Source: Table 6–5 through 6–12.

Table 6–55. Langley Value of Real Property Components as a Percentage of Total (total real property value in thousands of dollars)

Component	1979	1980	1981	1982	1983	1984	1985	1986	1987	1988
Land	0.04	0.04	0.04	0.04	0.04	0.04	0.03	0.03	0.03	0.03
Buildings	40.4	34.7	31.9	31.4	31.8	32.3	32.1	32.7	33.6	33.0
Other Structures and Facilities	59.6	65.6	68.0	68.6	68.2	67.7	67.9	67.2	66.4	67.0
Total Real Property Value	365,178	394,545	425,093	437,162	447,534	455,664	489,258	514,025	509,576	552,379

Source: Tables 6–8 and 6–13 through 6–15.

Table 6–56. Langley Personnel (at end of fiscal year)

Category	1979	1980	1981	1982	1983	1984	1985	1986	1987	1988
Paid Employees										
Permanent	3,005	2,966	2,895	2,801	2,904	2,821	2,827	2,814	2,851	2,840
Temporary	120	83	133	115	128	131	122	118	128	126
Total Paid Employees	3,125	3,094	3,028	2,916	3,032	2,952	2,165	2,949	2,979	2,966
Occupational Code Groups (permanent only)										
200, 700, and 900	1,338	1,306	1,305	1,281	1,343	1,328	1,296	1,293	1,304	1,310
600 and 500	554	556	538	508	547	531	547	556	575	562
300	1,056	1,010	963	927	957	920	926	904	899	909
100	57	94	89	85	57	42	58	61	73	59
Excepted and Supergrade	32	20	23	28	28	29	29	32	35	34
Minority Permanent Employees	275	319	311	305	331	321	338	329	341	355
Female Permanent Employees	508	541	531	515	570	553	579	601	633	642

Source: Tables 7–14 through 7–17, 7–22, and 7–27.

FACILITIES AND INSTALLATIONS

Table 6–57. Langley Funding by Fiscal Year (in millions of dollars)

Appropriation Title	1979	1980	1981	1982	1983	1984	1985	1986	1987	1988
Research and Development	138.8	170.9	142.9	129.6	131.0	140.1	176.0	173.0	217.6	197.1
Space Flight Control and Data Communications	—	—	—	—	—	0.2	0.1	0.1	0.1	0.1
Research and Program Management	106.6	114.0	120.8	126.6	132.7	140.0	147.1	145.6	154.3	177.9
Construction of Facilities	6.5	8.0	20.8	3.0	16.2	9.5	13.8	4.7	17.9	7.2
Total	251.9	292.9	284.5	259.2	279.9	289.8	337.0	223.4	389.9	382.3

Source: NASA Budget Estimates.

Table 6–58. Langley Total Procurement Activity by Fiscal Year (in millions of dollars)

	1979	1980	1981	1982	1983	1984	1985	1986	1987	1988
Net Value of Contract Awards	225.0	235.5	218.0	181.7	216.5	214.6	248.4	261.4	289.5	297.3
Percentage of NASA Total	5.4	4.9	4.0	3.1	3.2	2.9	3.0	3.2	3.4	3.1

Source: Annual Procurement Reports.

Table 6–59. Lewis in-House and Contractor-Held Property (at end of fiscal year; money amounts in thousands of dollars)

Category	1979	1980	1981	1982	1983	1984	1985	1986	1987	1988
In-House and Contractor-Held Property										
Land (acres)	6,204.6	6,204.6	6,204.6	8,356.7	8,356.7	8,356.7	6,804.8	6,804.8	6,804.8	6,804.8
Number of Buildings	250	250	250	259	258	256	256	256	256	258
Area of Buildings (square feet)	3,105,064	3,176,921	3,178,851	3,203,085	3,205,185	3,212,567	3,213,383	3,213,383	3,213,383	3,215,691
Value of in-House and Contractor-Held Property										
Land	2,230	2,230	2,230	3,651	3,651	3,651	2,621	2,621	2,621	2,621
Buildings	209,128	215,619	215,746	226,576	227,958	230,067	231,269	233,927	235,022	235,953
Other Structures and Facilities	76,377	76,489	77,169	80,103	80,742	81,591	81,929	83,653	87,417	87,960
Total Real Property Value	287,871	294,474	295,281	310,466	312,487	315,445	315,955	320,337	325,196	326,670
Capitalized Equipment Value	143,194	153,298	153,463	160,603	171,350	159,139	182,380	189,082	150,367	163,617

Source: Table 6–5 through 6–12.

Table 6–60. Lewis Value of Real Property Components as a Percentage of Total (total real property value in thousands of dollars)

Component	1979	1980	1981	1982	1983	1984	1985	1986	1987	1988
Land	0.8	0.8	0.8	1.2	1.2	1.2	0.8	0.8	0.8	0.8
Buildings	61.8	73.2	73.1	73.0	72.9	72.9	73.2	73.0	72.3	72.2
Other Structures and Facilities	32.8	26.0	26.1	25.8	25.8	25.7	25.9	26.1	26.9	26.9
Total Real Property Value	287,871	294,474	295,281	310,466	312,487	315,445	315,955	320,337	325,196	326,670

Source: Tables 6–8 and 6–13 through 6–15.

Table 6–61. *Lewis Personnel (at end of fiscal year)*

Category	1979	1980	1981	1982	1983	1984	1985	1986	1987	1988
Paid Employees										
Permanent	2,840	2,822	2,138	2,485	2,632	2,624	2,715	2,598	2,663	2,649
Temporary	67	79	644	182	119	78	67	44	53	67
Total Paid Employees	2,907	2,901	2,782	2,667	2,751	2,702	2,782	2,642	2,716	2,716
Occupational Code Groups (permanent only)										
200, 700, and 900	1,302	1,238	1,183	1,144	1,296	1,277	1,337	1,279	1,373	1,392
600 and 500	487	522	515	458	464	484	539	496	497	499
300	217	225	210	200	213	220	217	214	220	228
100	834	837	782	683	659	643	632	609	573	530
Excepted and Supergrade	29	25	22	26	28	28	24	26	30	40
Minority Permanent Employees	155	228	227	213	236	250	265	260	271	284
Female Permanent Employees	373	429	417	384	433	452	501	459	484	488

Source: Tables 7–14 through 7–17, 7–22, and 7–27.

Table 6–62. Lewis Funding by Fiscal Year (in millions of dollars)

Appropriation Title	1979	1980	1981	1982	1983	1984	1985	1986	1987	1988
Research and Development	149.2	170.7	163.4	175.1	281.1	281.9	311.4	259.9	277.8	254.8
Space Flight Control and Data Communications	—	—	—	—	—	2.0	3.4	3.0	5.1	3.7
Research and Program Management	87.5	94.8	99.9	106.4	118.8	128.7	137.4	143.4	151.7	181.9
Construction of Facilities	6.1	5.7	10.4	1.2	3.9	10.6	—	—	12.1	22.7
Total	242.8	271.2	273.7	282.7	403.1	423.2	452.2	406.3	446.7	463.1

Source: NASA Budget Estimates.

Table 6–63. Lewis Total Procurement Activity by Fiscal Year (in millions of dollars)

	1979	1980	1981	1982	1983	1984	1985	1986	1987	1988
Net Value of Contract Awards	274.1	319.8	338.0	427.2	467.3	628.0	657.7	625.5	381.8	417.9
Percentage of NASA Total	6.5	6.6	6.3	7.3	6.9	8.5	8.2	7.6	4.4	4.4

Source: Annual Procurement Reports.

FACILITIES AND INSTALLATIONS 453

Table 6–64. Marshall in-House and Contractor-Held Property
(at end of fiscal year; money amounts in thousands of dollars)

Category	1979	1980	1981	1982	1983	1984	1985	1986	1987	1988
In-House and Contractor-Held Property										
Land (acres)	409.5	409.5	409.5	409.5	409.5	409.5	409.5	409.5	409.5	409.5
Number of Buildings	176	168	166	166	166	165	167	174	175	194
Area of Buildings (square feet)	3,823,813	3,814,283	3,814,817	3,820,069	3,820,069	3,819,288	3,820,375	3,839,696	3,756,193	3,783,472
Value of in-House and Contractor-Held Property										
Land	7,157	7,160	7,164	7,164	7,164	7,164	7,164	7,171	7,171	7,171
Buildings	200,979	213,544	215,161	219,216	241,842	244,620	253,259	272,412	288,168	297,646
Other Structures and Facilities	130,147	134,279	133,201	137,255	141,076	143,985	145,778	147,431	151,813	154,213
Total Real Property Value	338,283	354,983	355,526	363,635	390,082	395,769	395,153	427,014	447,152	459,030
Capitalized Equipment Value	396,707	421,361	430,850	418,113	416,751	397,902	430,488	484,959	465,347	493,784

Source: Table 6–5 through 6–12.

Table 6–65. Marshall Value of Real Property Components as a Percentage of Total
(total real property value in thousands of dollars)

Component	1979	1980	1981	1982	1983	1984	1985	1986	1987	1988
Land	2.1	2.0	2.0	2.0	1.9	1.8	2.1	1.7	1.6	1.6
Buildings	59.4	60.2	60.5	60.3	62.0	61.8	74.9	63.8	64.4	64.8
Other Structures and Facilities	38.5	37.8	37.5	37.7	36.2	36.4	43.1	34.5	34.0	33.6
Total Real Property Value	338,283	354,983	355,526	363,635	390,082	395,769	395,153	427,014	447,152	459,030

Source: Tables 6–8 and 6–13 through 6–15.

Table 6–66. Marshall Personnel (at end of fiscal year)

Category	1979	1980	1981	1982	1983	1984	1985	1986	1987	1988
Paid Employees										
Permanent	3,598	3,563	3,385	3,332	3,351	3,223	3,284	3,260	3,384	3,340
Temporary	79	83	94	108	113	63	102	101	94	89
Total Paid Employees	3,677	3,646	3,479	3,440	3,464	3,286	3,386	3,361	3,478	3,429
Occupational Code Groups (permanent only)										
200, 700, and 900	1,977	1,982	1,900	1,924	1,980	1,910	1,996	2,016	2,146	2,106
600 and 500	1,069	1,085	1,055	1,001	999	981	949	987	1,021	1,040
300	525	474	419	393	286	264	244	219	195	194
100	27	22	11	14	86	68	50	38	22	—
Excepted and Supergrade	57	52	43	47	51	51	54	48	50	50
Minority Permanent Employees	126	173	198	183	215	233	276	279	318	333
Female Permanent Employees	628	692	710	695	722	735	813	858	955	983

Source: Tables 7–14 through 7–17, 7–22, and 7–27.

FACILITIES AND INSTALLATIONS 455

Table 6–67. Marshall Funding by Fiscal Year (in millions of dollars)

Appropriation Title	1979	1980	1981	1982	1983	1984	1985	1986	1987	1988
Research and Development	747.8	863.8	996.0	1,200.0	1,624.2	447.1	500.0	538.4	721.1	744.4
Space Flight Control and Data Communications	—	—	—	—	—	1,272.9	1,223.5	1,543.7	1,580.1	1,261.2
Research and Program Management	149.0	155.9	165.0	172.1	184.3	189.9	198.1	194.2	212.0	237.5
Construction of Facilities	—	3.5	—	—	17.8	11.7	1.6	—	10.1	19.2
Total	896.8	1,023.2	1,161.0	1,372.1	1,826.3	1,921.6	1,923.2	2,276.3	2,523.3	2,262.3

Source: NASA Budget Estimates.

Table 6–68. Marshall Total Procurement Activity by Fiscal Year (in millions of dollars)

	1979	1980	1981	1982	1983	1984	1985	1986	1987	1988
Net Value of Contract Awards	774.4	917.2	1,050.7	1,225.5	1,677.5	1,834.6	1,999.1	1,891.8	2,059.8	2,428.3
Percentage of NASA Total	18.4	19.0	19.4	20.8	24.7	24.9	24.1	23.1	23.9	25.5

Source: Annual Procurement Reports.

Table 6–69. National Space Technology Laboratories/Stennis in-House and Contractor-Held Property
(at end of fiscal year; money amounts in thousands of dollars)

Category	1979	1980	1981	1982	1983	1984	1985	1986	1987	1988
In-House and Contractor-Held Property										
Land (acres)	20,642.2	20,642.2	20,642.2	20,642.2	20,642.2	20,642.2	20,642.2	20,642.2	20,642.2	20,642.2
Number of Buildings	120	120	115	111	104	106	111	117	118	120
Area of Buildings (square feet)	1,178,177	1,212,226	1,220,982	1,247,031	1,250,970	1,272,435	1,361,392	1,456,829	1,458,841	1,546,685
Value of in-House and Contractor-Held Property										
Land	18,061	18,061	18,061	18,061	18,061	18,061	18,061	18,061	18,061	18,061
Buildings	68,629	69,893	68,775	69,802	70,909	72,844	80,566	88,733	90,094	102,344
Other Structures and Facilities	194,294	194,664	194,718	193,093	193,269	193,199	193,071	194,088	194,635	200,955
Total Real Property Value	280,984	282,618	281,554	280,956	282,239	284,104	291,698	300,882	302,790	321,360
Capitalized Equipment Value	29,296	24,134	25,475	28,680	30,306	31,252	33,356	36,923	27,719	29,876

Source: Table 6–5 through 6–12.

Table 6–70. National Space Technology Laboratories/Stennis Value of Real Property Components as a Percentage of Total
(total real property value in thousands of dollars)

Component	1979	1980	1981	1982	1983	1984	1985	1986	1987	1988
Land	6.4	6.4	6.4	6.4	6.4	6.4	6.2	6.0	6.0	5.6
Buildings	24.4	24.7	24.4	24.8	25.1	25.6	27.6	29.5	29.8	31.8
Other Structures and Facilities	69.1	68.9	69.2	68.7	68.5	68.0	66.2	64.5	64.3	62.5
Total Real Property Value	280,984	282,618	281,554	280,956	282,239	284,104	291,698	300,882	302,790	321,360

Source: Tables 6–8 and 6–13 through 6–15.

Table 6–71. *National Space Technology Laboratories/Stennis Personnel (at end of fiscal year)*

Category	1979	1980	1981	1982	1983	1984	1985	1986	1987	1988
Paid Employees										
Permanent	98	103	105	103	106	108	122	123	137	147
Temporary	10	8	8	16	22	21	13	14	10	12
Total Paid Employees	108	111	113	119	128	129	135	137	147	159
Occupational Code Groups (permanent only)										
200, 700, and 900	46	47	51	50	54	52	56	59	67	75
600 and 500	52	56	54	53	51	55	65	63	68	70
300	—	—	—	—	1	1	1	1	2	2
100	—	—	—	—	—	—	—	—	—	—
Excepted and Supergrade	3	2	3	3	3	3	2	3	3	3
Minority Permanent Employees	7	7	8	7	8	10	14	14	15	18
Female Permanent Employees	25	27	28	31	31	34	42	41	43	48

Source: Tables 7–14 through 7–17, 7–22, and 7–27.

Table 6–72. National Space Technology Laboratories/Stennis Funding by Fiscal Year (in millions of dollars)

Appropriation Title	1979	1980	1981	1982	1983	1984	1985	1986	1987	1988
Research and Development	15.4	9.4	8.5	9.9	8.3	9.6	10.8	9.6	11.8	19.1
Space Flight Control and Data Communications	—	—	—	—	—	0.8	6.3	8.5	15.8	16.2
Research and Program Management	4.5	4.9	4.9	5.5	6.3	10.2	10.7	11.2	12.0	20.5
Construction of Facilities	—	—	—	—	—	—	3.3	—	6.2	4.9
Total	19.9	14.3	13.4	15.4	14.6	20.6	31.1	29.3	45.8	60.7

Source: NASA Budget Estimates.

Table 6–73. National Space Technology Laboratories/Stennis Total Procurement Activity by Fiscal Year (in millions of dollars)

	1979	1980	1981	1982	1983	1984	1985	1986	1987	1988
Net Value of Contract Awards	33.7	36.7	38.4	39.6	40.4	49.3	60.0	57.7	71.4	91.5
Percentage of NASA Total	0.8	0.8	0.7	0.7	0.6	0.7	0.7	0.7	0.8	1.0

Source: Annual Procurement Reports.

*Table 6–74. Wallops in-House and Contractor-Held Property
(at end of fiscal year; money amounts in thousands of dollars)*

Category	1979	1980	1981	1982
In-House and Contractor-Held Property				
Land (acres)	6,165.8	6,065.8	6,165.8	6,165.8
Number of Buildings	282	284	258	257
Area of Buildings (square feet)	1,057,344	1,064,064	1,068,312	1,058,045
Value of in-House and Contractor-Held Property a				
Land	1,305	1,391		
Buildings	26,585	27,180		
Other Structures and Facilities	52,889	53,808		
Total Real Property Value	80,789	82,379		
Capitalized Equipment Value	56,015	57,025		

a   Value of property included with Goddard beginning in FY 1981.
Source: Table 6–5 through 6–12.

*Table 6–75. Wallops Value of Real Property Components as a
Percentage of Total
(total real property value in thousands of dollars)*

Component	1979	1980
Land	0.0	0.0
Buildings	68.0	68.9
Other Structures and Facilities	32.0	31.1
Total Real Property Value	80,789	82,379

Source: Tables 6–8 and 6–13 through 6–15.

Table 6–76. Wallops Personnel (at end of fiscal year)

Category	1979	1980
Paid Employees		
Permanent	391	382
Temporary	18	24
Total Paid Employees	409	406
Occupational Code Groups (permanent only)		
200, 700, and 900	107	103
600 and 500	108	109
300	151	150
100	25	20
Excepted and Supergrade	5	5
Minority Permanent Employees	32	37
Female Permanent Employees	71	72

Note: Personnel for Wallops were included with Goddard beginning in 1981.
Source: Tables 7–14 through 7–17, 7–22, and 7–27.

Table 6–77. Wallops Funding by Fiscal Year (in millions of dollars)

Appropriation Title	1979	1980
Research and Development	16.6	17.5
Research and Program Management	15.8	17.7
Construction of Facilities	—	1.1
Total	32.4	36.3

Note: Beginning in FY 1981, Wallops funding amounts were included with Goddard.
Source: NASA Budget Estimates.

Table 6–78. Wallops Total Procurement Activity by Fiscal Year (in millions of dollars)

	1979	1980	1981
Net Value of Contract Awards	26.6	29.0	32.9
Percentage of NASA Total	0.6	0.6	0.6

Source: Annual Procurement Reports.

CHAPTER SEVEN
NASA PERSONNEL

CHAPTER SEVEN
NASA PERSONNEL

Introduction

From 1979 to 1988, NASA's civil service workforce underwent no major changes. The number of civil servants (both permanent and temporary) dropped by a total of 537, a 2.2-percent decrease. The permanent civil service workforce decreased by 2.8 percent. The major change was in the size of the contractor workforce, which grew by 9,449 people between 1979 and 1988, an increase of 47.4 percent. Combined with the reduction in the civil service workforce, NASA experienced a total growth of 20.6 percent from 1979 to 1988, from 43,312 people in 1979 to 52,224 in 1988 (Figure 7–1). Kennedy Space Center, the National Space Technology Laboratories, and NASA Headquarters were the only NASA installations that grew in permanent civil service workforce size.

The composition of the workforce shifted toward a more professional makeup (Figure 7–2). The proportion of scientists and engineers and professional administrative personnel increased, and the proportion of staff in the other occupational codes decreased. The educational level of the workforce also increased. More NASA staff had college degrees, and fewer had only high school education (Figure 7–3).

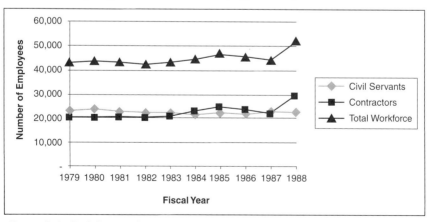

Figure 7–1. Civil Service, Contractor, and Total Workforce Trend (at end of fiscal year)

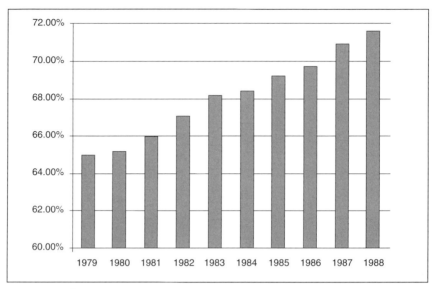

Figure 7–2. Growth in Professional Occupational Groups From 1979 to 1988 (Occupational Groups 600, 700, and 900)

In 1978, the federal government established the Senior Executive Service (SES) as part of the Civil Service Reform Act of 1978. In July 1979, NASA converted 465 of its executives into the newly established SES. Of the 465, 425 were converted from NASA Excepted, Public Law 313, or Executive Pay Act positions and forty from General Schedule (GS) positions (see Tables 7–12 and 7–13).

Salaries increased each year. In general, the annual increase was between 2.5 percent and 5.5 percent. However, salaries in FY 1981 grew by more than 9.7 percent, perhaps reflecting campaign promises made by the newly installed President Reagan. FY 1986 saw smaller than average salary increases, in line with an effort to curb federal spending (see Figure 7–4).

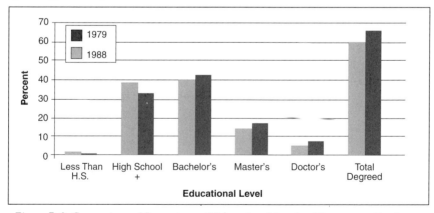

Figure 7–3. Comparison of Percentage of Educational Levels of Permanent Employees (1979 and 1988)

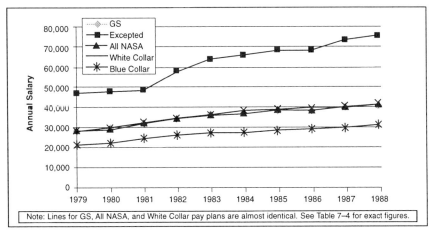

Figure 7–4. Average Salary Trends of Permanent Employees by Pay Plan

NASA's workforce became more diverse during the decade. The number of minority employees grew by more than 1,000 people, from 9.7 percent of the total permanent workforce in 1979 to 14.3 percent at the end of FY 1988 (Figure 7–5). This held for all racial categories, in all occupational codes, at all NASA installations, and in all grades. However, the average grade for minorities in all occupational categories still remained below the average grade for nonminority employees, although the size of the gap between the two groups decreased.

The number of women also grew by more than 1,500 during this period. The proportion of women in the permanent NASA workforce increased from 19.6 percent in FY 1979 to 27.4 percent in FY 1988 (Figure 7–6). This increase was for all occupational code groups, at all NASA installations, and in all grades. Women were, however, concentrated in the lower grades. They comprised approximately 90 percent of GS-1–6 employees but only 30 percent of GS-7–12 employees, 4 percent of GS-13–15 employees, and 3 percent of supergrade and excepted employees. Within each occupational code group, the average grade held by women remained lower than the average grade held by males, but the gap between the two groups decreased in size during the decade.

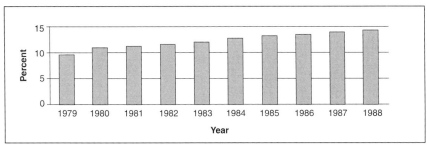

Figure 7–5. Growth in Minority Employment as a Percentage of NASA Total Employees (1979–1988)

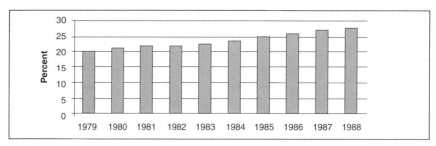

Figure 7–6. Growth in Percentage of Female Permanent Employees (1979–1988)

The average age of the workforce remained approximately the same during the ten-year period.

The information in the tables of this chapter (Tables 7–1 through 7–29) comes from *Civil Service Work Force: A Report to Management*, which is published each year by NASA's Personnel and General Management Office. The reader should note that beginning with FY 1982, figures for Dryden Flight Research Center are included with the Ames Research Center figures, and the Wallops Flight Center figures are included with the Goddard Space Flight Center figures, reflecting the two center consolidations.

Definition of Terms

The following terms are used when discussing NASA personnel.

Occupational Code Groups

NASA organized its jobs into the following principal job groups:

- 100—Wage System (Trade and Labor Positions): includes trade, craft, and general laboring positions (nonsupervisory, leader, and supervisory), compensated on the basis of prevailing locality wage rates.

- 200—Support Engineering and Related Positions: includes professional physical science, engineering, and mathematician positions in work situations not identified with aerospace technology.

- 300—Technical Support Positions: includes scientific and engineering aid, technician, drafting, photography, illustrating, salaried shop superintendents, quality assurance specialist production planning, and inspecting positions.

- 500—Clerical and Non-Professional Administrative Positions: includes secretarial, specialized and general clerical, and administrative specialist positions, the qualification requirements for which are clerical training and experience or specialized nonprofessional experience in supply, fiscal, procurement, and similar or related activities.

- 600—Professional Administrative Positions: includes professional management positions in research and development administration in such activities as financial management, contracting, personnel security, administration, law, public affairs, and the like for which a college degree, or the equivalent, and specialized training and experience are required.

- 700—Aero-Space Technology (AST) Scientific and Engineering Positions: includes professional scientific and engineering positions requiring AST qualifications, as well as professional positions engaged in aerospace research, development, operations, and related work, including the development and operation of specialized facilities and support equipment.

- 900—Life Science Support Positions: includes life science professional positions not requiring AST qualifications, as well as medical officers and other positions performing professional work in psychology, the biological sciences, and professions that support the science of medicine, such as nursing and medical technology.

- Professional Staff: includes all Group 200, 600, 700, and 900 positions.

- Support Staff: includes all Group 100, 300, and 500 positions.

Pay Grade

Civil service positions were classified in a variety of pay plans and grades based on the complexity, difficulty, and level of responsibility. The most common pay plan was the GS, which consisted of fifteen pay grades from GS-1 at the low end to GS-15. GS-16–18 were classified as supergrades. NASA also used the Senior Executive Schedule and various Federal Wage System pay plans.

Table 7–1. Total NASA Workforce (at end of fiscal year)

Personnel	1979	1980	1981	1982	1983	1984	1985	1986	1987	1988
Civil Service	23,360	22,470	22,736	22,310	22,534	21,870	22,316	21,960	22,646	22,823
Permanent	22,633	21,613	21,844	21,186	21,505	21,050	21,423	21,228	21,831	21,991
Other	727	857	892	1,124	1,029	820	893	732	815	832
Contractor	19,952	20,294	21,700	20,089	20,662	22,865	24,468	23,687	21,733	29,401
Total	43,312	43,764	44,436	42,399	43,196	44,735	46,784	45,647	44,379	52,224
Net Change, Permanent Only	–536	–20	–769	–658	319	–455	373	–195	603	160
Percentage Change, Permanent Only	–2.2	a	–3.4	–3.1	1.5	–2.1	1.8	–0.9	2.8	0.7

a Less than 0.05 percent.

Table 7–2. Accessions and Separations of Permanent Employees

Activity of Employee	1979	1980	1981	1982	1983	1984	1985	1986	1987	1988
Accessions	1,366	2,325	1,150	959	1,519	1,119	1,984	1,415	2,036	1,761
Separations	1,908	2,363	1,925	1,616	1,251	1,574	1,602	1,600	1,428	1,590
Net Accessions	–542	–38	–775	–657	286	–455	382	–185	608	171
Percentage Change	–2.3	a	–3.4	–3.0	1.3	–2.1	1.8	–0.9	2.9	0.8

a Less than 0.05 percent.

Table 7–3. *Permanent Employees by NASA Occupational Code Group: Number on Board (at end of fiscal year; percentage of NASA total in parentheses)*

Occupational Code Group	1979	1980	1981	1982	1983	1984	1985	1986	1987	1988
200, 700, 900 (Scientists/Engineers)	11,284 (49.7)	11,210 (49.6)	10,923 (50.0)	10,746 (50.7)	11,094 (51.6)	10,879 (51.7)	11,144 (52.0)	11,147 (52.5)	11,679 (53.5)	11,866 (54.0)
300 (Technical Support)	3,306 (14.6)	3,134 (13.9)	2,968 (13.6)	2,764 (13.0)	2,631 (12.2)	2,550 (12.1)	2,475 (11.6)	2,400 (11.3)	2,342 (10.7)	2,372 (10.8)
600 (Professional Admin.)	3,421 (15.1)	3,539 (15.7)	3,495 (16.0)	3,462 (16.3)	3,562 (16.6)	3,518 (16.7)	3,675 (17.2)	3,656 (17.2)	3,803 (17.4)	3,881 (17.6)
500 (Clerical)	3,228 (14.3)	3,292 (14.6)	3,098 (14.2)	2,905 (13.7)	2,889 (13.4)	2,850 (13.5)	2,880 (13.4)	2,810 (13.2)	2,854 (13.1)	2,801 (12.7)
100 (Trades/Labor)	1,394 (6.2)	1,438 (6.4)	1,360 (6.2)	1,309 (6.2)	1,329 (6.2)	1,253 (6.0)	1,249 (5.8)	1,215 (5.7)	1,153 (5.3)	1,071 (4.9)
Total Professional	14,705 (65.0)	14,749 (65.2)	14,418 (66.0)	14,208 (67.1)	14,656 (68.2)	14,397 (68.4)	14,819 (69.2)	14,803 (69.7)	15,482 (70.9)	15,747 (71.6)
Total	22,597	22,613	21,844	23,186	21,505	21,050	21,423	21,228	20,849	21,991

Table 7–4. Average Annual Salaries of Permanent Employees by Pay Plan
(at end of fiscal year; percentage increase from previous year in parentheses)

	1979	1980	1981	1982	1983	1984	1985	1986	1987	1988
GS	27,910 (5.2)	29,382 (5.3)	32,108 (9.3)	34,283 (6.8)	35,754 (4.3)	37,287 (4.3)	38,443 (3.1)	38,624 (0.5)	39,805 (3.1)	40,942 (2.9)
Excepted	47,167 (1.0)	47,715 (1.2)	48,602 (1.9)	58,106 (19.6)	63,790 (9.8)	66,040 (3.5)	68,198 (3.3)	68,127 (−0.1)	73,438 (7.8)	75,583 (2.9)
Total White Collar	28,328 (5.3)	29,776 (5.1)	32,460 (9.0)	34,819 (7.3)	36,392 (4.5)	38,000 (4.4)	39,137 (3.0)	33,337 (2.1)	40,576 (3.1)	41,728 (2.8)
Total Blue Collar	21,236 (4.4)	22,173 (4.4)	24,419 (10.1)	25,501 (4.4)	26,901 (5.5)	27,533 (2.3)	28,454 (3.3)	29,170 (2.5)	29,789 (2.1)	30,887 (3.7)
All NASA	27,891 (5.3)	29,293 (5.0)	31,959 (9.1)	34,243 (7.1)	35,795 (4.5)	37,377 (4.4)	38,544 (3.1)	38,755 (0.5)	40,006 (3.2)	41,200 (3.0)

Table 7–5. *Educational Profile of Permanent Employees: Number on Board (at end of fiscal year)*

Educational Level	1979	1980	1981	1982	1983	1984	1985	1986	1987	1988
Less High School	523	450	395	340	310	274	243	204	177	160
High School Plus	8,667	8,619	8,236	7,838	7,713	7,505	7,531	7,336	7,331	7,260
Bachelor's	9,014	8,987	8,717	8,538	8,735	8,563	8,824	8,779	9,152	9,278
Master's	3,243	3,314	3,265	3,238	3,401	3,354	3,420	3,460	3,651	3,739
Doctoral	1,186	1,243	1,231	1,232	1,346	1,354	1,405	1,446	1,520	1,554
Total Degreed Employees	13,443	13,544	13,213	13,008	13,482	13,271	13,649	13,685	14,323	14,571

Table 7–6. *Educational Profile of Permanent Employees: Percentage (at end of fiscal year)*

Educational Level	1979	1980	1981	1982	1983	1984	1985	1986	1987	1988
Less High School	1.7	2.0	1.8	1.6	1.4	1.3	1.1	1.0	0.8	0.7
High School Plus	38.3	38.1	37.7	37.0	35.9	35.7	35.2	34.6	33.6	33.0
Bachelor's	39.8	39.7	39.9	40.3	40.6	40.7	41.2	41.4	41.9	42.2
Master's	14.3	14.7	14.9	15.3	15.8	15.9	16.0	16.3	16.7	17.0
Doctoral	5.2	5.5	5.6	5.8	6.3	6.4	6.6	6.8	7.0	7.1
Total Degreed Employees	59.4	59.9	60.5	61.4	62.7	63.0	63.7	64.5	65.6	66.3

Table 7–7. Paid Employees by NASA Installation: Number on Board (Permanent and Other) (at end of fiscal year)

Installation	1979	1980	1981	1982	1983	1984	1985	1986	1987	1988
Ames Research Center	1,713	1,713	1,652	2,164	2,138	2,145	2,159	2,153	2,161	2,169
Dryden Flight Research Center [a]	498	499	491	—	—	—	—	—	—	—
Goddard Space Flight Center	3,562	3,535	3,431	3,746	3,794	3,647	3,738	3,785	3,746	3,727
Johnson Space Center	3,563	3,616	3,498	3,445	3,411	3,352	3,449	3,362	3,463	3,498
Kennedy Space Center	2,264	2,291	2,224	2,199	2,180	2,131	2,165	2,120	2,278	2,330
Langley Research Center	3,125	3,094	3,028	2,916	3,032	2,952	2,949	2,932	2,979	2,966
Lewis Research Center	2,907	2,901	2,782	2,667	2,751	2,702	2,782	2,642	2,716	2,716
Marshall Space Flight Center	3,677	3,646	3,479	3,440	3,464	3,286	3,386	3,361	3,478	3,429
National Space Technology Laboratories [b]	108	111	113	119	128	129	135	137	147	159
Wallops Flight Center (Facility) [c]	409	406	400	—	—	—	—	—	—	—
NASA Headquarters	1,534	1,658	1,638	1,614	1,636	1,526	1,553	1,468	1,648	1,829
Total	23,360	23,470	22,736	22,310	22,534	21,870	22,316	21,960	22,646	22,823

[a] Included with totals of Ames Research Center beginning in FY 1982.
[b] Renamed Stennis Space Center in 1988.
[c] Included with totals of Goddard Space Flight Center beginning in FY 1982.

Table 7–8. Paid Employees by NASA Installation: Percentage of NASA Total
(at end of fiscal year)

Installation	1979	1980	1981	1982	1983	1984	1985	1986	1987	1988
Ames Research Center	7.3	7.3	7.3	9.7	9.5	9.8	9.7	9.8	9.5	9.5
Dryden Flight Research Center	2.1	2.1	2.2	—	—	—	—	—	—	—
Goddard Space Flight Center	15.3	14.6	15.1	16.8	16.6	16.7	16.8	17.2	16.5	16.3
Johnson Space Center	15.3	15.4	15.4	15.4	15.2	15.3	15.5	15.3	15.3	15.3
Kennedy Space Center	9.7	9.8	9.8	9.9	9.7	9.7	9.7	9.7	10.1	10.2
Langley Research Center	13.5	13.2	13.3	13.6	13.5	13.5	13.2	13.4	13.2	13.0
Lewis Research Center	12.4	12.4	12.2	12.0	12.2	12.4	12.5	12.0	12.0	11.9
Marshall Space Flight Center	15.7	15.5	15.3	15.4	15.4	15.0	15.2	15.3	15.4	15.0
National Space Technology Laboratories	0.5	0.5	0.5	0.5	0.6	0.6	0.6	0.6	0.7	0.7
Wallops Flight Center (Facility)	1.8	1.7	1.8	—	—	—	—	—	—	—
NASA Headquarters	6.6	7.1	7.2	7.2	7.3	7.0	7.0	6.7	7.3	8.0
Total a	100.2	99.6	100.1	100.5	100.0	100.0	100.2	100.0	100.0	99.9

a Totals may not equal 100 percent due to rounding.

Table 7–9. Paid Employees by NASA Installation: Changes in Number on Board

Installation	1979	1980	1981	1982	1983	1984	1985	1986	1987	1988
Ames Research Center	22	0	-61	512	-26	7	14	-6	8	8
Dryden Flight Research Center	-16	1	-8	—	—	—	—	—	—	—
Goddard Space Flight Center	-79	-27	-104	315	48	-147	91	47	-39	-19
Johnson Space Center	-54	53	-118	-53	-34	11	97	-87	104	35
Kennedy Space Center	30	27	-1,177	1,085	-19	-49	39	-45	158	52
Langley Research Center	-42	-31	-66	-112	116	-80	-3	-17	47	0
Lewis Research Center	-57	-6	-119	-115	84	-49	80	-140	74	-13
Marshall Space Flight Center	-131	-31	-167	-39	24	-178	100	-25	117	-49
National Space Technology Laboratories	0	3	2	6	9	1	6	2	10	12
Wallops Flight Center (Facility)	-20	-3	-6	—	—	—	—	—	—	—
NASA Headquarters	-72	124	-20	-24	22	-110	27	-85	180	181
Total	-419	110	-734	-426	224	-664	446	-356	686	177

Table 7–10. Permanent Employees by NASA Installation: Number on Board
(at end of fiscal year)

Installation	1979	1980	1981	1982	1983	1984	1985	1986	1987	1988
Ames Research Center	1,664	1,651	1,606	2,041	2,033	2,043	2,052	2,072	2,079	2,101
Dryden Flight Research Center	468	465	446	—	—	—	—	—	—	—
Goddard Space Flight Center	3,482	3,436	3,319	3,621	3,668	3,541	3,629	3,679	3,648	3,626
Johnson Space Center	3,481	3,508	3,380	3,268	3,235	3,227	3,330	3,269	3,349	3,399
Kennedy Space Center	2,192	2,201	2,138	2,104	2,084	2,067	2,081	2,051	2,188	2,236
Langley Research Center	3,005	2,966	2,895	2,801	2,904	2,821	2,827	2,814	2,851	2,840
Lewis Research Center	2,840	2,822	2,138	2,485	2,632	2,624	2,715	2,598	2,663	2,649
Marshall Space Flight Center	3,598	3,563	3,385	3,332	3,351	3,223	3,284	3,260	3,384	3,340
National Space Technology Laboratories	98	103	105	103	106	108	122	123	137	147
Wallops Flight Center (Facility)	391	382	376	—	—	—	—	—	—	—
NASA Headquarters	1,414	1,516	1,504	1,431	1,492	1,396	1,383	1,362	1,532	1,653
Total	22,633	22,613	21,292	21,186	21,505	21,050	21,423	21,228	21,831	21,991

Table 7-11. Temporary Employees by NASA Installation: Number on Board (at end of fiscal year)

Installation	1979	1980	1981	1982	1983	1984	1985	1986	1987	1988
Ames Research Center	49	62	46	123	105	102	107	81	82	68
Dryden Flight Research Center	30	34	45	—	—	—	—	—	—	—
Goddard Space Flight Center	80	99	112	125	126	106	109	106	98	101
Johnson Space Center	82	108	118	177	176	125	119	93	114	99
Kennedy Space Center	72	90	86	95	96	64	84	69	90	94
Langley Research Center	120	83	133	115	128	131	122	118	128	126
Lewis Research Center	67	79	92	182	119	78	67	44	53	67
Marshall Space Flight Center	79	83	94	108	113	63	102	101	94	89
National Space Technology Laboratories	10	8	8	16	22	21	13	14	10	12
Wallops Flight Center (Facility)	18	18	24	—	—	—	—	—	—	—
NASA Headquarters	120	142	134	183	144	130	170	106	116	176
Total	727	806	892	1,124	1,029	820	893	732	785	832

Table 7–12. *NASA Excepted and Supergrade Employees by NASA Installation: Number on Board (at end of fiscal year)* [a]

Installation	1979	1980	1981	1982	1983	1984	1985	1986	1987	1988
Ames Research Center	29	22	24	36	35	34	33	35	38	37
Dryden Flight Research Center	9	11	12	—	—	—	—	—	—	—
Goddard Space Flight Center	45	47	45	49	51	49	46	41	45	50
Johnson Space Center	59	57	54	52	56	55	50	48	50	50
Kennedy Space Center	29	29	25	26	26	64	66	71	32	31
Langley Research Center	32	20	23	28	28	29	29	32	35	34
Lewis Research Center	29	25	22	26	28	28	24	26	30	40
Marshall Space Flight Center	57	52	43	47	51	51	54	48	49	50
National Space Technology Laboratories	3	2	3	3	3	3	2	3	3	3
Wallops Flight Center (Facility)	5	5	4	—	—	—	—	—	—	—
NASA Headquarters	173	194	187	180	194	188	173	189	198	225
Total	470	464	442	447	472	501	477	493	480	520

[a] Includes Senior Executive Service (SES), NASA Excepted Public Law 3104(a) Executive Pay Act, and GS-16 and above employees.

Table 7–13. NASA Excepted and Supergrade Employees by NASA Installation: Percentage of NASA Total
(at end of fiscal year) [a]

Installation	1979	1980	1981	1982	1983	1984	1985	1986	1987	1988
Ames Research Center	6.2	4.7	5.4	8.1	7.4	6.8	7.0	7.1	7.9	7.3
Dryden Flight Research Center [b]	1.9	2.4	2.7	—	—	—	—	—	—	—
Goddard Space Flight Center	9.6	10.1	10.2	11.0	10.8	9.8	9.6	8.3	9.4	9.8
Johnson Space Center	12.6	12.3	12.2	11.6	11.9	11.0	10.5	9.7	10.4	9.8
Kennedy Space Center	6.2	6.3	5.7	5.8	5.5	12.8	13.8	14.4	6.7	6.1
Langley Research Center	6.8	4.3	5.2	6.3	5.9	5.8	6.1	6.5	7.3	6.7
Lewis Research Center	6.2	5.4	5.0	5.8	5.9	5.6	5.0	5.3	6.3	7.8
Marshall Space Flight Center	12.1	11.2	9.7	10.5	10.8	10.2	11.3	9.7	10.2	9.8
National Space Technology Laboratories [c]	0.6	0.4	0.7	0.7	0.6	0.6	0.4	0.6	0.6	0.6
Wallops Flight Center (Facility) [d]	1.1	1.1	0.9	—	—	—	—	—	—	—
NASA Headquarters	36.8	41.8	42.3	40.3	41.1	37.5	36.3	38.3	41.3	44.1
Total [e]	100.1	100.0	100.0	100.1	99.9	100.1	100.0	99.9	100.1	102.0

[a] Includes SES, NASA Excepted Public Law 3104(a) Executive Pay Act, and GS-16 and above employees.
[b] Included with Ames Research Center beginning in FY 1982.
[c] Renamed Stennis Space Center in 1988.
[d] Included with Goddard Space Flight Center beginning in FY 1982.
[e] Totals may not equal 100 percent due to rounding.

Table 7–14. Scientific and Technical Permanent Employees (Occupational Code Group 200, 700, and 900) by NASA Installation: Number on Board (at end of fiscal year)

Installation	1979	1980	1981	1982	1983	1984	1985	1986	1987	1988
Ames Research Center	847	856	823	1,011	1,037	1,052	1,047	1,061	1,085	1,102
Dryden Flight Research Center a	183	186	181	—	—	—	—	—	—	—
Goddard Space Flight Center	1,726	1,707	1,675	1,725	1,774	1,738	1,830	1,895	1,893	1,901
Johnson Space Center	2,185	2,212	2,157	2,087	2,078	2,013	2,086	2,069	2,135	2,204
Kennedy Space Center	1,205	1,191	1,165	1,159	1,169	1,169	1,182	1,166	1,289	1,314
Langley Research Center	1,338	1,306	1,305	1,281	1,343	1,328	1,296	1,293	1,304	1,310
Lewis Research Center	1,302	1,238	1,183	1,144	1,296	1,277	1,337	1,279	1,373	1,392
Marshall Space Flight Center	1,977	1,982	1,900	1,924	1,980	1,910	1,996	2,016	2,146	2,106
National Space Technology Laboratories b	46	47	51	50	54	52	56	59	67	75
Wallops Flight Center (Facility) c	107	103	101	—	—	—	—	—	—	—
NASA Headquarters	368	382	382	365	363	340	314	309	387	462
Total	11,284	11,210	10,923	10,746	10,094	10,879	10,144	11,147	11,679	11,866

a   Included with Ames Research Center beginning in FY 1982.
b   Renamed Stennis Space Center in 1988.
c   Included with Goddard Space Flight Center beginning in FY 1982.

Table 7–15. *Technical Support Permanent Employees (Occupational Code Group 300) by NASA Installation: Number on Board (at end of fiscal year)*

Installation	1979	1980	1981	1982	1983	1984	1985	1986	1987	1988
Ames Research Center	149	105	106	216	175	177	166	153	146	142
Dryden Flight Research Center	185	192	183	—	—	—	—	—	—	—
Goddard Space Flight Center	497	463	443	563	557	536	513	500	478	468
Johnson Space Center	297	287	273	259	245	239	228	220	205	202
Kennedy Space Center	224	228	217	202	193	188	176	185	194	223
Langley Research Center	1,056	1,010	963	927	957	920	926	904	899	909
Lewis Research Center	217	225	210	200	213	220	217	214	220	228
Marshall Space Flight Center	525	474	419	393	286	264	244	219	195	194
National Space Technology Laboratories	—	—	—	—	1	1	1	1	2	2
Wallops Flight Center (Facility)	151	150	150	—	—	—	—	—	—	—
NASA Headquarters	5	4	4	4	4	5	4	4	3	4
Total	3,306	3,138	2,968	2,764	2,631	2,550	2,475	2,400	2,342	2,372

Table 7–16. *Trades and Labor Permanent Employees (Occupational Code Group 100) by NASA Installation: Number on Board (at end of fiscal year)*

Installation	1979	1980	1981	1982	1983	1984	1985	1986	1987	1988
Ames Research Center	267	290	286	352	363	345	353	356	342	342
Dryden Flight Research Center	3	—	4	—	—	—	—	—	—	—
Goddard Space Flight Center	147	140	138	142	133	125	128	132	127	123
Johnson Space Center	23	24	24	22	19	17	15	7	7	7
Kennedy Space Center	3	4	4	3	4	5	5	5	4	5
Langley Research Center	57	94	89	85	57	42	58	61	73	59
Lewis Research Center	834	837	782	683	659	643	632	609	573	530
Marshall Space Flight Center	27	22	11	14	86	68	50	38	22	—
National Space Technology Laboratories	—	—	—	—	—	—	—	—	—	—
Wallops Flight Center (Facility)	25	20	14	—	—	—	—	—	—	—
NASA Headquarters	8	7	8	8	8	8	8	7	5	5
Total	1,394	1,438	1,360	1,309	1,329	1,253	1,249	1,215	1,153	1,071

Table 7–17. *Clerical and Professional Administrative Permanent Employees (Occupational Code Group 600 and 500) by NASA Installation: Number on Board (at end of fiscal year)*

Installation	1979	1980	1981	1982	1983	1984	1985	1986	1987	1988
Ames Research Center	301	300	391	462	458	469	486	502	506	369
Dryden Flight Research Center	97	87	78	—	—	—	—	—	—	—
Goddard Space Flight Center	1,112	1,126	1,063	1,191	1,204	1,123	1,158	1,152	1,150	1,134
Johnson Space Center	976	989	926	900	893	958	1,001	573	1,002	986
Kennedy Space Center	760	778	752	740	718	705	718	695	701	694
Langley Research Center	554	556	538	508	547	531	547	556	575	562
Lewis Research Center	487	522	515	458	464	484	539	496	497	499
Marshall Space Flight Center	1,069	1,085	1,055	1,001	999	981	994	987	1,021	1,040
National Space Technology Laboratories	52	56	54	53	51	55	65	63	68	70
Wallops Flight Center (Facility)	160	109	111	—	—	—	—	—	—	—
NASA Headquarters	1,033	1,123	1,110	1,054	1,117	1,043	1,057	1,042	1,137	1,182
Total	6,601	6,731	6,593	6,367	6,451	6,368	6,565	6,066	6,657	6,536

Table 7–18. *Minority Permanent Employees: Number on Board (at end of fiscal year; percentage of NASA total in parentheses)*

Minority	1979	1980	1981	1982	1983	1984	1985	1986	1987	1988
Black	1,400 (6.2)	1,585 (7.0)	1,553 (7.1)	1,512 (7.1)	1,614 (7.5)	1,640 (7.8)	1,694 (7.9)	1,704 (8.0)	1,780 (8.2)	1,824 (8.3)
Hispanic	427 (1.9)	509 (2.3)	501 (2.3)	499 (2.4)	537 (2.5)	528 (2.5)	581 (2.7)	587 (2.8)	614 (2.8)	632 (2.9)
Asian	303 (1.3)	334 (1.5)	352 (1.6)	355 (1.7)	388 (1.8)	402 (1.9)	449 (2.1)	481 (2.3)	554 (2.5)	598 (2.7)
American Indian	57 (0.3)	66 (0.3)	66 (0.3)	67 (0.3)	75 (0.3)	79 (0.4)	82 (0.4)	84 (0.4)	94 (0.4)	98 (0.4)
Total	2,187 (9.7)	2,494 (11.0)	2,472 (11.3)	2,433 (11.5)	2,614 (12.2)	2,649 (12.6)	2,806 (13.1)	2,856 (13.5)	3,042 (13.9)	3,152 (14.3)

Table 7–19. Minority Permanent Employees by NASA Occupational Code Group: Number on Board (at end of fiscal year; percentage of NASA total in parentheses)

Occupational Code Group	1979	1980	1981	1982	1983	1984	1985	1986	1987	1988
500 (Clerical)	696 (21.6)	776 (23.6)	738 (23.8)	696 (24.0)	728 (25.2)	730 (25.6)	747 (25.9)	716 (25.5)	745 (26.1)	738 (26.3)
100 (Trades/labor)	224 (16.1)	265 (18.4)	253 (18.6)	259 (19.8)	240 (13.3)	244 (19.5)	249 (19.9)	243 (20.0)	233 (20.2)	221 (20.6)
600 (Professional Admin.)	355 (10.4)	395 (11.2)	406 (11.6)	423 (12.2)	473 (13.3)	480 (13.6)	524 (14.3)	550 (15.0)	275 (11.7)	294 (12.4)
200, 700, 900 (Scientists/Engineers)	697 (6.2)	795 (7.1)	812 (7.4)	814 (7.6)	919 (8.3)	939 (8.6)	1,025 (9.2)	1,079 (9.7)	565 (14.9)	589 (15.2)
300 (Technical Support)	215 (6.5)	263 (8.4)	263 (8.9)	241 (8.7)	254 (9.7)	256 (10.0)	261 (10.5)	268 (11.2)	1,224 (10.5)	1,310 (11.0)
Total	2,187 (9.7)	2,494 (11.0)	2,472 (11.3)	2,433 (11.5)	2,614 (12.2)	2,649 (12.6)	2,806 (13.1)	2,856 (13.5)	3,042 (13.9)	3,152 (14.3)

Table 7–20. *Minority Permanent Employees by Grade Range: Number on Board (at end of fiscal year; percentage of NASA total in parentheses)*

Grade of Employee	1979	1980	1981	1982	1983	1984	1985	1986	1987	1988
GS-1–6	688 (23.2)	786 (25.4)	720 (25.8)	646 (25.7)	655 (26.8)	630 (26.8)	659 (26.9)	617 (26.1)	630 (27.1)	594 (27.0)
GS-7–12	855 (10.9)	987 (12.6)	1,000 (13.1)	1,009 (13.7)	1,137 (14.8)	1,156 (15.5)	1,215 (16.0)	1,273 (16.7)	1,389 (17.3)	1,453 (18.0)
GS-13–15	406 (4.1)	441 (4.5)	485 (5.0)	504 (5.3)	561 (5.8)	599 (6.3)	664 (6.9)	701 (7.3)	771 (7.8)	864 (8.5)
GS-16–Excepted	14 (3.0)	15 (3.2)	15 (3.4)	15 (3.3)	21 (4.4)	20 (4.0)	19 (3.9)	22 (4.5)	19 (4.0)	20 (4.0)
Wage System	224 a	265 a	253 a	259 a	240 a	244 a	249 a	243 a	233 (20.2)	221 (20.6)
Total	2,187 (9.7)	2,494 (11.0)	2,473 (11.3)	2,433 (11.5)	2,614 (12.2)	2,649 (12.6)	2,806 (13.1)	2,856 (13.5)	3,042 (13.9)	3,152 (14.3)

a The percentage is included in the NASA total.

Table 7–21. Average GS Grade Level of Minority and Nonminority Permanent Employees by NASA Occupational Code Group (1979–1988) (at end of fiscal year)

Occupational Code Group	1979 Minority	1979 Non-minority	1980 Minority	1980 Non-minority	1981 Minority	1981 Non-minority	1982 Minority	1982 Non-minority	1983 Minority	1983 Non-minority
Scientists/Engineers	11.9	13.2	11.6	13.1	11.9	13.1	12.1	13.2	11.8	13.0
Professional Admin.	10.5	11.7	10.7	11.7	10.9	11.8	11.1	11.9	11.2	11.9
Technical Support	7.8	10.1	7.5	10.0	7.9	10.1	8.3	10.3	8.5	10.3
Clerical	4.6	5.4	4.6	5.4	4.9	5.5	5.1	5.6	5.2	5.7
All NASA	8.6	11.4	8.5	11.3	8.5	11.3	9.2	11.3	9.3	11.5

Occupational Code Group	1984 Minority	1984 Non-minority	1985 Minority	1985 Non-minority	1986 Minority	1986 Non-minority	1987 Minority	1987 Non-minority	1988 Minority	1988 Non-minority
Scientists/Engineers	12.0	13.0	12.0	13.0	12.1	13.0	12.0	12.9	12.2	13.0
Professional Admin.	11.3	12.0	11.3	12.0	11.3	12.0	11.4	12.0	11.4	12.1
Technical Support	8.6	10.5	8.7	10.4	8.7	10.4	8.8	10.4	8.9	10.4
Clerical	5.2	5.7	5.2	5.6	5.4	5.7	5.5	5.7	5.6	5.8
All NASA	9.4	11.6	9.5	11.6	9.7	11.6	9.8	11.6	10.1	11.7

Table 7–22. *Minority Permanent Employees by NASA Installation: Number on Board (1979–1988) (at end of fiscal year)*

Installation	1979	1980	1981	1982	1983	1984	1985	1986	1987	1988
Ames Research Center	263	285	284	372	377	391	409	419	425	425
Dryden Flight Research Center	65	70	75	—	—	—	—	—	—	—
Goddard Space Flight Center	385	431	410	455	476	453	469	506	527	541
Johnson Space Center	395	463	431	320	425	458	475	476	510	533
Kennedy Space Center	186	169	171	177	192	195	211	222	257	268
Langley Research Center	275	319	311	305	331	321	338	329	341	355
Lewis Research Center	155	228	227	213	236	250	265	260	271	284
Marshall Space Flight Center	126	173	198	183	215	233	276	279	318	333
National Space Technology Laboratories	6	7	8	7	8	10	14	14	15	18
Wallops Flight Center (Facility)	32	37	40	—	—	—	—	—	—	—
NASA Headquarters	299	312	317	301	354	338	349	351	378	395
Total	2,187	2,494	2,472	2,333	2,614	2,649	2,806	2,856	3,042	3,152

Table 7-23. Minorities as a Percentage of Permanent Employees by NASA Installation (1979–1988) (at end of fiscal year)

Installation	1979	1980	1981	1982	1983	1984	1985	1986	1987	1988
Ames Research Center	15.8	17.3	17.7	18.2	18.5	19.1	19.9	20.2	20.4	20.2
Dryden Flight Research Center	13.9	15.1	16.8	—	—	—	—	—	—	—
Goddard Space Flight Center	11.3	12.5	12.4	12.6	13.0	12.8	12.9	13.8	14.4	14.9
Johnson Space Center	11.1	13.2	12.8	12.9	13.1	14.2	14.3	14.6	15.2	15.7
Kennedy Space Center	7.1	7.7	8.0	8.4	9.2	9.4	10.1	10.8	11.7	12.0
Langley Research Center	9.2	10.8	10.7	10.9	11.4	11.4	12.0	11.7	12.0	12.5
Lewis Research Center	6.5	8.1	8.4	8.6	9.0	9.5	9.8	10.0	10.2	10.7
Marshall Space Flight Center	3.5	4.9	5.8	5.5	6.4	7.2	8.4	8.6	9.4	10.0
National Space Technology Laboratories	6.1	6.8	7.6	6.8	7.5	9.3	11.5	11.4	10.9	12.2
Wallops Flight Center (Facility)	8.2	9.7	10.6	—	—	—	—	—	—	—
NASA Headquarters	21.1	20.6	20.6	21.1	23.7	24.2	25.2	25.8	24.7	23.9
Total	9.7	11.0	11.3	11.5	12.2	12.6	13.1	13.5	13.9	14.3

Table 7–24. *Female Permanent Employees by NASA Occupational Code Group: Number on Board (at end of fiscal year; percentage of NASA total in parentheses)*

Occupational Code Group	1979	1980	1981	1982	1983	1984	1985	1986	1987	1988
500 (Clerical)	2,992 (92.7)	3,079 (93.5)	2,906 (93.8)	2,760 (95.0)	2,756 (95.4)	2,723 (95.5)	2,768 (96.1)	2,716 (96.7)	2,765 (96.9)	2,719 (97.1)
100 (Trades/labor)	23 (1.6)	24 (1.7)	21 (1.5)	25 (1.9)	27 (2.0)	38 (3.0)	43 (3.4)	42 (3.5)	37 (3.2)	33 (3.1)
600 (Professional Admin.)	869 (25.4)	1,018 (28.8)	1,077 (30.8)	1,114 (32.2)	1,229 (34.5)	1,308 (37.2)	1,445 (39.3)	1,503 (41.1)	1,631 (42.9)	1,749 (46.8)
200, 700, 900 (Scientists/Engineers)	463 (4.1)	578 (5.2)	596 (5.5)	620 (5.8)	756 (6.8)	808 (7.4)	969 (8.7)	1,086 (9.7)	1,265 (10.8)	1,352 (11.4)
300 (Technical Support)	91 (2.8)	113 (3.6)	111 (3.7)	101 (3.7)	106 (4.0)	108 (4.2)	124 (5.0)	132 (5.5)	151 (6.4)	162 (6.8)
Total	4,438 (19.6)	4,812 (21.3)	4,711 (21.6)	4,620 (21.8)	4,874 (22.7)	4,985 (23.7)	5,349 (25.0)	5,479 (25.8)	5,849 (26.8)	6,015 (27.4)

Table 7–25. Female Permanent Employees by Grade Range:
Number on Board (at end of fiscal year; percentage of NASA total in parentheses)

Grade of Employee	1979	1980	1981	1982	1983	1984	1985	1986	1987	1988
GS-1–6	2,592 (87.5)	2,689 (86.9)	2,461 (88.3)	2,269 (90.4)	2,228 (91.2)	2,184 (93.0)	2,252 (92.0)	2,179 (92.3)	2,136 (91.0)	2,006 (91.1)
GS-7–12	1,605 (20.4)	1,840 (23.5)	1,929 (25.3)	1,987 (27.0)	2,222 (29.4)	2,314 (31.1)	2,498 (32.8)	2,627 (32.8)	2,932 (36.6)	3,074 (38.0)
GS-13–15	210 (2.1)	248 (2.5)	287 (3.0)	315 (3.3)	382 (3.9)	433 (4.6)	538 (5.6)	609 (6.4)	727 (7.4)	882 (8.7)
GS-16–Excepted	8 (1.7)	11 (2.4)	13 (2.9)	14 (3.1)	15 (3.2)	16 (3.2)	18 (3.7)	22 (4.5)	17 (3.6)	20 (3.9)
Wage System	23 a	24 a	21 a	25 a	27 a	38 a	43 a	42 a	37 a	33 (6.8)
Total	4,438 (19.6)	4,812 (21.3)	4,711 (21.6)	4,610 (21.8)	4,874 (22.7)	4,985 (23.7)	5,349 (25.0)	5,479 (25.8)	5,849 (26.8)	6,015 (27.4)

a   The percentage is included in the NASA total.

Table 7–26. *Average GS Grade Level of Male and Female Permanent Employees by NASA Occupational Code Group (1979–1988) (at end of fiscal year)*

Occupational	1979		1980		1981		1982		1983	
Code Group	Males	Females	Males	Females	Males	Females	Males	Females	Males	Females
Scientists/Engineers	13.1	11.2	13.1	10.8	13.1	11.1	13.1	11.4	13.0	11.0
Professional Admin.	12.3	9.8	12.3	10.0	12.4	10.2	12.5	10.5	12.5	10.6
Technical Support	10.1	6.7	9.9	6.4	10.0	6.7	10.2	7.2	10.3	7.5
Clerical	6.2	5.2	6.1	5.1	5.9	5.4	6.1	5.4	6.2	5.5
All NASA	12.3	6.7	12.3	6.9	13.1	11.1	13.1	11.4	12.4	7.7

Occupational	1984		1985		1986		1987		1988	
Code Group	Males	Females	Males	Females	Males	Females	Males	Females	Males	Females
Scientists/Engineers	13.1	11.2	13.0	11.2	13.1	11.3	13.0	11.3	13.1	11.7
Professional Admin.	12.6	10.7	12.6	10.8	12.6	11.0	12.6	11.1	12.7	11.2
Technical Support	10.4	7.6	10.4	7.6	10.4	7.8	10.4	8.0	10.4	8.2
Clerical	6.1	5.5	6.2	5.5	6.2	5.6	6.0	5.6	6.4	5.8
All NASA	12.5	7.9	12.5	8.0	12.5	8.2	12.5	8.4	12.6	8.8

Table 7–27. Female Permanent Employees by NASA Installation: Number on Board (1979–1988) (at end of fiscal year)

Installation	1979	1980	1981	1982	1983	1984	1985	1986	1987	1988
Ames Research Center	329	244	404	402	412	435	450	478	487	494
Dryden Flight Research Center	66	67	67	—	—	—	—	—	—	—
Goddard Space Flight Center	750	803	821	834	875	865	914	962	973	990
Johnson Space Center	687	755	717	705	724	824	910	930	996	1,034
Kennedy Space Center	442	474	474	486	495	511	555	563	604	618
Langley Research Center	508	541	531	515	570	553	579	601	633	642
Lewis Research Center	373	429	417	384	433	452	501	459	484	488
Marshall Space Flight Center	628	692	710	695	722	735	813	858	955	983
National Space Technology Laboratories	25	27	28	31	31	34	42	41	43	48
Wallops Flight Center (Facility)	71	72	74	—	—	—	—	—	—	—
NASA Headquarters	559	608	609	568	612	576	585	587	674	718
Total	4,438	4,712	4,852	4,620	4,874	4,985	5,349	5,479	5,849	6,015

Table 7–28. *Females as a Percentage of Permanent Employees by NASA Installation (1979–1988) (at end of fiscal year)*

Installation	1979	1980	1981	1982	1983	1984	1985	1986	1987	1988
Ames Research Center	19.8	20.8	21.0	19.7	20.3	21.3	21.9	23.1	23.4	23.5
Dryden Flight Research Center	14.1	14.4	15.0	—	—	—	—	—	—	—
Goddard Space Flight Center	21.5	32.4	22.5	23.0	23.9	24.4	25.2	26.1	31.4	32.7
Johnson Space Center	19.7	21.5	21.2	21.6	22.4	25.5	27.3	28.4	29.7	30.4
Kennedy Space Center	20.2	21.5	22.2	23.1	23.8	24.7	26.7	27.5	27.6	27.6
Langley Research Center	16.9	18.2	18.3	18.4	19.6	19.6	20.5	21.4	22.2	22.6
Lewis Research Center	13.1	15.2	15.5	15.5	16.5	17.2	18.5	17.7	18.2	18.4
Marshall Space Flight Center	17.5	19.4	21.0	20.9	21.5	22.8	24.8	26.3	28.2	29.4
National Space Technology Laboratories	25.5	26.2	26.7	30.1	29.2	24.7	34.4	33.3	31.4	32.7
Wallops Flight Center (Facility)	18.2	18.8	19.7	—	—	—	—	—	—	—
NASA Headquarters	39.5	40.1	40.5	39.7	41.0	41.3	42.3	43.1	44.0	43.4
Total	19.6	21.3	21.6	21.8	22.7	23.7	25.0	25.8	26.8	27.4

Table 7–29. Age Profile of Permanent Employees by Grade Range:
Number on Board (at end of fiscal year; percentage of NASA total in parentheses)

Age	1979	1980	1981	1982	1983	1984	1985	1986	1987	1988
Under 25	783 (3.4)	1,031 (4.6)	844 (3.9)	825 (3.9)	1,116 (3.8)	1,072 (5.1)	1,181 (3.7)	1,051 (5.0)	1,139 (3.4)	999 (4.5)
25–29	1,428 (6.3)	1,707 (7.6)	1,736 (7.9)	1,650 (7.8)	1,708 (6.1)	1,722 (8.2)	1,977 (5.5)	2,191 (10.3)	2,436 (6.3)	2,558 (10.6)
30–34	1,800 (7.9)	1,965 (8.7)	1,932 (8.8)	1,739 (8.2)	1,856 (10.5)	1,809 (8.6)	1,998 (9.0)	2,086 (9.8)	2,212 (7.9)	2,341 (10.9)
35–39	2,895 (12.8)	2,744 (12.1)	2,546 (11.7)	2,410 (11.4)	2,209 (15.2)	2,138 (10.2)	2,184 (13.8)	2,236 (10.5)	2,259 (12.8)	2,397 (10.9)
40–44	3,862 (17.1)	3,798 (16.8)	3,596 (16.5)	3,406 (16.1)	3,294 (18.9)	3,057 (14.5)	2,859 (17.6)	2,699 (12.7)	2,745 (17.1)	2,659 (12.1)
45–49	4,501 (19.9)	4,474 (19.8)	4,289 (19.6)	4,074 (19.2)	3,954 (17.0)	3,828 (18.2)	3,712 (19.1)	3,540 (16.7)	3,483 (19.9)	3,441 (15.6)
50–54	3,548 (15.7)	3,601 (15.9)	3,610 (16.5)	3,788 (17.9)	3,940 (17.0)	3,885 (18.5)	3,900 (16.3)	3,808 (17.9)	3,759 (15.7)	3,731 (17.0)
55–59	2,717 (12.0)	2,259 (10.0)	2,185 (10.0)	2,096 (9.9)	2,160 (8.5)	2,222 (10.6)	2,268 (11.3)	2,307 (10.9)	2,467 (12.0)	2,515 (11.4)
60 Plus	1,099 (4.9)	1,034 (4.6)	1,106 (5.1)	1,198 (5.7)	1,263 (3.0)	1,317 (6.3)	1,344 (3.8)	1,310 (6.2)	1,331 (4.9)	1,350 (6.1)
Average Age	44.3	43.5	43.8	44.1	43.9	44.0	43.7	43.6	43.3	43.3

CHAPTER EIGHT
FINANCES AND PROCUREMENT

CHAPTER EIGHT
FINANCES AND PROCUREMENT

Introduction

This chapter provides an overview of NASA's budgetary and procurement activities, primarily in tabular form. The first part of the chapter covers NASA's budget; the second part addresses its procurement activities.

Finances

The tables covering NASA's budget at the end of this chapter provide top-level budget data for the years 1979–1988. More detailed budget data on each of NASA's programs are provided in Chapters 2 through 4 of Volume V of the *NASA Historical Data Book* and Chapters 2 through 5 of this volume. As shown in Table 8–1, NASA's outlays comprised only a small percentage of the total federal budget during this decade—remaining less than 1 percent for the entire period. The amount appropriated by Congress grew annually at a rate close to the rate of inflation in six of the years (FY 1981, FY 1982, FY 1983, FY 1984, FY 1985, and FY 1986) and at a greater rate in three of the years (FY 1979, FY 1980, and FY 1987). The one decrease occurred in FY 1988 but could really be considered an adjustment to the vast increase that took place in FY 1987, when NASA awarded the contract and provided funding for a replacement orbiter after the *Challenger* accident.

When looked at by individual appropriation, Congress authorized and appropriated amounts higher than NASA's budget request about half the time. However, NASA may have revised its budget request downward before submitting the budget to Congress to receive Office of Management and Budget approval.

NASA had four appropriation categories (see Figure 8–1). Research and Development (R&D) funded most of NASA's flight projects and basic research activities. The Space Flight, Control, and Data Communications (SFC&DC) appropriation funded Shuttle-related and most tracking and data acquisition activities. Research and Program Management (R&PM) financed civil service salaries regardless of the project or the office in

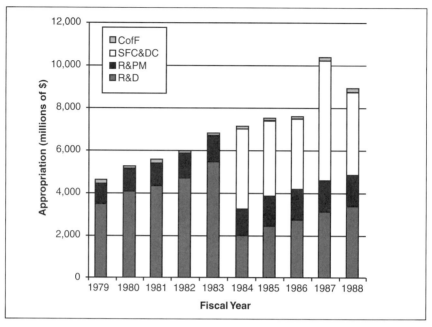

Figure 8–1. NASA Appropriations

which an individual worked, related expenses such as benefits and training, and employee travel. The Construction of Facilities (CofF) appropriation funded design, construction, purchase of land, modernization of facilities, and design of facilities planned for future authorization. By 1979, most of NASA's facilities had been acquired or built, and the amount allocated for this appropriation remained relatively low. Congress allocated funds for each specific appropriation category. By law, NASA may not reallocate funds among the appropriation categories without congressional approval.

Congress appropriated new funds for NASA each year. However, the funds in each appropriations category were available for use for a varying number of years. In addition, occasionally Congress specified in its appropriations legislation the length of time that money could be used for a particular project or the start date (other than the beginning of the fiscal year) for the availability of funds.

The tables that follow (Tables 8–2 through 8–14) use some budget-related terms unfamiliar to the reader. These are:

- **Budget authority** gives NASA permission (or authorization) to make an outlay or spend up to a specific amount. This authorization may be limited to a particular number of fiscal years and is usually for a particular purpose.
- An **appropriation** is the amount that Congress has made available for an agency's use.
- **Funding** is the amount that an entity has actually spent. It is computed at the end of the fiscal year.

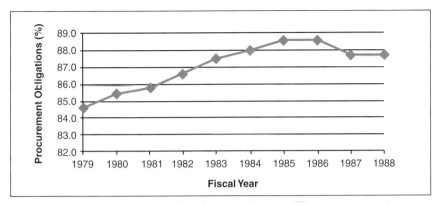

Figure 8–2. Percentage of NASA Obligations Devoted to Procurement Actions

- An **obligation** is a legal promise to pay for a specified item or service.
- The government **fiscal year** runs from October 1 through September 30. The fiscal year is designated by the year in which it ends—that is, FY 1988 begins October 1, 1987, and ends September 30, 1988.
- A **supplemental appropriation** is enacted after the regular annual appropriation act. Supplemental appropriations provide additional budget authority for programs or activities (including new programs authorized after the date of the original appropriations act) when the need for funds is too urgent to be postponed.

NASA's budget process is complex. An overview of this process is located in Chapter 1 in Volume V of the *NASA Historical Data Book*.

Procurement

The Space Act of 1958 established NASA's policy of using procurements for a large proportion of its activities. From 1979 to 1988, NASA used procurements to acquire more than 80 percent of its items, services, and systems. The proportion of obligations the agency devoted to procurement actions ranged from a low of 84.6 percent in 1979 to a high of 88.6 percent in FY 1985 and FY 1986 (Figure 8–2). During these years, NASA awarded procurements to large and small businesses, nonprofit and educational institutions, the Jet Propulsion Laboratory (JPL), other government agencies, and contractors outside the United States. The great majority (87 percent) was awarded to business firms, with the larger number awarded to small firms, but procurements with greater monetary value were awarded to large firms (Figure 8–3). During the decade, the total number of procurements reached a peak of 175,900 in 1981 and then began its decrease to a low of 108,800 in 1988 (Figure 8–4). Figure 8–5 shows the percentage of award value by type of contractor. The total value of the awards grew through the decade, with a small dip occurring only in 1986 (Figure 8–6). The combination of greater total award value and fewer awards meant that the average value of awards increased.

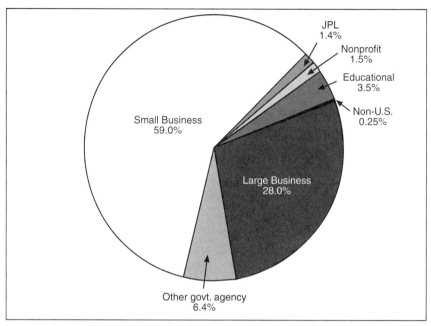

*Figure 8–3. Total Percentage of Procurement Actions by **Kind of Contractor** (1979–1988)*

NASA used a variety of contract pricing provisions in its awards. These included fixed-price, incentive, cost-plus-award-fee, cost-plus-fixed-fee, and a small number of other types of pricing provisions. Awards with the greatest monetary value used incentive or cost-plus-award-fee pricing provisions. Most awards used fixed-price provisions (FY 1979, 1983, 1984, 1985, 1986, 1987, and 1988) or incentive pricing provisions (FY 1980, 1981, and 1982).

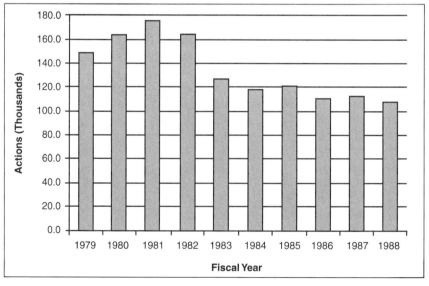

*Figure 8–4. **Total Number of Procurement Actions by Fiscal Year***

FINANCES AND PROCUREMENT

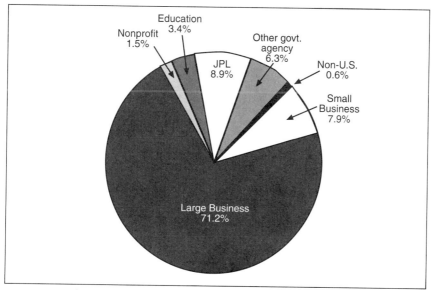

Figure 8–5. Percentage of Procurement Award Value by Kind of Contractor

Every state in the United States and the District of Columbia received NASA contracts during the decade. The greatest number of awards, with the greatest value, consistently went to California. Every geographic region increased the value of contracts it received except for New England, which ended the decade with awards of less value than it began (Figure 8–7).

Johnson Space Center and Marshall Space Flight Center awarded contracts with the greatest value throughout the decade. Rockwell International continued throughout the decade as NASA's top contractor, a position it had held since FY 1973. In all, NASA's top fifty contractors received contracts valued between 82 percent and 86 percent of the total value of NASA contracts. Of the nonprofit and educational institutions receiving NASA awards, the European Space Agency and

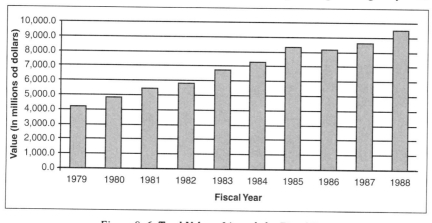

Figure 8–6. Total Value of Awards by Fiscal Year

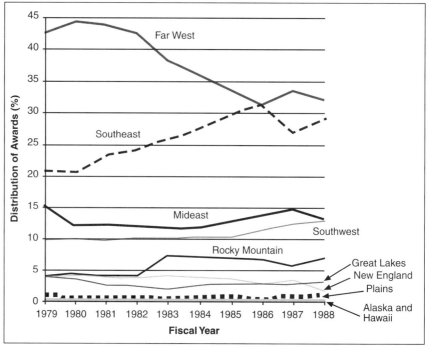

Figure 8–7. Distribution of Prime Contract Awards by Region

Stanford University led the category with each receiving awards with the greatest value for four years. Figure 8–8 shows a breakdown, by fiscal year, of NASA's total value of business and educational/nonprofit awards.

NASA's Procurement Process

The Federal Acquisition Regulation (FAR), supplemented by the NASA FAR Supplement, governs all stages of the procurement process. For a competitive procurement (from 1979 to 1988, approximately 80 percent of all NASA procurements were competitive), NASA's procurement process consisted of six steps:

1. Preparation of a Procurement Request (PR)
2. Preparation of a procurement plan
3. Solicitation
4. Bid or proposal evaluation
5. Selection, negotiation, and contract award
6. Contract management or administration

Prior to the preparation of a PR, a Mission Need Statement is prepared and approved. In addition, a Work Breakdown Structure (WBS) is prepared for most projects; it provides the basic framework for the pro-

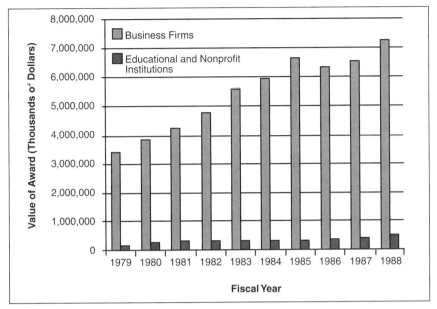

Figure 8–8. Total Value of Business and Educational and Nonprofit Awards by Fiscal Year

ject and presents the specific elements of a project to as great a level of detail as possible. It also often forms the structure of the Request for Proposal (RFP).

Procurement Request

The PR document describes NASA's requirements to procure goods or services over a particular dollar level in response to mission needs. It can include technical specifications for a material, product, or service (used primarily for procuring hardware); a statement of the work to be performed (used primarily for research and development, study, and service-oriented requirements); or a purchase description that includes the essential characteristics and functions of an item to be procured (used for off-the-shelf items). For procurements over a certain dollar amount, the PR also includes an in-house cost estimate that provides the government's best estimate of the actual price of the procurement.

Procurement Plan

A procurement plan describes each step to be taken to fulfill the procurement action. The contract negotiator, with the assistance of the PR initiator, develops the plan. Such a plan describes the items or systems to be procured in as much detail as possible, the schedule for completing the acquisition of the items or systems, possible sources for materials or services, and the recommended type of contract that will be used. It specifies whether the items will be procured through formal advertising for competitive bids or through a negotiation process, as well as the recommended type of contract that will be awarded.

Solicitation of Proposals

If the items to be procured can be accurately defined, then formal advertising for competitive bids takes place, and the government will release an Invitation for Bid (IFB) to interested suppliers. These are usually fixed-price contracts, in which the contractor is paid a fixed price for completing the required work within the allotted time.

More often, the items, systems, or services cannot be completely defined, and a cost reimbursement contract is awarded. This type of contract allows for government payment of reasonable allowable and otherwise allowable costs as defined in the FAR or in the contract provisions. Cost-plus-award-fee and cost-plus-fixed-fee contracts are two commonly used types of cost reimbursement contracts.

For a cost reimbursement contract, the government uses an RFP. The RFP contains a complete and specific description of the items or services to be procured, all applicable specifications, quantities, time and place of delivery, method of shipment, and other requirements. To reduce the government's time and expense needed to prepare major RFPs and evaluate proposals, the solicitation is normally restricted to those contractors that NASA has determined can adequately fulfill the mission needs objectives and that have an interest in the procurement. However, the release of RFPs is also announced in the *Commerce Business Daily* publication, and others may choose to submit a proposal.

The contracting officer is responsible for selecting the contract type. This person must be satisfied that the type selected is appropriate for the PR. The key element is the degree of uncertainty in the technical performance. The less uncertainty there is, the greater the likelihood that the government will choose a fixed-price contract.

Bid or Proposal Evaluation

The evaluation of bids is a straightforward process. As long as NASA is assured that the bidder can provide the items or services specified in the IFB, the lowest price receives the award.

The evaluation of proposals submitted in response to RFPs is more complex. The government evaluates proposals in accordance with the procedures set forth in the FAR. Bidders are usually evaluated in three areas: technical competence, managerial competence, and the quality of the proposal. The relative weight given to each of these categories and to the subareas within these categories is spelled out completely in the RFP. A Source Evaluation Board, through a formal process, evaluates contracts over a certain dollar level. Members of the organization that will administer the contract and contract specialists within the Procurement Office evaluate smaller procurements.

Selection, Negotiation, and Award

For a fixed-price contract, the process ends with the selection of the least expensive, technically qualified firm. For a cost reimbursement contract, once a firm has been selected, additional details still need to be negotiated before contract award. Any cost-related provision can be negotiated, including, but not limited to, the number of labor hours, travel expenses, overhead rates, and other direct costs. The question of fee, which is an amount in addition to the basic contract cost, needs to be settled. The fee can be a fixed amount specified in the contract, providing the contract requirements are met, or can vary depending on regular evaluations of contractor performance. The method of addressing inflation or escalating costs is another area that is negotiated. Following the completion of negotiations, the award takes place and work can begin on the contract.

Contract Management and Administration

Contract administration consists of all activities that take place from contract award until contract retirement. NASA has in place a large number of procedures that ensure that its contracts are managed efficiently and legally. For a procurement of an item, the delivery of that item may be the only major requirement. However, for administering the procurement of complicated systems or services, the government provides considerable oversight, and the contractor is responsible for providing regular information to the government regarding cost and progress of work. The government reviews and evaluates this information and may find that revisions to the contract or changes in project plans are necessary as a result of what these progress reports reveal. NASA employs a number of individuals who are responsible for contract administration.

Definition of Terms

The following information is from NASA Annual Procurement Reports:

Direct Actions (Direct Awards). Procurement actions placed directly with business firms or educational and nonprofit institutions or organizations. The term excludes procurement actions placed with or through other government agencies.

Intergovernmental. Procurement actions placed with or through other government agencies, except orders placed under Government Services Administration (GSA) Federal Supply Schedule contracts and awards to minority enterprises through the Small Business Administration under Section 8(a) of the Small Business Act.

Modification. Any written alteration in the specifications, delivery point, rate of delivery, contract period, price, quantity, or other contract provision of an existing contract, whether accomplished by unilateral action in accordance with a contract provision or by mutual action of the parties to the contract. It includes (a) bilateral actions, such as supplemental agreements, and (b) unilateral actions, such as change orders, notices of termination, and notices of the exercise of an option.

Competitive. Procurements in which offers are received from at least two responsible offerors capable of satisfying the government's requirements wholly or partially, and the award or awards are made on the basis of price, design, or technical competition.

Other Than Competitive. Procurements in which an offer was received from only one responsible offeror capable of satisfying the government's requirements wholly or partially (includes contracts resulting from unsolicited proposals).

Net Value. The net amount of obligations resulting from debit and credit procurement actions.

Procurement Action (Award). Any contractual action to obtain supplies, services, or construction that obligates or de-obligates funds, including:

1. Letter contracts or other preliminary notices of negotiated awards
2. Definitive contracts, including purchase orders
3. Orders under GSA Federal Supply Schedule contracts, basic order agreements, and against indefinite delivery type contracts
4. Intragovernmental
5. Grants
6. Cooperative and Space Agreements
7. Supplemental agreements, change orders, administrative changes, and terminations to existing procurements

Small Business. For purposes of government procurement, a profit-making concern, including its affiliates, which is independently owned and operated, is not dominant in its field, and further qualifies under the size standards criteria of the Small Business Administration that are published under Title 13 of the Code of Federal Regulations and in the FAR, Part 19, Subpart 19.1. The applicable size standard is prescribed in each NASA procurement solicitation.

The procurement tables (Tables 8–15 through 8–47) at the end of this chapter provide information on NASA's procurement activities. The data have been taken from the agency's Annual Procurement Reports.

Table 8–1. NASA's Budget Authority as a Percentage of the Total Federal Budget

FY	Percentage
1979	0.9
1980	0.8
1981	0.8
1982	0.8
1983	0.8
1984	0.8
1985	0.8
1986	0.7
1987	0.8
1988	0.9

Source: Historical Tables, Budget of the United States Government, Fiscal Year 1990.

Table 8–2. NASA Appropriations by Appropriation Title and Fiscal Year and Percentage Change (in millions of dollars)

Fiscal Year	R&D	R&PM	SFC&DC	CofF	Total	Percentage Change
1979	3,477.2	934.1	—	147.5	4,558.8	12.2
1980	4,091.1	996.2	—	156.1	5,243.4 a	15.0
1981	4,336.3	1,071.4	—	115.0	5,522.7 b	5.3
1982	4,740.9	1,103.3	—	95.8	5,940.0	7.6
1983	5,542.8	1,197.4	—	97.5	6,837.7 c	15.1
1984	2,011.9	1,238.5	3,791.6	135.5	7,177.5	5.0
1985	2,468.1	1,332.3	3,601.8	150.0	7,552.2 d	5.2
1986	2,756.8	1,362.0	3,397.9	139.3	7,756.0	2.7
1987	3,127.7	1,425.0	5,715.0	166.3	10,434.0	34.5
1988	3,374.2	1,495.7	3,908.3	178.3	8,956.5	−14.2
Total	35,927.0	12,155.9	20,414.6	1,381.3	27,146.5	

a Reflects supplemental appropriation.
b Reflects supplemental appropriation.
c Reflects supplemental appropriation.
d Reflects supplemental appropriation.

Source: NASA Chronological Budget History (1979–1988).

Table 8–3. *NASA's Budget History (1979–1988) (in millions of dollars)*

Fiscal Year/ Appropriation Title	Request	Authorization	Appropriation
1979			
R&PM	942.6	940.0	934.1
R&D	3,490.1	3,522.6	3,477.2
CofF	185.0	150.0	147.5
Total	4,585.2 *a*	4,612.6 *b*	4,558.8
1980			
R&PM	1,011.2	1,001.2	996.2
R&D	4,122.5	3,838.5	4,091.1
CofF	157.6	157.6	156.1
Total	5,291.3 *c*	4,997.3 *d*	5,243.4 *e*
1981			
R&PM	1,033.2	1,933.2	1,071.4
R&D	4,364.5	4,436.8	4,336.3
CofF	120.0	118.0	115.0
Total	5,517.7 *f*	5,587.9	5,522.7 *g*
1982			
R&PM	1,114.3	1,114.2	1,103.3
R&D	4,903.1	4,953.8	4,740.9
CofF	104.8	104.1	95.8
Total	6,122.2 *h*	6,172.2	5,940.0
1983			
R&PM	1,178.9	1,168.9	1,197.4
R&D	5,334.0	5,504.0	5,542.8
CofF	100.0	100.0	97.5
Total	6,612.9	6,772.9	6,837.7 *i*
1984			
R&PM	1,247.5	1,242.5	1,238.5
R&D	5,708.5	5,833.0	2,011.9
SFC&DC	—	—	3,791.6
CofF	150.5	142.1	135.5
Total	7,106.5	7,267.6	7,177.5
1985			
R&PM	1,331.0	1,316.0	1,332.3
R&D	2,400.1	2,475.1	2,468.1
SFC&DC	3,600.3	3,585.3	3,601.8
CofF	160.0	150.0	150.0
Total	7,491.4	7,526.4	7,552.2 *j*

Table 8–3 continued

Fiscal Year/ Appropriation Title	Request	Authorization	Appropriation
1986			
R&PM	1,345.0	1,367.0	1,362.0
R&D	2,881.8	2,786.8	2,756.8
SFC&DC	3,509.9	3,372.9	3,397.9
CofF	149.3	139.3	139.3
Total	7,886.0	7,666.0	7,756.0
1987			
R&PM	1,425.0	1,425.0	1,425.0
R&D	3,032.1	3,112.7	3,127.7
SFC&DC	3,343.0	3,899.0	5,715.0
CofF	166.3	161.3	166.3
Total	7,996.4 k	8,598.0	10,434.0
1988			
R&PM	1,598.0	1,593.0	1,495.7
R&D	3,623.2	3,719.0	3,374.2
SFC&DC	4,092.3	4,045.3	3,908.3
CofF	195.5	216.5	178.3
Total	9,509.0 l	9,573.8	8,956.5

a   Includes supplemental request and recission.
b   Reflects supplemental authorization.
c   Amended budget request; reflects supplemental request.
d   Reflects supplemental authorization.
e   Reflects supplemental appropriation.
f   Amended budget request.
g   Reflects supplemental appropriation.
h   Amended budget request.
i   Reflects supplemental appropriation.
j   Reflects supplemental appropriation.
k   Amended budget request.
l   Amended budget request.
Source: Table 8–1 and *NASA Chronological Budget History (1979–1988).*

Table 8–4. Authorizations and Appropriations Compared With Budget Requests (in millions of dollars)

Amounts and Percentages Cut (or Added) by Congress

	Research and Program Management		Research and Development		Space Flight, Control, and Data Communications		Construction of Facilities		Total	
	Amount	Percentage	Amount	Percentage	Amount	Percentage	Amount	Percentage	Amount	Percentage
FY 1979										
Auth.	-2.6	-0.3	32.5	0.9	—	—	-2.5	-1.4	27.4	0.6
Appr.	-8.5	-0.9	-12.9	-0.4	—	—	-5.0	-2.7	-26.4	-0.6
FY 1980										
Auth.	-10.0	-1.0	-284.0	-6.9	—	—	0.0	0.0	-294.0	-5.6
Appr.	-15.0	-1.5	-31.4	-0.8	—	—	-1.5	-1.0	-47.9	-0.9
FY1981										
Auth.	0.0	0.0	72.3	1.7	—	—	-2.0	-1.7	70.3	1.3
Appr. a	0.0	0.0	0.0	0.0	—	—	0.0	0.0	0.0	0.0
FY 1982										
Auth.	0.0	0.0	50.7	1.2	—	—	-0.7	-0.7	50.0	0.8
Appr.	-11.0	-1.0	-162.2	-3.3	—	—	-9.0	-8.6	-182.2	-2.3
FY 1983										
Auth.	-10.0	-0.8	170.0	3.2	—	—	0.0	0.0	160.0	2.4
Appr.	-10.0	-0.8	208.0	3.9	—	—	-2.5	-2.5	196.3	3.0
FY 1984										
Auth.	-5.0	-0.4	174.5	3.1	—	—	-8.4	-5.6	161.1	2.3
Appr.	-9.0	-0.7	-3,696.6	-64.8	3,791.6	66.4	-15.0	-10.0	-90.1	3.0

Table 8-4 continued
Amounts and Percentages Cut (or Added) by Congress

	Research and Program Management		Research and Development		Space Flight, Control, and Data Communications		Construction of Facilities		Total	
	Amount	Percentage	Amount	Percentage	Amount	Percentage	Amount	Percentage	Amount	Percentage
FY 1985										
Auth.	-15.0	-1.1	75.0	3.1	-15.0	-0.4	-10.0	-6.3	35.0	0.5
Appr.	1.3	0.1	68.0	2.8	1.5	b	-10.0	-6.3	60.8	0.8
FY 1986										
Auth.	22.0	1.6	-95.0	-3.3	-137.0	-3.9	-10.0	-6.6	-220.0	-2.8
Appr.	17.0	1.2	-125.0	-4.3	-112.0	-3.2	-10.0	-6.6	-130.0	-1.6
FY 1987										
Auth.	0.0	0.0	80.6	2.7	556.0	16.6	-5.0	-3.0	631.6	7.9
Appr.	0.0	0.0	95.6	3.2	2,372.0	70.1	0.0	0.0	2,467.6	73.8
FY 1988										
Auth.	-5.0	—	95.8	2.3	-47.0	-1.1	21.0	10.7	64.8	0.7
Appr.	-102.3	-6.4	-249.0	-6.9	-184.0	-4.5	-17.2	-8.8	-552.5	-5.8

a Difference from revised budget estimate.
b Less than 0.05 percent.
Source: NASA Chronological History (1979–1988).

Table 8–5. Budget Requests, Authorizations, Appropriations, and Obligations (in millions of dollars)

Fiscal Year	Budget Request	Authorization	Appropriation	Obligation	Expenditure
1979	4,585.2	4,612.6	4,558.8	4,557.5	4,196.5
1980	5,291.3	4,997.3	5,243.4	5,098.1	4,851.6
1981	5,517.7	5,587.9	5,522.7	5,606.2	5,421.2
1982	6,122.2	6,172.2	5,940.0	5,946.7	6,035.4
1983	6,612.9	6,772.9	6,837.7	6,732.9	6,663.9
1984	7,106.5	7,267.6	7,177.5	7,135.2	7,047.6
1985	7,491.4	7,526.4	7,552.2	7,638.4	7,317.7
1986	7,886.0	7,666.0	7,756.0	7,463.7	7,403.5
1987	7,996.4	8,598.0	10,434.0	8,603.7	7,591.4
1988	9,509.0	9,573.8	8,956.5	9,914.7	9,091.6
Total	63,348.6	68,774.7	69,978.8	68,688.1	65,620.4

Source: NASA Pocket Statistics, January 1989, and Table 8–3.

Table 8–6. Research and Program Management Funding by Installation (in millions of dollars; at end of fiscal year)

Installation	1979	1980	1981	1982	1983	1984	1985	1986	1987	1988
Ames Research Center	62.7	67.4	94.8	101.1	107.2	113.9	120.3	123.3	133.6	165.2
Dryden Flight Research Center	19.1	20.4	—	—	—	—	—	—	—	—
Goddard Space Flight Center	127.9	133.5	162.4	169.1	180.6	186.8	196.9	199.5	213.9	242.8
Johnson Space Center	152.9	164.7	176.1	186.5	195.2	201.1	214.8	206.0	227.9	282.0
Kennedy Space Center	123.3	133.2	150.2	156.0	161.3	172.6	184.5	192.2	200.2	242.8
Langley Research Center	106.6	114.0	120.8	126.6	132.7	140.0	147.1	145.6	154.3	177.9
Lewis Research Center	87.5	94.8	99.9	106.4	118.8	128.7	137.4	143.4	151.7	181.9
Marshall Space Flight Center	149.0	155.9	165.0	172.1	184.3	189.9	198.1	194.2	212.0	237.5
National Space Technology Laboratories/Stennis	4.5	4.9	5.5	6.6	6.3	10.2	10.7	11.2	12.0	20.5
Wallops Flight Center (Facility)	15.8	17.7	—	—	—	—	—	—	—	—
NASA Headquarters	84.5	89.6	96.4	109.8	111.0	108.2	116.9	74.0	139.4	212.5
R&PM Applied to Shuttle Development				49.0						
Inspector General						4.7	5.2			
Total R&PM	933.8	996.0	1,071.01	1,183.0	1,197.4	1,256.1	1,331.9	1,341.5	1,451.5	1,762.2

Table 8–7. Research and Development Funding by Installation (in millions of dollars; at end of fiscal year)

Installation	1979	1980	1981	1982	1983	1984	1985	1986	1987	1988
Ames Research Center	141.5	147.9	160.4	175.3	180.1	189.7	217.3	245.8	282.1	261.2
Dryden Flight Research Center	13.1	16.6	—	—	—	—	—	—	—	—
Goddard Space Flight Center	519.3	552.0	511.2	744.0	730.1	361.6	399.0	461.6	483.9	508.0
Jet Propulsion Laboratory	235.1	315.4	258.7	316.2	304.9	248.3	338.7	412.7	452.4	486.3
Johnson Space Center	1,151.2	1,388.0	1,523.3	1,598.5	1,550.0	175.8	239.7	244.7	325.1	327.5
Kennedy Space Center	235.5	300.6	365.1	486.1	529.3	53.0	45.7	69.0	53.8	86.0
Langley Research Center	138.8	170.9	142.9	129.6	131.0	140.1	176.0	173.0	217.6	197.1
Lewis Research Center	149.2	170.7	163.4	175.1	281.1	281.9	311.4	259.9	277.8	254.8
Marshall Space Flight Center	747.8	863.8	996.0	1,200.0	1,624.2	447.1	500.0	538.4	721.1	744.4
National Space Technology Laboratories/Stennis	15.4	9.4	8.5	9.9	8.3	9.6	10.8	9.6	11.8	19.1
Wallops Flight Center (Facility)	16.6	17.5	—	—	—	—	—	—	—	—
NASA Headquarters	115.7	135.9	144.5	137.2	239.7	160.7	179.6	204.5	328.1	374.6
Total R&D	3,477.2	4,088.1	4,335.5	4,738.0	5,539.0	2,064.2	2,468.1	2,619.3	3,153.7	3,279.9

Table 8–8. Space Flight Control and Data Communications Funding by Installation (in millions of dollars; at end of fiscal year)

Installation	1984	1985	1986	1987	1988
Ames Research Center	9.8	12.3	15.4	16.3	15.4
Dryden Flight Research Center	—	—	—		
Goddard Space Flight Center	431.0	431.1	331.2	416.9	464.3
Jet Propulsion Laboratory	97.2	111.0	116.2	124.3	131.5
Johnson Space Center	1,303.2	1,159.7	988.3	2,337.9	908.8
Kennedy Space Center	439.9	383.1	448.5	642.1	719.7
Langley Research Center	0.2	0.1	0.1	0.1	0.1
Lewis Research Center	2.0	3.4	3.0	5.1	3.7
Marshall Space Flight Center	1,272.9	1,223.5	1,543.7	1,580.1	1,261.2
National Space Technology Laboratories/Stennis	0.8	6.3	8.5	15.8	16.2
Wallops Flight Center (Facility)	—	—	—	—	—
NASA Headquarters	240.4	263.7	211.0	860.9	202.9
Total SFC&DC	3,772.3	3,544.2	3,665.9	5,999.5	3,805.7

Source: NASA Budget Estimates (1979–1988).

Table 8–9. Construction of Facilities Funding by Facility (in millions of dollars; at end of fiscal year)

Facility	1979	1980	1981	1982	1983	1984	1985	1986	1987	1988
Ames Research Center	9.7	2.9	13.9	18.5	3.5	4.7	13.6	7.8	22.2	23.4
Dryden Flight Research Center	5.6	—	—	—	—	—	—	—	—	—
Goddard Space Flight Center	—	—	—	—	4.7	—	2.2	3.6	15.2	19.8
Jet Propulsion Laboratory	4.6	—	3.5	1.0	—	4.3	12.2	9.4	16.3	7.2
Johnson Space Center	—	—	—	0.7	—	2.3	3.2	—	18.4	11.1
Kennedy Space Center	—	5.8	0.8	1.7	10.7	58.6	36.9	—	11.9	26.9
Langley Research Center	6.5	8.0	20.8	3.0	16.2	9.5	13.8	4.7	17.9	7.2
Lewis Research Center	6.1	5.7	10.4	1.2	3.9	10.6	—	—	12.1	22.7
Marshall Space Flight Center	—	3.5	—	—	17.8	11.7	1.6	—	10.1	19.2
National Space Technology Laboratories/Stennis	—	—	—	—	—	—	3.3	—	6.2	4.9
Wallops Flight Center (Facility)	—	1.1	—	—	—	—	—	—	—	—
Large Aeronautical Facilities	56.1	45.9	—	—	—	—	—	—	—	—
Various Locations	—	—	3.4	9.8	—	—	13.8	16.7	22.0	8.6
Space Shuttle Facilities	31.1	27.8	10.1	35.1	—	—	—	—	—	—
Space Shuttle Payload Facilities	—	4.3	1.6	—	—	—	—	—	—	—
Facilities Planning and Design	10.7	14.0	10.0	10.0	—	—	—	—	17.0	16.0
Repair	—	12.0	15.0	12.8	—	—	—	—	—	—
Minor Construction	4.2	3.5	4.0	2.3	—	—	—	—	—	—
Rehabilitation and Modification	12.8	19.8	19.0	17.7	—	—	—	—	—	—
Total CofF	147.5	156.1	117.0	113.7	101.3	155.5	150.0	133.3	169.3	178.3

Source: Annual NASA Budget Estimates.

FINANCES AND PROCUREMENT 517

Table 8–10. Research and Development Appropriation by Program (in millions of dollars)

Program	1979 a	1980 b	1981	1982	1983	1984	1985	1986	1987	1988
Space Station							155.5	205.0	410.0	425.0
Industrial Space Facility										25.0
Space Shuttle c			1,995.0	d	1,769.0	427.4	407.4	439.0	507.5	549.6
Space Flight Operations			679.2	e	1,796.0	f				
Expendable Launch Vehicles			54.4	31.2	42.8	g				
Physics and Astronomy			323.7	323.5	461.7	578.6	680.2	605.4	528.5	577.1
Lunar and Planetary h			175.6	205.0	180.4	205.4	293.9	354.0	374.3	332.3
Life Sciences			42.2	39.5	55.7	59.0	63.3	68.0	69.7	74.6
Space Applications			331.6	328.2	341.3	293.0	384.1	519.8	578.1	641.3
Technology Utilization			8.8	8.0	4.0	9.0	9.5	11.1	15.7	18.3
Commercial Use of Space								17.0	25.6	31.7
Aeronautical Research and Technology			272.2	264.8	280.0	302.3	342.4	354.0	376.0	377.0
Transatmospheric Research and Technology									40.0	53.0
Space Research and Technology			110.7	111.0	123.0	138.0	154.0	168.0	185.2	235.0
Safety, Reliability, and Quality Assurance										16.2
Advanced Systems									17.1	18.1
Energy Technology Applications			1.9	—	i				j	
Tracking and Data Acquisition			341.1	402.1	508.9	14.2 k	15.3	16.2		

Table 8–10 continued

Program	1979 a	1980 b	1981	1982	1983	1984	1985	1986	1987	1988
Supplemental	185.0	285.0	l							
Recission		-1.4								
Undistributed					-20.0	-15.0	-37.5	-1.0		
Total	3,477.2	4,091.1	4,336.3	4,740.9	5,542.8	2,011.9	2,468.1	2,756.8	3,127.7	3,374.2

a FY 1979 appropriations were undistributed (no amounts specified for particular programs).
b FY 1980 appropriations were undistributed (no amounts specified for particular programs).
c Renamed Space Transportation Capability Development beginning in FY 1984. See Table 8–11 for SFC&DC Shuttle appropriations.
d Appropriated amount for Space Shuttle budget category not specified.
e Appropriated amount for Space Flight Operations budget category not specified.
f Budget category moved to SFC&DC appropriation.
g Budget category eliminated.
h Renamed Planetary Exploration beginning in FY 1981.
i Energy Technology Applications budget category eliminated.
j Budget category eliminated.
k Remaining Tracking and Data Acquisition appropriation moved to SFC&DC appropriation.
l Supplemental appropriation amounts incorporated into individual budget categories.

Table 8–11. Space Flight Control and Data Communications Appropriation by Program (in millions of dollars)

Installation	1984	1985	1986	1987	1988
Space Transportation Capability Development (STCD) a	1,500.0	1,510.6	976.5	2,984.4	1,100.6
STCD Reserve	45.0	—			
Space Transportation System Operations	1,570.6	1,339.0	1,725.1	1,867.7	1,885.8
Tracking and Data Acquisition b	676.0	795.7	701.3	862.9	893.9
General Reduction		–43.5	–5.0		
Total	3,791.6	3,601.8 c	3,397.9	5,715.0	3,908.3

a Renamed Shuttle Production and Operational Capability beginning in FY 1985.

b Renamed Space and Ground Network, Communications and Data Systems beginning in FY 1986.

c Supplemental appropriation amounts incorporated into individual budget categories.

Table 8–12. Research and Development Funding by Program (in millions of dollars)

Program	1979	1980	1981	1982	1983	1984	1985	1986	1987	1988
Space Station							155.5	184.7	420.0	489.5
Industrial Space Facility										
Space Transportation Systems a	2,011.6	2,681.1	2,728.6	3,089.9	3,563.0	431.7	391.4	403.9	491.1	593.4
Space Shuttle/Capability										
Development	1,638.3	1943.0	1,995.0	2,623.4	2,141.3	—				
Operations	299.6	683.7	679.2	466.5	1,421.7	—				
Expendable Launch Vehicles	73.6	54.4	54.4	—b						
Space Science c	505.4	562.5	881.8	896.2	1,060.1	1,157.0	1,404.5	1,476.5	1,547.6	1,581.8
Physics and Astronomy	282.9	344.7	323.7	322.4	470.3	567.6	677.2	569.3	554.0	614.4
Life Sciences	40.1	42.2	42.2	39.5	55.7	58.0	62.3	66.1	71.8	72.0
Planetary Exploration	182.4	175.6	175.6	210.0	186.4	217.4	290.9	353.6	359.2	327.7
Space and Terrestrial Applications	283.9	365.3	—		—					
Space Applications	274.8	353.5	331.5	324.3						
Technology Transfer	9.1	11.8	8.8							
Solid Earth Observations					128.9	76.4	57.6	70.9	72.4	d
Environmental Observations					156.9	162.0	212.7	271.6	318.3	389.2
Materials Processing					22.0	25.6	27.0	31.0	47.3	62.7
Communications					32.4	41.1	60.6	96.4	103.4	94.8
Information Systems					7.5	8.9	16.2	17.6	21.2	20.8
Commercial Programs						9.0	9.5	26.8	40.9	48.7
Technology Utilization				8.0	9.0	9.0	9.5	10.6	15.7	19.0
Commercial Use of Space								16.2	25.2	29.7

Table 8–12 continued

Program	1979	1980	1981	1982	1983	1984	1985	1986	1987	1988
Aeronautical and Space Technology	376.4	390.8	384.0	375.8	404.5	452.3	492.4	488.7	625.0	606.7
Aeronautical Research and Technology	264.1	276.2	271.4	264.8	280.0	315.3	342.4	337.3	374.0	332.9
Transatmospheric Research and Technology									45.0	52.5
Space Research and Technology	107.3	110.7	110.7	111.1	124.5	137.0	150.0	151.4	206.0	221.3
Energy Technology	5.0	3.9	1.9	— e	—					
Safety, Reliability, and Quality Assurance									12.0	
Tracking and Data Acquisition f, g	299.9	341.1	339.9	402.1	485.5	14.2	14.8	15.5	17.1	17.9
Total	3,477.2	4,340.8	4,334.3	4,772.0	5,519.1	2,064.2	2,468.1	2,596.1	3,153.7	3,279.9

a Renamed Space Transportation Capability Development. Remainder of funding moved to SFC&DC appropriation.
b Program eliminated.
c Merged with the Office of Space and Terrestrial Applications in FY 1983 to become the Office of Space Science and Applications.
d Programs included in Earth Science budget category.
e Program eliminated.
f Renamed Tracking and Data Advanced Systems in FY 1984. Remainder of funding moved to SFC&DC appropriation.
g Renamed Space and Ground Network, Communications and Data Systems beginning in FY 1986.

Table 8–13. *Space Flight Control and Data Communications Funding by Program (in millions of dollars)*

Installation	1984	1985	1986	1987	1988
Space Transportation Capability Development a	1,646.3	1,484.5	1,365.3	3,408.1	1,092.7
Space Transportation System Operations	1,452.0	1,314.0	1,640.2	1,746.0	1,833.6
Tracking and Data Acquisition b	674.0	795.7	660.4	845.4	879.4
Total	3,772.3	3,594.2	3,665.9	5,999.5	3,805.7

a Renamed Shuttle Production and Operational Capability beginning in FY 1985.
b Renamed Space and Ground Network, Communications and Data Systems beginning in FY 1986.
Source: NASA Budget Estimates (1986–1990).

Table 8–14. *NASA Budget Authority in Millions of Real-Year Dollars and in Equivalent FY 1996 Dollars*

Fiscal Year	NASA Total (Real-Year Dollars)	1996 Deflator	Total (Inflated 1996 Dollars)
1979	4,596	2.0456	9,401
1980	5,240	1.8763	9,832
1981	5,518	1.7031	9,398
1982	6,044	1.5851	9,580
1983	6,875	1.5218	10,463
1984	7,458	1.4577	10,871
1985	7,573	1.4041	10,633
1986	7,807	1.3636	10,645
1987	10,923	1.3243	14,465
1988	9,062	1.2779	11,581

Source: Aeronautics and Space Report of the President, 1996 (Washington, DC: U.S. Government Printing Office, 1997), Appendix E-1A and E-1B.

Table 8–15. *Total Number of Procurement Actions by Kind of Contractor: FY 1979–1988 (in thousands)*

Kind of Contractor	Number	Percentage
Business Firms	1,174.8	87.1
Small Business Firms	797.1	68.4
Large Business Firms	377.7	31.6
Nonprofit Institutions	20.3	1.5
Educational Institutions	47.5	3.5
Jet Propulsion Laboratory	19.0	1.4
Government Agencies	86.1	6.4
Contractors Outside the United States	3.4	0.25
Total	1,351.1	100.0

Table 8–16. *Number of Procurement Actions by Kind of Contractor and Fiscal Year (in thousands)*

Kind of Contractor	FY 1979		FY 1980		FY 1981		FY 1982		FY 1983	
	Number	Percentage	Number	Percentage	Number	Percentage	Number	Percentage	Number	Percentage
Business Firms	130.8	87.8	142.5	87.2	152.2	86.5	136.4	82.7	109.1	86.4
Nonprofit Institutions	2.0	1.3	1.9	1.2	2.1	1.2	2.2	1.3	2.2	1.7
Educational Institutions	3.6	2.2	4.1	2.5	4.3	2.4	4.1	2.5	4.4	3.5
Jet Propulsion Laboratory	1.6	1.1	1.6	1.0	1.8	1.1	2.1	1.3	4.7	3.7
Government Agencies	10.5	7.1	12.9	7.9	15.1	8.6	19.8	12.0	5.5	4.4
Contractors Outside the United States	0.4	0.3	0.4	0.2	0.4	0.2	0.3	0.2	0.3	0.2
Total	148.9	100.0	163.4	100.0	175.9	100.0	164.9	100.0	126.3	100.0

Kind of Contractor	FY 1984		FY 1985		FY 1986		FY 1987		FY 1988	
	Number	Percentage	Number	Percentage	Number	Percentage	Number	Percentage	Number	Percentage
Business Firms	104.7	88.6	107.7	89.1	98.3	88.1	98.4	87.8	94.7	87.0
Nonprofit Institutions	2.0	1.7	2.2	1.8	2.0	1.8	1.9	1.5	1.7	1.6
Educational Institutions	4.4	3.8	5.1	4.2	5.5	4.9	5.7	5.1	6.3	5.8
Jet Propulsion Laboratory	1.8	1.5	1.2	1.0	1.4	1.3	1.4	1.3	1.4	1.3
Government Agencies	5.0	4.2	4.4	3.6	4.1	3.7	4.3	3.8	4.5	4.1
Contractors Outside the United States	0.3	0.3	0.3	0.3	0.3	0.3	0.4	0.4	0.3	0.3
Total	118.2	100.0	120.9	100.0	111.6	100.0	112.1	100.0	108.9	100.0

Table 8–17. Number of Procurement Actions Awarded to Small and Large Business Firms by Fiscal Year (in thousands)

Kind of Business	FY 1979		FY 1980		FY 1981		FY 1982		FY 1983	
	Number	Percentage	Number	Percentage	Number	Percentage	Number	Percentage	Number	Percentage
Small Business Firms	82.9	63.4	87.8	61.6	98.9	65.0	90.7	66.5	77.4	70.9
Large Business Firms	47.9	36.6	54.7	38.4	53.3	35.0	45.7	33.5	31.7	29.1
Total	1,30.8	100.0	142.5	100.0	152.2	100.0	136.4	100.0	109.1	100.0

Kind of Business	FY 1984		FY 1985		FY 1986		FY 1987		FY 1988	
	Number	Percentage	Number	Percentage	Number	Percentage	Number	Percentage	Number	Percentage
Small Business Firms	74.1	70.8	75.3	69.9	69.8	71.0	71.5	72.7	68.7	72.5
Large Business Firms	30.6	29.2	32.4	30.1	28.5	29.0	26.9	27.3	26.0	27.5
Total	104.7	100.0	107.7	100.0	98.3	100.0	98.4	100.0	94.7	100.0

Table 8–18. Total Procurement Award Value by Kind of Contractor and Method of Procurement: FY 1979–1988 (in millions of dollars)

Kind of Contractor	Number	Percentage
Business Firms	1,174.8	87.1
Business Firms	54,746.8	79.2
Small Business Firms	5,491.7	10.0
Large Business Firms	49,255.1	90.0
Nonprofit Institutions	1,068.7	1.5
Educational Institutions	2,356.8	3.4
Jet Propulsion Laboratory	6,162.3	8.9
Government Agencies	4,365.01	6.3
Contractors Outside the United States	445.3	0.6
Total	69,144.9	100.0

Method of Procurement (Business)	Number	Percentage
Business Firms	40,997.6	79.3

Table 8–19. Value of Awards by Kind of Contractor and Fiscal Year (in millions of dollars)

Kind of Contractor	FY 1979 Number	FY 1979 Percentage	FY 1980 Number	FY 1980 Percentage	FY 1981 Number	FY 1981 Percentage	FY 1982 Number	FY 1982 Percentage	FY 1983 Number	FY 1983 Percentage
Business Firms	130.8	87.8	142.5	87.2	152.2	86.5	136.4	82.7	109.1	86.4
Business Firms	3,416.4	81.1	3,868.3	79.9	4,272.8	79.0	4,805.6	81.7	5,586.0	82.2
Nonprofit Institutions	50.8	1.2	82.2	1.7	155.1	2.9	108.8	1.9	102.5	1.5
Educational Institutions	147.2	3.5	177.0	3.7	192.5	3.6	187.0	3.2	211.3	3.1
Jet Propulsion Laboratory	338.6	8.0	397.2	8.2	410.8	7.6	426.3	7.3	454.9	6.7
Government Agencies	221.4	5.3	271.8	5.6	321.9	6.0	308.1	5.2	394.2	5.8
Contractors Outside the United States	37.4	0.9	46.1	1.0	55.2	1.0	47.9	0.8	47.9	0.7
Total	4,211.8	100.0	4,842.6	100.0	5,408.3	100.0	5,883.7	100.0	6,796.8	100.0

Kind of Contractor	FY 1984 Number	FY 1984 Percentage	FY 1985 Number	FY 1985 Percentage	FY 1986 Number	FY 1986 Percentage	FY 1987 Number	FY 1987 Percentage	FY 1988 Number	FY 1988 Percentage
Business Firms	5,967.4	81.1	6,652.9	80.2	6,356.0	77.7	6,540.5	76.0	7,274.9	76.2
Nonprofit Institutions	98.6	1.3	103.1	1.2	119.0	1.5	119.1	1.4	129.5	1.3
Educational Institutions	222.6	3.0	256.9	3.1	276.6	3.4	315.4	3.7	370.3	3.9
Jet Propulsion Laboratory	533.1	7.3	724.6	8.7	891.3	10.9	1,005.6	11.7	979.9	10.3
Government Agencies	494.3	6.7	525.1	6.3	489.7	6.0	594.9	6.9	734.6	7.7
Contractors Outside the United States	38.1	0.5	35.4	0.4	47.1	0.6	34.3	0.4	55.9	0.6
Total	7,354.1	100.0	8,298.0	100.0	8,179.7	100.0	8,609.8	100.0	9,545.1	100.0

FINANCES AND PROCUREMENT 527

Table 8–20. *Value of Awards to Small and Large Business Firms by Fiscal Year (in millions of dollars)*

Kind of Business	FY 1979		FY 1980		FY 1981		FY 1982		FY 1983	
	Number	Percentage	Number	Percentage	Number	Percentage	Number	Percentage	Number	Percentage
Small Business Firms	325.4	9.5	384.6	9.9	409.4	9.6	430.1	9.0	482.3	8.6
Large Business Firms	3,091.0	90.5	3,483.7	90.1	3,863.4	90.4	4,375.5	91.0	5,103.7	91.4
Total	3,416.4	100.0	3,868.3	100.0	4,272.8	100.0	4,805.6	100.0	5,586.0	100.0

Kind of Business	FY 1984		FY 1985		FY 1986		FY 1987		FY 1988	
	Number	Percentage	Number	Percentage	Number	Percentage	Number	Percentage	Number	Percentage
Small Business Firms	556.2 a	9.3	644.7 b	9.7	671.3 c	10.6	786.3 d	12.0	801.4 e	11.0
Large Business Firms	5,411.2	90.7	6,008.2	90.3	5,684.7	89.4	5,754.2	88.0	6,437.5	89.0
Total	5,967.4	100.0	6,652.9	100.0	6,356.0	100.0	6,540.5	100.0	7,274.9	100.0

a This includes $108.1 million awarded to small minority firms under authority of section 8(a) of the Small Business Act and $5 million through the Small Business Innovation Research program.
b This includes $127.1 million awarded to small minority firms under authority of section 8(a) of the Small Business Act and $29.5 million through the Small Business Innovation Research program.
c This includes $148.3 million awarded to small minority firms under authority of section 8(a) of the Small Business Act and $36 million through the Small Business Innovation Research program.
d This includes $172.5 million awarded to small minority firms under authority of section 8(a) of the Small Business Act and $32.2 million through the Small Business Innovation Research program.
e This includes $172.6 million awarded to small minority firms under authority of section 8(a) of the Small Business Act and $47.3 million through the Small Business Innovation Research program.

Table 8–21. Value of Awards to Business Firms by Kind of Procurement and Fiscal Year (in millions of dollars)

Kind of Procurement	FY 1979		FY 1980		FY 1981		FY 1982		FY 1983	
	Number	Percentage	Number	Percentage	Number	Percentage	Number	Percentage	Number	Percentage
Competitive	2,541.1	74.4	2,858.1	73.9	3,127.7	73.2	3,436.5	71.5	3,845.3	68.8
Noncompetitive	875.3	25.6	1,010.2	26.1	1,145.1	26.8	1,369.1	28.5	1,740.7	31.2
Total	3,416.4	100.0	3,868.3	100.0	4,272.8	100.0	4,805.6	100.0	5,586.0	100.0

Kind of Procurement	FY 1984		FY 1985		FY 1986		FY 1987		FY 1988	
	Number	Percentage	Number	Percentage	Number	Percentage	Number	Percentage	Number	Percentage
Competitive	4,286.6	71.8	5,030.2	77.1	4,950.1	67.7	5,031.7	76.5	5,890.3	80.1
Noncompetitive	1,680.8	28.2	1,495.7 a	22.9	2,143.1	29.3	1,445.7	22.0	1,279.4	17.4
Follow-on	—	—	—	—	217.1	3.0	97.9	1.5	180.7	2.5
Total	5,967.4	100.0	6,525.9	100.0	7,310.3 b	100.0	6,575.3 c	100.0	7,350.4 d	100.0

a   This does not include 8(a) awards.
b   This amount does not include $869.4 million in awards that were not available for competition.
c   The $6,575.3 million does not include $2,034.5 million in awards that were not available for competition.
d   The $7,350.4 million does not include $2,194.7 million in awards that were not available for competition.

Table 8–22. *Value and Percentage of Direct Awards to Business Firms by Contract Pricing Provision: 1979–1988 (in millions of dollars)* [a]

Pricing Provision	FY 1979		FY 1980		FY 1981		FY 1982		FY 1983	
	Number	Percentage	Number	Percentage	Number	Percentage	Number	Percentage	Number	Percentage
Firm-Fixed-Price	466.6	14	454.5	12	508.0	12	551.2	12	648.6	12
Incentive	2,244.1	72	2,680.4	72	214.2	5	277.2	6	378.4	7
Cost-Plus-Award-Fee [b]	—	—	—	—	2,887.2	70	3,219.7	69	3,625.0	67
Cost-Plus-Fixed-Fee	440.1	13	432.4	12	366.6	9	405.6	8	421.8	8
Other	149.5	5	171.1	4	170.2	4	221.5	5	322.1	6
Total	3,300.3	100	3,738.4	100	4,146.2	100	4,675.2	100	5,395.9	100

Pricing Provision	FY 1984		FY 1985		FY 1986		FY 1987		FY 1988	
	Number	Percentage	Number	Percentage	Number	Percentage	Number	Percentage	Number	Percentage
Firm-Fixed-Price	709.7	12	863.7	13	770.1	12	849.7	13	853.5	12
Incentive	710.5	12	1,650.1	26	1,981.7	32	1,425.7	22	1,347.2	19
Cost-Plus-Award-Fee	3,528.8	61	2,985.2	46	2,642.2	43	3,294.1	52	4,007.2	56
Cost-Plus-Fixed-Fee	472.3	8	587.8	9	647.8	10	696.1	11	766.0	11
Other	401.5	7	400.3	6	152.2	3	111.2	2	121.0	2
Total	5,822.8	100	6,487.1	100	6,194.0	100	6,376.8	100	7,094.9	100

[a] Excludes smaller procurements, generally those less than $10,000.
[b] Included with Incentive category through FY 1980.

Table 8–23. Number and Percentage of Procurement Actions in Direct Awards to Business Firms by Contract Pricing: 1979–1988

Pricing Provision	FY 1979 Number	FY 1979 Percentage	FY 1980 Number	FY 1980 Percentage	FY 1981 Number	FY 1981 Percentage	FY 1982 Number	FY 1982 Percentage	FY 1983 Number	FY 1983 Percentage
Firm-Fixed-Price	10,679	60	10,823	41	10,518	38	11,537	40	11,563	60
Incentive	2,314	13	11,351	43	13,010	47 [a]	12,402	43 [b]	3,469	18 [c]
Cost-Plus-Award-Fee	—	—	—	—	—	—	—	—	—	—
Cost-Plus-Fixed-Fee	3,560	20	3,432	13	3,598	13	3,750	13	3,469	18
Other	1,246	7	792	3	554	2	1,154	4	771	4
Total	17,799	100	26,398	100	27,680	100	28,843	100	19,272	100

Pricing Provision	FY 1984 Number	FY 1984 Percentage	FY 1985 Number	FY 1985 Percentage	FY 1986 Number	FY 1986 Percentage	FY 1987 Number	FY 1987 Percentage	FY 1988 Number	FY 1988 Percentage
Firm-Fixed-Price	10,220	49	9,377	53	8,217	61	7,981	60	8,114	58
Incentive	6,257	30 [d]	354	2	269	2	266	2	280	2
Cost-Plus-Award-Fee	—	—	3,185	18	1,212	9	1,330	10	1,539	11
Cost-Plus-Fixed-Fee	3,754	18	3,715	21	3,233	24	3,192	24	3,498	25
Other	626	3	1,026	6	539	4	532	4	560	4
Total	20,857	100	17,692	100	13,470	100	13,301	100	13,991	100

[a] Includes cost-plus-award-fee procurements.
[b] Includes cost-plus-award-fee procurements.
[c] Includes cost-plus-award-fee procurements.
[d] Includes cost-plus-award-fee procurements.

Table 8–24. Distribution of Prime Contract Awards by State: 1979–1988 (in thousands of dollars)

State	FY 1979 Number	FY 1979 Percentage	FY 1980 Number	FY 1980 Percentage	FY 1981 Number	FY 1981 Percentage	FY 1982 Number	FY 1982 Percentage	FY 1983 Number	FY 1983 Percentage
Alabama	77,813	2.2	81,093	2.1	94,071	2.1	106,352	2.2	117,122	2.1
Alaska	815	a	1,493	a	2,111	a	2,160	a	2,301	a
Arizona	18,854	0.5	32,169	0.8	27,856	0.6	30,213	0.6	29,123	0.5
Arkansas	88	a	189	a	315	a	512	a	47	a
California	1,438,969	41.3	1,721,269	43.5	1,875,403	42.8	2,056,491	42.0	2,131,242	37.6
Colorado	60,232	1.7	92,273	2.3	89,124	2.0	75,959	1.6	153,377	2.7
Connecticut	98,342	2.8	115,330	2.9	112,886	2.6	120,458	2.5	172,209	3.0
Delaware	680	a	584	a	1,072	a	4,461	0.1	1,647	a
District of Columbia	14,277	0.4	11,372	0.3	19,760	0.5	23,850	0.5	30,019	0.5
Florida	360,222	10.3	401,030	10.1	536,684	12.2	633,552	12.9	773,978	13.7
Georgia	7,433	0.2	7,479	0.2	9,132	0.2	7,844	0.2	11,874	0.2
Hawaii	2,632	0.1	4,195	0.1	3,420	0.1	4,467	0.1	5,469	0.1
Idaho	26	a	78	a	253	a	225	a	295	a
Illinois	12,165	0.3	10,670	0.3	11,954	0.3	11,217	0.2	12,658	0.2
Indiana	16,874	0.5	24,810	0.6	18,430	0.4	23,741	0.5	13,577	0.2
Iowa	3,349	0.1	4,516	0.1	3,738	0.1	3,920	0.1	4,076	0.1
Kansas	1,689	a	3,754	0.1	3,896	0.1	4,443	0.1	6,302	0.1
Kentucky	917	a	1,117	a	689	a	1,013	a	1,278	a
Louisiana	133,363	3.8	173,007	4.4	205,045	4.7	265,367	5.4	344,292	6.1
Maine	334	a	120	a	14	a	287	a	278	a
Maryland	257,178	7.4	272,409	6.9	309,600	7.1	360,077	7.3	407,970	7.2
Massachusetts	51,850	1.5	53,264	1.3	56,409	1.3	52,842	1.1	61,548	1.1
Michigan	14,589	0.4	10,524	0.3	14,265	0.3	15,549	0.3	7,317	0.1

532 NASA HISTORICAL DATA BOOK

Table 8–24 continued

State	FY 1979 Number	FY 1979 Percentage	FY 1980 Number	FY 1980 Percentage	FY 1981 Number	FY 1981 Percentage	FY 1982 Number	FY 1982 Percentage	FY 1983 Number	FY 1983 Percentage
Minnesota	10,679	0.3	8,028	0.2	6,547	0.1	8,186	0.2	6,294	0.1
Mississippi	26,341	0.8	29,083	0.7	30,770	0.7	33,393	0.7	38,451	0.7
Missouri	6,573	0.2	5,865	0.2	6,936	0.2	5,469	0.1	5,759	0.1
Montana	3	a	114	a	164	a	102	a	79	a
Nebraska	135	a	175	a	398	a	261	a	496	a
Nevada	933	a	724	a	1,101	a	1,389	a	568	a
New Hampshire	1,421	a	1,714	a	2,924	0.1	3,357	0.1	4,660	0.1
New Jersey	58,386	1.7	39,497	1.0	41,313	0.9	37,432	0.8	74,102	1.3
New Mexico	17,100	0.5	17,323	0.4	16,132	0.4	23,504	0.5	21,703	0.4
New York	69,360	2.0	52,935	1.3	56,319	1.3	50,063	1.0	51,838	0.9
North Carolina	3,117	0.1	3,773	0.1	4,133	0.1	4,786	0.1	5,701	0.1
North Dakota	—	—	—	—	13	a	10	a	—	—
Ohio	82,632	2.4	90,040	2.3	70,009	1.6	71,587	1.5	78,860	1.4
Oklahoma	1,207	a	2,554	0.1	1,936	a	2,086	a	4,353	0.1
Oregon	2,620	0.1	2,675	0.1	4,222	0.1	3,268	0.1	2,918	0.1
Pennsylvania	98,142	2.8	100,609	2.5	105,715	2.4	112,506	2.3	104,320	1.8
Rhode Island	639	a	1,349	a	1,148	a	1,315	a	1,243	a
South Carolina	550	a	621	a	393	a	204	a	413	a
South Dakota	227	a	273	a	90	a	194	a	324	a
Tennessee	3,898	0.1	4,233	0.1	4,563	0.1	6,953	0.1	7,891	0.1
Texas	306,838	8.8	344,792	8.7	384,300	8.8	448,117	9.1	526,331	9.3
Utah	71,243	2.0	71,898	1.8	85,352	1.9	124,016	2.5	253,518	4.5

FINANCES AND PROCUREMENT

Table 8–24 continued

State	FY 1979		FY 1980		FY 1981		FY 1982		FY 1983	
	Number	Percentage	Number	Percentage	Number	Percentage	Number	Percentage	Number	Percentage
Vermont	152	a	335	a	417	a	388	a	250	a
Virginia	104,436	3.0	114,444	2.9	124,461	2.8	127,904	2.6	157,184	2.8
Washington	41,321	1.2	37,205	0.9	34,786	0.8	22,910	0.5	21,909	0.4
West Virginia	11	a	—	—	34	a	9	a	92	a
Wisconsin	4,713	0.1	4,969	0.1	5,726	0.1	5,409	0.1	9,820	0.2
Wyoming	327	a	248	a	248	a	325	a	518	a
Total b	3,485,896	100.0	3,958,221	100.0	4,386,223	100.0	4,900,143	100.0	5,667,594	100.0

State	FY 1984		FY 1985		FY 1986		FY 1987		FY 1988	
	Number	Percentage	Number	Percentage	Number	Percentage	Number	Percentage	Number	Percentage
Alabama	291,507	4.8	377,117	5.5	411,466	6.3	463,511	6.8	546,789	7.2
Alaska	1,890	a	1,776	a	2,268	a	3,333	a	2,807	a
Arizona	30,448	0.5	28,993	0.4	25,184	0.3	24,035	0.4	21,061	0.3
Arkansas	214	a	299	a	174	a	393	a	239	a
California	2,149,986	35.1	2,236,531	32.7	1,999,674	30.4	2,224,687	32.8	2,411,400	31.8
Colorado	116,830	1.9	127,455	1.9	113,707	1.7	113,036	1.7	86,641	1.1
Connecticut	177,736	2.9	172,360	2.5	113,527	1.7	162,600	2.4	76,060	1.0
Delaware	2,737	a	1,697	a	1,322	a	2,334	a	3,621	a
District of Columbia	33,307	0.5	37,929	0.6	39,760	0.6	49,275	0.7	56,739	0.7
Florida	816,993	13.3	979,033	14.3	966,303	14.7	682,070	10.0	873,427	11.5
Georgia	15,959	0.3	20,305	0.3	34,813	0.5	26,931	0.4	15,700	0.2
Hawaii	4,780	0.1	5,746	0.1	5,434	0.1	5,507	0.1	6,890	0.1
Idaho	283	a	212	a	208	a	271	a	643	a

534 NASA HISTORICAL DATA BOOK

Table 8–24 continued

State	FY 1984 Number	FY 1984 Percentage	FY 1985 Number	FY 1985 Percentage	FY 1986 Number	FY 1986 Percentage	FY 1987 Number	FY 1987 Percentage	FY 1988 Number	FY 1988 Percentage
Illinois	16,619	0.3	17,067	0.2	17,445	0.3	16,173	0.2	15,299	0.2
Indiana	15,384	0.3	14,202	0.2	12,733	0.2	9,899	0.1	11,745	0.2
Iowa	4,729	0.1	4,321	0.1	4,373	0.1	6,914	0.1	7,861	0.1
Kansas	2,295	a	2,351	a	1,948	a	7,995	0.1	20,712	0.3
Kentucky	1,321	a	635	a	738	a	780	a	649	a
Louisiana	343,434	5.6	394,624	5.8	364,829	5.6	303,456	4.5	332,207	4.4
Maine	328	a	403	a	475	a	695	a	452	a
Maryland	477,645	7.8	587,567	8.6	560,074	8.5	620,268	9.1	661,290	8.7
Massachusetts	58,443	1.0	68,195	1.0	57,449	0.9	63,310	0.9	63,989	0.8
Michigan	8,849	0.1	14,587	0.2	17,999	0.3	16,166	0.2	17,642	0.2
Minnesota	10,463	0.2	10,289	0.2	8,621	0.1	14,641	0.2	8,964	0.1
Mississippi	46,269	0.8	53,875	0.8	55,477	0.8	68,341	1.0	84,976	1.1
Missouri	7,802	0.1	12,705	0.2	9,561	0.1	8,640	0.1	13,376	0.2
Montana	35	a	248	a	126	a	215	a	339	a
Nebraska	381	a	244	a	402	a	853	a	535	a
Nevada	535	a	616	a	672	a	505	a	511	a
New Hampshire	4,852	0.1	5,353	0.1	5,314	0.1	9,287	0.1	7,398	0.1
New Jersey	78,975	1.3	108,292	1.6	148,319	2.3	126,995	1.9	112,331	1.5
New Mexico	23,339	0.4	28,317	0.4	33,711	0.5	54,970	0.8	45,559	0.6
New York	58,744	1.0	58,182	0.9	55,875	0.9	68,182	1.0	54,992	0.7
North Carolina	8,275	0.1	7,510	0.1	5,845	0.1	7,561	0.1	6,254	0.1
North Dakota	26	a	15	a	—	—	106	a	90	a
Ohio	104,543	1.7	115,872	1.7	127,116	1.9	131,299	1.9	159,626	2.1

FINANCES AND PROCUREMENT 535

Table 8–24 continued

State	FY 1984 Number	FY 1984 Percentage	FY 1985 Number	FY 1985 Percentage	FY 1986 Number	FY 1986 Percentage	FY 1987 Number	FY 1987 Percentage	FY 1988 Number	FY 1988 Percentage
Oklahoma	1,148	a	2,607	a	4,452	0.1	3,118	a	3,406	a
Oregon	5,831	0.1	4,050	0.1	4,068	0.1	3,491	0.1	2,906	a
Pennsylvania	70,541	1.2	69,430	1.0	81,915	1.2	110,416	1.6	105,174	1.4
Rhode Island	1,166	a	1,481	a	1,869	a	2,585	a	2,544	a
South Carolina	254	a	551	a	721	a	1,049	a	896	a
South Dakota	314	a	829	a	476	a	444	a	485	a
Tennessee	8,931	0.1	7,736	0.1	10,505	0.2	15,946	0.2	21,494	0.3
Texas	587,124	9.6	657,512	9.6	693,020	10.6	764,386	11.3	913,120	12.1
Utah	317,690	5.2	335,823	4.9	316,717	4.8	289,036	4.3	429,151	5.7
Vermont	334	a	82	a	1,752	a	740	a	530	a
Virginia	161,050	2.6	187,973	2.8	188,622	2.9	240,589	3.5	318,312	4.2
Washington	31,808	0.5	42,813	0.6	39,304	0.6	42,944	0.6	17,547	0.2
West Virginia	93	a	351	a	291	a	819	a	260	a
Wisconsin	17,291	0.3	28,494	0.4	20,966	0.3	20,216	0.3	32,363	0.4
Wyoming	433	a	585	a	446	a	512	a	226	a
Total b	6,119,937	100.0	6,835,240	100.0	6,567,966	100.0	6,791,565	100.0	5,577,228	100.0

a Less than 0.05 percent.
b Excludes smaller procurements—generally those of $10,000 or less through FY 1984 and $25,000 beginning in FY 1985; also excludes other government agencies, awards outside the United States, and actions on the Jet Propulsion Laboratory contracts.

Table 8–25. Distribution of Prime Contract Awards by Region: FY 1979–1988

Region a	1979	1980	1981	1982	1983	1984	1985	1986	1987	1988	Total
Net Value of Awards (in thousands of dollars)											
New England	152,738	172,112	173,798	178,647	240,188	242,859	247,987	180,386	239,217	150,973	1,978,792
Mideast	525,023	477,406	533,779	589,389	669,896	721,949	863,097	887,265	977,470	994,147	7,239,421
Southeast	718,189	816,069	1,009,987	1,187,889	1,458,323	1,684,300	2,030,009	2,039,784	1,811,446	2,201,203	14,957,199
Great Plains	22,616	22,611	21,618	22,483	23,251	26,010	30,754	25,381	39,593	52,023	286,340
Great Lakes	130,973	141,013	120,384	127,503	118,637	162,686	190,222	196,259	193,753	236,675	1,618,105
Southwest	343,999	396,838	430,224	503,920	581,510	642,059	717,429	756,367	846,509	983,146	6,202,001
Rocky Mountain	131,831	164,641	175,141	200,627	407,787	435,271	464,323	431,204	403,070	517,000	3,330,895
Far West	1,483,843	1,761,873	1,915,512	2,084,058	2,156,637	2,188,160	2,284,010	2,043,718	2,271,627	2,432,364	20,621,802
Alaska and Hawaii	3,447	5,688	5,531	6,627	7,770	6,670	7,522	7,702	8,840	9,697	69,494
Total	3,485,896	3,958,221	4,386,223	4,900,143	5,667,594	6,119,937	6,835,240	6,567,966	6,791,565	7,577,288	56,290,073
Percentage of Total											
New England	4.4	4.3	4.0	3.7	4.2	4.0	3.6	2.8	3.5	2.0	3.5
Mideast	15.1	12.1	12.2	12.0	11.8	11.8	12.6	13.5	14.4	13.1	12.9
Southeast	20.5	20.6	23.0	24.2	25.7	27.5	29.7	31.0	26.7	29.1	26.6
Great Plains	0.7	0.6	0.5	0.5	0.4	0.4	0.5	0.4	0.6	0.7	0.5
Great Lakes	3.3	3.6	2.7	2.6	2.1	2.7	2.8	3.0	2.9	3.1	2.9
Southwest	9.9	10.0	9.8	10.3	10.3	10.5	10.5	11.5	12.5	13.0	11.0
Rocky Mountain	3.9	4.2	4.0	4.1	7.2	7.1	6.8	6.6	5.9	6.8	5.9
Far West	42.6	44.5	43.7	42.5	38.1	35.8	33.4	31.1	33.5	32.1	36.6
Alaska and Hawaii	0._	0.1	0.1	0.1	0.1	0.1	0.1	0.1	0.1	0.1	0.1
Total	100.0	100.0	100.0	100.0	100.0	100.0	100.0	100.0	100.0	100.0	100.0

Table 8–25 continued

Percentage Change of Prime Contract Award Value Over Previous Year

Region [a]	1979	1980	1981	1982	1983	1984	1985	1986	1987	1988	Total
New England	2.9	12.7	1.0	2.8	34.4	1.1	2.1	-27.2	32.6	-36.9	—
Mideast	23.5	-9.1	11.8	10.4	13.7	7.8	19.6	2.8	10.2	1.7	—
Southeast	21.3	13.6	23.8	17.6	22.8	13.4	20.5	0.5	-11.2	21.5	—
Great Plains	-22.0	[b]	-4.4	4.0	3.4	10.6	18.2	-17.5	56.0	31.4	—
Great Lakes	28.4	7.7	-14.6	5.9	-7.0	37.1	16.9	3.2	-1.3	22.2	—
Southwest	10.6	15.4	8.4	17.1	15.4	10.4	11.7	5.4	11.9	22.2	—
Rocky Mountain	10.8	24.9	6.4	14.6	103.3	6.7	6.7	-7.1	-6.5	28.3	—
Far West	5.0	18.7	8.7	8.8	3.5	1.5	4.4	-10.5	11.2	7.0	—
Alaska and Hawaii	15.0	65.0	-2.8	19.8	17.2	-14.2	12.8	2.4	14.8	9.7	—
United States	15.0	13.5	10.8	11.7	15.7	7.4	11.7	-3.9	3.4	11.6	—

[a] The New England region consists of Connecticut, Maine, Massachusetts, Rhode Island, New Hampshire, and Vermont. The Mideast region is Delaware, the District of Columbia, Maryland, New Jersey, New York, and Pennsylvania. The Southeast region consists of Alabama, Arkansas, Florida, Georgia, Kentucky, Louisiana, Mississippi, North Carolina, South Carolina, Tennessee, Virginia, and West Virginia. The Great Plains region is Iowa, Kansas, Minnesota, Missouri, Nebraska, North Dakota, and South Dakota. The Great Lakes region contains Illinois, Indiana, Michigan, Ohio, and Wisconsin. The Southwest is Arizona, New Mexico, Oklahoma, and Texas. The Rocky Mountain region consists of Colorado, Idaho, Montana, Utah, and Wyoming. The Far West region is California, Nevada, Oregon, and Washington.

[b] Less than 0.05 percent.

Table 8–26. Value and Percentage of Total of Awards by Installation (in millions of dollars)

Installation	FY 1979 Number	FY 1979 Percentage	FY 1980 Number	FY 1980 Percentage	FY 1981 Number	FY 1981 Percentage	FY 1982 Number	FY 1982 Percentage	FY 1983 Number	FY 1983 Percentage
Ames Research Center a	219.8	4.6	236.2	4.3	207.0	3.4	222.1	3.8	247.8	3.6
Goddard Space Flight Center b	634.2	14.4	694.0	13.7	790.4	14.0	864.0	14.7	952.4	14.0
Johnson Space Center	1,197.6	28.4	1,429.1	29.5	1,625.8	30.0	1,690.0	38.7	1,737.8	25.5
Kennedy Space Center	357.3	8.5	395.9	8.2	527.9	9.8	585.4	9.9	724.7	10.7
Langley Research Center	225.0	5.4	235.5	4.9	218.0	4.0	181.7	3.1	216.5	3.2
Lewis Research Center	274.1	6.5	319.8	6.6	338.0	6.3	427.2	7.3	467.3	6.9
Marshall Space Flight Center	774.4	18.4	917.2	19.0	1,050.7	19.4	1,225.5	20.8	1,677.5	24.7
NASA Resident Office/JPL	338.6	8.1	397.2	8.2	410.8	7.6	426.3	7.2	454.9	6.7
National Space Technology Laboratories/Stennis	33.7	0.8	36.7	0.8	38.4	0.8	39.6	0.7	40.4	0.6
NASA Headquarters	157.1	3.7	181.0	3.7	201.3	3.7	221.9	3.8	277.5	4.1
Total	4,211.8	100.0	4,842.6	100.0	5,408.3	100.0	5,883.7	100.0	6,796.8	100.0

Table 8–26 continued

Installation	FY 1979 Number	FY 1979 Percentage	FY 1980 Number	FY 1980 Percentage	FY 1981 Number	FY 1981 Percentage	FY 1982 Number	FY 1982 Percentage	FY 1983 Number	FY 1983 Percentage
Ames Research Center [a]	281.2	3.8	340.4	4.1	361.3	4.4	467.0	5.4	431.5	4.5
Goddard Space Flight Center [b]	953.8	13.0	1,076.8	13.0	1,265.4	15.5	1,313.3	15.3	1,356.3	14.2
Johnson Space Center	1,636.2	22.1	1,719.1	20.7	1,359.5	16.6	1,627.4	18.9	1,806.4	18.9
Kennedy Space Center	814.2	11.1	977.9	11.8	1,026.6	12.6	883.4	10.3	1,069.2	11.2
Langley Research Center	214.6	2.9	248.4	3.0	261.4	3.2	289.5	3.4	297.3	3.1
Lewis Research Center	628.0	8.5	675.7	8.2	625.5	7.6	381.8	4.4	417.9	4.4
Marshall Space Flight Center	1,824.6	24.9	1999.1	24.1	1891.8	23.1	2,059.8	23.9	2,428.3	25.5
NASA Resident Office/JPL	533.1	7.3	724.6	8.7	895.4	11.0	1,008.7	11.7	986.0	10.3
National Space Technology Laboratories/Stennis	49.3	0.7	60.0	0.7	57.7	0.7	71.4	0.8	91.5	1.0
NASA Headquarters	419.1	5.7	476.0	5.7	435.1	5.3	507.5	5.9	660.7 [c]	6.9
Total	7,354.1	100.0	8,298.0	100.0	8,179.7	100.0	8,609.8	100.0	9,545.1	100.0

[a] Includes Dryden procurements.
[b] Includes Wallops procurements.
[c] Includes $14.2 million in reimbursable funds to the U.S. Treasury for the Tracking and Data Relay Satellite System (TDRSS).

Table 8–27. Ranking of NASA's Top Ten Contractors by Fiscal Year

Contractor	1979	1980	1981	1982	1983	1984	1985	1986	1987	1988
Rockwell International Corp.	1	1	1	1	1	1	1	1	1	1
Martin Marietta Corp.	2	2	2	2	2	2	3	3	2	4
General Electric	3	4	6	10	—	—	—	—	6	7
McDonnell Douglas Corp.	4	3	3	3	4	6	7	5	5	5
Bendix Corp. *a*	5	6	7	8	7	8	8	10	—	—
IBM	6	7	8	9	9	9	—	—	—	—
Computer Sciences Corp.	7	5	4	5	6	—	—	—	—	—
Morton Thiokol Corp.	8	8	5	4	3	3	4	4	4	4
United Technologies Corp.	9	9	10	—	8	10	—	—	9	—
Hughes Aircraft Corp.	10	10	—	—	—	—	—	—	—	—
Boeing Services Int'l.	—	—	9	—	—	—	—	—	—	—
United Space Boosters, Inc. *b*	—	—	—	6	10	7	6	7	7	8
General Dynamics Corp.	—	—	—	7	5	5	5	8	—	—
Lockheed Space Ops. Co.	—	—	—	—	—	4	2	2	3	2
Lockheed Missiles & Space Co.	—	—	—	—	—	—	9	—	—	—
Lockheed Engineering Mgmt. Co. *c*	—	—	—	—	—	—	10	—	10	9
Ford Aerospace and Comm.	—	—	—	—	—	—	—	6	—	—
RCA Corp.	—	—	—	—	—	—	—	9	—	—
Boeing Co.	—	—	—	—	—	—	—	—	8	6
EG&G Florida, Inc.	—	—	—	—	—	—	—	—	—	10

a Also called Allied Bendix Aerospace.
b Includes contracted work performed at more than one location.
c Also called Lockheed Engineering and Science Co.

Table 8–28. Top Fifty Contractors: FY 1979 (in thousands of dollars) a

Contractor/Place of Contract Performance b	Rank in FY 1978	Net Value of Awards	
		Amount	Percentage
1. Rockwell International Corp. Downey, California	1	1,071,755	31.37
2. Martin Marietta Corp. New Orleans, Louisiana	2	178,391	5.22
3. General Electric Co. King of Prussia, Pennsylvania	8	120,941	3.54
4. McDonnell Douglas Corp. Huntington Beach, California	3	113,676	3.33
5. Bendix Corp. Columbia, Maryland	4	100,205	2.93
6. IBM Houston, Texas	6	93,485	2.74
7. Computer Sciences Corp. Kennedy Space Center, Florida	10	92,854	2.72
8. Thiokol Corp. Brigham City, Utah	9	77,724	2.27
9. United Technologies Corp. East Hartford, Connecticut	13	73,270	2.14
10. Hughes Aircraft Co. Los Angeles, California	7	71,106	2.08
11. Boeing Services International Kennedy Space Center, Florida	14	58,198	1.70
12. RCA Corp. Princeton, New Jersey	12	50,763	1.49
13. Lockheed Electronics Co. Houston, Texas	5	50,571	1.48
14. General Dynamics Corp. San Diego, California	11	46,928	1.37
15. Boeing Co. Seattle, Washington	15	43,311	1.27
16. Lockheed Missile & Space Co. Sunnyvale, California	22	36,020	1.05
17. Ford Aerospace and Communications Houston, Texas	17	35,189	1.03
18. Planning Research Corp. Kennedy Space Center, Florida	18	34,748	1.02
19. United Space Boosters, Inc. Kennedy Space Center, Florida	25	32,634	0.96
20. Perkin Elmer Corp. Danbury, Connecticut	27	30,564	0.89
21. TRW Inc. Redondo Beach, California	24	29,000	0.85

Table 8–28 continued

Contractor/Place of Contract Performance b	Rank in FY 1978	Net Value of Awards	
		Amount	Percentage
22. Pan American World Airways Bay St. Louis, Mississippi	33	27,257	0.80
23. Singer Co. Houston, Texas	23	27,248	0.80
24. Ball Corp. Boulder, Colorado	26	22,102	0.65
25. Northrop Services, Inc. Houston, Texas	28	19,509	0.59
26. Sperry Corp. Phoenix, Arizona	19	19,509	0.57
27. Lockheed Corp. Burbank, California	37	19,483	0.57
28. Air Products & Chemicals, Inc. Allentown, Pennsylvania	21	19,430	0.57
29. Vought Corp. Dallas, Texas	16	18,415	0.54
30. Computer Sciences Technicolor Associates Greenbelt, Maryland	31	16,143	0.47
31. Westinghouse Electric Corp. Baltimore Washington Int'l. Airport, Maryland	32	16,070	0.47
32. Honeywell Information Systems McLean, Virginia	39	15,794	0.46
33. Fairchild Industries, Inc. Germantown, Maryland	34	15,387	0.45
34. Briscoe Frank Co. Kennedy Space Center, Florida	20	14,129	0.41
35. Kentron International, Inc. Hampton, Virginia	65	13,733	0.40
36. Mechanical Technology, Inc. Latham, New York	54	13,700	0.40
37. Bostrom Bergen Metal Products c Mountain View, California	—	13,466	0.39
38. Teledyne Industries, Inc. Huntsville, Alabama	38	12,938	0.38
39. General Motors Corp. Indianapolis, Indiana	47	10,997	0.32
40. Raytheon Services Co. Halethorpe, Maryland	35	9,382	0.27
41. Control Data Corp. Minneapolis, Minnesota	43	9,108	0.27

Table 8–28 continued

Contractor/Place of Contract Performance b	Rank in FY 1978	Net Value of Awards	
		Amount	Percentage
42. *Modular Computer Systems, Inc.*			
Fort Lauderdale, Florida	74	8,879	0.26
43. *SDC Integrated Services, Inc.*			
Slidell, Louisiana	42	8,364	0.24
44. *Virginia Electric & Power Co.*			
Hampton, Virginia	44	7,464	0.22
45. *Metro Contract Services, Inc.* c			
Huntsville, Alabama	45	7,281	0.21
46. *Digital Equipment Corp.*			
Maynard, Massachusetts	64	6,923	0.20
47. *Wackenhut Services, Inc.*			
Kennedy Space Center, Florida	57	6,521	0.19
48. Klate Holt Co. c			
Hampton, Virginia	86	6,496	0.19
49. *Honeywell, Inc.*			
Minneapolis, Minnesota	40	6,403	0.19
50. *Textron, Inc.*			
Fort Worth, Texas	50	6,343	0.19
Other		576,491	16.88
Total Awards to Business Firms		3,416,927	100.00

a Excludes smaller procurements, generally those of less than $10,000.
b For contractors in italics, awards during the year represent awards on several contracts.
c Small business concern.

Table 8–29. Top Fifty Contractors: FY 1980 (in thousands of dollars) a

Contractor/Place of Contract Performance	Rank in FY 1979	Net Value of Awards	
		Amount	Percentage
1. Rockwell International Corp. Downey, California	1	1,273,345	32.92
2. Martin Marietta Corp. New Orleans, Louisiana	2	232,948	6.02
3. McDonnell Douglas Corp. Huntington Beach, California	4	159,747	4.13
4. General Electric Co. King of Prussia, Pennsylvania	3	114,286	2.95
5. Computer Sciences Corp. Kennedy Space Center, Florida	7	112,146	2.90
6. Bendix Corp. Columbia, Maryland	5	97,290	2.52
7. IBM Houston, Texas	6	84,336	2.18
8. Thiokol Corp. Brigham City, Utah	8	79,232	2.05
9. United Technologies Corp. East Hartford, Connecticut	9	74,639	1.93
10. Hughes Aircraft Co. El Segundo, California	10	68,449	1.77
11. Boeing Services International Kennedy Space Center, Florida	11	59,252	1.53
12. Lockheed Engineering and Mgmt. Corp. Houston, Texas	—	59,051	1.53
13. Ford Aerospace and Communications Houston, Texas	17	47,636	1.23
14. Lockheed Missile & Space Co. Sunnyvale, California	16	47,047	1.22
15. General Dynamics Corp. San Diego, California	14	46,181	1.19
16. Boeing Co. Seattle, Washington	15	44,613	1.15
17. Perkin Elmer Corp. Danbury, Connecticut	20	43,147	1.12
18. United Space Boosters, Inc. Kennedy Space Center, Florida	19	43,117	1.11
19. TRW, Inc. Redondo Beach, California	21	41,852	1.08
20. Planning Research Corp. Kennedy Space Center, Florida	18	37,952	0.98
21. RCA Corp. Princeton, New Jersey	12	31,656	0.82

FINANCES AND PROCUREMENT

Table 8–29 continued

Contractor/Place of Contract Performance	Rank in FY 1979	Net Value of Awards	
		Amount	Percentage
22. Pan American World Airways Bay St. Louis, Mississippi	22	31,585	0.82
23. Singer Co. Houston, Texas	23	18,035	0.72
24. Lockheed Corp. Burbank, California	27	25,401	0.66
25. Kentron International, Inc. Hampton, Virginia	35	23,339	0.60
26. Northrop Services, Inc. Houston, Texas	25	22,066	0.57
27. Ball Corp. Boulder, Colorado	24	20,870	0.54
28. Teledyne Industries, Inc. Huntsville, Alabama	38	19,631	0.51
29. General Motors Corp. Indianapolis, Indiana	39	19,514	0.50
30. Computer Sciences Technicolor Associates Greenbelt, Maryland	30	18,241	0.47
31. Sperry Corp. Phoenix, Arizona	26	17,651	0.46
32. Air Products & Chemicals, Inc. Allentown, Pennsylvania	28	17,410	0.45
33. Garrett Corp. Phoenix, Arizona	80	16,492	0.43
34. Mechanical Technology, Inc. Latham, New York	36	14,504	0.37
35. Honeywell Information Systems McLean, Virginia	32	14,280	0.37
36. Vought Corp. Dallas, Texas	29	13,667	0.35
37. Christie Willamette JV Mountain View, California	—	12,987	0.34
38. Raytheon Services Co. Halethorpe, Maryland	40	12,721	0.33
39. Fairchild Industries, Inc. Germantown, Maryland	33	10,426	0.27
40. Virginia Electric & Power Co. Hampton, Virginia	44	9,449	0.24
41. Westinghouse Electric Corp. Baltimore Washington Int'l. Airport, Maryland	31	9,044	0.23

Table 8–29 continued

Contractor/Place of Contract Performance	Rank in FY 1979	Net Value of Awards	
		Amount	Percentage
42. Pioneer Contract Services, Inc. *b* Huntsville, Alabama	—	9,030	0.23
43. Wackenhut Services, Inc. Kennedy Space Center, Florida	47	8,948	0.23
44. SDC Integrated Services, Inc. Slidell, Louisiana	43	8,827	0.23
45. Management & Technical Services Houston, Texas	60	8,648	0.22
46. Control Data Corp. Minneapolis, Minnesota	41	8,256	0.21
47. Klate Holt Co. *b* Hampton, Virginia	48	7,699	0.20
48. Northrop Worldwide Aircraft Houston, Texas	51	7,348	0.19
49. Briscoe Frank Co. Kennedy Space Center, Florida	34	7,192	0.19
50. Global Associates New Orleans, Louisiana	59	7,154	0.18
Other		639,960	16.56
Total Awards to Business Firms		3,868,297	100.00

a Excludes smaller procurements, generally those of less than $10,000.
b Small business concern.

Table 8–30. Top Fifty Contractors: FY 1981 (in thousands of dollars) a

Contractor/Place of Contract Performance	Rank in FY 1980	Net Value of Awards	
		Amount	Percentage
1. Rockwell International Corp. Downey, California	1	1,471,331	34.43
2. Martin Marietta Corp. New Orleans, Louisiana	2	261,308	6.11
3. McDonnell Douglas Corp. Huntington Beach, California	3	198,323	4.64
4. Computer Sciences Corp. Kennedy Space Center, Florida	5	129,442	3.03
5. Thiokol Corp. Brigham City, Utah	8	105,273	2.46
6. General Electric Co. King of Prussia, Pennsylvania	4	103,731	2.43
7. Bendix Corp. Columbia, Maryland	6	103,133	2.41
8. IBM Houston, Texas	7	95,013	2.22
9. Boeing Services International Kennedy Space Center, Florida	11	81,202	1.90
10. United Technologies Corp. Windsor Locks, Connecticut	9	71,370	1.67
11. General Dynamics Corp. San Diego, California	15	66,070	1.55
12. United Space Boosters, Inc. Kennedy Space Center, Florida	18	65,340	1.53
13. Ford Aerospace and Communications Houston, Texas	13	61,752	1.45
14. Lockheed Engineering and Mgmt. Corp. Houston, Texas	12	60,548	1.42
15. Hughes Aircraft Co. Los Angeles, California	10	53,148	1.24
16. Perkin Elmer Corp. Danbury, Connecticut	17	51,427	1.20
17. Planning Research Corp. Kennedy Space Center, Florida	20	44,394	1.04
18. Lockheed Missile & Space Co. Sunnyvale, California	14	43,022	1.01
19. Boeing Co. Seattle, Washington	16	40,132	0.94
20. TRW, Inc. Redondo Beach, California	19	36,629	0.86
21. Pan American World Airways Bay St. Louis, Mississippi	22	34,023	0.80

Table 8–30 continued

Contractor/Place of Contract Performance	Rank in FY 1980	Net Value of Awards	
		Amount	Percentage
22. Ball Corp. Boulder, Colorado	27	30,342	0.71
23. Singer Co. Houston, Texas	23	29,918	0.70
24. RCA Corp. Princeton, New Jersey	21	27,334	0.64
25. Air Products & Chemicals, Inc. Allentown, Pennsylvania	32	25,151	0.59
26. Sperry Corp. Phoenix, Arizona	31	23,895	0.56
27. Northrop Services, Inc. Houston, Texas	26	23,737	0.56
28. Teledyne Industries, Inc. Huntsville, Alabama	28	23,181	0.54
29. Computer Sciences Technicolor Associates Greenbelt, Maryland	30	19,559	0.49
30. Kentron International, Inc. Hampton, Virginia	25	16,292	0.38
31. Lockheed Corp. Burbank, California	24	16,051	0.38
32. Raytheon Services Co. Halethorpe, Maryland	38	14,918	0.35
33. Vought Corp. Dallas, Texas	36	14,455	0.34
34. Mechanical Technology, Inc. Latham, New York	34	13,755	0.32
35. General Motors Corp. Indianapolis, Indiana	29	13,710	0.32
36. Wackenhut Services, Inc. Kennedy Space Center, Florida	43	13,254	0.31
37. Honeywell Information Systems McLean, Virginia	35	12,972	0.30
38. Management & Technical Services Houston, Texas	45	11,758	0.28
39. Control Data Corp. Hampton, Virginia	46	11,391	0.27
40. SDC Integrated Services, Inc. Slidell, Louisiana	44	11,201	0.26
41. Virginia Electric & Power Co. Hampton, Virginia	40	10,193	0.24
42. Honeywell, Inc. St. Petersburg, Florida	59	9,853	0.23

Table 8-30 continued

Contractor/Place of Contract Performance	Rank in FY 1980	Net Value of Awards	
		Amount	Percentage
43. Garrett Corp.			
Phoenix, Arizona	33	9,204	0.22
44. Eaton Corp.			
Farmingdale, New York	—	8,880	0.21
45. Digital Equipment Corp.			
Landover, Maryland	52	8,728	0.20
46. Klate Holt Co. *b*			
Hampton, Virginia	47	8,375	0.20
47. W&J Construction Corp. *b*			
Kennedy Space Center, Florida	—	7,846	0.18
48. SF&G Inc. DBA Mercury *b*			
Huntsville, Alabama	—	7,795	0.18
49. Northrop Worldwide Aircraft			
Houston, Texas	48	7,766	0.18
50. Informatics, Inc.			
Mountain View, California	60	7,297	0.17
Other		657,384	15.35
Total Awards to Business Firms		4,272,806	100.00

a Excludes smaller procurements, generally those of less than $10,000.
b Small business concern.

Table 8–31. Top Fifty Contractors: FY 1982 (in thousands of dollars) a

Contractor/Place of Contract Performance	Rank in FY 1981	Net Value of Awards	
		Amount	Percentage
1. Rockwell International Corp. Downey, California	1	1,564,210	32.55
2. Martin Marietta Corp. New Orleans, Louisiana	2	309,896	6.45
3. McDonnell Douglas Corp. Huntington Beach, California	3	220,312	4.58
4. Thiokol Corp. Brigham City, Utah	5	152,413	3.17
5. Computer Sciences Corp. Kennedy Space Center, Florida	4	138,334	2.88
6. United Space Boosters, Inc. Kennedy Space Center, Florida	12	127,055	2.64
7. General Dynamics Corp. San Diego, California	11	113,882	2.37
8. Bendix Corp. Columbia, Maryland	7	108,720	2.26
9. IBM Houston, Texas	8	106,512	2.22
10. General Electric Co. King of Prussia, Pennsylvania	6	97,003	2.02
11. United Technologies Corp. Windsor Locks, Connecticut	10	89,855	1.87
12. Lockheed Engineering and Mgmt. Corp. Houston, Texas	14	88,872	1.85
13. Boeing Services International Kennedy Space Center, Florida	9	81,574	1.70
14. Ford Aerospace and Communications Houston, Texas	13	74,057	1.54
15. Lockheed Missile & Space Co. Sunnyvale, California	18	68,592	1.43
16. Planning Research Corp. Kennedy Space Center, Florida	17	54,665	1.14
17. Perkin Elmer Corp. Danbury, Connecticut	16	44,394	0.92
18. TRW, Inc. Redondo Beach, California	20	43,859	0.91
19. Boeing Co. Seattle, Washington	19	40,986	0.85
20. Hughes Aircraft Co. Los Angeles, California	15	39,535	0.82
21. Pan American World Airways Bay St. Louis, Mississippi	21	34,792	0.72

FINANCES AND PROCUREMENT 551

Table 8–31 continued

Contractor/Place of Contract Performance	Rank in FY 1981	Net Value of Awards	
		Amount	Percentage
22. Singer Co.			
Houston, Texas	23	33,097	0.69
23. Space Communications Co.			
Gaithersburg, Maryland	81	30,191	0.63
24. Teledyne Industries, Inc.			
Northridge, California	28	29,175	0.61
25. Ball Corp.			
Boulder, Colorado	22	26,420	0.55
26. Sperry Corp.			
Phoenix, Arizona	26	25,989	0.54
27. Northrop Services, Inc.			
Houston, Texas	27	25,261	0.53
28. Air Products & Chemicals, Inc.			
Allentown, Pennsylvania	25	25,036	0.52
29. RCA Corp.			
Princeton, New Jersey	24	23,682	0.49
30. Raytheon Services Co.			
Greenbelt, Maryland	32	19,586	0.41
31. Honeywell Information Systems			
McLean, Virginia	37	18,453	0.38
32. Santa Barbara Research Center			
Goleta, California	—	16,362	0.34
33. Management & Technical Services			
Houston, Texas	38	15,565	0.32
34. Kentron International, Inc.			
Hampton, Virginia	30	14,774	0.31
35. SF&G Inc. DBA Mercury b			
Huntsville, Alabama	48	14,256	0.30
36. Wackenhut Services, Inc.			
Kennedy Space Center, Florida	36	13,948	0.29
37. Systems Development Corp.			
Slidell, Louisiana	—	13,406	0.28
38. Control Data Corp.			
Minneapolis, Minnesota	45	12,789	0.27
39. Westinghouse Electric Corp.			
Large, Pennsylvania	51	12,149	0.25
40. International Telephone and Telegraph (ITT)			
Fort Wayne, Indiana	100	11,701	0.24
41. Mechanical Technology, Inc.			
Latham, New York	34	10,275	0.21
42. General Motors Corp.			
Indianapolis, Indiana	35	10,275	0.21

Table 8–31 continued

Contractor/Place of Contract Performance	Rank in FY 1981	Net Value of Awards	
		Amount	Percentage
43. Modular Computer Systems, Inc. Fort Lauderdale, Florida	71	10,109	0.21
44. Digital Equipment Corp. Mountain View, California	45	10,036	0.21
45. Virginia Electric & Power Co. Hampton, Virginia	41	9,622	0.20
46. Vought Corp. Dallas, Texas	33	9,510	0.20
47. Honeywell, Inc. Largo, Florida	42	9,221	0.19
48. Northrop Worldwide Aircraft Houston, Texas	49	9,002	0.19
49. Garrett Corp. Phoenix, Arizona	43	8,967	0.19
50. Global Associates *b* New Orleans, Louisiana	52	8,961	0.19
Other		728,252	15.16
Total Awards to Business Firms		4,805,588	100.00

a   Excludes smaller procurements, generally those of less than $10,000.
b   Small business concern.

Table 8–32. Top Fifty Contractors: FY 1983 (in thousands of dollars) a

Contractor/Place of Contract Performance	Rank in FY 1982	Net Value of Awards	
		Amount	Percentage
1. Rockwell International Corp. Downey, California	1	1,568,303	28.07
2. Martin Marietta Corp. New Orleans, Louisiana	2	466,074	8.34
3. Morton Thiokol Corp. Brigham City, Utah	4	268,457	4.81
4. McDonnell Douglas Corp. Huntington Beach, California	3	236,717	4.24
5. General Dynamics Corp. San Diego, California	7	156,336	2.80
6. Computer Sciences Corp. Kennedy Space Center, Florida	5	147,135	2.63
7. Bendix Corp. Columbia, Maryland	8	137,038	2.45
8. United Technologies Corp. Windsor Locks, Connecticut	11	116,152	2.08
9. IBM Houston, Texas	9	116,055	2.08
10. United Space Boosters, Inc. Kennedy Space Center, Florida	6	115,450	2.07
11. Ford Aerospace and Communications Houston, Texas	14	106,804	1.91
12. Lockheed Engineering and Mgmt. Corp. Houston, Texas	12	100,733	1.80
13. Lockheed Missile & Space Co. Sunnyvale, California	15	96,419	1.73
14. Boeing Services International Kennedy Space Center, Florida	13	85,943	1.54
15. General Electric Co. King of Prussia, Pennsylvania	10	85,492	1.53
16. Perkin Elmer Corp. Danbury, Connecticut	17	70,093	1.25
17. EG&G Florida Inc. Kennedy Space Center, Florida	—	67,571	1.21
18. Planning Research Corp. Kennedy Space Center, Florida	16	57,148	1.02
19. RCA Corp. Princeton, New Jersey	29	56,786	1.02
20. Hughes Aircraft Co. Los Angeles, California	20	51,434	0.92
21. TRW, Inc. Redondo Beach, California	18	49,427	0.88

Table 8–32 continued

Contractor/Place of Contract Performance	Rank in FY 1982	Net Value of Awards	
		Amount	Percentage
22. Teledyne Industries, Inc. Huntsville, Alabama	24	47,330	0.85
23. Boeing Co. Seattle, Washington	19	43,606	0.78
24. Ball Corp. Boulder, Colorado	25	38,898	0.70
25. Pan American World Airways Bay St. Louis, Mississippi	21	35,877	0.64
26. Singer Co. Houston, Texas	22	35,316	0.63
27. Northrop Services, Inc. Houston, Texas	27	28,281	0.52
28. Sperry Corp. Houston, Texas	26	26,509	0.47
29. Air Products & Chemicals, Inc. Allentown, Pennsylvania	28	25,117	0.45
30. Honeywell Information Systems McLean, Virginia	31	23,069	0.41
31. Raytheon Services Co. Greenbelt, Maryland	30	21,414	0.38
32. Lockheed Space Operations Co. Kennedy Space Center, Florida	—	19,381	0.35
33. Honeywell, Inc. Clearwater, Florida	47	19,030	0.34
34. Management & Technical Services Houston, Texas	33	17,761	0.32
35. Kentron International, Inc. Hampton, Virginia	34	17,701	0.32
36. Digital Equipment Corp. Landover, Maryland	44	17,116	0.31
37. Mercury Consolidated Inc. Huntsville, Alabama	—	16,621	0.30
38. Systems Development Corp. Slidell, Louisiana	37	14,198	0.25
39. Control Data Corp. Greenbelt, Maryland	38	13,309	0.24
40. Vought Corp. Dallas, Texas	46	11,915	0.21
41. Mechanical Technology, Inc. Latham, New York	41	11,838	0.21
42. Informatics General Corp. Mountain View, California	52	10,689	0.19

Table 8–32 continued

Contractor/Place of Contract Performance	Rank in FY 1982	Net Value of Awards	
		Amount	Percentage
43. Westinghouse Electric Corp. Large, Pennsylvania	39	10,455	0.19
44. Fairchild Industries, Inc. Germantown, Maryland	53	10,450	0.19
45. Virginia Electric & Power Co. Hampton, Virginia	45	10,087	0.18
46. Garrett Corp. Phoenix, Arizona	49	9,689	0.17
47. Northrop Worldwide Aircraft Houston, Texas	48	9,324	0.17
48. Klate Holt Co. *b* Hampton, Virginia	51	9,281	0.17
49. Lockheed Corp. Burbank, California	55	9,171	0.16
50. Cleveland Electric Illuminating Cleveland, Ohio	57	8,710	0.16
Other		875,695	15.36
Total Awards to Business Firms		5,585,985	100.00

a Excludes smaller procurements, generally those of less than $10,000.
b Small business concern.

Table 8–33. Top Fifty Contractors: FY 1984 (in thousands of dollars) a

Contractor/Place of Contract Performance	Rank in FY 1983	Net Value of Awards	
		Amount	Percentage
1. Rockwell International Corp. Downey, California	1	1,401,411	23.50
2. Martin Marietta Corp. New Orleans, Louisiana	2	427,788	7.17
3. Morton Thiokol Corp. Brigham City, Utah	3	322,362	5.40
4. Lockheed Space Operations Co. Kennedy Space Center, Florida	32	301,357	5.05
5. General Dynamics Corp. San Diego, California	5	252,515	4.23
6. McDonnell Douglas Corp. Huntington Beach, California	4	199,763	3.35
7. United Space Boosters, Inc. Huntsville, Alabama	10	196,520	3.29
8. Bendix Corp. Columbia, Maryland	7	162,643	2.73
9. IBM Houston, Texas	9	134,408	2.25
10. United Technologies Corp. Windsor Locks, Connecticut	8	117,924	1.98
11. EG&G Florida Inc. Kennedy Space Center, Florida	17	109,357	1.83
12. Ford Aerospace and Communications Houston, Texas	11	105,663	1.77
13. Lockheed Engineering and Mgmt. Corp. Houston, Texas	12	105,116	1.76
14. Lockheed Missile & Space Co. Sunnyvale, California	13	102,096	1.71
15. Computer Sciences Corp. Houston, Texas	6	89,465	1.50
16. TRW, Inc. Redondo Beach, California	21	82,291	1.38
17. Perkin Elmer Corp. Danbury, Connecticut	16	79,349	1.33
18. RCA Corp. Princeton, New Jersey	19	67,979	1.14
19. Planning Research Corp. Kennedy Space Center, Florida	18	57,366	0.96
20. Teledyne Industries, Inc. Huntsville, Alabama	22	51,528	0.89
21. Boeing Co. Seattle, Washington	23	44,188	0.74

FINANCES AND PROCUREMENT

Table 8–33 continued

Contractor/Place of Contract Performance	Rank in FY 1983	Net Value of Awards	
		Amount	Percentage
22. Singer Co.			
Houston, Texas	26	43,936	0.74
23. General Electric Co.			
King of Prussia, Pennsylvania	15	43,553	0.73
24. Hughes Aircraft Co.			
El Segundo, California	20	42,163	0.71
25. Pan American World Services, Inc.			
Bay St. Louis, Mississippi	25	40,296	0.68
26. Ball Corp.			
Boulder, Colorado	24	39,088	0.65
27. Northrop Services, Inc.			
Houston, Texas	27	32,193	0.54
28. Raytheon Services Co.			
Greenbelt, Maryland	31	27,764	0.47
29. Westinghouse Electric Corp.			
Large, Pennsylvania	43	25,889	0.43
30. Sperry Corp.			
Houston, Texas	28	24,874	0.42
31. Boeing Services International			
Kennedy Space Center, Florida	14	24,797	0.42
32. Air Products & Chemicals, Inc.			
Allentown, Pennsylvania	29	21,773	0.36
33. Kentron International, Inc.			
Hampton, Virginia	35	20,937	0.35
34. Control Data Corp.			
Greenbelt, Maryland	39	20,221	0.34
35. Management & Technical Services			
Houston, Texas	34	19,776	0.33
36. Lockheed Corp.			
Marietta, Georgia	49	18,210	0.30
37. Digital Equipment Corp.			
Mountain View, California	36	16,091	0.27
38. Honeywell, Inc.			
Clearwater, Florida	33	14,643	0.25
39. Systems Development Corp.			
Slidell, Louisiana	38	14,637	0.25
40. Fairchild Industries, Inc.			
Germantown, Maryland	44	13,553	0.23
41. Bechtel National, Inc.			
Kennedy Space Center, Florida	—	13,228	0.22
42. Mechanical Technology, Inc.			
Latham, New York	41	12,972	0.22

Table 8–33 continued

Contractor/Place of Contract Performance	Rank in FY 1983	Net Value of Awards	
		Amount	Percentage
43. Grumman Aerospace Corp. Bethpage, New York	57	12,559	0.21
44. Informatics General Corp. Mountain View, California	42	12,316	0.21
45. Honeywell Information Systems McLean, Virginia	30	12,013	0.20
46. Northrop Worldwide Aircraft Houston, Texas	47	11,906	0.20
47. Sauer Mechanical, Inc. Kennedy Space Center, Florida	79	11,768	0.20
48. Space Communications Co. Gaithersburg, Maryland	—	11,614	0.19
49. RMS Technologies, Inc. *b, c* Greenbelt, Maryland	—	11,427	0.19
50. Klate Holt Co. *b* Hampton, Virginia	48	10,842	0.18
Other		930,299	14.55
Total Awards to Business Firms		5,967,427	100.00

a   Excludes smaller procurements, generally those of less than $10,000.
b   Small business concern.
c   Disadvantaged/minority business firm.

Table 8–34. Top Fifty Contractors: FY 1985 (in thousands of dollars)

Contractor/Place of Contract Performance	Rank in FY 1984	Net Value of Awards	
		Amount	Percentage
1. Rockwell International Corp. Downey, California	1	1,345,265	20.22
2. Lockheed Space Operations Co. Kennedy Space Center, Florida	4	551,235	8.29
3. Martin Marietta Corp. New Orleans, Louisiana	2	482,520	7.25
4. Morton Thiokol Corp. Brigham City, Utah	3	334,151	5.02
5. General Dynamics Corp. San Diego, California	5	300,284	4.51
6. United Space Boosters, Inc. Huntsville, Alabama	7	207,336	3.12
7. McDonnell Douglas Corp. Huntington Beach, California	6	193,728	2.91
8. Allied Bendix Aerospace a Columbia, Maryland	8	150,229	2.26
9. Lockheed Missile & Space Co. Sunnyvale, California	14	136,679	2.05
10. Lockheed Engineering and Mgmt. Corp. Houston, Texas	13	124,869	1.88
11. IBM Houston, Texas	9	124,224	1.87
12. Ford Aerospace and Communications Houston, Texas	12	120,287	1.81
13. United Technologies Corp. Windsor Locks, Connecticut	10	110,067	1.65
14. EG&G Florida Inc. Kennedy Space Center, Florida	11	108,064	1.62
15. TRW, Inc. Redondo Beach, California	16	103,181	1.55
16. Computer Sciences Corp. Houston, Texas	15	102,273	1.54
17. RCA Corp. Princeton, New Jersey	18	102,088	1.53
18. Space Communications Co. Gaithersburg, Maryland	48	98,389	1.48
19. Boeing Co. Seattle, Washington	21	69,176	1.04
20. Planning Research Corp. Kennedy Space Center, Florida	19	65,285	0.98
21. Perkin Elmer Corp. Danbury, Connecticut	17	63,659	0.96

Table 8-34 continued

Contractor/Place of Contract Performance	Rank in FY 1984	Net Value of Awards	
		Amount	Percentage
22. Pan American World Services, Inc. Bay St. Louis, Mississippi	25	49,269	0.74
23. Teledyne Industries, Inc. Huntsville, Alabama	20	45,837	0.69
24. General Electric Co. King of Prussia, Pennsylvania	23	43,471	0.65
25. Singer Co. Houston, Texas	22	42,552	0.64
26. Boeing Technical Operations Inc. Kennedy Space Center, Florida	—	39,304	0.59
27. Northrop Services, Inc. Houston, Texas	27	39,127	0.59
28. Hughes Aircraft Co. El Segundo, California	24	38,134	0.57
29. Lockheed Corp. Marietta, Georgia	36	30,451	0.46
30. Ball Corp. Boulder, Colorado	26	30,123	0.45
31. Management & Technical Services Houston, Texas	35	26,444	0.40
32. Sperry Corp. Houston, Texas	30	25,239	0.38
33. Raytheon Services Co. Greenbelt, Maryland	28	24,999	0.38
34. Air Products & Chemicals, Inc. Allentown, Pennsylvania	32	24,353	0.37
35. Fairchild Industries, Inc. Germantown, Maryland	40	20,176	0.30
36. Control Data Corp. Mountain View, California	34	19,471	0.29
37. Systems Development Corp. Slidell, Louisiana	39	18,753	0.28
38. Cray Research Inc. Chippewa Falls, Wisconsin	62	18,656	0.28
39. LTV Aerospace & Defense Co. Dallas, Texas	68	17,393	0.26
40. American Telephone & Telegraph (AT&T) Greenbelt, Maryland	65	16,570	0.25
41. Informatics General Corp. Mountain View, California	44	15,607	0.23
42. Digital Equipment Corp. Huntsville, Alabama	37	15,266	0.23

Table 8–34 continued

Contractor/Place of Contract Performance	Rank in FY 1984	Net Value of Awards	
		Amount	Percentage
43. Honeywell Information Systems McLean, Virginia	45	14,330	0.22
44. International Fuel Cells Corp. South Windsor, Connecticut	—	14,286	0.21
45. Bamsi, Inc. b, c Huntsville, Alabama	73	13,410	0.20
46. Management Services, Inc. b Huntsville, Alabama	71	13,353	0.20
47. Honeywell, Inc. Clearwater, Florida	38	13,089	0.20
48. Westinghouse Electric Corp. Large, Pennsylvania	29	12,819	0.19
49. Northrop Worldwide Aircraft Houston, Texas	46	12,651	0.19
50. Mechanical Technology, Inc. Latham, New York	42	12,619	0.19
Other		1,052,177	14.83
Total Awards to Business Firms		6,652,918	100.00

a   In FY 1984, this was the Bendix Corporation.
b   Small business concern.
c   Disadvantaged/minority business firm.

Table 8–35. *Top Fifty Contractors: FY 1986 (in thousands of dollars)*

Contractor/Place of Contract Performance	Rank in FY 1985	Net Value of Awards	
		Amount	Percentage
1. Rockwell International Corp. Downey, California	1	1,155,806	18.18
2. Lockheed Space Operations Co. Kennedy Space Center, Florida	2	559,159	8.80
3. Martin Marietta Corp. New Orleans, Louisiana	3	427,393	6.72
4. Morton Thiokol Corp. Brigham City, Utah	4	319,888	5.03
5. McDonnell Douglas Corp. Huntington Beach, California	7	265,828	4.18
6. Ford Aerospace and Communications Huntsville, Alabama	12	207,726	3.27
7. USBI Booster Production Co. Huntsville, Alabama	—	196,441	3.09
8. General Dynamics Corp. San Diego, California	5	194,386	3.06
9. RCA Corp. Princeton, New Jersey	17	140,909	2.22
10. Allied Bendix Aerospace Columbia, Maryland	8	137,955	2.17
11. Lockheed Engineering and Mgmt. Corp. Houston, Texas	10	123,899	1.95
12. Lockheed Missile & Space Co. Sunnyvale, California	9	121,095	1.91
13. EG&G Florida Inc. Kennedy Space Center, Florida	14	117,411	1.85
14. Boeing Co. Seattle, Washington	19	112,603	1.77
15. United Technologies Corp. Windsor Locks, Connecticut	13	96,876	1.52
16. Computer Sciences Corp. Houston, Texas	16	96,202	1.51
17. IBM Houston, Texas	11	93,543	1.47
18. TRW, Inc. Redondo Beach, California	15	84,920	1.34
19. Contel Corp. Gaithersburg, Maryland	—	68,531	1.08
20. General Electric Co. King of Prussia, Pennsylvania	24	66,076	1.04
21. Planning Research Corp. Kennedy Space Center, Florida	20	51,336	0.81

Table 8–35 continued

Contractor/Place of Contract Performance	Rank in FY 1985	Net Value of Awards Amount	Percentage
22. Teledyne Industries, Inc. Huntsville, Alabama	23	48,075	0.76
23. Pan American World Services, Inc. Bay St. Louis, Mississippi	22	46,508	0.73
24. Northrop Services, Inc. Houston, Texas	27	41,081	0.65
25. Boeing Technical Operations Inc. Kennedy Space Center, Florida	26	35,580	0.56
26. Raytheon Services Co. Greenbelt, Maryland	33	28,449	0.45
27. Lockheed Corp. Marietta, Georgia	29	27,393	0.43
28. Management & Technical Services Houston, Texas	31	27,065	0.43
29. Ball Corp. Boulder, Colorado	30	26,916	0.42
30. Fairchild Industries, Inc. Germantown, Maryland	35	26,013	0.41
31. Perkin Elmer Corp. Danbury, Connecticut	21	23,520	0.37
32. Singer Co. Houston, Texas	25	20,252	0.32
33. Aerojet General Corp. Azusa, California	62	19,589	0.31
34. Bamsi, Inc. a, b Huntsville, Alabama	45	19,029	0.30
35. Informatics General Corp. Mountain View, California	41	19,010	0.30
36. Sperry Corp. Houston, Texas	32	18,382	0.29
37. Westinghouse Electric Corp. Large, Pennsylvania	48	18,095	0.28
38. Air Products & Chemicals, Inc. Allentown, Pennsylvania	34	17,253	0.27
39. Systems Development Corp. Slidell, Louisiana	37	17,066	0.27
40. Harris Corp. Melbourne, Florida	—	16,988	0.27
41. Management Services, Inc. a Huntsville, Alabama	46	15,036	0.24
42. Sverdrup Technology, Inc. Middleburgh Heights, Ohio	74	14,842	0.23

Table 8–35 continued

Contractor/Place of Contract Performance	Rank in FY 1985	Net Value of Awards	
		Amount	Percentage
43. Analex Corp. *a* Cleveland, Ohio	53	14,617	0.23
44. Honeywell, Inc. Clearwater, Florida	47	14,586	0.23
45. Control Data Corp. Mountain View, California	36	13,934	0.22
46. Honeywell Information Systems McLean, Virginia	43	13,776	0.22
47. Mechanical Technology, Inc. Latham, New York	50	13,634	0.21
48. LTV Aerospace & Defense Co. Dallas, Texas	39	13,391	0.21
49. RMS Technologies, Inc. *a, b* Greenbelt, Maryland	59	13,373	0.21
50. Digital Equipment Corp. Landover, Maryland	42	13,111	0.21
Other		1,081,496	17.00
Total Awards to Business Firms		6,356,043	100.00

a   Small business concern.
b   Disadvantaged/minority business firm.

Table 8–36. Top Fifty Contractors: FY 1987 (in thousands of dollars)

Contractor/Place of Contract Performance	Rank in FY 1986	Net Value of Awards	
		Amount	Percentage
1. Rockwell International Corp. Downey, California	1	1,610,806	24.62
2. Martin Marietta Corp. New Orleans, Louisiana	3	325,879	4.98
3. Lockheed Space Operations Co. Kennedy Space Center, Florida	2	323,290	4.94
4. Morton Thiokol Corp. Brigham City, Utah	4	286,017	4.37
5. McDonnell Douglas Corp. Huntington Beach, California	5	285,017	4.36
6. General Electric Co. a Princeton, New Jersey	20	225,369	3.45
7. USBI Booster Production Co. Huntsville, Alabama	7	183,052	2.80
8. Boeing Co. Seattle, Washington	14	174,872	2.67
9. United Technologies Corp. Stratford, Connecticut	15	165,616	2.53
10. Lockheed Engineering and Mgmt. Corp. Houston, Texas	11	162,507	2.48
11. Allied Bendix Aerospace Columbia, Maryland	10	142,010	2.17
12. EG&G Florida Inc. Kennedy Space Center, Florida	13	130,743	2.00
13. TRW, Inc. Redondo Beach, California	18	124,473	1.90
14. Ford Aerospace and Communications Palo Alto, California	6	119,978	1.83
15. Lockheed Missile & Space Co. Sunnyvale, California	12	107,967	1.65
16. Computer Sciences Corp. Greenbelt, Maryland	16	90,013	1.38
17. Contel Corp. Gaithersburg, Maryland	19	80,900	1.24
18. IBM Houston, Texas	17	72,017	1.10
19. Pan American World Services, Inc. Bay St. Louis, Mississippi	23	59,849	0.92
20. Orbital Sciences Corp. b Denver, Colorado	—	41,598	0.64
21. Teledyne Industries, Inc. Huntsville, Alabama	22	38,378	0.59

Table 8–36 continued

Contractor/Place of Contract Performance	Rank in FY 1986	Net Value of Awards	
		Amount	Percentage
22. Planning Research Corp.			
Hampton, Virginia	21	36,947	0.56
23. Perkin Elmer Corp.			
Danbury, Connecticut	31	34,915	0.53
24. Raytheon Services Co.			
Greenbelt, Maryland	26	32,254	0.49
25. Bamsi, Inc. b, c			
Huntsville, Alabama	34	31,195	0.48
26. Northrop Services, Inc.			
Mountain View, California	24	29,950	0.46
27. Boeing Technical Operations Inc.			
Houston, Texas	25	27,218	0.42
28. Unisys Corp.			
Greenbelt, Maryland	—	27,148	0.42
29. Sverdrup Technology, Inc.			
Middleburgh Heights, Ohio	42	26,958	0.41
30. Aerojet General Corp.			
Azusa, California	33	26,214	0.40
31. General Dynamics Corp.			
San Diego, California	8	25,433	0.39
32. Wyle Laboratories			
Huntsville, Alabama	53	24,487	0.37
33. Sterling Federal Systems, Inc.			
Mountain View, California	—	24,103	0.37
34. Fairchild Industries, Inc.			
Germantown, Maryland	30	23,889	0.37
35. Grumman Aerospace Corp.			
Bethpage, New York	67	23,243	0.36
36. Lockheed Corp.			
Marietta, Georgia	27	21,555	0.33
37. Control Data Corp.			
Greenbelt, Maryland	45	20,076	0.31
38. Air Products & Chemicals, Inc.			
Allentown, Pennsylvania	38	19,309	0.30
39. Digital Equipment Corp.			
Landover, Maryland	50	18,871	0.29
40. RMS Technologies, Inc. b, c			
Greenbelt, Maryland	49	18,613	0.28
41. Singer Co.			
Houston, Texas	32	17,868	0.27
42. Ball Corp.			
Boulder, Colorado	29	17,416	0.27

Table 8–36 continued

Contractor/Place of Contract Performance	Rank in FY 1986	Net Value of Awards	
		Amount	Percentage
43. Management Services, Inc. *b*			
Huntsville, Alabama	41	17,387	0.27
44. Klate Holt Co. *b*			
Hampton, Virginia	52	17,163	0.26
45. Westinghouse Electric Corp.			
Large, Pennsylvania	37	15,908	0.24
46. Grumman Data Systems Corp.			
Huntsville, Alabama	58	15,246	0.23
47. Northrop Worldwide Aircraft			
Houston, Texas	51	14,529	0.22
48. Harris Corp.			
Melbourne, Florida	40	13,252	0.20
49. Argee Corp.			
Las Cruces, New Mexico	—	13,172	0.20
50. Micro Craft, Inc. *b*			
Tullahoma, Tennessee	92	12,559	0.19
Other		1,143,877	17.49
Total Awards to Business Firms		6,540,541	100.00

a Includes awards to RCA Corp.
b Small business concern.
c Disadvantaged/minority business firm.

Table 8–37. Top Fifty Contractors: FY 1988 (in thousands of dollars)

Contractor/Place of Contract Performance	Rank in FY 1987	Net Value of Awards	
		Amount	Percentage
1. Rockwell International Corp. Downey, California	1	1,714,205	23.56
2. Lockheed Space Operations Co. Kennedy Space Center, Florida	3	474,338	6.52
3. Morton Thiokol Corp. Brigham City, Utah	4	422,798	5.81
4. Martin Marietta Corp. New Orleans, Louisiana	2	341,047	4.69
5. McDonnell Douglas Corp. Huntington Beach, California	5	299,103	4.11
6. Boeing Co. Marshall Space Flight Center, Alabama	8	260,333	3.58
7. General Electric Co. Princeton, New Jersey	6	211,157	2.90
8. USBI Booster Production Co. Huntsville, Alabama	7	190,700	2.62
9. Lockheed Engineering & Science Corp. Houston, Texas	—	177,776	2.44
10. EG&G Florida Inc. Kennedy Space Center, Florida	12	155,505	2.14
11. AlliedSignal Aerospace Co. Columbia, Maryland	—	152,060	2.09
12. Computer Sciences Corp. Greenbelt, Maryland	16	151,449	2.08
13. TRW, Inc. Redondo Beach, California	13	142,715	1.96
14. Lockheed Missile & Space Co. Sunnyvale, California	15	140,823	1.94
15. Ford Aerospace Corp. Palo Alto, California	14	137,372	1.89
16. United Technologies Corp. West Palm Beach, Florida	9	91,463	1.26
17. IBM Houston, Texas	18	87,262	1.20
18. Contel Corp. Gaithersburg, Maryland	17	75,658	1.04
19. Grumman Aerospace Corp. Reston, Virginia	35	74,134	1.02
20. Pan American World Services, Inc. Stennis Space Center, Mississippi	19	69,977	0.96
21. Planning Research Corp. Hampton, Virginia	22	46,785	0.64

Table 8–37 continued

Contractor/Place of Contract Performance	Rank in FY 1987	Net Value of Awards	
		Amount	Percentage
22. Boeing Technical Operations Inc. Houston, Texas	27	41,830	0.57
23. Teledyne Industries, Inc. Marshall Space Flight Center, Alabama	21	39,972	0.55
24. Bamsi, Inc. a Marshall Space Flight Center, Alabama	25	39,878	0.55
25. Sverdrup Technology, Inc. Middleburgh Heights, Ohio	29	37,964	0.52
26. Raytheon Services Co. Greenbelt, Maryland	24	37,725	0.52
27. Perkin Elmer Corp. Danbury, Connecticut	23	30,972	0.43
28. Cray Research, Inc. Chippewa Falls, Wisconsin	57	30,715	0.42
29. Orbital Sciences Corp. b Denver, Colorado	20	26,038	0.36
30. NSI Technology Services Corp. c Moffett Field, California	26	25,123	0.35
31. Sterling Federal Systems, Inc. Moffett Field, California	33	24,930	0.34
32. Fairchild Industries, Inc. Germantown, Maryland	34	24,151	0.33
33. General Dynamics Corp. San Diego, California	31	22,817	0.31
34. Unisys Corp. Greenbelt, Maryland	28	22,645	0.31
35. Aerojet General Corp. Sacramento, California	30	22,159	0.30
36. Klate Holt Co. b Hampton, Virginia	44	20,274	0.28
37. Wyle Laboratories Hampton, Virginia	32	20,259	0.28
38. Ball Corp. Boulder, Colorado	42	19,611	0.27
39. W&J Construction Corp. Kennedy Space Center, Florida	—	19,301	0.27
40. Lockheed Corp. Marietta, Georgia	36	18,519	0.25
41. LTV Aerospace & Defense Co. Dallas, Texas	58	16,789	0.23
42. Continental Construction Corp. b Edwards, California	—	16,491	0.23

570 NASA HISTORICAL DATA BOOK

Table 8–37 continued

Contractor/Place of Contract Performance	Rank in FY 1987	Net Value of Awards	
		Amount	Percentage
43. Krug International Corp.			
Houston, Texas	55	16,240	0.22
44. Grumman Data Systems Corp.			
Marshall Space Flight Center, Alabama	46	16,166	0.22
45. Northrop Worldwide Aircraft			
Houston, Texas	47	15,954	0.22
46. Engineering and Economic Research *a, b*			
Beltsville, Maryland	68	15,047	0.21
47. Air Products & Chemicals, Inc.			
Allentown, Pennsylvania	38	14,876	0.20
48. Singer Co.			
Houston, Texas	41	14,272	0.20
49. Zero One Systems, Inc. *a, b*			
Moffett Field, California	53	13,111	0.18
50. ST System Corp. *a, b*			
Hyattsville, Maryland	52	12,777	0.18
Other		1,261,600	15.25
Total Awards to Business Firms		7,274,866	100.00

a Disadvantaged/minority business firm.
b Small business concern.
c Formerly Northrop Services, Inc.

Table 8–38. *Top Fifty Educational and Nonprofit Institutions: FY 1979 (in thousands of dollars)*

Contractor/Place of Contract Performance	Rank in FY 1978	Net Value of Awards	
		Amount	Percentage
1. University of California at San Diego La Jolla, California	7	9,501	4.80
2. Massachusetts Institute of Technology Cambridge, Massachusetts	1	8,951	4.52
3. Smithsonian Institution a Cambridge, Massachusetts	2	8,924	4.51
4. Charles Stark Draper Laboratory a Cambridge, Massachusetts	4	7,047	3.56
5. National Academy of Sciences a Washington, D.C.	6	5,388	2.72
6. Stanford University Stanford, California	5	5,107	2.58
7. University of Michigan Ann Arbor, Michigan	3	4,845	2.45
8. Battelle Memorial Institute a Columbus, Ohio	15	4,513	2.28
9. University of Chicago Chicago, Illinois	8	4,458	2.25
10. University of California at Berkeley Berkeley, California	10	4,319	2.18
11. Harvard University Cambridge, Massachusetts	11	4,314	2.18
12. California Institute of Technology Pasadena, California	12	3,955	2.00
13. Universities Space Research a Washington, D.C.	19	3,640	1.84
14. University of Chile Santiago, Chile	9	3,591	1.81
15. University of Maryland at College Park College Park, Maryland	20	3,476	1.76
16. University of Arizona Tucson, Arizona	23	3,324	1.68
17. University of Wisconsin at Madison Madison, Wisconsin	18	3,257	1.65
18. Purdue University West Lafayette, Indiana	24	3,022	1.53
19. SRI International Corp. a Menlo Park, California	22	2,826	1.43
20. University of Iowa Iowa City, Iowa	13	2,816	1.42
21. University of Hawaii Honolulu, Hawaii	25	2,589	1.31

Table 8–38 continued

Contractor/Place of Contract Performance	Rank in FY 1978	Net Value of Awards	
		Amount	Percentage
22. University of Colorado at Boulder Boulder, Colorado	16	2,563	1.29
23. University of California at Los Angeles Los Angeles, California	26	2,562	1.29
24. Princeton University Princeton, New Jersey	34	2,478	1.25
25. University of Texas at Dallas Richardson, Texas	21	2,327	1.18
26. Old Dominion University Norfolk, Virginia	27	2,207	1.12
27. Columbia University New York, New York	17	2,069	1.05
28. University of Texas at Austin Austin, Texas	14	1,913	0.97
29. American Institute of Aeronautics and Astronautics [a] New York, New York	29	1,911	0.97
30. Virginia Polytechnic Institute Blacksburg, Virginia	31	1,787	0.90
31. Cornell University Ithaca, New York	40	1,716	0.87
32. New Mexico State University at Las Cruces Las Cruces, New Mexico	30	1,606	0.81
33. University of Minnesota at Minneapolis St. Paul Minneapolis, Minnesota	32	1,545	0.78
34. University of Southern California Los Angeles, California	65	1,528	0.77
35. George Washington University Washington, D.C.	33	1,511	0.76
36. University of Alabama at Huntsville Huntsville, Alabama	63	1,480	0.75
37. Pennsylvania State University State College, Pennsylvania	53	1,407	0.71
38. Southwest Research Institute [a] San Antonio, Texas	58	1,322	0.67
39. University of Pittsburgh Pittsburgh, Pennsylvania	51	1,266	0.64
40. Case Western Reserve University Cleveland, Ohio	49	1,256	0.63
41. Ohio State University Columbus, Ohio	62	1,238	0.63

Table 8–38 continued

Contractor/Place of Contract Performance	Rank in FY 1978	Net Value of Awards	
		Amount	Percentage
42. Environmental Research Institute *a* Ann Arbor, Michigan	47	1,233	0.62
43. University of Illinois at Urbana Urbana, Illinois	46	1,220	0.62
44. Research Triangle Institute *a* Durham, North Carolina	59	1,185	0.60
45. Johns Hopkins University Baltimore, Maryland	37	1,177	0.59
46. San Jose State University San Jose, California	48	1,165	0.59
47. Georgia Institute of Technology Atlanta, Georgia	50	1,162	0.59
48. California State University at Chico Chico, California	44	1,122	0.57
49. University of Washington Seattle, Washington	39	1,112	0.56
50. Utah State University Logan, Utah	45	1,102	0.56
Other		51,907	26.20
Total Awards to Educational and Nonprofit Institutions		197,940	100.00

a Nonprofit institution.

Table 8–39. *Top Fifty Educational and Nonprofit Institutions: FY 1980*
(in thousands of dollars)

Contractor/Place of Contract Performance	Rank in FY 1979	Net Value of Awards	
		Amount	Percentage
1. European Space Agency a			
Paris, France	52	31,100	12.00
2. Smithsonian Institution a			
Cambridge, Massachusetts	3	10,748	4.15
3. Massachusetts Institute of Technology			
Cambridge, Massachusetts	2	10,089	3.89
4. University of California at San Diego			
La Jolla, California	1	9,670	3.73
5. Charles Stark Draper Laboratory a			
Cambridge, Massachusetts	4	7,847	3.03
6. Stanford University			
Stanford, California	6	7,428	2.87
7. University of Chile			
Santiago, Chile	14	5,484	2.12
8. California Institute of Technology			
Pasadena, California	12	5,392	2.08
9. National Academy of Sciences a			
Washington, D.C.	5	4,661	1.80
10. University of Chicago			
Chicago, Illinois	9	4,602	1.78
11. University of California at Berkeley			
Berkeley, California	10	4,472	1.72
12. University of Michigan			
Ann Arbor, Michigan	7	4,234	1.63
13. University of Wisconsin at Madison			
Madison, Wisconsin	17	4,176	1.61
14. University of Hawaii			
Honolulu, Hawaii	21	4,155	1.60
15. Harvard University			
Cambridge, Massachusetts	11	4,145	1.60
16. Battelle Memorial Institute a			
Columbus, Ohio	8	4,086	1.58
17. University of Maryland at College Park			
College Park, Maryland	15	4,048	1.56
18. University of Texas at Austin			
Austin, Texas	28	3,981	1.54
19. University of Iowa			
Iowa City, Iowa	20	3,614	1.39
20. University of Arizona			
Tucson, Arizona	16	3,522	1.36
21. Purdue University			
West Lafayette, Indiana	18	3,458	1.33

Table 8–39 continued

Contractor/Place of Contract Performance	Rank in FY 1979	Net Value of Awards	
		Amount	Percentage
22. University of Colorado at Boulder Boulder, Colorado	22	3,321	1.28
23. Universities Space Research a Columbia, Maryland	13	3,106	1.20
24. University of California at Los Angeles Los Angeles, California	23	2,684	1.04
25. Princeton University Princeton, New Jersey	24	2,461	0.95
26. Virginia Polytechnic Institute Blacksburg, Virginia	30	2,355	0.91
27. University of Washington Seattle, Washington	49	2,265	0.87
28. American Institute of Aeronautics and Astronautics a New York, New York	29	2,162	0.83
29. Cornell University Ithaca, New York	31	2,161	0.83
30. University of Alabama at Huntsville Huntsville, Alabama	36	2,066	0.80
31. Columbia University New York, New York	27	2,062	0.80
32. Old Dominion University Norfolk, Virginia	26	1,957	0.75
33. Texas A&M University College Station, Texas	56	1,921	0.74
34. SRI International Corp. a Menlo Park, California	19	1,904	0.73
35. Case Western Reserve University Cleveland, Ohio	40	1,872	0.72
36. Oklahoma State University Stillwater, Oklahoma	67	1,818	0.70
37. University of Minnesota at Minneapolis St. Paul Minneapolis, Minnesota	33	1,780	0.69
38. Pennsylvania State University State College, Pennsylvania	37	1,699	0.66
39. University of Texas at Dallas Richardson, Texas	25	1,655	0.64
40. University of Southern California Los Angeles, California	34	1,585	0.61
41. New Mexico State University at Las Cruces Las Cruces, New Mexico	32	1,579	0.61

Table 8–39 continued

Contractor/Place of Contract Performance	Rank in FY 1979	Net Value of Awards	
		Amount	Percentage
42. Research Triangle Institute [a] Durham, North Carolina	44	1,572	0.61
43. Franklin Institute [a] Philadelphia, Pennsylvania	—	1,475	0.57
44. University of Kansas Lawrence, Kansas	—	1,474	0.57
45. Environmental Research Institute [a] Ann Arbor, Michigan	42	1,464	0.56
46. Ohio State University Columbus, Ohio	41	1,436	0.55
47. Washington University at St. Louis St. Louis, Missouri	50	1,423	0.55
48. Northeast Radio Observatory [a] Chico, California	82	1,415	0.55
49. Georgia Institute of Technology Atlanta, Georgia	47	1,381	0.53
50. San Jose State University Mountain View, California	46	1,291	0.50
Other		62,930	24.28
Total Awards to Educational and Nonprofit Institutions		259,186	100.00

[a] Nonprofit institution.

Table 8–40. Top Fifty Educational and Nonprofit Institutions: FY 1981 (in thousands of dollars)

Contractor/Place of Contract Performance	Rank in FY 1980	Net Value of Awards Amount	Percentage
1. European Space Agency a Paris, France	1	97,100	27.94
2. Massachusetts Institute of Technology Cambridge, Massachusetts	3	12,406	3.57
3. Smithsonian Institution a Cambridge, Massachusetts	2	9,752	2.81
4. University of California at San Diego La Jolla, California	4	9,453	2.72
5. National Academy of Sciences a Washington, D.C.	9	7,482	2.15
6. Stanford University Stanford, California	6	7,368	2.12
7. University of Chile Santiago, Chile	7	7,000	2.01
8. Charles Stark Draper Laboratory a Cambridge, Massachusetts	5	6,455	1.86
9. California Institute of Technology Pasadena, California	8	5,526	1.59
10. University of California at Berkeley Berkeley, California	11	5,240	1.51
11. University of Wisconsin at Madison Madison, Wisconsin	13	5,216	1.50
12. University of Chicago Chicago, Illinois	10	4,777	1.38
13. University of Maryland at College Park College Park, Maryland	17	4,755	1.37
14. Johns Hopkins University Baltimore, Maryland	51	4,539	1.31
15. University of Arizona Tucson, Arizona	20	4,527	1.30
16. University of Michigan Ann Arbor, Michigan	12	4,227	1.22
17. University of Colorado at Boulder Boulder, Colorado	22	3,701	1.07
18. Battelle Memorial Institute a Columbus, Ohio	16	3,653	1.05
19. Purdue University West Lafayette, Indiana	21	3,511	1.01
20. Harvard University Cambridge, Massachusetts	15	3,496	1.01
21. Universities Space Research a Columbia, Maryland	23	3,436	0.99

Table 8–40 continued

Contractor/Place of Contract Performance	Rank in FY 1980	Net Value of Awards	
		Amount	Percentage
22. Association of University Research and Astronomy a Baltimore, Maryland	—	3,368	0.97
23. University of Iowa Iowa City, Iowa	19	3,357	0.97
24. University of Hawaii Honolulu, Hawaii	14	3,347	0.96
25. Princeton University Princeton, New Jersey	25	3,031	0.87
26. University of Washington Seattle, Washington	27	2,924	0.84
27. University of California at Los Angeles Los Angeles, California	24	2,802	0.81
28. Texas A&M University College Station, Texas	33	2,372	0.68
29. American Institute of Aeronautics and Astronautics a New York, New York	28	2,364	0.68
30. Cornell University Ithaca, New York	29	2,362	0.68
31. Pennsylvania State University State College, Pennsylvania	38	2,343	0.67
32. University of New Hampshire Durham, New Hampshire	56	2,328	0.67
33. Environmental Research Institute a Ann Arbor, Michigan	45	2,196	0.63
34. Southwest Research Institute a San Antonio, Texas	54	2,127	0.61
35. University of Texas at Austin Austin, Texas	18	2,036	0.59
36. New Mexico State University at Las Cruces Las Cruces, New Mexico	41	2,019	0.58
37. George Washington University Washington, D.C.	60	1,981	0.57
38. SRI International Corp. a Menlo Park, California	34	1,967	0.57
39. Case Western Reserve University Cleveland, Ohio	35	1,923	0.55
40. Virginia Polytechnic Institute Blacksburg, Virginia	26	1,874	0.54
41. Columbia University New York, New York	31	1,868	0.54

Table 8–40 continued

Contractor/Place of Contract Performance	Rank in FY 1980	Net Value of Awards	
		Amount	Percentage
42. Hampton City *a*			
Hampton, Virginia	—	1,826	0.53
43. Washington University at St. Louis			
St. Louis, Missouri	47	1,785	0.51
44. University of Minnesota at Minneapolis St. Paul			
Minneapolis, Minnesota	37	1,753	0.50
45. Oklahoma State University			
Stillwater, Oklahoma	36	1,704	0.49
46. University of Southern California			
Los Angeles, California	40	1,702	0.49
47. University of Texas at Dallas			
Richardson, Texas	39	1,701	0.49
48. University of Kansas			
Lawrence, Kansas	44	1,673	0.48
49. University of Alabama at Huntsville			
Huntsville, Alabama	30	1,648	0.47
50. Old Dominion University			
Norfolk, Virginia	32	1,642	0.47
Other		69,935	20.12
Total Awards to Educational and Nonprofit Institutions		347,578	100.00

a Nonprofit institution.

Table 8–41. Top Fifty Educational and Nonprofit Institutions: FY 1982 (in thousands of dollars)

Contractor/Place of Contract Performance	Rank in FY 1981	Net Value of Awards	
		Amount	Percentage
1. European Space Agency a			
Paris, France	1	55,074	18.62
2. Massachusetts Institute of Technology			
Cambridge, Massachusetts	2	10,025	3.39
3. National Academy of Sciences a			
Washington, D.C.	5	8,839	3.01
4. Stanford University			
Stanford, California	6	8,663	2.93
5. University of California at San Diego			
La Jolla, California	4	8,049	2.72
6. University of Chile			
Santiago, Chile	7	7,774	2.63
7. Smithsonian Institution a			
Cambridge, Massachusetts	3	6,761	2.29
8. Charles Stark Draper Laboratory a			
Cambridge, Massachusetts	8	6,488	2.19
9. California Institute of Technology			
Pasadena, California	9	4,946	1.67
10. University of Maryland at College Park			
College Park, Maryland	13	4,875	1.64
11. Universities Space Research a			
Columbia, Maryland	21	4,770	1.61
12. University of California at Berkeley			
Berkeley, California	10	4,643	1.57
13. University of Chicago			
Chicago, Illinois	12	4,561	1.54
14. Harvard University			
Cambridge, Massachusetts	20	4,373	1.48
15. University of Hawaii			
Honolulu, Hawaii	24	4,244	1.44
16. University of Michigan			
Ann Arbor, Michigan	16	3,983	1.35
17. University of Arizona			
Tucson, Arizona	15	3,814	1.29
18. University of Wisconsin at Madison			
Madison, Wisconsin	11	3,577	1.21
19. University of Colorado at Boulder			
Boulder, Colorado	17	3,464	1.17
20. University of California at Los Angeles			
Los Angeles, California	27	3,450	1.17
21. University of Iowa			
Iowa City, Iowa	23	3,206	1.08

FINANCES AND PROCUREMENT

Table 8–41 continued

Contractor/Place of Contract Performance	Rank in FY 1981	Net Value of Awards Amount	Net Value of Awards Percentage
22. Purdue University West Lafayette, Indiana	19	3,162	1.07
23. Texas A&M University College Station, Texas	28	2,784	0.94
24. University of Texas at Austin Austin, Texas	35	2,750	0.93
25. University of New Hampshire Durham, New Hampshire	32	2,709	0.92
26. Case Western Reserve University Cleveland, Ohio	39	2,626	0.89
27. American Institute of Aeronautics and Astronautics [a] New York, New York	29	2,595	0.88
28. Battelle Memorial Institute [a] Columbus, Ohio	18	2,541	0.86
29. Princeton University Princeton, New Jersey	25	2,479	0.84
30. Virginia Polytechnic Institute Blacksburg, Virginia	40	2,187	0.74
31. Association of University Research and Astronomy [a] Baltimore, Maryland	22	2,158	0.73
32. Pennsylvania State University State College, Pennsylvania	31	2,097	0.71
33. Cornell University Ithaca, New York	30	2,007	0.68
34. New Mexico State University at Las Cruces Las Cruces, New Mexico	36	1,986	0.67
35. Columbia University New York, New York	41	1,978	0.67
36. University of Washington Seattle, Washington	26	1,927	0.65
37. SRI International Corp. [a] Menlo Park, California	38	1,904	0.64
38. Southwest Research Institute [a] San Antonio, Texas	24	1,860	0.63
39. Hampton City [a] Hampton, Virginia	42	1,844	0.62
40. Oklahoma State University Stillwater, Oklahoma	45	1,832	0.62
41. Washington University at St. Louis St. Louis, Missouri	43	1,749	0.59

Table 8–41 continued

Contractor/Place of Contract Performance	Rank in FY 1981	Net Value of Awards	
		Amount	Percentage
42. Research Triangle Institute *a* Durham, North Carolina	53	1,732	0.59
43. University of Kansas Lawrence, Kansas	48	1,718	0.58
44. University of Alabama at Huntsville Huntsville, Alabama	49	1,685	0.57
45. University of Southern California Los Angeles, California	46	1,685	0.57
46. University of Minnesota at Minneapolis St. Paul Minneapolis, Minnesota	44	1,599	0.54
47. Old Dominion University Norfolk, Virginia	50	1,509	0.51
48. University of Texas at Dallas Richardson, Texas	47	1,492	0.50
49. Rensselaer Polytechnic Institute Troy, New York	57	1,477	0.50
50. Arizona State University Tempe, Arizona	60	1,471	0.50
Other		70,624	23.88
Total Awards to Educational and Nonprofit Institutions		295,746	100.00

a Nonprofit institution.

Table 8–42. Top Fifty Educational and Nonprofit Institutions: FY 1983
(in thousands of dollars)

Contractor/Place of Contract Performance	Rank in FY 1982	Net Value of Awards	
		Amount	Percentage
1. European Space Agency [a] Paris, France	1	24,135	7.69
2. Massachusetts Institute of Technology Cambridge, Massachusetts	2	11,862	3.78
3. Stanford University Stanford, California	4	11,170	3.56
4. Smithsonian Institution [a] Cambridge, Massachusetts	7	10,976	3.50
5. University of California at San Diego La Jolla, California	5	10,659	3.40
6. National Academy of Sciences [a] Washington, D.C.	3	10,359	3.30
7. Charles Stark Draper Laboratory [a] Cambridge, Massachusetts	8	7,903	2.52
8. Association of University Research and Astronomy [a] Baltimore, Maryland	31	7,669	2.44
9. California Institute of Technology Pasadena, California	9	6,211	1.98
10. University of Chile Santiago, Chile	6	5,710	1.82
11. Universities Space Research [a] Columbia, Maryland	11	5,563	1.77
12. University of Maryland at College Park College Park, Maryland	10	5,515	1.76
13. University of Chicago Chicago, Illinois	13	5,514	1.76
14. University of California at Berkeley Berkeley, California	12	5,385	1.75
15. University of Hawaii Honolulu, Hawaii	15	5,362	1.71
16. University Corporation for Atmospheric Research [a] Palestine, Texas	—	4,710	1.50
17. University of Wisconsin at Madison Madison, Wisconsin	18	4,610	1.47
18. University of New Hampshire Durham, New Hampshire	25	4,436	1.41
19. University of Michigan Ann Arbor, Michigan	16	4,389	1.40
20. Battelle Memorial Institute [a] Columbus, Ohio	28	4,323	1.38

Table 8–42 continued

Contractor/Place of Contract Performance	Rank in FY 1982	Net Value of Awards	
		Amount	Percentage
21. University of Arizona Tucson, Arizona	17	4,244	1.35
22. University of Colorado at Boulder Boulder, Colorado	19	4,044	1.29
23. Harvard University Cambridge, Massachusetts	14	3,587	1.14
24. University of Iowa Iowa City, Iowa	21	3,544	1.13
25. Oklahoma State University Stillwater, Oklahoma	40	3,334	1.06
26. University of California at Los Angeles Los Angeles, California	20	3,277	1.04
27. Southwest Research Institute [a] San Antonio, Texas	38	3,215	1.02
28. Case Western Reserve University Cleveland, Ohio	26	3,073	0.98
29. American Institute of Aeronautics and Astronautics [a] New York, New York	27	2,949	0.94
30. Purdue University West Lafayette, Indiana	22	2,788	0.89
31. Texas A&M University College Station, Texas	23	2,715	0.86
32. University of Texas at Austin Austin, Texas	24	2,631	0.84
33. Virginia Polytechnic Institute Blacksburg, Virginia	30	2,569	0.82
34. Johns Hopkins University Baltimore, Maryland	58	2,538	0.81
35. Hampton City [a] Hampton, Virginia	39	2,376	0.76
36. University of Washington Seattle, Washington	36	2,332	0.74
37. Old Dominion University Norfolk, Virginia	47	2,285	0.73
38. Cornell University Ithaca, New York	33	2,261	0.72
39. SRI International Corp. [a] Menlo Park, California	37	2,222	0.71
40. New Mexico State University at Las Cruces Las Cruces, New Mexico	34	2,155	0.69

Table 8-42 continued

Contractor/Place of Contract Performance	Rank in FY 1982	Net Value of Awards	
		Amount	Percentage
41. Research Triangle Institute *a*			
Research Triangle Park, North Carolina	42	2,153	0.69
42. Northeast Radio Observatory			
Westford, Massachusetts	59	2,107	0.67
43. Columbia University			
New York, New York	35	1,941	0.62
44. Princeton University			
Princeton, New Jersey	29	1,933	0.62
45. Howard University			
Washington, D.C.	61	1,915	0.61
46. Washington University at St. Louis			
St. Louis, Missouri	41	1,828	0.58
47. George Washington University			
Washington, D.C.	55	1,815	0.58
48. Ohio State University			
Columbus, Ohio	53	1,770	0.56
49. Pennsylvania State University			
State College, Pennsylvania	32	1,660	0.53
50. San Jose State University			
Mountain View, California	57	1,647	0.52
Other		80,407	25.63
Total Awards to Educational and Nonprofit Institutions		313,776	100.00

a Nonprofit institution.

Table 8–43. Top Fifty Educational and Nonprofit Institutions: FY 1984
(in thousands of dollars)

Contractor/Place of Contract Performance	Rank in FY 1983	Net Value of Awards	
		Amount	Percentage
1. Stanford University Stanford, California	3	13,109	4.08
2. Massachusetts Institute of Technology Cambridge, Massachusetts	2	11,835	3.68
3. Smithsonian Institution [a] Cambridge, Massachusetts	4	11,803	3.67
4. Association of University Research and Astronomy [a] Baltimore, Maryland	8	11,025	3.43
5. National Academy of Sciences [a] Washington, D.C.	6	10,793	3.36
6. University of California at San Diego La Jolla, California	5	9,750	3.04
7. Charles Stark Draper Laboratory [a] Cambridge, Massachusetts	7	8,404	2.62
8. European Space Agency [a] Paris, France	1	8,303	2.58
9. University Corporation for Atmospheric Research [a] Palestine, Texas	16	8,075	2.51
10. Johns Hopkins University Baltimore, Maryland	34	7,793	2.43
11. Universities Space Research [a] Columbia, Maryland	11	7,356	2.29
12. California Institute of Technology Pasadena, California	9	6,782	2.11
13. University of Chicago Chicago, Illinois	13	6,280	1.95
14. University of Colorado at Boulder Boulder, Colorado	22	5,644	1.76
15. University of Wisconsin at Madison Madison, Wisconsin	17	5,599	1.74
16. University of Michigan Ann Arbor, Michigan	19	5,214	1.62
17. University of Maryland at College Park College Park, Maryland	12	5,180	1.61
18. University of Chile Santiago, Chile	10	5,144	1.60
19. University of Arizona Tucson, Arizona	21	5,051	1.57
20. University of California at Berkeley Berkeley, California	14	4,964	1.55

Table 8–43 continued

Contractor/Place of Contract Performance	Rank in FY 1983	Net Value of Awards	
		Amount	Percentage
21. University of Hawaii Honolulu, Hawaii	15	4,516	1.41
22. University of Iowa Iowa City, Iowa	24	4,260	1.33
23. University of New Hampshire Durham, New Hampshire	18	4,256	1.32
24. Southwest Research Institute *a* San Antonio, Texas	27	3,724	1.16
25. University of California at Los Angeles Los Angeles, California	26	3,712	1.16
26. Case Western Reserve University Cleveland, Ohio	28	3,446	1.07
27. Hampton City *a* Hampton, Virginia	35	3,338	1.04
28. Battelle Memorial Institute *a* Columbus, Ohio	20	3,332	1.04
29. University of Texas at Austin Austin, Texas	32	3,223	1.00
30. Harvard University Cambridge, Massachusetts	23	2,934	0.91
31. American Institute of Aeronautics and Astronautics *a* New York, New York	29	2,743	0.85
32. University of Washington Seattle, Washington	36	2,687	0.84
33. New Mexico State University at Las Cruces Las Cruces, New Mexico	40	2,576	0.80
34. SRI International Corp. *a* Menlo Park, California	39	2,485	0.77
35. Virginia Polytechnic Institute Blacksburg, Virginia	33	2,269	0.71
36. Columbia University New York, New York	43	2,167	0.67
37. Cornell University Ithaca, New York	38	2,137	0.67
38. Mitre Corp. *a* McLean, Virginia	58	2,079	0.65
39. Old Dominion University Norfolk, Virginia	37	2,055	0.64
40. Utah State University Logan, Utah	62	2,018	0.63

Table 8–43 continued

Contractor/Place of Contract Performance	Rank in FY 1983	Net Value of Awards	
		Amount	Percentage
41. Research Triangle Institute *a*			
Research Triangle Park, North Carolina	41	1,979	0.62
42. Washington University at St. Louis			
St. Louis, Missouri	46	1,950	0.61
43. Princeton University			
Princeton, New Jersey	44	1,946	0.61
44. Pennsylvania State University			
State College, Pennsylvania	49	1,912	0.60
45. Texas A&M University			
College Station, Texas	31	1,887	0.59
46. Northeast Radio Observatory *a*			
Westford, Massachusetts	42	1,880	0.59
47. George Washington University			
Washington, D.C.	47	1,875	0.58
48. Purdue University			
West Lafayette, Indiana	30	1,863	0.58
49. University of Minnesota at Minneapolis St. Paul			
Minneapolis, Minnesota	55	1,836	0.57
50. University of Illinois at Urbana			
Urbana, Illinois	54	1,798	0.56
Other		84,264	26.22
Total Awards to Educational and Nonprofit Institutions		321,251	100.00

a Nonprofit institution.

Table 8–44. *Top Fifty Educational and Nonprofit Institutions: FY 1985*
(*in thousands of dollars*)

Contractor/Place of Contract Performance	Rank in FY 1984	Net Value of Awards	
		Amount	Percentage
1. Stanford University Stanford, California	1	17,177	4.77
2. Association of University Research and Astronomy [a] Baltimore, Maryland	4	15,581	4.33
3. Massachusetts Institute of Technology Cambridge, Massachusetts	2	14,411	4.00
4. National Academy of Sciences [a] Washington, D.C.	5	11,750	3.26
5. Charles Stark Draper Laboratory [a] Cambridge, Massachusetts	7	11,163	3.10
6. University of California at San Diego La Jolla, California	6	9,462	2.63
7. Universities Space Research [a] Columbia, Maryland	11	8,843	2.46
8. Smithsonian Institution [a] Cambridge, Massachusetts	3	8,092	2.25
9. Southwest Research Institute [a] San Antonio, Texas	24	7,672	2.13
10. University of Michigan Ann Arbor, Michigan	16	7,622	2.12
11. University of Wisconsin at Madison Madison, Wisconsin	15	7,466	2.07
12. Harvard University Cambridge, Massachusetts	30	7,457	2.07
13. University of Colorado at Boulder Boulder, Colorado	14	7,413	2.06
14. California Institute of Technology Pasadena, California	12	7,280	2.02
15. University of Maryland at College Park College Park, Maryland	17	6,520	1.81
16. University of Arizona Tucson, Arizona	19	6,233	1.73
17. University Corporation for Atmospheric Research [a] Palestine, Texas	9	5,841	1.62
18. University of Hawaii Honolulu, Hawaii	21	5,651	1.57
19. University of Chicago Chicago, Illinois	13	5,372	1.49
20. University of California at Berkeley Berkeley, California	20	5,216	1.45

Table 8–44 continued

Contractor/Place of Contract Performance	Rank in FY 1984	Net Value of Awards	
		Amount	Percentage
21. Johns Hopkins University Baltimore, Maryland	10	4,530	1.26
22. University of New Hampshire Durham, New Hampshire	23	4,511	1.25
23. University of California at Los Angeles Los Angeles, California	25	4,309	1.20
24. Battelle Memorial Institute [a] Columbus, Ohio	28	4,250	1.18
25. University of Iowa Iowa City, Iowa	22	3,989	1.11
26. Hampton City [a] Hampton, Virginia	27	3,806	1.06
27. New Mexico State University at Las Cruces Las Cruces, New Mexico	33	3,558	0.99
28. University of Chile Santiago, Chile	18	3,519	0.98
29. Case Western Reserve University Cleveland, Ohio	26	3,505	0.97
30. Cornell University Ithaca, New York	37	3,440	0.96
31. Research Triangle Institute [a] Research Triangle Park, North Carolina	41	3,284	0.91
32. American Institute of Aeronautics and Astronautics [a] New York, New York	31	2,907	0.81
33. University of Texas at Austin Austin, Texas	29	2,905	0.81
34. University of Washington Seattle, Washington	32	2,895	0.80
35. University of Illinois at Urbana Urbana, Illinois	50	2,845	0.79
36. Old Dominion University Norfolk, Virginia	39	2,757	0.77
37. Virginia Polytechnic Institute Blacksburg, Virginia	35	2,682	0.74
38. Princeton University Princeton, New Jersey	43	2,656	0.74
39. Columbia University New York, New York	36	2,640	0.73
40. University of Alabama at Huntsville Huntsville, Alabama	55	2,554	0.71

Table 8–44 continued

Contractor/Place of Contract Performance	Rank in FY 1984	Net Value of Awards	
		Amount	Percentage
41. Pennsylvania State University State College, Pennsylvania	44	2,505	0.70
42. Washington University at St. Louis St. Louis, Missouri	42	2,446	0.68
43. Georgia Institute of Technology Atlanta, Georgia	62	2,310	0.64
44. Oklahoma State University Stillwater, Oklahoma	—	2,214	0.61
45. George Washington University Washington, D.C.	47	2,187	0.61
46. SRI International Corp. [a] Menlo Park, California	34	2,185	0.61
47. University of Minnesota at Minneapolis St. Paul Minneapolis, Minnesota	49	2,112	0.59
48. Howard University Washington, D.C.	56	2,073	0.58
49. Purdue University West Lafayette, Indiana	48	1,975	0.55
50. Arizona State University Tempe, Arizona	59	1,922	0.53
Other		94,341	26.01
Total Awards to Educational and Nonprofit Institutions		360,034	100.00

[a] Nonprofit institution.

Table 8–45. Top Fifty Educational and Nonprofit Institutions: FY 1986 (in thousands of dollars)

Contractor/Place of Contract Performance	Rank in FY 1985	Net Value of Awards	
		Amount	Percentage
1. Association of University Research and Astronomy a Baltimore, Maryland	2	19,728	4.99
2. Stanford University Stanford, California	1	18,324	4.63
3. Massachusetts Institute of Technology Cambridge, Massachusetts	3	12,512	3.16
4. National Academy of Sciences a Washington, D.C.	4	12,228	3.09
5. Charles Stark Draper Laboratory a Cambridge, Massachusetts	5	12,034	3.04
6. Universities Space Research a Houston, Texas	7	12,000	3.03
7. University of Michigan Ann Arbor, Michigan	10	11,356	2.87
8. Smithsonian Institution a Cambridge, Massachusetts	8	10,179	2.57
9. University of California at Berkeley Berkeley, California	20	9,906	2.51
10. University of Colorado at Boulder Boulder, Colorado	13	9,708	2.45
11. University Corporation for Atmospheric Research a Palestine, Texas	17	8,966	2.27
12. University of Wisconsin at Madison Madison, Wisconsin	11	8,906	2.25
13. University of California at San Diego La Jolla, California	6	8,609	2.18
14. University of Maryland at College Park College Park, Maryland	15	7,873	1.99
15. Southwest Research Institute a San Antonio, Texas	9	7,868	1.99
16. University of Alabama at Huntsville Huntsville, Alabama	40	6,549	1.66
17. University of Arizona Tucson, Arizona	16	6,549	1.66
18. California Institute of Technology Pasadena, California	14	6,364	1.61
19. University of Chicago Chicago, Illinois	19	5,873	1.49
20. University of Hawaii Honolulu, Hawaii	18	5,318	1.35

Table 8-45 continued

Contractor/Place of Contract Performance	Rank in FY 1985	Net Value of Awards Amount	Net Value of Awards Percentage
21. Case Western Reserve University Cleveland, Ohio	29	5,192	1.31
22. New Mexico State University at Las Cruces Las Cruces, New Mexico	27	5,033	1.27
23. University of California at Los Angeles Los Angeles, California	23	4,480	1.13
24. Oklahoma State University Stillwater, Oklahoma	44	4,239	1.07
25. University of Iowa Iowa City, Iowa	25	4,141	1.05
26. Battelle Memorial Institute *a* Columbus, Ohio	24	3,992	1.01
27. University of Chile Santiago, Chile	28	3,815	0.97
28. Hampton City *a* Hampton, Virginia	26	3,801	0.96
29. University of Washington Seattle, Washington	34	3,687	0.93
30. Johns Hopkins University Baltimore, Maryland	21	3,477	0.88
31. University of Texas at Austin Austin, Texas	33	3,466	0.88
32. University of New Hampshire Durham, New Hampshire	22	3,451	0.87
33. University of Houston at Clear Lake Houston, Texas	96	3,145	0.80
34. American Institute of Aeronautics and Astronautics *a* New York, New York	32	3,139	0.79
35. Mitre Corp. *a* McLean, Virginia	65	3,098	0.78
36. Princeton University Princeton, New Jersey	38	2,980	0.75
37. Columbia University New York, New York	39	2,824	0.71
38. Cornell University Ithaca, New York	30	2,795	0.71
39. Pennsylvania State University State College, Pennsylvania	41	2,626	0.66
40. Ohio State University Columbus, Ohio	51	2,623	0.66

Table 8–45 continued

Contractor/Place of Contract Performance	Rank in FY 1985	Net Value of Awards	
		Amount	Percentage
41. University of Illinois at Urbana Urbana, Illinois	35	2,523	0.64
42. University of Houston Houston, Texas	78	2,423	0.61
43. Washington University at St. Louis St. Louis, Missouri	42	2,402	0.61
44. Research Triangle Institute [a] Research Triangle Park, North Carolina	31	2,367	0.60
45. Virginia Polytechnic Institute Blacksburg, Virginia	37	2,327	0.59
46. Old Dominion University Norfolk, Virginia	36	2,270	0.57
47. Harvard University Cambridge, Massachusetts	12	2,250	0.57
48. University of Alaska at Fairbanks Fairbanks, Alaska	52	2,182	0.55
49. Georgia Institute of Technology Atlanta, Georgia	43	2,113	0.54
50. George Washington University Washington, D.C.	45	2,052	0.52
Other		99,819	25.22
Total Awards to Educational and Nonprofit Institutions		395,582	100.00

[a] Nonprofit institution.

Table 8–46. *Top Fifty Educational and Nonprofit Institutions: FY 1987*
(in thousands of dollars)

Contractor/Place of Contract Performance	Rank in FY 1986	Net Value of Awards	
		Amount	Percentage
1. Stanford University Stanford, California	2	23,207	5.34
2. Association of University Research and Astronomy a Baltimore, Maryland	1	17,737	4.08
3. University of California at Berkeley Berkeley, California	9	13,482	3.10
4. National Academy of Sciences a Washington, D.C.	4	13,337	3.07
5. Charles Stark Draper Laboratory a Cambridge, Massachusetts	5	13,238	3.05
6. Massachusetts Institute of Technology Cambridge, Massachusetts	3	12,012	2.76
7. Universities Space Research a Houston, Texas	6	11,498	2.65
8. University of Colorado at Boulder Boulder, Colorado	10	10,930	2.52
9. Smithsonian Institution a Cambridge, Massachusetts	8	10,667	2.49
10. University of California at San Diego La Jolla, California	13	9,687	2.23
11. University of Michigan Ann Arbor, Michigan	7	9,470	2.18
12. University of Arizona Tucson, Arizona	17	9,440	2.17
13. Southwest Research Institute a San Antonio, Texas	15	9,277	2.14
14. University of Maryland at College Park College Park, Maryland	14	8,300	1.91
15. University of Wisconsin at Madison Madison, Wisconsin	12	7,195	1.66
16. California Institute of Technology Pasadena, California	18	7,108	1.64
17. University of Alabama at Huntsville Huntsville, Alabama	16	6,537	1.50
18. University of Iowa Iowa City, Iowa	25	6,375	1.47
19. University of Chicago Chicago, Illinois	19	6,257	1.44
20. University Corporation for Atmospheric Research a Palestine, Texas	11	6,255	1.44

Table 8–46 continued

Contractor/Place of Contract Performance	Rank in FY 1986	Net Value of Awards	
		Amount	Percentage
21. New Mexico State University at Las Cruces			
Las Cruces, New Mexico	22	5,748	1.32
22. University of California at Los Angeles			
Los Angeles, California	23	5,422	1.25
23. University of Hawaii			
Honolulu, Hawaii	20	5,355	1.23
24. University of New Hampshire			
Durham, New Hampshire	32	5,283	1.22
25. Case Western Reserve University			
Cleveland, Ohio	21	5,121	1.18
26. University of Texas at Austin			
Austin, Texas	31	4,854	1.12
27. Battelle Memorial Institute [a]			
Columbus, Ohio	26	4,419	1.02
28. University of Washington			
Seattle, Washington	29	4,118	0.95
29. Harvard University			
Cambridge, Massachusetts	47	4,085	0.94
30. Pennsylvania State University			
State College, Pennsylvania	39	3,817	0.88
31. Mitre Corp. [a]			
McLean, Virginia	35	3,807	0.88
32. Hampton City [a]			
Hampton, Virginia	28	3,779	0.87
33. Cornell University			
Ithaca, New York	38	3,667	0.84
34. Johns Hopkins University			
Baltimore, Maryland	30	3,561	0.82
35. Old Dominion University			
Norfolk, Virginia	46	3,552	0.82
36. Columbia University			
New York, New York	37	3,498	0.81
37. American Institute of Aeronautics and Astronautics [a]			
New York, New York	34	3,357	0.77
38. University of Chile			
Santiago, Chile	27	3,342	0.77
39. University of Alaska at Fairbanks			
Fairbanks, Alaska	48	3,287	0.76
40. Princeton University			
Princeton, New Jersey	36	3,248	0.75

Table 8-46 continued

Contractor/Place of Contract Performance	Rank in FY 1986	Net Value of Awards	
		Amount	Percentage
41. Research Triangle Institute *a*			
Research Triangle Park, North Carolina	44	3,032	0.70
42. Ohio State University			
Columbus, Ohio	40	2,969	0.68
43. Texas A&M University			
College Station, Texas	56	2,877	0.66
44. San Jose State University			
Mountain View, California	67	2,679	0.62
45. University of Illinois at Urbana			
Urbana, Illinois	41	2,561	0.59
46. Oklahoma State University			
Stillwater, Oklahoma	24	2,521	0.58
47. George Washington University			
Washington, D.C.	50	2,482	0.57
48. Georgia Institute of Technology			
Atlanta, Georgia	49	2,402	0.55
49. Virginia Polytechnic Institute			
Blacksburg, Virginia	45	2,398	0.55
50. University of Houston at Clear Lake			
Houston, Texas	33	2,263	0.52
Other		112,967	25.94
Total Awards to Educational and Nonprofit Institutions		434,480	100.00

a Nonprofit institution.

Table 8–47. *Top Fifty Educational and Nonprofit Institutions: FY 1988 (in thousands of dollars)*

Contractor/Place of Contract Performance	Rank in FY 1987	Net Value of Awards	
		Amount	Percentage
1. Stanford University Stanford, California	1	27,674	5.54
2. Association of University Research and Astronomy [a] Baltimore, Maryland	2	23,696	4.74
3. New Mexico State University at Las Cruces Las Cruces, New Mexico	21	19,231	3.85
4. Universities Space Research [a] Columbia, Maryland	7	16,957	3.39
5. Massachusetts Institute of Technology Cambridge, Massachusetts	6	14,279	2.86
6. University of California at Berkeley Berkeley, California	3	13,633	2.73
7. National Academy of Sciences [a] Washington, D.C.	4	12,866	2.57
8. University of Arizona Tucson, Arizona	12	12,001	2.40
9. University of Colorado at Boulder Boulder, Colorado	8	11,847	2.37
10. Smithsonian Institution [a] Cambridge, Massachusetts	9	11,447	2.29
11. Southwest Research Institute [a] San Antonio, Texas	13	11,361	2.27
12. Charles Stark Draper Laboratory [a] Cambridge, Massachusetts	5	10,604	2.12
13. University of Maryland at College Park College Park, Maryland	14	9,718	1.94
14. University of Alabama at Huntsville Huntsville, Alabama	17	9,527	1.91
15. University of Wisconsin at Madison Madison, Wisconsin	15	9,232	1.85
16. University of Michigan Ann Arbor, Michigan	11	8,996	1.80
17. University of California at San Diego La Jolla, California	10	8,605	1.72
18. California Institute of Technology Pasadena, California	16	7,890	1.58
19. Case Western Reserve University Cleveland, Ohio	25	7,802	1.56
20. University of Iowa Iowa City, Iowa	18	7,019	1.40

Table 8–47 continued

Contractor/Place of Contract Performance	Rank in FY 1987	Net Value of Awards Amount	Percentage
21. University of Hawaii Honolulu, Hawaii	23	6,496	1.30
22. University of California at Los Angeles Los Angeles, California	22	5,873	1.18
23. Pennsylvania State University State College, Pennsylvania	30	5,365	1.07
24. University of Chicago Chicago, Illinois	19	5,362	1.07
25. University of New Hampshire Durham, New Hampshire	24	5,011	1.00
26. University of Houston at Clear Lake Houston, Texas	50	5,004	1.00
27. Harvard University Cambridge, Massachusetts	29	4,791	0.96
28. University of Washington Seattle, Washington	28	4,681	0.94
29. University of Texas at Austin Austin, Texas	26	4,441	0.89
30. University of Southern California Los Angeles, California	55	4,343	0.87
31. Columbia University New York, New York	36	3,733	0.75
32. Old Dominion University Norfolk, Virginia	35	3,731	0.75
33. Johns Hopkins University Baltimore, Maryland	34	3,722	0.74
34. Ohio State University Columbus, Ohio	42	3,705	0.74
35. University of Chile Santiago, Chile	38	3,670	0.73
36. Cornell University Ithaca, New York	33	3,323	0.67
37. Texas A&M University College Station, Texas	43	3,287	0.66
38. Virginia Polytechnic Institute Blacksburg, Virginia	49	3,272	0.65
39. Hampton City [a] Hampton, Virginia	32	3,166	0.63
40. American Institute of Aeronautics and Astronautics [a] New York, New York	37	3,109	0.62
41. Princeton University Princeton, New Jersey	40	2,959	0.59

Table 8-47 continued

Contractor/Place of Contract Performance	Rank in FY 1987	Net Value of Awards	
		Amount	Percentage
42. Oklahoma State University Stillwater, Oklahoma	46	2,904	0.58
43. San Jose State University Mountain View, California	44	2,776	0.56
44. University of Alaska at Fairbanks Fairbanks, Alaska	39	2,755	0.55
45. Georgia Institute of Technology Atlanta, Georgia	48	2,746	0.55
46. George Washington University Washington, D.C.	47	2,698	0.54
47. University of Houston Houston, Texas	69	2,657	0.53
48. Battelle Memorial Institute [a] Moffett Field, California	27	2,580	0.52
49. University of Illinois at Urbana Urbana, Illinois	45	2,502	0.50
50. Cleveland State University Cleveland, Ohio	54	2,236	0.52
Other		132,599	26.47
Total Awards to Educational and Nonprofit Institutions		499,882	100.00

[a] Nonprofit institution.

INDEX

A

Abex Corporation, 375
Active Cavity Irradiance Monitor II, ACRIM II, 103
Advanced Communications Technology Satellite, ACTS, 17, 18, 47, 96, 254
Advanced Digital Engine Control System, ADECS, 213
Advanced Digital Synthetic Aperture Radar (SAR) Processor, ADSP, 257
Advanced Fighter Technology Integration, AFTI, 198
Advanced short takeoff and vertical landing, ASTOVL, 181, 210
Advanced technology blades, SV-15/ATB program, 210
Advanced Transport Operating System, ATOPS, 229
Advanced Very High Resolution Radiometer, AVHRR, 35
Aerojet General Corp., 563, 566, 569
Aeronautics, Balloons, and Sounding Rockets, AB&SR, 312
Aero-Space Technology, AST, 467
Aetna Life and Casualty, Inc., 52, 53
Agency for International Development, 16
Agriculture, U.S. Department of, 13, 16–17, 392; and AgRISTARS, 16–17, 81
Aiken, William S., Jr., 186, 187
Air Force, U.S., 17, 35, 60, 62, 70, 198, 200, 204, 214, 217, 224–225, 227, 237, 252, 306; and Air Force Satellite Control Facility, Sunnyvale, California, 305
Air Products & Chemicals, Inc., 551, 554, 557, 560, 563, 566, 570
Airborne Windshear Detection and Avoidance Program, 234
Aircraft Energy Efficiency program, ACEE, 181, 189, 190–197, 198
Alabama, University of, 360, 572, 575, 579, 582, 590, 592, 595, 598
Alaska, 51, 55, 57; and Fairbanks, 306, 307, 342; and University of, 594, 596, 600
Allen, H. Julian, 390
Allen, Lew, Jr., 394
Aller, Robert O., 301
AlliedSignal Aerospace Co., 568
Allison Corp., 210

American Institute of Aeronautics and Astronautics, 572, 575, 578, 581, 584, 587, 590, 593, 596, 599
American Satellite Company, ASC, 50, 74, 128
American Society of Mechanical Engineers, 237
American Telephone and Telegraph, AT&T, 50, 51, 560
Ames, Joseph S., 391
Ames Research Center, 181, 182, 186, 190, 202, 203, 205, 208, 220, 222, 224, 289, 321, 383, 389, 390–391, 392, 404, 407, 414–426, 428, 432–434, 466, 472–482, 487–488, 492–493, 513–516, 538–539; and Numerical Aerodynamic Simulation Facility, 183
Anelex Corp., 564
Anik, 17, 51, 54, 64–65, 66, 67, 72, 159–161; and Telesat Canada Corporation, 17, 51, 64, 159–161
Announcement of Opportunity, 15, 25
Antigua, 306
Apollo program, 211, 397, 402; and Apollo-Soyuz Test Project, 28
Applications Explorer Mission, AEM, 14, 16
Applications Technology Satellites, ATS, 12, 13, 17,
Arabsat, 17, 65, 66, 67, 73, 162
Argee Corp., 567
ARGOS, 48
Ariane, 54, 67
Arizona, University of, 571, 574, 577, 580, 584, 586, 589, 592, 595, 598
Arizona State University, 582, 591
Army, U.S., 14, 205, 208, 392, 396, 401
Arnold, Ray, 19
Ascension Island, 305, 307, 342
Asia, 57
Asia Satellite, AsiaSat, 55
Assembly Concept for Construction of Erectable Space Structures, ACCESS, 184, 243, 248, 253
Association of University Research and Astronomy, 578, 581, 583, 586, 589, 592, 595, 598
Atlantic Missile Range, 396
Atlantic Ocean, 39, 57, 63
Atlantis, 138, 163, 166
Atlas, 35, 37; and Atlas-Centaur, 58, 148–156
Atomic Energy Commission, 180

Auburn University, 369
Aussat, 17, 66–67, 74, 162–163; and Aussat Proprietary, Ltd., 66, 67
Australia, 57, 66, 74, 312; and Alice Springs, 312; and Canberra, 300, 306, 307, 308, 311, 320, 343; and Orroral Valley, 306, 347; and Parkes, 311, 312; and Tidbinbilla, 311, 312, 343; and Yarragardee, 305, 306, 349
Auter, Harry, 402

B

Bahamas, Grand, 306
Balch, Jackson M., 401
Ball Corp., 551, 553, 554, 557, 560, 563, 566, 569
Ballhaus, William F., Jr., 187, 390
Bamsi, Inc., 561, 563, 566, 569
Battelle Columbus Laboratories, 369, 577
Battelle Memorial Institute, 571, 574, 581, 583, 587, 590, 593, 596, 600
Beattie, Donald A., 186
Bechtel National, Inc., 557
Beeler, D.E., 391
Beggs, James M., 355, 358, 388
Bell, 209
Bendix Corporation, Allied, 540, 550, 553, 556, 559, 562, 565
Benson, Robert, 20
Bermuda, 305, 306, 307, 314, 342
Bikle, Paul F., 391
Boeing Aerospace Corporation, Services International, 196, 198, 201, 203, 209, 212, 229, 231, 238, 373, 540, 550, 553, 554, 556, 557, 559, 560, 562, 565, 566, 568
Botswana, Africa, 305, 344; and Botswana National Museum, 344
Bowles, Roland, 235
Boyd, John, 391
Brazil, 17
British Aerospace, 210
Bryant, Frederick B., 301
Bulgaria, 49
Buckhorn, California, 305, 307
Bush, George, 228

C

California, University of, at Berkeley, 571, 574, 577, 580, 583, 586, 589, 592, 595, 598; and at Los Angeles, 572, 575, 578, 580, 584, 587, 590, 593, 596, 599; and at San Diego, 571, 574, 577, 580, 583, 586, 589, 592, 595, 598
California Institute of Technology, 394, 571, 574, 577, 580, 583, 586, 589, 592, 595, 598
California State University at Chico, 573
Calio, Anthony J., 18, 47
Cape Canaveral, 306, 320, 395
Caribbean, 14, 57
Carruthers, John, 18
Carter, Jimmy, 16
Case Western Reserve University, 369, 572, 575, 578, 581, 584, 587, 590, 593, 596, 598
Centaur, 400; and Atlas-Centaur, 58, 148–156
Centers for the Commercial Development of Space, 360
Central America, 39
Centre Nationale d'Études Spatiales, CNES, 34
Challenger, 101, 144, 160, 165, 166, 167, 247, 290, 319; and *Challenger* accident, 3, 32, 235, 242, 247, 251, 252, 290, 291, 307, 307, 313, 319, 331, 350, 361, 389, 497
Charles Stark Draper Laboratory, 571, 574, 577, 580, 583, 586, 589, 592, 595, 598
Charlesworth, Charles E., 395
Cheney, Richard, 228
Chicago, University of, 571, 574, 577, 580, 583, 586, 589, 592, 595, 599
Chile, University of, 571, 574, 577, 580, 583, 586, 590, 593, 596, 599
Civil Service Reform Act of 1978, 464
Clark, John F., 393
Clarks, Henry J., 355, 356
Clarkson University, 369
Clean Air Act, Amendments, 15
Clear Lake, Texas, 394
Cleveland Electric Illuminating, 555
Cleveland Hopkins International Airport, 399
Cleveland State University, 600
Cohen, Aaron, 395
Colladay, Raymond S., 186, 187, 224, 225
Collier, Robert J., Trophy, 4, 182, 191, 398, 400; and National Aeronautic Association, 4
Colorado, University of, 369, 372, 572, 575, 577, 580, 584, 586, 589, 592, 595, 598
Columbia, 22, 138, 140, 160, 290
Columbia University, 572, 575, 578, 581, 585, 587, 590, 593, 596, 599
Commerce Business Daily, 504
Commerce, U.S. Department of, 13, 32
Commercial Use of Space, 1984 National Policy, 355
Commonwealth Scientific and Industrial Research Organization, Australia, 311
Communications Satellite Corporation, Comsat, 47, 52, 53, 58
Communications Technology Satellites, CTS, 12
Compton, Dale L., 390
Computer Sciences Corporation, 540, 550, 553, 556, 559, 562, 565, 568

Comstar, 17, 50, 72, 128; and Comsat General Corporation, 50
Congress, 11, 12, 15, 16, 31, 32, 46, 47, 175, 176, 188, 189, 226, 228, 357, 358, 361, 396, 497, 498
Construction of Facilities, CofF, 498, 516
Continental Construction Corp., 569
Continental Telecom, Inc., Contel, 50, 314, 562, 565, 568
Continental Telephone, 314
Control Data Corporation, 551, 554, 557, 560, 564, 566
Cooper, Robert, 393
Cornell University, 572, 575, 578, 581, 584, 587, 590, 593, 596, 599
Cortright, Edgar M., 397
COSMOS, 48
COSPAS, 17, 37, 48, 49
Costa, S. Richard, 302
Cray Research Inc., 560, 569
Cryogenic Limb Array Etalon Spectrometer, CLAES, 102
Culbertson, Philip E., 357

D

Data Capture Facility, 321
Data Collection System, 34
Debus, Kurt H., 396
Deep Space Network, DSN, 5, 299, 304, 307, 308, 309, 311, 312, 313, 325, 336, 337, 343, 345, 346, 347, 394
Deere, John, & Co., 370
Defense Systems, Inc., 370
Defense, U.S. Department of, DOD, 61, 70, 73, 176, 189, 192, 204, 205, 207, 208, 210, 213, 221, 224, 225, 226, 227, 228, 247, 299, 305, 348; and Defense Advanced Research Projects Agency, DARPA, 189, 205, 207, 208, 213, 217, 222, 223, 224, 227; and Defense Space Communications System, DSCS, 63
De France, Smith J., 390
Delmarva Peninsula, 402
Delta, 40, 51, 52, 67, 68
Denmark, 17, 247
Denver, Colorado, 235
Deutsch, George C., 187
Digital Autonomous Terminal Access Communication, DATAC, 231
Digital Electronic Engine Control, DEEC, 212
Digital Equipment Corporation, 552, 554, 557, 560, 564, 566
Digital fly-by-wire, DFBW, 210, 211, 212, 216
Discovery, 106, 128, 129, 130, 141, 157, 158, 161, 162, 165

District of Columbia, Washington, D.C., 501
Donlan, Charles J., 397
Draper Laboratory, 211
Dryden Flight Research Center, 182, 186, 190, 193, 197, 198, 201, 202, 211, 212, 213, 214, 217, 218, 219, 220, 221, 238, 241, 242, 302, 305, 306, 307, 312, 342, 344, 355, 383, 388, 390, 391, 392, 404, 407, 414–426, 435–436, 466, 472–482, 487–488, 492–493, 513–516
DuPont, 370
Dynamic Augmentation Experiment, DAE, 246

E

Earth, 5, 12, 16, 22, 26, 29, 30–38, 41, 42, 45, 48, 50, 55, 59, 60, 61, 62, 67, 68, 69, 178, 182, 183, 184, 222, 240, 247, 248, 249, 250, 251, 299, 300, 304, 309, 313, 315, 344, 394, 401, 403
Earth Data Corporation, 375
Earth Observation Satellite Company, EOSAT, 16, 42
Earth Observing System, EOS, 256
Earth Radiation Budget Experiment, ERBE, 15, 30, 31, 37
Earth Radiation Budget Satellite, ERBS, 14, 15, 28, 30, 31, 37, 73, 101
Earth Resources Technology Satellite, ERTS, 3, 13
Eastern Test Range, Cape Canaveral, 396
Eaton, Peter T., 356
ECHO, 345
Ecuador, Quito, 305, 306
Edelson, Burton, 19
Edwards Air Force Base, California, 207, 214, 223, 241, 306, 391, 392
EG&G Florida, Incorporated, 540, 553, 556, 559, 562, 565, 568
Ellington Air Force Base, 394
Elms, James C., 395
Emergency Locator Transmitter, ELT, 47–49
Emergency Position Indicating Radio Beacon, EPIRB, 47–49
Endeavour, 241
Energy, U.S. Department of, 180, 250, 400
Engineering and Economic Research, 570
England, 69; and Winkfield, 305
Enterprise, 211, 241, 392
Environmental Research Institute, of Michigan, 369, 573, 576, 578
Estess, Roy S., 402
Europe, 47, 57
European Space Agency, ESA, 312, 501, 574, 577, 580, 583, 586

Evans, L.J., Jr., 357
Experimental Assembly of Structures in Extravehicular Activity, EASE, 184, 243, 248
Extravehicular activity, EVA, 249, 258
Extremely high frequency, EHF, 61

F

Fairchild Industries, Inc., 50, 314, 371, 555, 557, 560, 563, 566, 569
Feature Identification and Landmark Experiment, FILE, 26–27
Federal Acquisition Regulation, FAR, 502, 504, 506
Federal Aviation Administration, FAA, 176, 189, 192, 210, 229, 234, 235, 238
Federal Communications Commission, 18, 50
Federal Financing Bank, U.S. Department of Treasury, 314, 331, 332
Federal Technology Transfer Act of 1986, 359
Federal Wage System, 467
Ferrick, Eugene, 302
Finland, 17, 49
Fisk, Lennard A., 19
Fleet Satellite Communications, Fltsatcom, 17, 60, 72, 154–156
Fletcher, James C., 388
Force, Charles T., 301, 302
Ford, Aerospace, Communications Corporation, 59, 62, 67, 540, 550, 553, 556, 559, 562, 565, 568
Fort Irwin, California, 306
Fosque, Hugh S., 301
France, 17, 34, 37, 48, 57, 247; and Modane, 195
Franklin Institute, 576
Frosch, Robert A., 356, 357, 360, 388

G

Gabris, Edward A., 188
Galapagos Islands, 14
Galaxy, 51, 52, 55, 67, 73, 131
Garrett Corporation, 552, 555
Gemini, Project, 396
General Dynamics Corp., 197, 198, 210, 212, 540, 550, 553, 556, 559, 562, 566, 569
General Electric, GE, 31–32, 191, 195, 196, 206, 210, 216, 317, 540, 541, 544, 550, 553, 557, 560, 562, 565, 568
General Motors, 194, 551
General Schedule, GS, 464, 465
General Sciences Corporation, 375
General Services Administration, GSA, 505, 506
Geological Survey, U.S., 13, 46
George Washington University, 572, 578, 585, 588, 591, 594, 597, 600
Georgia Institute of Technology, 573, 576, 591, 594, 597, 600
Geosat, 16
Geostar Corporation, 374
Geostationary Operational Environmental Satellite, GOES, 12, 14, 28, 32, 38, 39, 40, 41, 72, 116–121
Germany, 25, 57, 219, 247
Get Away Special, GAS, 25–26
Gillam, Isaac T., Ike, IV, 355, 391
Gilruth, Robert R., 395
Giotto, 312
Glennan, T. Keith, 388
Global Associates, 552
Global positioning system, GPS, 229
Global Weather Experiment, 15
Goddard, Robert H., 393
Goddard Space Flight Center, 18, 31, 51, 257, 300, 302, 305, 306, 307, 315, 320, 329, 346, 383, 389, 392, 393, 403, 404, 407, 414–426, 428, 437–439, 466, 472–482, 487–488, 492–493, 513–516, 538–539; and Goddard Institute of Space Studies, 392; and National Science Space Data Center, 18
Goett, Harry J., 393
Goetz, Robert, 395
Goldstone, California, 306, 307, 308, 309, 311, 313, 344
Goodyear Aerospace Corporation, 257
Graham, William R., 178, 226, 388
Graves, Randolph A., 187
Greenbelt, Maryland, 51, 300, 304, 392
Greenwood, Lawrence, 19
Griffin, Gerald, 395, 396
Grumman Aircraft Corporation, 199, 210, 217, 373, 558, 566, 567, 568, 570
GTE, 373
GTI Corporation, 370
Guam, 50, 306, 307, 346
Guastaferro, Angelo, 390, 391
Gulfstream Aerospace Corporation, 194, 196, 197

H

Halley's Comet, 312
Halogen Occultation Experiment, HALOE, 33, 86, 102
Halpern, Richard, 20
Hamilton Standard, 192, 195
Hampton, Virginia, 397, 579, 581, 584, 587, 590, 593, 596, 599
Harrier, aircraft, 205

Harris Corporation, 314, 563, 567
Harris, Leonard A., 187
Hart, Terry, 247
Harvard University, 571, 574, 577, 580, 584, 587, 589, 594, 596, 599
Hawaii, 39, 50, 51, 55, 57, 306, 571
Hawaii, University of, 574, 578, 580, 583, 587, 589, 592, 596, 599
Hearth, Donald P., 393, 397
Heflex Bioengineering Test, 23
Hercules Corporation, 375
High Angle of Attack Research Vehicle, HARV, 220
High Resolution Doppler Imager, HRDI, 102
Highly Integrated Digital Electronic Control, HIDEC, 213
Highly Maneuverable Aircraft Technology, HiMAT, 214, 215, 216, 217, 286
Hinners, Noel W., 393
Hlass, Jerry I., 401
Holcomb, Lee B., 187
Honeywell Corporation, Inc., 371, 551, 554, 557, 558, 560, 561, 564
Hornstein, Robert M., 302
Houston, Texas, 394
Houston, University of, 369, 593, 594, 597, 599, 600
Howard University, 585, 591
Hubble Space Telescope, 321, 401
Hughes, Communications, Aircraft Corporation, 51, 53, 55, 61, 62, 66, 67, 68, 540, 550, 553, 557, 560
Huntsville, Alabama, 400

I

Ice accretion code, Lewis Research Center, LEWICE, 238
Icing Research Tunnel, IRT, 237, 238
Illinois, University of, Urbana, 573, 588, 590, 593, 597, 600
Improved Stratospheric and Mesospheric Sounder, ISAMS, 102
INCO, 370
Indonesia, 17, 51
Induced Environment Contamination Monitor, IECM, 245, 289
Industrial Applications Centers, IACs, 359
Industrial Guest Investigator (IGI) Agreement, 360, 361
Inertial Upper Stage, IUS, 317, 318
Informatics General Corporation, 554, 558, 560, 563
Insat, 17, 67–68, 164–165
Institute for Technology Development, ITD, 369, 375

Instrumentation Technology Associates, 373
Interior, U.S. Department of, 13, 16, 398
International Business Machines, IBM, 52, 53, 211, 540, 550, 553, 556, 559, 562, 565, 568
International Fuel Cells Corp., 561
International Telecommunications Satellite Organization, Intelsat, 13, 50, 56–60, 68, 72, 145–153
International Telephone and Telegraph, ITT, 551
Invitation for Bid, IFB, 504
Iowa, University of, 571, 574, 578, 580, 584, 587, 590, 593, 595, 598
Ireland, 247
Italy, 57

J

Jacksonville, Florida, 396
Japan, 47, 57
Jet Propulsion Laboratory, JPL, 5, 246, 257, 300, 302, 303, 307, 311, 313, 320, 326, 328, 383, 389, 394, 404, 407, 414–426, 428, 440–441, 499, 526, 538–539
Johannes, Robert P., 391
Johns Hopkins University, 70, 391, 573, 577, 584, 586, 590, 593, 596, 599
Johnson, Harry W., 356
Johnson, Lyndon B., 396
Johnson Space Center/Manned Spacecraft Center, 197, 306, 315, 320, 389, 394, 395, 398, 404, 407, 414–426, 428, 442–444, 472, 501, 513–516, 538–539; and Johnson Mission Control Center, 306
Joint Endeavor Agreements, JEA, 360, 361, 370, 371, 372, 373
Jones, Robert T., 202
Jupiter, 3, 5, 309, 311

K

Kansas, University of, 576, 579, 582
Kayten, Gerald G., 187
Kennedy Space Center, 51, 223, 242, 290, 291, 306, 314, 347, 350, 351, 383, 389, 395, 396, 397, 404, 407, 414–426, 428, 444–446, 463, 472–482, 487–488, 492–493, 513–516, 538–539
Kentron International, Inc., 551, 554, 557
Kerrebrock, Jack, 186
Keyworth, G.A., 177
Kimball, Harold G., 301, 302
Klate Holt Co., 555, 558, 567, 569
Klineberg, John M., 399
Kraft, Christopher C., Jr., 395
Kramer, James, 186

Kreiger, Robert, 402
Krier, Gary E., 356
Krings, John, 227
Krug International Corporation, 570
Kutler, Paul, 187
Kutyna, Donald J., Major General, 224

L

Lamberth, Horace, 396
Landsat, 3, 12, 13, 16, 32, 33, 42, 43, 44, 79, 122–124, 312, 327, 329, 340
Langley, Samuel Pierpont, 397
Langley Research Center, 183, 184, 185, 190, 197, 198, 200, 201, 204, 205, 208, 209, 220, 221, 222, 223, 224, 229, 231, 232, 234, 235, 237, 242, 247, 248, 250, 251, 290, 291, 296, 383, 389, 392, 397, 398, 403, 404, 407, 414–426, 428, 447–449, 472–482, 487–488, 492–493, 513–516, 538–539; and Langley Memorial Aeronautical Laboratory, 397, 398; and National Transonic Facility, 183
Large-scale Advanced Propfan, LAP, 195, 196
Laser Geodynamics Satellite, LAGEOS, 14
La Soufriere, volcano, 14
Leasat/Syncom, 17, 61, 62, 73, 157–158
Lee, Thomas J., 400
Lemkey, Frank, 20
Levine, Jack, 187
Lewis, George W., 399
Lewis Research Center, 4, 46, 47, 58, 186, 190, 191, 192, 193, 194, 195, 196, 220, 221, 224, 237, 238, 383, 389, 399, 400, 404, 407, 414–426, 428, 450–452, 472–482, 487–488, 492–493, 513–516, 538–539; and Glenn Research Center, 58, 237
Light Detecting and Ranging, LIDAR, 235, 257, 258
Little, Arthur D., Inc., 371
Local User Terminal, LUT, 48
Lockheed Company, Space Operations, Missiles, Engineering, 194, 195, 196, 197, 210, 540, 550, 553, 554, 555, 556, 557, 559, 560, 562, 563, 565, 566, 568, 569
Long Duration Exposure Facility, LDEF, 243, 246, 247, 248, 296, 398
Louisiana, New Orleans, 400, 401
Lovelace, Alan M., 388
Lovell, Robert, 19
Low, George M., 388, 395
LTV Aerospace & Defense Co., 560, 564, 569
Lubarsky, Bernard, 399
Lucas, William R., 400
Lunar Roving Vehicle, 401

Lundin, Bruce T., 399
Luxenberg, Barbara A., 356
Lyman, Peter T., 394

M

Madrid, Spain, 300, 307, 308, 313, 320, 346
Magnetic Field Satellite, Magsat, 14, 16, 42, 44, 46, 72, 125–127
Magnetoplasmadynamic, MPD, 252
Malaysia, 68
Management & Technical Services, 551, 554, 557, 560, 563, 567
Management Services, Inc., 561, 563
Manganiello, Eugene J., 399
Manned Spaceflight Network, MSFN, 300
Mariner, 309
Mark, Hans, 388, 390
Mars, 240, 309, 345, 398
Marshall, General George C., 400
Marshall Space Flight Center, 21, 25, 184, 248, 289, 383, 389, 396, 400, 401, 404, 407, 414–426, 428, 453–455, 472–482, 487–488, 492–493, 501, 513–516, 538–539; and Michoud Assembly Facility, 400; and Slidell Computer Complex, 400
Martin, John J., 186, 187
Martin Marietta Corporation, 372, 540, 541, 544, 547, 550, 553, 556, 559, 562, 565, 568
Martin Thiokol Corporation, 52, 540, 550, 556, 559, 562, 565, 568
Maryland, University of, College Park, 571, 574, 577, 580, 583, 586, 589, 592, 595, 598
Massachusetts Institute of Technology, 248, 571, 574, 577, 580, 583, 586, 589, 592, 595, 598
Massively Parallel Processor, MPP, 257
Materials Experiment Assembly, MEA, 25, 26
Materials Processing in Space, MPS, 360, 361
Materialwissenschaftliche Autonome Experimente unter Schwerelosigkeit, MAUS, 25, 26
McCarthy, John F., 399
McCartney, Forrest, 396
McCoy, Caldwell, Jr., 19
McDonnell Douglas Corporation, 51, 52, 196, 197, 204, 210, 221, 361, 370, 540, 541, 544, 547, 550, 553, 556, 559, 562, 565, 568
McElroy, John, 19, 393
MCI, 53
McIver, Duncan E., 188
Measurement of Air Pollution From Satellites, MAPS, 26, 27

INDEX 607

Mechanical Technology, Inc., 551, 554, 557, 561, 564
Memorandum of Agreement, MOA, 374, 375
Memorandum of Understanding, MOU, 371, 372, 373, 374, 375
Mercury, Project, 309, 396, 398, 401
Mercury Consolidated Inc., 554
Meredith, Leslie H., 393
Merritt Island, Florida, 306, 307, 314, 347, 396
Meteoroid and Exposure Module, MEM, 247, 248
Miami, Florida, 396
Michigan, University of, 571, 574, 577, 580, 583, 586, 589, 592, 595, 598
Micro Craft, Inc., 567
Micro-Gravity Research Associates, 371
Microwave Limb Sounder, MLS, 102
Middle East, 57, 65
Minnesota, University of, 572, 575, 579, 582, 588, 591
Mission Adaptive Wing, MAW, 198
Mission Needs Statement, 502
Mission Peculiar Equipment Support Structure, MPESS, 23, 24, 27
Mississippi, Hancock County, 402
Mitre Corp., 587, 593, 596
Modular Computer Systems, Inc., 552
Mojave Desert, California, 196, 300, 307, 344
Moon, 240, 396, 401
Moore, Jesse W., 395
Morelos, Mexico, 17, 68, 73, 166
Mouat, David A., 375
Mount St. Helens, 14, 29
Multimission Modular Spacecraft, MMS, 31, 32, 33, 43
Multispectral Scanner, MSS, 42–44
Muroc, California, 392
Murray, Bruce C., 394
Myers, Dale D., 388

N

NASA Communications, NASCOM, 303, 305, 307, 320
NASA End-to-End Data System, NEEDS, 256
NASA Headquarters, 428, 429–431, 472–482, 487–488, 492–493, 538–539
NASA Inspector General, 513
National Academy of Sciences, 571, 574, 577, 580, 583, 586, 589, 592, 595, 598
National Advisory Committee for Aeronautics, NACA, 175, 237, 383, 390, 392, 398, 399, 403
National Aeronautics and Space Act, 4, 31, 175, 356, 358, 499

National Aeronautics and Space Administration, NASA, 3, 4, 6, 7, 11–19, 21, 23, 25, 28, 30, 31–38, 40, 41, 43, 45, 46, 47, 48, 50, 54, 55, 58, 60, 63, 65, 67, 68, 69, 70, 175, 176, 177, 178, 179, 180, 183, 184, 185, 186, 188, 189, 190, 191, 192, 194, 195, 197, 198, 199, 201, 202, 204, 205, 207, 208, 209, 210, 211, 216, 220, 222, 223, 224, 225, 227, 228, 229, 231, 232, 235, 236, 237, 238, 239, 241, 242, 243, 247, 252, 255, 256, 257, 296, 299, 300, 301, 302, 303, 304, 305, 307, 308, 309, 311, 312, 313, 314, 315, 319, 320, 321, 351, 355, 356, 357, 358, 359, 360, 362, 383, 386, 387, 389, 394, 396, 397, 398, 400, 401, 402, 403, 463, 465, 466, 497, 498, 499, 502, 503, 504, 505, 506
National Aerospace Plane, NASP, 178, 182, 221–228
National Meteorological Center, 34, 35
National Oceanic and Atmospheric Administration, NOAA, 12, 14, 15, 16, 17, 28, 30, 32, 33, 34, 35, 36, 37, 38, 42, 43, 44, 48, 49, 72, 86, 107–115, 307
National Science Foundation, 311
National Scientific Balloon Facility, 312
National Space Policy, 5, 240, 357, 358, 359
National Transportation Safety Board, 234
Naumann, Robert, 20
Naval Research Laboratory, 393
Navy, U.S., 17, 60, 70, 199, 204, 402
Neptune, 3, 308, 311, 343
Netherlands, The, 247
Network Control Center, 307, 313, 333
New England, 501, 537
New Guinea, Papua, 67
New Hampshire, University of, 578, 581, 583, 587, 590, 593, 596, 599
New Mexico State University, 572, 575, 578, 581, 584, 587, 590, 593, 596, 598
Nicks, Oran W., 397
Nighttime/Daylight Optical Survey of Thunderstorm Lightning, NOSL, 22
Nimbus, 12, 14, 15, 28, 32, 33, 37, 48, 86, 340, 341
North America, 39, 48, 57
North American Aircraft, 214
North Atlantic Treaty Organization, NATO, 62, 63, 159
Northeast Radio Observatory, 576, 585, 588
Northrop Corporation, Worldwide Aircraft, 200, 551, 552, 554, 555, 557, 558, 560, 561, 563, 566, 567, 570
Norway, 17
NOVA, 70, 169–170
NSI Technology Services Corp., 569

O

Oak Ridge National Laboratory, 253
Office of Advanced Research and Technology, OART, 179
Office of Management and Budget, 497
Office of Science and Technology Policy, 177
Ohio State University, 369, 572, 576, 585, 593, 597, 599
Oklahoma State University, 575, 579, 581, 584, 591, 593, 597, 600
Old Dominion University, 572, 575, 579, 582, 584, 587, 590, 594, 596, 599
Olstad, Walter B., 186
Ontario, Canada, 17, 37, 48, 57, 64, 247
Orbital Sciences Corporation, 371, 565, 569
Orbital Transfer Vehicle, 251
Orbiting Satellite Carrying Amateur Radio, Oscar, 69
Organization of the Petroleum Exporting Countries, OPEC, 190
Orient Express, 182
Orlando, Florida, 235
Ott, Richard H., 356

P

Page, George F., 396
Paine, Thomas O., 388
Palapa, 17, 51, 67, 68, 69, 73, 166–167
Pan Am Pacific Satellite Corporation, 55
Pan American World Airways, Services, 550, 554, 557, 560, 563, 565, 568
Parkes, Australia, 311, 312
Parks, Robert J., 394
Particle Environment Monitor, PEM, 103
Pasadena, California, 300, 303, 313, 320, 394
Patrick Air Force Base, 306
Payload Assist Module, PAM, 17, 51, 52
Pegasus, 401
Pennsylvania State University, 369, 572, 575, 578, 581, 585, 588, 591, 593, 596, 599
Perkin Elmer Corporation, 550, 553, 556, 559, 563, 566, 569
Petersen, Richard H., 397
Petrone, Rocco A., 400
Philippines, 68
Phobos, 312
Pickering, William H., 394
Pioneer, 309, 311, 345
Pittsburgh, University of, 572
Planning Research Corp., 550, 553, 556, 559, 562, 566, 568
Plum Brook Station, 399
Poker Flats Research Facility, 312

Polar Orbiting Geophysical Observatory, POGO, 45
Ponce de Leon, 306, 307, 314, 347
Povinelli, Frederick P., 187
Pratt & Whitney, 191, 210, 212
Princeton University, 572, 575, 578, 581, 585, 588, 590, 593, 596, 599
Procurement Request, PR, 502, 503, 504
Program Communications Support Network, PCSN, 320, 321, 339
Propfan Test Assessment, PTA, 195, 197
Propulsion Systems Laboratory, 399
Pseudorandom noise, PN, 350
Puerto Rico, 50, 55
Purdue University, 571, 574, 577, 581, 584, 588, 591

Q

Quality Short-haul Research Aircraft, QSRA, 181, 203, 204
Quann, John J., 393
Quiet, Clean, Short-haul Experimental Engine, QCSEE, 204

R

Rantek, 374
Raytheon Services Company, 551, 554, 557, 560, 563, 566, 569
RCA, Satcom, Americom, 17, 35, 47, 50, 51, 52, 53, 70, 72, 132–138, 540, 551, 553, 559, 562
Reagan, Ronald, 3, 6, 11, 47, 182, 225, 355, 357, 464
Reck, Gregory, 188
Redstone Arsenal, 400, 401
Rees, Eberhard F.M., 400
Reis, Victor, 177
Rensselaer Polytechnic Institute, 582
Request for Proposal, RFP, 502, 504
Rescue Coordination Center, 48
Research and Program Management, R&PM, 497, 513
Research Triangle Institute, 573, 576, 582, 585, 588, 590, 594, 597
RMS Technologies, Inc., 558, 564, 566
Robertson, Floyd, 19
Rockwell International, 214, 372, 540, 541, 544, 547, 550, 553, 556, 559, 562, 565, 568
Roeder, John H., 302
Rogers Commission, 236, 389
Rogers Dry Lake, 391
Rohr Industries, 194, 195
Rolls Royce, 210
Rose, James T., 187, 355
Rosen, Cecil C., III, 187

INDEX

Rosen, Robert, 187, 188
Rosman, North Carolina, 305, 348
Ross, Lawrence J., 399
Ross, Miles, 396
Rotor Systems Research Aircraft, RSRA, Sikorsky Aircraft Division, 181, 205–208

S

Salisbury, Maryland, 402
Sander, Michael, 20
Sandusky, Ohio, 399
San Jose State University, California, 573, 576, 585, 597, 600
Santa Barbara Research Center, 551
Santiago, Chile, 306, 307, 348
Satellite Business Systems, SBS, 17, 51–54, 72, 139–141
Satellite Control Center, 34
Saturn, 3, 309, 310, 311, 396, 397, 401
Saudi Arabia, 65
Sauer Mechanical, Inc., 558
Scherer, Lee R., 391, 396
Schmoll, Kathryn, 20
Schneider, William, 301
Scott, David R., 391
Scott Science and Technology, Inc., 374
Seamans, Robert C., 388
Search and Rescue, SAR, 36, 37, 95, 257,
Search and Rescue Satellite Aided Tracking, SARSAT, 17, 37, 47, 48, 49
Seasat, 14
Senate, U.S., 227, 228
Senegal, Dakar, 305, 307, 344
Sharp, Edward R., 399
Shepard, Alan B., 401
Short takeoff and landing, STOL, 180, 181, 203, 204, 210
Short takeoff and vertical landing, STOVL, 181, 203
Shuttle Carrier Aircraft, SCA, 241
Shuttle Imaging Radar, SIR-A, 22
Shuttle Multispectral Infrared Radiometer, 22
Sierra Negra, 14
Silverstein, Abe, 399
Singapore, 68
Singer Company, 551, 554, 557, 560, 563, 566, 570
Sjoberg, Sigurd A., 395
Skylab, 13, 396, 401
Slone, Henry O., 187
Small Business Administration, 506
Small Business Innovation Development Act, 361
Small Business Innovation Research, SBIR, 356, 359, 361, 362, 377, 378, 379

Small Business Technology Transfer, STTR, 362
Smith, Richard G., 396, 400
Smithsonian Institution, 571, 574, 577, 580, 583, 586, 589, 592, 595, 598
Smylie, Robert E., 301, 393
Socorro, New Mexico, 311
Solar Array Experiment, SAE, 246
Solar Backscatter Ultraviolet, SBUV, 37
Solar Cell Calibration Facility, SCCF, 246
Solar Mesospheric Explorer, SME, 69
Solar Stellar Irradiance Comparison Experiment, SOLSTICE, 103
Solar Ultraviolet Spectral Irradiance Monitor, SUSIM, 103
Source Evaluation Board, 504
South America, 39, 57
South Korea, 57
Southern California, University of, 572, 575, 579, 582
Southwest Research Institute, 572, 578, 581, 584, 587, 589, 592, 595, 598
Soviet Union, 17, 37, 48, 49, 397
Space Communications Company, 551, 558, 559
Space Flight, Control, and Data Communications, SFC&DC, 303, 304, 497
Space Industries, Inc., 372
Space Services, Inc., 373
Space Shuttle program, 3, 11, 14, 17, 20, 21–28, 30, 31, 35, 47, 50, 51, 52, 62, 64, 65, 66, 67, 75, 79, 85, 89, 91, 104, 133, 140, 144, 153, 157, 175, 176, 179, 180 183–184, 211, 224, 228, 239, 240, 241–248, 251, 253, 288, 289, 290, 300, 305, 306, 307, 309, 313, 314, 315, 317, 320, 321, 342, 344, 347, 348, 357, 361, 370–375, 389, 391, 392, 393, 395, 397, 398, 401, 402, 404, 497, 513, 516, 517, 520; and Orbiter Processing Facility, 397; and Shuttle Payload Engineering Division, 20; and Shuttle Landing Facility, 397; and Space Shuttle Main Engine, SSME, 180, 402
Space Systems Development Agreement, SSDA, 361, 373, 374,
Space Tracking and Data Acquisition Network, STADAN, 300
Space Tracking and Data Network, STDN, 299, 300, 304, 305, 306, 307, 309, 313, 322, 325, 328, 338, 339, 343, 345, 347, 393
Space Transportation System, STS, 5, 6, 7, 14, 21, 55, 69, 348, 351
Spaceco, Ltd., 371
SPACEHAB, Inc., 374
Spacelab, 321, 329, 341

Spacelab Data Processing Facility, 321
Sperry Corporation, 373, 551, 554, 557, 560, 563
Spinak, Abraham D., 402, 403
SRI International Corp., 571, 575, 578, 581, 584, 587, 591
Stacked Oscars on Scout, SOOS, 70, 71, 74, 171–172
Stanford University, 502, 571, 574, 577, 580, 583, 586, 589, 592, 595, 598
Stennis, John C., Senator, 402
Stennis Space Center, 383, 389, 401, 402, 404, 428; and Mississippi Test Facility, 402; and National Space Technology Laboratories, 383, 389, 401, 402, 407, 414–426, 456–458, 463, 472–482, 487–488, 492–493, 513–516, 538–539
Sterling Federal Systems, Inc., 566, 569
Stevenson-Wydler Technology Act of 1980, 359
Stofan, Andrew, 19, 399
Stone, Barbara A., 356
Strategic Air Command, 60
Stratospheric Aerosol and Gas Experiment, SAGE, 14, 28, 29, 30, 31, 72, 100
Stratospheric Aerosol Measurement II, 28
Suitland, Maryland, 35
Sun, 29, 32, 33, 69, 318
Superhigh frequency, SHF, 69
Surrey, University of, England, 69
Sverdrup Technology, Inc., 563, 566, 569
Sweden, 17
Switzerland, 247
Synchronous Meteorological Satellites, SMS, 12, 38
Systems Development Corporation, 551, 554, 557, 560, 563
Syvertson, Clarence A., 390

T

Tasmania, University of, 312, 347
Taylor, Charles A., 301
Technical Exchange Agreement, TEA, 360, 361, 370, 371, 372, 375, 376
Technology Transfer, 92, 360
Technology Utilization, 363, 364, 366
Teledyne Industries, Inc., 551, 553, 556, 560, 563, 565, 568
Television Infrared Observation Satellite, TIROS, 12, 14, 33–38, 48
Telstar, 17, 50, 51, 73, 129–130
Tennessee, University of, Space Institute, 369
Terhune, Charles H., Jr., 394
Terminal-Configured Vehicle, TCV, 229, 232
Testardi, Louis R., 18, 20

Texas A&M University, 369, 575, 578, 581, 584, 588, 597, 599
Texas, University of, at Austin, 574, 578, 581, 584, 587, 590, 593, 596, 599; and at Dallas, 572, 575, 579, 582
Thematic Mapper, TM, 42, 43
Thome, Pitt, 19
Thompson, James R., Jr., 400
3M Company, 201, 372
Tidbinbilla, Australia, 311, 312, 343
Tile Gap Heating, TGH, 245, 289
Tilford, Shelby G., 19
Total Ozone Monitoring System, 15
Tokyo, 182
Total Energy Control System, TECS, 231
Townsend, John W., Jr., 393
Tracking and Data Relay Satellite System, TDRSS, 28, 30, 42, 43, 44, 257, 299, 300, 301, 302, 304, 307, 309, 313, 314, 315, 316, 317, 318, 319, 320, 322, 325, 329, 330, 331, 332, 333, 334, 336, 339, 350, 351, 352, 393, 395
Transonic Aircraft Technology, TACT, 198
Transport Systems Research Vehicle, TSRV, 229, 230, 231, 232
Transportation, U.S. Department of, 176, 192
Trimble, George S., 395
TRW, 47, 314, 352, 550, 553, 556, 559, 562, 565, 568
Tula Peak, 306

U

Ultrahigh frequency, UHF, 60, 62, 69
Union Oil Company, 375
Unisys Corp., 566, 569
United Kingdom, 17, 34, 48, 57, 210, 247
United Space Boosters, Inc., 540, 550, 553, 556, 559
United States, U.S., 3, 5, 6, 7, 11, 13, 14, 17, 18, 34, 40, 47, 48, 50, 51, 55, 56, 59, 68, 176, 177, 178, 210, 234, 237, 247, 356, 357, 358, 359, 391, 499, 501
United Technologies Corporation, 192, 205, 540, 550, 553, 556, 559, 562, 565, 568
Universities Space Research, 571, 575, 577, 580, 583, 586, 589, 592, 595, 598
University Corporation for Atmospheric Research, 583, 586, 589, 592, 595
UoSAT, 69, 73, 168–169
Upper Atmospheric Research Satellites, UARS, 15, 28, 31, 32, 33, 84, 87, 102, 104
Uranus, 3, 184, 255, 311, 312, 345
USBI Booster Production Co., 562, 565, 568
Utah State University, 573, 587
Utsman, Thomas E., 396

V

Vandenberg Air Force Base, California, 305, 306, 320, 397
Vanderbilt University, 369
Vanguard, Project, 393
Vega, Soviet spacecraft, 312
Venneri, Samuel L., 187
Venus, 309, 312, 345
Vernamonti, Len, Colonel, 226
Vertical short takeoff and landing, VSTOL, 181, 203, 205, 208
Vertical takeoff and landing, VTOL, 181, 203, 210
Very high frequency, VHF, 67, 69
Very Large Array, 311
Viking, project, 5, 398
Virgin Islands, 55
Virginia Electric & Power Company, 552, 555
Virginia Polytechnic Institute, 572, 575, 578, 581, 584, 587, 590, 594, 597, 599
Visible/Infrared Spin Scan Radiometer, VISSR, 39, 40
von Braun, Wernher, 400, 401
Vought Corporation, 552, 554
Voyager, 3, 5, 184, 255, 308, 309, 310, 311, 312, 343, 345, 347

W

Wallops Flight Facility/Wallops Flight Center, 14, 302, 306, 307, 312, 326, 346, 348, 383, 393, 402, 403, 404, 407, 414–426, 459–460, 466, 472–482, 487–488, 492–493, 513–516; and Wallops Research Airport, 403
W&J Construction Corp., 569
Washington, University of, 573, 575, 578, 581, 584, 587, 590, 593, 596, 599
Washington University, St. Louis, 576, 579, 581, 585, 588, 591, 594
Webb, James E., 388
Weitz, Paul J., 395
Westar Satellite System, 17, 50, 51, 54, 55, 56, 72, 142–144
Western Union, Spacecom, 51, 54, 55, 301, 314, 351
Westinghouse Electric Corporation, 551, 555, 557, 561, 563, 567
Whitcomb, Richard T., 197
White House, 358
White Sands, Las Cruces, New Mexico, 42, 44, 300, 305, 306, 307, 313, 395
White Sands Ground Terminal, 307, 315, 316, 318, 350
White Sands Missile Range, 312
White Sands Test Facility, 395
Whitten, Raymond, 355
Wild, Jack W., 301, 359
Williams, Dell P., III, 187
Williams, Walter C., 391
Wind Imaging Interferometer, WINDII, 102
Wisconsin, University of, 369, 571, 574, 577, 580, 583, 586, 589, 592, 595, 598
Wood, H. William, 301
Work Breakdown Structure, WBS, 502
World Meteorological Organization, 15
World War II, 219, 237, 392, 399
Wright, Linwood C., 187, 188
Wright-Patterson Air Force Base, 195, 214
Wyle Laboratories, 566, 569

Y

Young, A. Thomas, 390, 393

Z

Zero One Systems, Inc., 570

ABOUT THE COMPILER

Judy A. Rumerman is a professional technical writer who has written or contributed to numerous documents for the National Aeronautics and Space Administration. She has been the author of documents covering various spaceflight missions, the internal workings of NASA's Goddard Space Flight Center, and other material used for training. She was also the compiler of *U.S. Human Spaceflight: A Record of Achievement, 1961–1998,* a monograph for the NASA History Office detailing NASA's human spaceflight missions.

Ms. Rumerman has degrees from the University of Michigan and George Washington University. She grew up in Detroit and presently resides in Silver Spring, Maryland.

THE NASA HISTORY SERIES

Reference Works, NASA SP-4000

Grimwood, James M. *Project Mercury: A Chronology* (NASA SP-4001, 1963).
Grimwood, James M., and Hacker, Barton C., with Vorzimmer, Peter J. *Project Gemini Technology and Operations: A Chronology* (NASA SP-4002, 1969).
Link, Mae Mills. *Space Medicine in Project Mercury* (NASA SP-4003, 1965).
Astronautics and Aeronautics, 1963: Chronology of Science, Technology, and Policy (NASA SP-4004, 1964).
Astronautics and Aeronautics, 1964: Chronology of Science, Technology, and Policy (NASA SP-4005, 1965).
Astronautics and Aeronautics, 1965: Chronology of Science, Technology, and Policy (NASA SP-4006, 1966).
Astronautics and Aeronautics, 1966: Chronology of Science, Technology, and Policy (NASA SP-4007, 1967).
Astronautics and Aeronautics, 1967: Chronology of Science, Technology, and Policy (NASA SP-4008, 1968).
Ertel, Ivan D., and Morse, Mary Louise. *The Apollo Spacecraft: A Chronology, Volume I, Through November 7, 1962* (NASA SP-4009, 1969).
Morse, Mary Louise, and Bays, Jean Kernahan. *The Apollo Spacecraft: A Chronology, Volume II, November 8, 1962–September 30, 1964* (NASA SP-4009, 1973).
Brooks, Courtney G., and Ertel, Ivan D. *The Apollo Spacecraft: A Chronology, Volume III, October 1, 1964–January 20, 1966* (NASA SP-4009, 1973).
Ertel, Ivan D., and Newkirk, Roland W., with Brooks, Courtney G. *The Apollo Spacecraft: A Chronology, Volume IV, January 21, 1966–July 13, 1974* (NASA SP-4009, 1978).
Astronautics and Aeronautics, 1968: Chronology of Science, Technology, and Policy (NASA SP-4010, 1969).
Newkirk, Roland W., and Ertel, Ivan D., with Brooks, Courtney G. *Skylab: A Chronology* (NASA SP-4011, 1977).
Van Nimmen, Jane, and Bruno, Leonard C., with Rosholt, Robert L. *NASA Historical Data Book, Volume I: NASA Resources, 1958–1968* (NASA SP-4012, 1976, rep. ed. 1988).
Ezell, Linda Neuman. *NASA Historical Data Book, Volume II: Programs and Projects, 1958–1968* (NASA SP-4012, 1988).
Ezell, Linda Neuman. *NASA Historical Data Book, Volume III: Programs and Projects, 1969–1978* (NASA SP-4012, 1988).
Gawdiak, Ihor Y., with Fedor, Helen. Compilers. *NASA Historical Data Book, Volume IV: NASA Resources, 1969–1978* (NASA SP-4012, 1994).
Rumerman, Judy A. Compiler. *NASA Historical Data Book, Volume V: NASA Launch Systems, Space Transportation, Human Spaceflight, and Space Science, 1979–1988* (NASA SP-4012, 1999).

Astronautics and Aeronautics, 1969: Chronology of Science, Technology, and Policy (NASA SP-4014, 1970).
Astronautics and Aeronautics, 1970: Chronology of Science, Technology, and Policy (NASA SP-4015, 1972).
Astronautics and Aeronautics, 1971: Chronology of Science, Technology, and Policy (NASA SP-4016, 1972).
Astronautics and Aeronautics, 1972: Chronology of Science, Technology, and Policy (NASA SP-4017, 1974).
Astronautics and Aeronautics, 1973: Chronology of Science, Technology, and Policy (NASA SP-4018, 1975).
Astronautics and Aeronautics, 1974: Chronology of Science, Technology, and Policy (NASA SP-4019, 1977).
Astronautics and Aeronautics, 1975: Chronology of Science, Technology, and Policy (NASA SP-4020, 1979).
Astronautics and Aeronautics, 1976: Chronology of Science, Technology, and Policy (NASA SP-4021, 1984).
Astronautics and Aeronautics, 1977: Chronology of Science, Technology, and Policy (NASA SP-4022, 1986).
Astronautics and Aeronautics, 1978: Chronology of Science, Technology, and Policy (NASA SP-4023, 1986).
Astronautics and Aeronautics, 1979–1984: Chronology of Science, Technology, and Policy (NASA SP-4024, 1988).
Astronautics and Aeronautics, 1985: Chronology of Science, Technology, and Policy (NASA SP-4025, 1990).
Noordung, Hermann. *The Problem of Space Travel: The Rocket Motor.* Stuhlinger, Ernst, and Hunley, J.D., with Garland, Jennifer. Editors (NASA SP-4026, 1995).
Astronautics and Aeronautics, 1986–1990: A Chronology (NASA SP-4027, 1997).
Gawdiak, Ihor Y., and Shetland, Charles. Compilers. *Astronautics and Aeronautics, 1991–1995: A Chronology* (NASA SP-2000-4028, 2000).

Management Histories, NASA SP-4100

Rosholt, Robert L. *An Administrative History of NASA, 1958–1963* (NASA SP-4101, 1966).
Levine, Arnold S. *Managing NASA in the Apollo Era* (NASA SP-4102, 1982).
Roland, Alex. *Model Research: The National Advisory Committee for Aeronautics, 1915–1958* (NASA SP-4103, 1985).
Fries, Sylvia D. *NASA Engineers and the Age of Apollo* (NASA SP-4104, 1992).
Glennan, T. Keith. *The Birth of NASA: The Diary of T. Keith Glennan.* Hunley, J.D. Editor (NASA SP-4105, 1993).
Seamans, Robert C., Jr. *Aiming at Targets: The Autobiography of Robert C. Seamans, Jr.* (NASA SP-4106, 1996).

THE NASA HISTORY SERIES

Project Histories, NASA SP-4200

Swenson, Loyd S., Jr., Grimwood, James M., and Alexander, Charles C. *This New Ocean: A History of Project Mercury* (NASA SP-4201, 1966; rep. ed. 1998).

Green, Constance McL., and Lomask, Milton. *Vanguard: A History* (NASA SP-4202, 1970; rep. ed. Smithsonian Institution Press, 1971).

Hacker, Barton C., and Grimwood, James M. *On Shoulders of Titans: A History of Project Gemini* (NASA SP-4203, 1977).

Benson, Charles D. and Faherty, William Barnaby. *Moonport: A History of Apollo Launch Facilities and Operations* (NASA SP-4204, 1978).

Brooks, Courtney G., Grimwood, James M., and Swenson, Loyd S., Jr. *Chariots for Apollo: A History of Manned Lunar Spacecraft* (NASA SP-4205, 1979).

Bilstein, Roger E. *Stages to Saturn: A Technological History of the Apollo/Saturn Launch Vehicles* (NASA SP-4206, 1980, rep. ed. 1997).

SP-4207 not published.

Compton, W. David, and Benson, Charles D. *Living and Working in Space: A History of Skylab* (NASA SP-4208, 1983).

Ezell, Edward Clinton, and Ezell, Linda Neuman. *The Partnership: A History of the Apollo-Soyuz Test Project* (NASA SP-4209, 1978).

Hall, R. Cargill. *Lunar Impact: A History of Project Ranger* (NASA SP-4210, 1977).

Newell, Homer E. *Beyond the Atmosphere: Early Years of Space Science* (NASA SP-4211, 1980).

Ezell, Edward Clinton, and Ezell, Linda Neuman. *On Mars: Exploration of the Red Planet, 1958–1978* (NASA SP-4212, 1984).

Pitts, John A. *The Human Factor: Biomedicine in the Manned Space Program to 1980* (NASA SP-4213, 1985).

Compton, W. David. *Where No Man Has Gone Before: A History of Apollo Lunar Exploration Missions* (NASA SP-4214, 1989).

Naugle, John E. *First Among Equals: The Selection of NASA Space Science Experiments* (NASA SP-4215, 1991).

Wallace, Lane E. *Airborne Trailblazer: Two Decades with NASA Langley's Boeing 737 Flying Laboratory* (NASA SP-4216, 1994).

Butrica, Andrew J. Editor. *Beyond the Ionosphere: Fifty Years of Satellite Communication* (NASA SP-4217, 1997).

Butrica, Andrew J. *To See the Unseen: A History of Planetary Radar Astronomy* (NASA SP-4218, 1996).

Mack, Pamela E. Editor. *From Engineering Science to Big Science: The NACA and NASA Collier Trophy Research Project Winners* (NASA SP-4219, 1998).

Reed, R. Dale. With Lister, Darlene. *Wingless Flight: The Lifting Body Story* (NASA SP-4220, 1997).

Heppenheimer, T.A. *The Space Shuttle Decision: NASA's Search for a Reusable Space Vehicle* (NASA SP-4221, 1999).

Hunley, J.D. Editor. *Toward Mach 2: The Douglas D-558 Program* (NASA SP-4222, 1999).

Swanson, Glen E. Editor. *"Before This Decade Is Out...": Personal Reflections on the Apollo Program* (NASA SP-4223, 1999).

Center Histories, NASA SP-4300

Rosenthal, Alfred. *Venture into Space: Early Years of Goddard Space Flight Center* (NASA SP-4301, 1985).
Hartman, Edwin, P. *Adventures in Research: A History of Ames Research Center, 1940–1965* (NASA SP-4302, 1970).
Hallion, Richard P. *On the Frontier: Flight Research at Dryden, 1946–1981* (NASA SP- 4303, 1984).
Muenger, Elizabeth A. *Searching the Horizon: A History of Ames Research Center, 1940–1976* (NASA SP-4304, 1985).
Hansen, James R. *Engineer in Charge: A History of the Langley Aeronautical Laboratory, 1917–1958* (NASA SP-4305, 1987).
Dawson, Virginia P. *Engines and Innovation: Lewis Laboratory and American Propulsion Technology* (NASA SP-4306, 1991).
Dethloff, Henry C. *"Suddenly Tomorrow Came . . .": A History of the Johnson Space Center* (NASA SP-4307, 1993).
Hansen, James R. *Spaceflight Revolution: NASA Langley Research Center from Sputnik to Apollo* (NASA SP-4308, 1995).
Wallace, Lane E. *Flights of Discovery: 50 Years at the NASA Dryden Flight Research Center* (NASA SP-4309, 1996).
Herring, Mack R. *Way Station to Space: A History of the John C. Stennis Space Center* (NASA SP-4310, 1997).
Wallace, Harold D., Jr. *Wallops Station and the Creation of the American Space Program* (NASA SP-4311, 1997).
Wallace, Lane E. *Dreams, Hopes, Realities: NASA's Goddard Space Flight Center, The First Forty Years* (NASA SP-4312, 1999).

General Histories, NASA SP-4400

Corliss, William R. *NASA Sounding Rockets, 1958–1968: A Historical Summary* (NASA SP-4401, 1971).
Wells, Helen T., Whiteley, Susan H., and Karegeannes, Carrie. *Origins of NASA Names* (NASA SP-4402, 1976).
Anderson, Frank W., Jr. *Orders of Magnitude: A History of NACA and NASA, 1915–1980* (NASA SP-4403, 1981).
Sloop, John L. *Liquid Hydrogen as a Propulsion Fuel, 1945–1959* (NASA SP-4404, 1978).
Roland, Alex. A *Spacefaring People: Perspectives on Early Spaceflight* (NASA SP-4405, 1985).
Bilstein, Roger E. *Orders of Magnitude: A History of the NACA and NASA, 1915–1990* (NASA SP-4406, 1989).
Logsdon, John M. Editor. With Lear, Linda J., Warren-Findley, Jannelle, Williamson, Ray A., and Day, Dwayne A. *Exploring the Unknown: Selected Documents in the History of the U.S. Civil Space Program, Volume I: Organizing for Exploration* (NASA SP-4407, 1995).

Logsdon, John M. Editor. With Day, Dwayne A., and Launius, Roger D. *Exploring the Unknown: Selected Documents in the History of the U.S. Civil Space Program, Volume II: External Relationships* (NASA SP-4407, 1996).

Logsdon, John M. Editor. With Launius, Roger D., Onkst, David H., and Garber, Stephen E. *Exploring the Unknown: Selected Documents in the History of the U.S. Civil Space Program, Volume III: Using Space* (NASA SP-4407, 1998).

Logsdon, John M. Editor. With Williamson, Ray A., Launius, Roger D., Acker, Russell J., Garber, Stephen J., and Friedman, Jonathan L. *Exploring the Unknown: Selected Documents in the History of the U.S. Civil Space Program, Volume IV: Accessing Space* (NASA SP-4407, 1999).

ISBN 0-16-050266-7